Springer Asia Pacific Mathematics Series

Volume 8

The Springer Asia Pacific Mathematics Series promotes high-quality scientific research connected to Asia-Pacific countries. Monographs, edited volumes, lecture notes, and textbooks will cover topics related to pure and applied mathematics as well as interdisciplinary research.

All material will be rigorously peer-reviewed and should be related to mathematical research conducted at or with institutions in the APAC region. This series is open to subseries, which may be topical or exclusively supplied by a single research institution.

Takashi Suzuki

Liouville's Theory in Linear and Nonlinear Partial Differential Equations

Springer

Takashi Suzuki
Center for Mathematical Modeling and Data
Science
Osaka University
Toyonaka, Osaka, Japan

ISSN 3091-2555 ISSN 3091-2563 (electronic)
Springer Asia Pacific Mathematics Series
ISBN 978-981-95-1974-3 ISBN 978-981-95-1975-0 (eBook)
https://doi.org/10.1007/978-981-95-1975-0

Mathematics Subject Classification: Primary: 35B33, 35J60, 35J61, 35K55, 37L60, 53A05, 70F45, 70S15, 74H40, 80A30, 82C31, 92C17
Secondary: 35B12, 35B35, 35B40, 35B44, 35J20, 35K57, 35K58, 35Q72, 35Q80, 37L45, 53A10, 70K20, 70K40, 74G35, 74H35, 74H55, 80A25

This Springer imprint is published by the registered company Springer Nature Singapore Pte Ltd.
The registered company address is: 152 Beach Road, #21-01/04 Gateway East, Singapore 189721, Singapore

Preface

Beginners in mathematical analysis encounter several of Liouville's theorems, or theorems of his type. These theorems are common in the fields of analysis, geometry, and physics. The same is true for specialists in these fields, because several concepts underlying them are often used in modern approaches to partial differential equations, linear and nonlinear. However, recent studies are revealing strong connections between them, looking independent of each other.

Joseph Liouville (1809–1882) is thought of as a master of pure and applied mathematics [1]. This book describes some aspects of this profile, through his five theorems, for the intended readers, graduate students, and researchers in mathematics and sciences; each chapter thus deals with each of these theorems.

The first theorem is concerned with the spectral theory, that is, the Sturm–Liouville property of nodal points of one-dimensional elliptic operators. This property is used efficiently in prescribing iso-spectral symmetric potentials. A related property of decreasing of the number of zero points of solutions to parabolic equations is also given, with its significant applications. Discrete structures contained in continuously distributed materials are thus noticed. The first principle of singularity cancellation, however, arises from a different point of view; symmetric interaction of particles, referred to as the action–reaction law. Rather than this symmetry, then, duality hidden behind differential operators with different boundary conditions induces new insight for this classical problem in one-space dimension. A remarkable application is the assignment of total eigenvalues of the operator using the explicit symmetric potential of elementary functions. These methods and results are summarized at the beginning of Chap. 1.

The second theorem is the Liouville integral for surfaces with constant Gaussian curvature. The structure of the equation in the complex analysis induces a geometric profile of its nonlinearity. That is, the one between a flat plane and a sphere, which results in the quantized blowup mechanism observed in the family of solutions to the Boltzmann–Poisson (BP) equation. This equation arises in statistical mechanics for many point vortices. Isoperimetric features of the associated spectral problem derived from linearization are observed, which results in uniqueness of the solution.

It is now well known that this uniqueness justifies the propagation of chaos of many point vortices assumed in their canonical ensembles. The Liouville integral, however, leads to a blowup mechanism other than quantization, for a family of solutions to BP equation: Hamiltonian control. More precisely, climbing up the material hierarchy, particles cluster to several spots. Location of these spots, then, is subject to the Hamiltonian governing the first hierarchy of the original particles. This scientific principle, called recursive hierarchy, is observed in several selected models in physics. Its realization in dynamical systems is to be described in the final chapter.

The third theorem is the classification with bounded entire solutions to the Laplace equation. This property is extended to semilinear elliptic equations, and then, the blowup analysis reveals the structure of total set of solutions. The quantized blowup mechanism observed in the BP equation in the previous chapter is then reproduced from this point of view: the scaling limit of the solution and Liouville property of the limit equation. The method of moving planes is efficient in achieving this Liouville property, if the nonlinearity is critical or sub-critical. In Chap. 3, we thus gain several real analytic tools efficient in scaling invariant models, called blowup analysis based on Liouville's third theorem.

The fourth theorem is the time derivative of the volume integral on deforming bodies. Among many applications of this theorem, there are elliptic regularity, derivation of fundamental equations from physical laws, and problems in engineering such as the free boundary problem and shape optimization. Hadamard's variational formula arises in this context of applications. Here, the second volume derivative is associated with the second fundamental form of the boundary of the domain. This transformation theory, then, ensures blowup of irrotational incompressible Euler flow and enthalpy formulation of free boundary problems of Stefan type. It is, however, a tool other than scaling, to solve dynamical recursive hierarchy, again shown in the last chapter.

The fifth theorem is the integrability of Hamilton systems under transformation of variables in accordance with the Poisson bracket. This Poisson structure is seen in several biological models where interaction of many species makes the system stable. This structure casts the second principle of singularity cancellation different from the one discovered in Chap. 1; skew-symmetric interaction of many species. Gierer–Meinhardt (GM) systems with specific parameters and Lotka–Volterra (LV) system with or without diffusion terms are such examples. In contrast, there arises also the first principle of singularity cancellation in Chap. 1; symmetry of interaction of many particles, through the gradient of field distribution created by them. This principle lies inside the Smoluchowski–Poisson (SP) equation other than the above models, accompanied by quantized blowup mechanism and recursive hierarchy. The BP equation is actually realized as a stationary state of the SP equation. Consequently, we reach three levels of blowup quantization in the SP equation; in stationary, finite time, and infinite time, via scaling and volume derivatives. A dynamical recursive hierarchy is also observed, in the scaling and translation limit of the solution. We conclude this chapter with a topic in mathematical oncology, the chemotactic paradox in connection with the Keller–Segel (KS) system, where methods of multi-scale modeling of reaction–diffusion, that of numerical simulation based on detailed mass balance, and that of the Turing paradigm are applied.

This book thus presents seven concepts governing physical and biological phenomena associated with partial differential equations. The first is a *quantized blowup mechanism*. Here we describe this property in the context of SP and BP equations, where Liouville's theorems cast analytic tools: parametric representation of a surface with constant scalar curvature, classification of entire solutions, and the first volume derivatives. The second is *duality between fields and particles*. Although this phenomenon is connected with convex analysis developed in later ages, some of its origin is observed in the Sturm–Liouville theory. The third is *recursive hierarchy*. Here we notice the role of point vortex Hamiltonians in the study of SP and BP equations. *Nonlinear spectral mechanics* is the fourth, where a stationary state is realized as a nonlinear eigenvalue problem involving a conserved quantity as a parameter, which controls the dynamical system including blowup mechanism. Again, it is typically observed in SP and BP equations, again. The fifth is *space homogenization of isolated systems*. GM systems with specific parameters or LV systems are such examples. The sixth is *emergence from networks*. Some features are described in Chap. 5 in the context of the modeling of reaction and integrability of the fundamental process. The seventh is the *cancellation of singularities in skew-symmetric species*. This phenomenon is sometimes realized in the Poisson structure of dependent variables, differently from cancellation of singularities via the action–reaction law, which results in the second concept, duality between fields and particles. It is remarkable that all of these concepts have emerged from the classical Liouvile's theorems.

This book deals with fundamental, classical, and modern theories in linear and nonlinear partial differential equations, motivated and technically originated from Liouville's five theorems. The reader will be presented with inverse problems, semilinear elliptic equations, free boundary problems, biological models, fluid mechanics, statistical mechanics, thermodynamics, and mathematical oncology. New insights emerging from these observations are the linearization of the iso-spectral deformation, quantized blowup mechanisms, recursive hierarchy observed in nonlinear systems, and singularity cancellation through the skew-symmetric structure of curvature tensors and interaction of many particles under symmetric forces.

At this moment, the author is standing in front of a landscape: a panoramic, bright, hard, and deep world written in mathematical formulae. This world is summarized as a fusion of five of Liouville's theorems, which has passed through the author during his 50 years. It would be his pleasure if the reader could share this unexpected experience.

Finally, the author thanks Prof. Takuya Tsuchiya for discussions and comments on frames. Thanks are also due to Masayuki Nakamura in Springer-Verlag, for his kind supports.

Osaka, Japan Takashi Suzuki
March, 2021

Reference

1. J. Lutzen, *Joseph Liouville 1809–1882, Master of Pure and Applied Mathematics* (Springer, New York, 1990)

Acknowledgement This work is supported by JSPS core-to-core research project, Kakenhi 16H06576, and Kakenhi 19H01799.

Acknowledgement. This work is supported by JSPS core-to-core research project, Kakenhi 16H06570, and Kakenhi 19H01709.

Contents

Chapter 1
Spectral Theory

Abstract The first of Liouville's theorems in this monograph is concerned with the linear eigenvalue problem in mathematical physics. Having a summary in the first section, we begin with the applied inverse problem in the second section, to introduce boundary identification of parabolic equations and its ill-posedness to reach the inverse Sturm–Liouville problem. In the third section, the Gel'fand–Levitan– Marchenko (GLM) equation is used to reduce the problem to a study on Goursat problems for hyperbolic equations. We are thus lead to eigenvalue problems in the fourth section, and the intersection property as a striking tool in the theory for one space dimension in the fifth section. We then come back to the inverse Sturm–Liouville problem in the sixth section; uniqueness, reconstruction, and linearization of the Hochstadt equation. The τ-function emerges as a Fredholm determinant in the final section, from symmetry of the problem and duality between the boundary conditions; those of Dirichlet and Robin.

1.1 Introduction

1.1.1 Intersection Property

Liouville's intersection property is realized in both parabolic and spectral problems. It is concerned with the problem in one space dimension. The former result says that the number of zeros of the solution to the linear parabolic equation provided with the boundary condition does not increase. The latter result is on the number of zeros of eigenfunctions. This chapter deals mostly with the latter.

For the Dirichlet boundary condition this eigenvalue problem takes the form

$$-\varphi'' + p(x)\varphi = \lambda\varphi, \quad 0 \leq x \leq 1, \qquad \varphi|_{x=0,1} = 0 \tag{1.1}$$

under normalization, where $p = p(x)$ is a regular coefficient. Its significance lies in the fact that the number of zeros of this eigenfunction increase one by one with its order, intersecting each other. Reviewing the proof of this fundamental fact is still

T. Suzuki, *Liouville's Theory in Linear and Nonlinear Partial Differential Equations*, Springer Asia Pacific Mathematics Series 8,
https://doi.org/10.1007/978-981-95-1975-0_1

exciting; one sees interesting communication between continuity and discreteness, which results in a huge number of applications in mathematical analysis. Our point of view here is the observation of the underlying symmetric structure of the problem.

1.1.2 Reverse Theory

A properly defined linear operator possesses a spectrum. Given the domain $\Omega \subset \mathbf{R}^2$, for example, eigenvalues of

$$-\Delta = \frac{\partial^2}{\partial x_1^2} + \frac{\partial^2}{\partial x_2^2}$$

provided with the Dirichlet boundary condition constitute a set of spectra. It describes the tone delivered from Ω if this Ω stands for a vibrating membrane with fixed boundary. Then, what do we know of Ω from this tone? This is the question posed by M. Kac [12].

In the one-dimensional case of (1.1), the eigenvalues determine the operator within the framework of symmetric operators [21]. The following property is one of the results shown in this chapter, where C^1 regularity of coefficients is assumed:

$$C_s^1 = \{p \in C^1[0, 1] \mid p(1-x) = p(x),\ 0 \leq x \leq 1\}.$$

Denote the eigenvalues of (1.1) by $\lambda_n^*(p)$, $n = 1, 2, \cdots$. Then, Theorem 1.33 below assures that the condition

$$\lambda_n^*(q) = \lambda_n^*(p),\ \forall n \in \mathbf{N}^* = \{1, 2, \cdots\}, \quad q, p \in C_s^1$$

implies

$$q \equiv p.$$

Let $p \in C_s^1$, and put

$$\lambda_n^* = \lambda_n(p), \quad n \in \mathbf{N}^*$$

to proceed to the following question on the reconstruction of coefficient: Given

$$n_1 \in \mathbf{N}^*, \quad \mu_1 \in (\lambda_{n_1 - 1}, \lambda_{n_1 + 1})$$

with $\lambda_0 = -\infty$, is there any $q \in C_s^1$ satisfying

$$\lambda_n(q) = \lambda_n^*,\ \forall n \neq n_1, \quad \lambda_{n_1} = \mu_1? \tag{1.2}$$

From the above result, this q is unique if it exists.

Actually, the answer to the above existence question is positive, and this $q \in C_s^1$ is given by

$$q = p - 2\frac{d^2}{dx^2}\log\tau, \quad \tau = W(f_1, \varphi_{n_1}^*), \tag{1.3}$$

where

$$W(f, g) = f'g - fg'$$

denotes the *Wronskian*,

$$\varphi_{n_1}^* = \varphi_{n_1}^*(p)$$

is the eigenfunction to (1.1) for $\lambda = \lambda_{n_1}^*$, and $f = f_1(x)$ is a solution to

$$-f'' + p(x)f = \lambda f \quad \text{in } 0 \leq x \leq 1 \tag{1.4}$$

for $\lambda = \mu_1$ with suitable symmetry or anti-symmetry, according to whether n_1 is odd or even. This property is formulated as Theorem 1.36 below.

By this result, any one of $\{\lambda_n^*\}$ is moved arbitrarily within the category of symmetric functions. In the simplest case,

$$p = 0 \quad \Rightarrow \quad \lambda_n^*(p) = (n\pi)^2, \ n \in \mathbf{N}^*,$$

therefore, the coefficient $q \in C_s^1$ satisfying (1.2), denoted by $q_1 = q_1(x)$, takes the form of (1.3) with an elementary function $\tau(x) = W(f_1, \varphi_{n_1}^*)$, where

$$\varphi_{n_1}^*(x) = \sin n_1\pi x$$

and $f_1 = f_1(x)$ is a solution to (1.4) for $p(x) = 0$ and $\mu = \mu_1$.

We now remove one of the eigenvalues $\{\lambda_n^*(q_1)\}$, say $\tilde{\lambda}_{n_2}^* = \lambda_{n_2}^*(q_1)$ for $n_2 \in \mathbf{N}^* \setminus \{n_1\}$, to

$$\mu_2 \in (\lambda_{n_2-1}^*(q_1), \lambda_{n_2+1}^*(q_1)).$$

Let q_2 be the renewed coefficient obtained in this way. This $q_2 = q_2(x)$ is actually given by

$$q_2 = q_1 - 2\frac{d^2}{dx^2}W(f_2, \tilde{\varphi}_{n_2}^*),$$

where $f = f_2(x)$ is a solution to

$$-f'' + q_1(x)f = \mu f \quad \text{in } 0 \le x \le 1 \tag{1.5}$$

for $\mu = \mu_2$ and $\tilde{\varphi}_{n_2}^* = \varphi_{n_2}^*(q_1)$. Since $\tilde{\varphi}_{n_2}^*$ is also a solution to (1.5) for $\mu = \tilde{\lambda}_{n_2}^*$, if any solution to (1.5) for any $\mu \in \mathbf{R}$ is represented by that to (1.3) for some $\lambda \in \mathbf{R}$ and $p(x) = 0$, we can represent $q_2 = q_2(x)$ by the solutions to (1.3). This property is actually valid, and is achieved by an elementary calculation of integration described in Theorem 1.38 in Sect. 1.6.2.

We thus establish the method of *superposition*, which ensures, in particular, that $p \in C_s^1$ satisfying

$$\lambda_n^*(p) = (n\pi)^2$$

except for a finite number of $n \in \mathbf{N}^*$, is an *elementary* function in the form of

$$p = -2\frac{d^2}{dx^2}\log \tau$$

(Theorem 1.39 in Sect. 1.6.2).

The function $\tau = \tau(x)$ in (1.3), however, is represented by a *Fredholm determinant* associated with the Gel'fand–Levitan–Marchenko (GLM) equation (Theorem 1.46 in Sect. 1.7.3).

1.1.3 Organization of Chap. 1

We begin with *applied inverse problems*, introducing the notion of identification and identifiability in Sect. 1.2. Method of spectral expansion is introduced and then the key structure, the Gel'fand–Levitan theory is described there.

Most applied inverse problems are not provided with stability. This property is called *ill-posedness*. A typical example is the *quasi-inverse matrix* for the finite dimensional under-determined linear equation, where the method of regularization is used to detect unstable solutions produced by the quasi-inverse matrix.

Section 1.3 is devoted to uniqueness and non-uniqueness of the solution to a parabolic inverse problem [19], although a sharp degree of ill-posedness is detected in an inverse hyperbolic problem [23]. Taking preparations on Dirichlet series and Goursat problems for hyperbolic equations, we reduce the problems to a GLM equation, to reach the Hochstadt equation by an elementary method of *separation of variables*. In Sects. 1.4 and 1.5 we develop a standard theory on general eigenvalue problems and the above described Sturm–Liouville property of (1.1), respectively. Decreasing of zeros in the time of the solution to parabolic equations in one space dimension is also illustrated. Finding a discrete structure in continuously distributed objects becomes one of the key concerns in later studies developed in this book.

Inverse spectral theory is treated in Sect. 1.6, via the study of the Gel'fand–Levitan–Marchenko equation. We show that the symmetric Robin boundary condition ensure the other category than the above described Dirichlet condition, uniqueness, existence, construction, and superposition. Finally, the operator theoretical structure of GLM equations is revealed in accordance with the Fredholm determinant in Sect. 1.7.

1.2 Applied Inverse Problems

1.2.1 *Identification of Conductivity*

To confirm the derivation of the heat equation, let $\Omega \subset \mathbf{R}^3$ be a domain composed of a heat conductive material and u be the temperature. Write $x = (x_1, x_2, x_3) \in \Omega$ and $t \in \mathbf{R}_+ \equiv (0, +\infty)$ for space and time variables, respectively, and take an arbitrary subdomain $\omega \subset \Omega$. Let c and ρ be the specific heat and the material density, respectively. Since

$$\int_\omega c\rho u(x,t)dx$$

stands for the heat quantity contained in ω,

$$\frac{d}{dt}\int_\omega c\rho u(x,t)\,dx = \int_\omega \frac{\partial}{\partial t}(c\rho u(x,t))\,dx$$

denotes the rate of its change, while

$$\int_{\partial\omega} \kappa\frac{\partial u}{\partial n}(x,t)\,d\sigma_x$$

indicates the heat quantity incoming ω, where κ, n, and $d\sigma$ denote the conductivity, the outer unit normal vector, and the surface element of $\partial\omega$, respectively. Then we obtain

$$\int_\omega \frac{\partial}{\partial t}(c\rho)\,dx = \int_{\partial\omega} \kappa\frac{\partial u}{\partial n}\,d\sigma = \int_\omega \mathrm{div}\,(\kappa\ \mathrm{grad}\ u)\,dx$$

by the divergence formula, and hence

$$\frac{\partial}{\partial t}(c\rho u) = \mathrm{div}\,(\kappa\ \mathrm{grad}\ u), \quad x \in \Omega,\ 0 < t < \infty, \tag{1.6}$$

because $\omega \subset \Omega$ is arbitrary. Equality (1.6) is the heat equation to be provided with the initial-boundary condition. Thus the initial condition

$$u|_{t=0} = u_0(x), \quad x \in \Omega \tag{1.7}$$

indicates the state of the initial time, while the boundary condition prescribes that on the boundary.

The first boundary condition,

$$u = \alpha(\xi, t), \quad \xi \in \partial\Omega, \ 0 < t < \infty \tag{1.8}$$

describes the temperature on the boundary itself. The second boundary condition

$$\kappa \frac{\partial u}{\partial n} = \beta(\xi, t), \quad \xi \in \partial\Omega, \ 0 < t < \infty \tag{1.9}$$

denotes the heat flux on the boundary. Under the presence of the boundary radiation, the condition

$$\kappa \frac{\partial u}{\partial n} + \mu u = \gamma(\xi, t), \quad \xi \in \partial\Omega, \ 0 < t < \infty \tag{1.10}$$

is taken, where μ is a function of $\xi \in \partial\Omega$ and $t \in (0, \infty)$. These conditions (1.8)–(1.10) are called the *Dirichlet, Neumann, and Robin* boundary condition, respectively.

In *direct problems*, the coefficients c, ρ, κ, μ, initial value $u_0(x)$, and boundary values α, β, γ, are known, and the fundamental question of the well-posedness is concerned with the unique existence of the solution $u = u(x, t)$ and its continuous dependence on the data; coefficients, and initial, and boundary values. By this property we can predict the temperature anywhere and anytime, provided that the data are actually given. If the data are not completely known, however, there arises a new problem to determine the unknown data by taking conditions for the solution, which is called the *inverse problem.*

In practical problems the coefficients may be important rather more than the initial value. These unknown coefficients can be constants or piecewise constants, where the inverse problem forms a profile of parameter identification. This kind of problem arises in science, engineering, and medicine. Nevertheless, this problem has a mathematical difficulty, that is, ill-posedness, lack of uniqueness and stability. Several techniques have been developed to overcome it.

Here we consider

$$\begin{aligned} &\frac{\partial u}{\partial t} = a(t)\frac{\partial^2 u}{\partial x^2}, && 0 < x < \infty, \ 0 < t < T \\ &u|_{x=0} = f(t), && 0 < t < T \\ &u|_{t=0} = 0, && 0 \le x < \infty, \end{aligned} \tag{1.11}$$

where the material is assumed to be a semi-infinite interval of a radioactive medium and the conductivity $a = a(t)$ varies with the time. The temperature $u = u(x, t)$ is supposed to be zero at the initial state $t = 0$, and $f = f(t)$ at the end point $x = 0$. Then the direct problem is to determine u for given a and f [11].

To establish the well-posedness of this problem, compatibility conditions at $x = 0$ and prescribed behavior as $x \to \infty$ of u are necessary.

Theorem 1.1 *Suppose the following:*

(a) $0 < a = a(t) \in C[0, T)$.
(f1) $f = f(t) \in C[0, T)$ *and* $f(0) = 0$.

Then there is a unique solution u to (1.11) such that

(b) $u = u(x, t) \in C([0, \infty) \times (0, T))$.
(c) $u_x,\ u_{xx},\ u_t \in C((0, \infty) \times (0, T))$.
(d) $\sup_{0\le t\le T'}\{|u(x, t)| + |u_x(x, t)|\} \le C \exp(x^\alpha)$, $x \uparrow \infty$, *for each given* $T' < T$ *with* $\alpha < 2$.

Proof See that $(f1)$ follows from (b), that is, the compatibility condition at $x = 0$, and introduce the transformation

$$\theta = \int_0^t a(\tau)\, d\tau.$$

Since $\theta(0) = 0$ and $\theta'(t) > 0$, $0 \le t < T$, the inverse transformatic $t = t(\theta)$ is well-defined as $t'(\theta) = 1/a(t)$. Then $U(x, \theta) = u(x, t(\theta))$ satisfies

$$\begin{aligned} &\frac{\partial U}{\partial \theta} = \frac{\partial^2 U}{\partial x^2}, \quad 0 < x < \infty,\ 0 < \theta < \Theta \\ &U|_{x=0} = F(\theta),\ 0 \le \theta < \Theta \\ &U|_{\theta=0} = 0, \qquad 0 \le x < \infty, \end{aligned} \tag{1.12}$$

where $F(\theta) = f(t(\theta))$ and $\Theta = \theta(T)$. It is a classical problem provided with the fundamental solution

$$K(x, \theta) = \frac{1}{\sqrt{4\pi\theta}} \exp(-\frac{x^2}{4\theta}), \quad 0 \le x < \infty,\ 0 < \theta < \infty. \tag{1.13}$$

Then it holds that

$$U(x, \theta) = -2 \int_0^\theta \frac{\partial K}{\partial x}(x, \theta - \tau) F(\tau)\, d\tau, \quad 0 < x < \infty,\ 0 < \theta < \Theta. \tag{1.14}$$

In fact, since K in (1.14) is the fundamental solution, the function

$$\tilde{V}(x, \theta) = \int_{-\infty}^{+\infty} K(x - y, \theta) \tilde{V}_0(y)\, dy, \quad -\infty < x < \infty,\ 0 < \theta < \infty$$

satisfies

$$
\begin{aligned}
&(\frac{\partial}{\theta\theta} - \frac{\partial^2}{\partial x^2})\tilde{V} = 0, \quad -\infty < x < \infty,\ 0 < \theta < \infty \\
&\lim_{\theta \downarrow 0} \tilde{V}(x,\theta) = \tilde{V}_0(x), \quad -\infty < x < \infty
\end{aligned} \tag{1.15}
$$

for appropriate $\tilde{V}_0(x)$. The solution to

$$
\begin{aligned}
&(\frac{\partial}{\partial\theta} - \frac{\partial^2}{\partial x^2})V = 0, \quad 0 < x < \infty,\ 0 < \theta < \infty \\
&V|_{x=0} = 0,\ 0 < \theta < \infty, \quad V|_{\theta=0} = V_0(x),\ 0 < x < \infty,
\end{aligned} \tag{1.16}
$$

on the other hand, is given by the reflection,

$$
\tilde{V}(x,\theta) = \begin{cases} V(x,\theta), & 0 < x \leq \infty,\ 0 < \theta < \infty \\ -V(-x,\theta), & 0 < x < \infty,\ 0 < \theta < \infty. \end{cases} \tag{1.17}
$$

This $\tilde{V}$ satisfies (1.15) for

$$
\overline{V}_0(x) = \begin{cases} V_0(x), & 0 < x < \infty \\ -V_0(-x), & -\infty < x < 0. \end{cases}
$$

Hence there arises

$$
V(x,\theta) = \int_0^\infty G(x,y;\theta)V_0(y)dy, \quad 0 < x < \infty,\ 0 < \theta < \infty,
$$

for

$$
G(x,y;\theta) = K(x-y,\theta) - K(x+y,\theta),
$$

called the Green's function in (1.16).

The inhomogeneous problem

$$
\begin{aligned}
&(\frac{\partial}{\partial\theta} - \frac{\partial^2}{\partial x^2})V = F(x,\theta), && 0 < x < \infty,\ 0 < \theta < \infty \\
&V|_{x=0} = 0, && 0 < \theta < \infty \\
&V|_{\theta=0} = 0, && 0 < x < \infty
\end{aligned}
$$

is solved by the Duhamel principle,

$$
V(x,\theta) = \int_0^\theta d\tau \int_0^\infty G(x,y;\theta-\tau)F(y,\tau)dy.
$$

Similarly, the solution to the inhomogeneous boundary value problem (1.12) is given by

$$V(x,\theta) = \int_0^\theta -\frac{\partial}{\partial x}G(x,y;\theta-\tau)\Big|_{y=0} F(\tau)\,d\tau, \tag{1.18}$$

that is, (1.14). Equality (1.14) then implies the theorem by the inverse transformation,

$$u(x,t) = U(x,\theta(t)),$$

and the proof is complete. □

We turn to an inverse problem for (1.11), that is,

$$\begin{aligned} &\frac{\partial u}{\partial t} = a(t)\frac{\partial^2 u}{\partial x^2}, && 0 < x < \infty,\ 0 < t < T \\ &u|_{x=0} = f(t), && 0 \le t < T \\ &u|_{t=0} = 0, && 0 \le x < \infty. \end{aligned} \tag{1.19}$$

Here we assume that the conductivity $a = a(t)$ is unknown as well as the temperature $u = u(x,t)$, and instead, the heat flux at the endpoint,

$$-a(t)\frac{\partial u}{\partial x}\Big|_{x=0} = g(t), \quad 0 < t < \infty \tag{1.20}$$

is known besides the boundary temperature $f = f(t)$. We study unique solvability of $a = a(t)$ and $u = u(x,t)$ in (1.19) for given $(f,g) = (f(t),g(t))$.

Theorem 1.2 (Jones) *Assume that* $a = a(t)$ *and* $f = f(t)$ *satisfy (a) in Theorem 1.1 and*

(f2) $f \in W^{1,1}_{loc}[0,\ T) \cap W^{1,\infty}_{loc}(0,\ T)$ *with* $f(0) = 0$,

respectively. Then the boundary flux

$$g(t) = -a(t)\frac{\partial u}{\partial x}\Big|_{x=0}$$

is continuous on $[0,\ T)$ *in Theorem 1.1.*

Note that $(f2)$ implies $(f1)$.

Theorem 1.3 *Under the assumption*

[$(f3)$] $f \in W^{1,1}_{loc}[0,\ T) \cap W^{1,\infty}_{loc}(0,T)$, $f(0) = 0$, $f' \ge 0,\ f \not\equiv 0,\ a.e.$, *there exist $a = a(t)$ and $u = u(x,t)$ satisfying (1.19)–(1.20) with (a)–(d), if and only if the given continuous function $g = g(t)$ satisfies*

$$[(K)] \quad K(t) = \frac{1}{\sqrt{\pi} g(t)} \int_0^t \frac{f'(\tau)}{(t-\tau)^{1/2}} d\tau > 0, \quad 0 \leq t < T.$$

Theorem 1.4 *Under the assumptions of $(f3)$ and*

$$[(K_\varepsilon)] \quad K_\varepsilon \in L^\infty_{loc}[0, \; T),$$

the solution $(a, u) = (a(t), u(x,t))$ satisfying (1.19)–(1.20) and (a)–(d) is unique, where $\varepsilon > 0$ and

$$K_\varepsilon(t) = \frac{1}{\sqrt{\pi} g(t)} \int_0^t \frac{f'(\tau)}{(l-\tau)^{1/2+\varepsilon}} d\tau.$$

Proof *(Theorem 1.2)* We use (1.13)–(1.14). Since

$$\partial^2 K / \partial x^2 = \partial K / \partial \theta, \quad \theta > 0$$

we obtain

$$\begin{aligned}
\frac{\theta U}{\partial x}(x, \theta) &= -2 \int_0^\theta \frac{\partial^2 K}{\partial x^2}(x, \theta - \tau) F(\tau)\, d\tau \\
&= 2 \int_0^\theta \frac{\partial K}{\partial \tau}(x, \theta - \tau) F(\tau)\, d\tau \\
&= -2 \int_0^\theta K(x, \theta - \tau) F'(\tau) d\tau, \quad 0 < x < \infty, \; 0 < \theta < \Theta.
\end{aligned}$$

Then $(f2)$ implies

$$\begin{aligned}
\frac{\partial U}{\partial x}(0, \theta) &= -2 \int_0^\theta K(0, \theta - \tau) F'(\tau)\, d\tau \\
&= -\frac{1}{\sqrt{\pi}} \int_0^\theta \frac{f'(\tau)}{\sqrt{\theta - \tau}}\, d\tau, \; 0 < \theta < \Theta,
\end{aligned}$$

or

$$g(t) = -a(t) \frac{\partial u}{\partial x}\bigg|_{x=0} = \frac{a(t)}{\sqrt{\pi}} \int_0^t \frac{f'(r)}{(\int_r^t a(r') dr')^{1/2}} dr,$$

which is continuous on $[0, \; T)$ by

$$f \in W^{1,1}_{loc}[0, T) \cap W^{1,\infty}_{loc}(0, T),$$

and the proof is complete. □

From the above proof we can verify the following property.

Proposition 1.1 *Given continuous functions a, f, and $g \not\equiv 0$ satisfying (a) and $(f2)$, we have $u = u(x,t)$ such that (1.18) and (b)–(d) hold if and only if*

$$a(t) = \sqrt{\pi} g(t) / \int_0^t \frac{f'(t)}{(\int_r^t a(r)dr)^{1/2}} d\tau, \quad 0 \le t < T. \tag{1.21}$$

Proof *(Theorem 1.3)* Suppose that $0 < a = a(t) \in C[0, T)$ satisfies (1.21). Then $g \ge 0$ follows from $(f3)$ so that

$$0 < a(t) \le a_S(t)^{1/2} \sqrt{\pi} g(t) / \int_0^t \frac{f'(\tau)}{(t-\tau)^{1/2}} d\tau, \quad 0 \le t < T,$$

where

$$a_S(t) = \sup_{0 \le t' \le t} a(t').$$

Hence it holds that

$$K(t) = \frac{1}{\sqrt{\pi} g(t)} \int_0^t \frac{f'(\tau)}{(t-\tau)^{1/2}} d\tau > 0, \quad 0 \le t < T.$$

Suppose, conversely, $(f3)$ and (K). Setting

$$(La)(t) = \sqrt{\pi} g(t) / \int_0^t \frac{f'(\tau)}{(\int_\tau^t a(r)dr)^{1/2}},$$

we show the existence of its fixed point. Let

$$a_I(t) = \inf_{0 \le t' \le t} a(t').$$

Since $f', g \ge 0$ we obtain

$$a_I(t)^{1/2} K(t) \le (La)(t) \le a_S(t)^{1/2} K(t), \quad 0 < \forall a = a(t) \in C[0, T).$$

In particular, $\underline{a} = K_I^2$ and $\overline{a} = K_S^2$ are sub- and super-solutions to $a = La$, respectively, which means

$$\underline{a} \le L\underline{a}, \quad \overline{a} \ge L\overline{a}.$$

It holds also that

$$0 < \underline{a} \leq \overline{a},$$

while the mapping L: $L^1_+(0, T) \to C[0, T)$ is order-preserving, that is,

$$a \leq b \quad \Rightarrow \quad La \leq Lb,$$

where $L^1_+(0, T)$ denotes the set of non-negative L^1 functions. Therefore,

$$\{L^n \underline{a} | n = 1, 2, 3, \cdots\}$$

forms a monotone increasing family of continuous functions bounded above by $\overline{a}$. By the Beppo–Levi theorem the limit function

$$a_\infty = \lim_{n\to\infty} L^n a \in L^\infty(0, T)$$

satisfies (1.21) with $a_\infty \geq \underline{a} > 0$. It belongs to $C[0, T)$ and is a fixed point of L. We thus obtain a desired a and hence that of u. □

Proof *(Theorem 1.4)* To show the uniqueness of $a \in C[0, T)$ for (1.21), let a_1 and a_2 be fixed points of L. It holds that

$$\begin{aligned}
a_1(t) &- a_2(t) = (La_1)(t) - (La_2)(t) \\
&= \sqrt{\pi}\{\frac{1}{g(t)}\int_0^t \frac{f'(\tau)}{(\int_\tau^t a_1(r)dr)^{1/2}}d\tau\}^{-1} \\
&\cdot\{\frac{1}{g(t)}\int_0^t \frac{f'(\tau)}{(\int_\tau^t a_2(r)dr)^{1/2}}d\tau\}^{-1}\frac{1}{g(t)}\int_0^t f'(\tau)d\tau\cdot(\int_\tau^t a_1(r)dr)^{-1/2} \\
&\cdot(\int_\tau^t a_2(r)dr)^{-1/2}\{(\int_\tau^t a_1(r)dr)^{1/2} + (\int_\tau^t a_2(r)dr)^{1/2}\}^{-1} \\
&\cdot\int_\tau^t (a_1(r) - a_2(r))\, dr,
\end{aligned}$$

and hence,

$$\begin{aligned}
|a_1(t) - a_2(t)| &\leq C\frac{\sqrt{\pi}}{g(t)}\int_0^t \tau f'(\tau)(t-\tau)^{-\frac{3}{2}}\int_\tau^t |a_1(r) - a_2(r)|\, dr \\
&\leq C\|K_\varepsilon\|_{L^\infty(0,t)}\int_0^t \frac{|a_1(r) - a_2(r)|}{(t-r)^{1-\varepsilon}}dr, \quad 0 \leq t < T,
\end{aligned}$$

which implies $a_1 \equiv a_2$ by Gronwall's inequality, as desired. □

1.2.2 Optimization in Seismography

Here we study a one-dimensional model of elastic wave,

$$\begin{aligned} \rho(z)\frac{\partial^2 y}{\partial t^2} &= \frac{\partial}{\partial z}(\mu(z)\frac{\partial y}{\partial z}), && 0 < z < \infty,\ 0 < t < T \\ -\mu(0)\frac{\partial y}{\partial z}\bigg|_{z=0} &= g(t), && 0 < t < T \\ y|_{t=0} = \frac{\partial y}{\partial t}\bigg|_{t=0} &= 0, && 0 < z < \infty, \end{aligned} \tag{1.22}$$

where $y = y(z, t)$ denotes the displacement of the rock at the depth z and the time t. The coefficients $\rho = \rho(z)$ and $\mu = \mu(z)$ stand for the density and the elastic coefficient of the rock, which are assured to be the functions of z. In practice they may be piecewise constant, but here we consider the general setting of $p, \mu \in L^\infty_+(\mathbf{R}_+)$, where $L^\infty_+(\mathbf{R}_+)$ denotes the set of positive L^∞ functions on $R_+ \equiv [0, \infty)$.

The boundary condition indicates that the displacement is zero and changes with the velocity zero at the initial time, while the function $g = g(t)$ indicates the excitation communicated on the medium surface. Hence the quantity

$$Y(t) = y|_{z=0}, \quad 0 < t < T, \tag{1.23}$$

denotes a seismogram, the recording of the vibrating state of the surface.

It is an inverse problem of reflection seismology, to determine the unknown coefficients ρ and μ as well as the unknown displacement $y = y(z, t)$ through the unknown data, the input function $g = g(t)$ and the output function $Y = Y(t)$. The most crucial difference from the previous problem is that the unknown coefficients ρ and μ are supposed to be the functions of z, not of t, and hence it is impossible to derive a closed relation between (p, μ) and (g, Y) as in the nonlinear integral equation (1.21).

There are several things that we should note. First, the hyperbolic equation preserves singularity of the initial value so that recording of the vibrating state of the surface is technically difficult, involving noise. Second, from the finite propagation property of the hyperbolic equation, it is hopeless to determine unknown coefficients (ρ, μ) up to $z \to \infty$, only by the observation in finite time of $(g(t), Y(t))$, $0 < t < T$. Finally, it is not reasonable to determine two unknown functions (ρ, μ) simultaneously, just by adding one more condition (1.23) other than (1.22). This ambiguity is resolved by the equivalent class below.

We are concentrating on the construction of the least squares approximate solution, called identification in applied sciences [4]. Set

$$Y(\rho, \mu; t) = y(0,\ t),$$

where $y = y(z,t)$ denotes the solution to (1.22) for given (ρ, μ). This solution is taken in the weak sense described later. Let Y_d be the observed seismogram, which may contain a noise but is supposed to be a known function.

Taking a suitable set $\Sigma \subset L^2(\mathbf{R}_+) \times L^2(\mathbf{R}_+)$, called the set of admissible parameters, we consider the optimization problem

$$\text{Minimize } \{J(\rho,\mu) \mid (\rho,\mu) \in \Sigma\}, \tag{1.24}$$

where

$$J(\rho,\mu) = \int_0^T \{Y(\rho,\mu;t) - Y_d(t)\}^2 dt$$

[4]. The question is the existence of a minimizer in (1.24), or the possibility of a topology in Σ, which is strong enough to make J continuous and simultaneously weak enough to make Σ compact.

The optimization (1.24) accepts the natural action of the group due to the transformation of variables. To see this property, taking $\varphi \in L^\infty_+(\mathbf{R}_+)$, we introduce the new variable z' by

$$\frac{dz'}{dz} = \varphi(z), \quad z'\big|_{z=0} = 0. \tag{1.25}$$

Then, the function $y'(z',t) = y(z(z'),t)$ satisfies

$$\begin{aligned}
\rho'(z')\frac{\partial^2 y'}{\partial t^2} &= \frac{\partial}{\partial z'}(\mu'(z')\frac{\partial y'}{\partial z'}), && 0 < z' < \infty, \ 0 < t < T \\
-\mu'(0)\frac{\partial y'}{\partial z'}\bigg|_{z=0} &= g(t), && 0 < t < T \\
y'\big|_{t=0} = \frac{\partial y'}{\partial t}\bigg|_{t=0} &= 0, && 0 < z' < \infty
\end{aligned}$$

whenever $y = y(z,t)$ solves (1.22), where

$$\rho'(z') = \frac{\rho(z(z'))}{\varphi(z(z'))}, \quad \mu'(z') = \frac{\mu(z(z'))}{\varphi(z(z'))}. \tag{1.26}$$

Then it holds that $Y(\rho',\mu';t) = Y(\rho,\mu;t)$ by $y|_{z=0} = y'\big|_{z=0}$. Regarding this property, we write

$$(\rho,\mu) \sim (\rho',\mu')$$

for $(\rho,\mu), (\rho',\mu') \in L^\infty_+(\mathbf{R}_+) \times L^\infty_+(\mathbf{R}_+)$ if there is $\varphi \in L^\infty_+(\mathbf{R}_+)$ such that (1.26) holds, where z and z' are related as in (1.25).

Then the optimization problem (1.24) may be replaced by

$$\text{Minimize}\ \{J(p, \mu) \mid (\rho, \mu) \in \Sigma/\sim\}. \tag{1.27}$$

If $\varphi(z) = \sqrt{\rho(z)/\mu(z)}$ in (1.25), we reach a physical quantity. In fact, $v(z) = 1/\varphi(z)$ denotes the velocity of the wave propagation and hence $z'(z)$ indicates the travelling time from $z = 0$ to $z = z$. Then we have $(\rho', \mu') = (\sigma, \sigma)$ with

$$\sigma(z') = \sqrt{\rho(z(z'))\mu(z(z'))},$$

called acoustical impedance. Given $\sigma\pm > 0$, we provide a topology to

$$\Sigma^X = \{\sigma \in L^\infty(0, X) \mid \sigma_- \leq \sigma(x) \leq \sigma_+ \text{ a.e. } x \in (0, X)\},$$

strong enough to make J Lipschitz continuous and weak enough to make Σ^T compact, which guarantees a solution to the optimization problem

$$\text{Minimize}\ \{J(\sigma, \sigma) \mid \sigma \in \Sigma^T\}. \tag{1.28}$$

To this end, we introduce the notion of weak solutions to (1.22). Since

$$X \equiv \int_0^Z \sqrt{\rho(z)/\mu(z)}dz$$

stands for the travelling time between $z = 0$ to $z = Z$, system (1.22) may be provided with the additional condition $y|_{z=Z} = 0$ in $(z, t) \in (0, Z) \times (0, T)$ for $X \geq T$:

$$\begin{aligned}
p(z)\frac{\partial^2 y}{\partial t^2} &= \frac{\partial}{\partial z}(\mu(z)\frac{\partial y}{\partial z}), && 0 < z < Z,\ 0 < t < T \\
-\mu(0)\frac{\partial y}{\partial z}\bigg|_{z=0} &= g(t),\ \ y|_{z=Z} = 0, && 0 < t < T \\
y|_{t=0} = \frac{\partial y}{\partial z}\bigg|_{t=0} &= 0, && 0 < z < Z.
\end{aligned} \tag{1.29}$$

We use this system, which is suitable for numerical computations.

Take a pair of Hilbert spaces,

$$H = L^2(0, Z), \quad V = \{v \in H^1(0, Z) \mid v(Z) = 0\},$$

where

$$H^1(0, Z) = W^{1,2}(0, Z) = \{v \in L^2(0, Z) \mid v' \in L^2(0, Z)\}$$

denotes the usual Sobolev space. Identifying H' with H through Riesz's representation theorem, we get a *Gel'fand triple*

$$V \hookrightarrow H \cong H' \hookrightarrow V'.$$

Given $\rho, \mu \in L^\infty_+(R_+)$, we put

$$((u, v))_\mu = \int_0^Z \mu(z)\frac{\partial u}{\partial z}\frac{\partial v}{\partial z}\, dz, \quad \|u\|_\mu^2 = (u, u)_\mu, \quad u, v \in V$$

$$(u, v)_\rho = \int_0^Z \rho(z) uv\, dz, \qquad |u|_\rho^2 = (u, u)_\rho, \quad u, v \in H.$$

If $y = y(z, t)$ is smooth enough and satisfies (1.22) in the classical sense, each $v \in H^1(0, Z)$ admits

$$\begin{aligned}\frac{d^2}{dt^2}(y(t), v)_\rho &= \int_0^Z \frac{\partial}{\partial z}(\mu(z)\frac{\partial y}{\partial z})v(z)\, dz \\ &= \mu(z)\frac{\partial y}{\partial z}v(z)\Big|_{z=0}^{z=Z} - \int_0^Z \mu(z)\frac{\partial y}{\partial z}\frac{\partial v}{\partial z}\, dz \\ &= g(t)v(0) - ((y(t), v))_\mu.\end{aligned}$$

Assuming $g \in W^{1,1}(0, T)$ for

$$W^{1,1}(0, T) = \{g \in L^1(0, T) \mid g' \in L^1(0, T)\},$$

we can define $L_g \in L^2(0, T; V')$ by

$$_{V'}\langle L_g(t), v\rangle_V = g(t)v(0), \quad v \in V \hookrightarrow C^0[0, Z]).$$

Then, the weak solution to (1.22) indicates a function

$$y = y(\cdot, t) \in L^2(0, T; V), \quad \frac{dy}{dt}(\cdot, t) \in L^2(0, T; H)$$

satisfying

$$\begin{aligned}&\frac{d^2}{dt^2}(y(t), v)_\rho + ((y(t), v))_\mu = \langle L_g(t), v\rangle \quad \text{a.e. } t \text{ for each } v \in V \\ &y(0) = \frac{dy}{dt}(0) = 0, \end{aligned} \tag{1.30}$$

We note that

$$I(t) = \frac{1}{2}\int_0^Z \rho(z)y_t^2 + \mu(z)y_z^2\, dz \in L^\infty(0,T)$$

stands for the energy, and equality (1.31) in the following theorem is formally derived from

$$\begin{aligned} I'(t) &= \int_0^Z \rho(z)y_t y_{tt} + \mu(z)y_z y_{zt}\, dz \\ &= \int_0^Z y_t \frac{\partial}{\partial z}(\mu(z)y_z) + \mu(z)y_z y_{zt}\, dz \\ &= y_t\mu(z)y_z|_{z=0}^{z=Z} - \int_0^Z y_{tz}\mu(z)y_z dz + \int_0^Z \mu(z)y_z y_{zt} dz = y_t(0,t)g(t). \end{aligned}$$

We note also that the relation $\frac{dy}{dt} \in L^2(0,T;H)$ implies $y \in C([0,\ T];\ H)$. It holds also that

$$\frac{dy}{dt} \in C([0,T];V')$$

by $\frac{d^2y}{dt^2} \in L^2(0,T;V')$, which is derived from (1.30) and $y \in L^2(0,T;V)$.

Theorem 1.5 *If $\rho \in L^\infty_+(\mathbf{R}_+)$ and $\mu \in L^\infty_+(\mathbf{R}_+)$ satisfy*

$$\rho_- \le \rho(z) \le \rho_+,\ \mu_- \le \mu(z) \le \mu_+, \quad a.e.\ z \in (0,\ Z)$$

for some constants $\rho_\pm > 0$ and $\mu_\pm > 0$, there is a unique weak solution $y = y(\cdot,t)$ to (1.22) such that

$$y \in L^\infty(0,T;V), \quad \frac{dy}{dt} \in L^\infty(0,T;H),$$

with the property that

$$g \in W^{1,1}(0,T) \mapsto (y, \frac{dy}{dt}) \in L^\infty(0,T;V) \times L^\infty(0,T;H)$$

is continuous and

$$I(t_2) - I(t_1) = g(t_2)y(0,t_2) - g(t_1)y(0,t_1) - \int_{t_1}^{t_2} g'(s)y(0,\ s)ds \tag{1.31}$$

for a.e. $t_1,\ t_2$.

Proof To show the uniqueness, assume $y \in L^2(0,T;V)$, $y' \in L^2(0,T;H)$, and

$$\frac{d^2}{dt^2}(y(t), v)_\rho + ((y(t), v))_\mu = 0, \quad \text{a.e. } t \in (0, T) \text{ for each } v \in V$$

$$y(0) = y'(0) = 0.$$

Since the inclusion $V \subset H$ is compact, we have a complete orthonormal system of eigenfunctions $\{\varphi_j\}$ satisfying

$$((\varphi_j, v))_\mu = \lambda_j(\varphi_j, v)_\rho, \ \forall v \in V, \quad (\varphi_i, \varphi_j)_\rho = \delta_{ij}, \ ((\varphi_i, \varphi_j))_\mu = \lambda_i \delta_{ij},$$

where $\{\lambda_j\}$ in $0 < \lambda_1 \leq \lambda_2 \leq \cdots$ stand for the eigenvalues. The function $Y_j(t) = (y(t), \varphi_j)_\rho$ satisfies

$$\frac{d^2}{dt^2}Y_j + \lambda_j Y_j = 0, \ \text{a.e. } t \in (0, T), \quad Y_j(0) = Y_j'(0) = 0,$$

and hence $Y_j \equiv 0$ for any j which implies $y(t) \equiv 0$.

The existence of the solution is verified by Galerkin's method. Let

$$V_m = \langle \varphi_1, \cdots \varphi_m \rangle \subset V, \quad m = 1, 2, \cdots,$$

and define the function $y_m = y_m(t) : [0, T] \to V_m$ by

$$\frac{d^2}{dl^2}(y_m(t), \varphi_j)_\rho + ((y_m(t), \varphi_j))_\mu = \langle L_g(t), \varphi_j \rangle, \quad 1 \leq j \leq m, \ 0 \leq t \leq T$$

$$y_m(0) = \frac{dy_m}{dt}(0) = 0.$$

Writing $y_m(t) = \sum_{j=1}^m Y_j^m(t)\varphi_j$, we obtain

$$\frac{d^2}{dt^2}Y_m^j + \lambda_j Y_j^m = g(t)\varphi_j(0), \quad 1 \leq j \leq m, \ 0 \leq l \leq T$$

$$Y_j^m(0) = g(t)\frac{dY_j^m}{dt}(0) = 0. \tag{1.32}$$

to define $y_m(t)$. This $Y_j^m = Y_j$ is independent of m by the uniqueness of the solution to the above system of ordinary differential equations.

Putting

$$I_m(t) = \frac{1}{2}\sum_{j=1}^m (\frac{dY_j}{dt})^2 + \frac{1}{2}\sum_{j=1}^m \lambda_j Y_j^2 = \frac{1}{2}\left|\frac{dy_m}{dt}\right|_\rho^2 + \frac{1}{2}\|y_m(t)\|_\mu^2,$$

we have

$$\frac{d}{dt} I_m(t) = \left. \frac{dy_m}{dt} \right|_{z=0}$$

by (1.32), and hence

$$I_m(t_2) - I_m(t_1) = g(t_2) y_m(0, t_2) - g(t_1) y_m(0, t_1) - \int_{t_1}^{t_2} g'(s) y_m(0, s) ds$$

for $y_m(\cdot, t) = y_m(t)$. Let $t_2 = t$ and $t_1 = 0$. Since $\varphi_j \in V$ implies $y_m(Z, s) = 0$, it holds that

$$-y_m(0, s) = \int_0^Z \frac{\partial y_m}{\partial z}(z, s)\, dz.$$

Then we obtain

$$y_m(0, s)^2 \le Z \int_0^Z \{\frac{\partial y_m}{\partial z}(z, s)\}^2 dz \le Z \mu_-^{-1} \|y_m(s)\|_\mu^2, \quad 0 \le s \le T, \tag{1.33}$$

and hence

$$I_m(t) \le C\{|g(t)| + \int_0^t |g'(s)|\, ds\} \cdot \max_{0 \le s \le t} I_m(t)^{\frac{1}{2}} \le C \|g\|_{W^{1,1}(0,T)} \cdot \max_{0 \le s \le t} I_m(t)^{\frac{1}{2}},$$

recalling $W^{1,1}(0, T) \subset L^\infty(0, T)$. There thus arises

$$\max_{0 \le t \le T} I_m(t) \le C \|g\|_{W^{1,1}(0,T)}^2, \tag{1.34}$$

which implies the boundedness of

$$\{y_m\} \subset L^\infty(0, T; V), \quad \{\frac{dy_m}{dt}\} \subset L^\infty(0, T; H),$$

and hence that of

$$\{\frac{d^2 y_m}{dt^2}\} \subset L^\infty(0, T; V').$$

Their suitable subsequence converges $*$-weakly, which assures the weak solution to (1.11) satisfying

$$y \in L^\infty(0, T; V), \quad \frac{dy}{dt} \in L^\infty(0, T; H).$$

Furthermore (1.34) implies

$$\max_{0\le t\le T} I(t) \le C\|g\|^2_{W^{1,1}(0,T)}$$

and hence

$$g \in W^{1,1}(0,T) \mapsto (y, \frac{dy}{dt}) \in L^\infty(0,T;V) \times L^\infty(0,T;H)$$

is continuous. Finally, using

$$I_m(t) \to I(t) \equiv \frac{1}{2}\sum_{j=1}^{\infty}(\frac{dY_j}{dt})^2 + \frac{1}{2}\sum_{j=1}^{\infty}\lambda_j y_j^2,$$

we have the strong convergence

$$y_m(t) \to y(t) = \sum_{j=1}^{\infty} Y_j(t)\varphi_j \qquad \text{in } V$$

$$\frac{dy_m}{dt}(t) \to \frac{dy}{dt}(t) = \sum_{j=1}^{\infty}\frac{dY_j}{dt}(t)\varphi_j q \quad \text{in } H$$

for each $t \in I$, and hence (1.31) for a.e. t_1, t_2. The proof is complete. □

Proposition 1.2 *If* $g(0) = 0$ *and* $g \in W^{1,2}(0,T)$ *in the previous theorem, it holds that*

$$y \in C^0([0,T];V)$$
$$\frac{dy}{dt} \in C^0([0,T];H) \cap L^\infty(0,T;V)$$
$$\frac{d^2y}{dt^2} \in C^0([0,T];V') \cap L^\infty(0,T;H).$$

The function $I = I(t)$ *is continuous and is equal to* $\int_0^t g(s)\frac{\partial y}{\partial t}(0,s)ds$.

Proof By the hypothesis, we can take a weak solution $x = x(z,t)$ to

$$\rho(z)\frac{\partial^2 x}{\partial t^2} - \frac{\partial}{\partial z}(\mu(z)\frac{\partial x}{\partial z}) = 0, \quad 0 < z < Z,\ 0 < t < T$$

$$-\mu(0)\frac{\partial x}{\partial z}\Big|_{z=0} = g'(t), \quad x|_{x=Z} = 0, \quad 0 < t < T$$

$$x|_{t=0} = \frac{\partial x}{\partial t}\Big|_{t=0} = 0, \quad 0 < z < Z$$

with the property described in Theorem 1.5. The function

$$\tilde{y}(t) = \int_0^t x(s)ds$$

satisfies (1.29) by $g(0) = 0$, and hence

$$y(t) = \int_0^t x(s)ds$$

follows from uniqueness of the solution. Hence we obtain $x(t) = y'(t)$ with desired properties, to complete the proof. □

We proceed to the existence of a minimizer to (1.28). Recall that $\mathrm{Y}_d = Y_d(t)$ is the observed seismogram and

$$J(\rho, \mu) = \int_0^T \{Y(\rho, \mu; t) - Y_d(t)\}^2 \, dt$$

$$\Sigma^X = \{\sigma \in L^\infty(0, X) \mid \sigma_- \le \sigma(x) \le \sigma_+, \quad \text{a.e. } x \in (0, X)\},$$

where $\sigma_\pm > 0$ are given constants and $X = T$. To verify the existence of the optimization problem

$$\text{Minimize } \{J(\sigma, \sigma) \mid \sigma \in \Sigma^X\}, \tag{1.35}$$

we introduce a topology to Σ^X. Given $\sigma \in \Sigma^X$, we define $r = r(z')$, $0 \le z' \le X'$, by

$$r(z') = \sigma(z(z'))^2, \quad X' = \int_0^X \frac{dz}{\sigma(z)}, \tag{1.36}$$

where z and z' are related as in (1.25) for $\varphi(z) = \frac{1}{\sigma(z)}$. Here we take

$$(1, r(z')), \quad 0 \le z' \le X',$$

equivalent to $(\sigma(z), \sigma(z))$, $0 \le z \le X$, to define

$$d^X(\sigma_1, \sigma_2) = \max_{0 \le z' \le X_1' \vee X_2} \left| \int_0^z \tilde{r}_1(\tau) - \tilde{r}_2(\tau) \, d\tau \right|, \quad \sigma_1, \sigma_2 \in \Sigma^X,$$

where

$$\tilde{r}(z') = \begin{cases} r(z'), & 0 \le z' \le X' \\ r(X'), & X' \le z'. \end{cases} \tag{1.37}$$

Let

$$R^S = \{r \in L^\infty(0,S) \mid \sigma_-^2 \le r(s) \le \sigma_+^2 \text{ a.e. } s \in (0,S)\}$$
$$D^S(r_1,r_2) = \max_{0\le s\le S}\left|\int_0^s r_1(s) - r_2(s)\,ds\right|, \quad r_1,\ r_2 \in \mathbf{R}^S$$

for $S > 0$.

Lemma 1.1 *This $D^S(\cdot,\cdot)$ is a metric on R^S, which is compact.*

Proof Given $r \in R^S$, put $\hat{r} = \int_0^s r(\tau)d\tau$. It holds that

$$D^S(r_1,r_2) = \|\hat{r}_1 - \hat{r}_2\|_{L^\infty(0,S)},$$

and hence $R^S \cong \mathscr{K}$ as a Banach space, where

$$\mathscr{K} = \{\hat{r} \in C^0[0,S] \mid \hat{r}(0) = 0 \text{ absolutely continuous on } [0,S],$$
$$\sigma_-^2 \le \hat{r}'(s) \le \sigma_+^2 \text{ a.e. } s \in (0,S))\}.$$

Hence R^S is compact by the Ascoli–Arzela theorem and the proof is complete. □

Lemma 1.2 *It holds that*

$$\|Y(1,r_1;\cdot) - Y(1,r_2;\cdot)\|_{C^0[0,T]} \le C\|g\|_{W^{3,1}(0,T)} D^S(r_1,r_2) \tag{1.38}$$

for $C = C(\sigma_-,T,S) > 0$, provided that $g \in \mathrm{W}^{3,1}(0,T)$ and $g'(0) = 0$.

Proof We have $Y(1,r_j;t) = y_j(0,t)$, $j = 1,2$, for

$$\begin{aligned}
&r_j(s)\frac{\partial^2 y_j}{\partial t^2} - \frac{\partial^2 y_j}{\partial s^2} = 0, && 0 < s < S,\ 0 < t < T\\
&-\left.\frac{\partial y_j}{\partial s}\right|_{s=0} = g(t),\ \ y_j\big|_{s=S} = 0, && 0 < t < T,\\
&y_j\big|_{t=0} = \left.\frac{\partial y_j}{\partial t}\right|_{t=0} = 0, && 0 < s < S.
\end{aligned} \tag{1.39}$$

As in the proof of Theorem 1.5, there arises

$$\left\|\frac{\partial^2 y_j}{\partial s\partial t}\right\|_{L^\infty(0,T;H)} + \left\|\frac{\partial^2 y_j}{\partial t^2}\right\|_{L^\infty(0,T;H)} \le C\|g\|_{W^{2,1}(0,T;H)}$$
$$\left\|\frac{\partial^2 y_j}{\partial s\partial t^2}\right\|_{L^\infty(0,T;H)} + \left\|\frac{\partial^2 y_j}{\partial t^3}\right\|_{L^\infty(0,T;H)} \le C\|g\|_{W^{3,1}(0,T;H)}$$

by $g \in W^{3,1}(0,T)$ and $g(0) = g'(0) = 0$.

Let

$$\Delta I = \frac{1}{2}\int_0^S r_1 \left(\frac{\partial \Delta y}{\partial t}\right)^2 + \left(\frac{\partial \Delta y}{\partial s}\right)^2 ds$$

for

$$\Delta y = y_2 - y_1, \quad \Delta r = r_2 - r_1, \quad \Delta \hat{r}(s) = \int_0^s \Delta r(\tau) d\tau.$$

We obtain

$$\begin{aligned}
&\Delta r \frac{\partial^2 y_2}{\partial t^2} + r_1 \frac{\partial^2 y}{\partial t^2} - \frac{\partial^2 \Delta y}{\partial s^2} = 0, && 0 < s < S,\ 0 < t < T \\
&\left.\frac{\partial \Delta y}{\partial s}\right|_{S=0} = \Delta y|_{s=S} = 0, && 0 < t < T \\
&\Delta y|_{t=0} = \left.\frac{\partial \Delta y}{\partial t}\right|_{t=0} = 0, && 0 < s < S
\end{aligned}$$

by (1.39), which implies

$$\begin{aligned}
\Delta I(t) &= \frac{1}{2}\int_0^S r_1 \left(\frac{\partial \Delta y}{\partial t}\right)^2 + \left(\frac{\partial \Delta y}{\partial s}\right)^2 ds \\
&= -\int_0^t dt \int_0^S ds\, \Delta r \frac{\partial^2 y_2}{\partial t^2}\frac{\partial \Delta y}{\partial t} \\
&= -\int_0^t dt \int_0^S ds\, (\Delta \hat{r})_s \frac{\partial^2 y_2}{\partial t^2}\frac{\partial \Delta y}{\partial t} \\
&= \int_0^t dt \int_0^S ds\, \Delta \hat{r}(s) \frac{\partial}{\partial s}\left(\frac{\partial^2 y_2}{\partial t^2}\frac{\partial \Delta y}{\partial t}\right) \\
&= \int_0^t dt \int_0^S ds\, \Delta \hat{r}(s) \left(\frac{\partial^3 y_2}{\partial s \partial t^2}\frac{\partial \Delta y}{\partial t} + \frac{\partial^2 y_2}{\partial t^2}\frac{\partial^2 \Delta y}{\partial t \partial s}\right) \\
&= \int_0^t dt \int_0^S ds\, \Delta \hat{r}(s) \left(\frac{\partial^3 y_2}{\partial s \partial t^2}\frac{\partial \Delta y}{\partial t} - \frac{\partial^3 y_2}{\partial t^3}\frac{\partial \Delta y}{\partial s}\right) \\
&\quad + \int_0^S ds\, \Delta \hat{r}(s)\, \frac{\partial^2 y_2}{\partial t^2} \cdot \left.\frac{\partial \Delta y}{\partial s}\right|_{t=t} \\
&\le C\|g\|_{W^{3,1}(0,T)}\|\Delta \hat{r}\|_{C^0[0,S]}\left\{\int_0^t \Delta I(t)^{\frac{1}{2}} dt + \Delta I(t)^{\frac{1}{2}}\right\},
\end{aligned}$$

and hence

$$\sup_{t\in[0,T]} \Delta I(t) \leq C\|g\|^2_{W^{3,1}(0,T)}\|\Delta\hat{r}\|^2_{C^0[0,S]}.$$

Then the desired (1.38) follows from (1.33) and the proof is complete. □

We conclude this section with the solvability of (1.35).

Theorem 1.6 (Bamberger–Chavent–Lailly) *The set Σ^X is equipped with a pseudo distance, denoted by $d^X(\cdot,\cdot)$, which does not satisfy the triangular inequality but induces a Hausdorff topology. This Σ^X is sequentially compact and $j(\sigma) \equiv J(\sigma,\sigma)$ is continuous, provided that $g \in W^{3,1}(0,T)$ and $g(0) = g'(0) = 0$.*

Proof Given $\sigma \in \Sigma$, we write r_σ and S_σ for r and X' defined by (1.36). There arises the mapping

$$\sigma \in \Sigma^X \mapsto r_\sigma \in R^{S_\sigma}$$

in this way. We have $S \equiv X/\sigma_- \geq S_\sigma$ for any $\sigma \in \Sigma$, and hence the mapping $\sigma \in \Sigma^X \mapsto \tilde{r}_\sigma \in R^S$ arises with the extension (1.37). For $T = X$ and

$$\tilde{\sigma}(z) = \begin{cases} \sigma(z), & 0 \leq z \leq X \\ \sigma(X), & X \leq z \end{cases}$$

we have $d^X(\sigma_1,\sigma_2) = D^S(r_{\tilde{\sigma}_1}, r_{\tilde{\sigma}_1})$ by the definition.

To verify the sequential compactness, let $\{\sigma_k\} \subset \Sigma^X$. Then its subsequence, denoted by the same symbol, satisfies

$$D^S(r_{\tilde{\sigma}_k}, r) \to 0$$

with some $r \in R^S$ by Lemma 1.1. Inverting the transformation, we find $\sigma \in \Sigma^X$ such that $r_{\tilde{\sigma}} = r$, which means $d^\Sigma(\sigma_k,\sigma) \to 0$. Thus Σ^X is sequentially compact.

We have, on the other hand,

$$\begin{aligned}
&\|Y(\sigma_1,\sigma_1;\cdot) - Y(\sigma_2,\sigma_2;\cdot)\|_{C^0[0,T]} \\
&= \|Y(\tilde{\sigma}_1,\tilde{\sigma}_1;\cdot) - Y(\tilde{\sigma}_2,\tilde{\sigma}_2;\cdot)\|_{C^0[0,T]} \\
&= \|Y(1,r_{\tilde{\sigma}_1};\cdot) - Y(1,r_{\tilde{\sigma}_2};\cdot)\|_{C^0[0,T]} \\
&\leq C\|g\|_{W^{1,3}(0,T)} d^S(r_{\tilde{\sigma}_1}\ r_{\tilde{\sigma}_2}) \\
&= C\|g\|_{W^{1,3}(0,T)} d^X(\sigma_1,\sigma_2)
\end{aligned}$$

by Lemma 1.2. Therefore,

$$J(\sigma,\sigma) = \|Y(\sigma,\sigma) - Y_d\|^2_{L^2(0,T)}$$

is continuous in $\sigma \in \Sigma^X$, and the proof is complete. □

The metric $D^S(\cdot, \cdot)$ on R^S is rather weak, admitting oscillating functions $\{r_k\}$ which converge to a flat function r_∞. The associated output $Y(1, r_\infty; t)$, however, detects the average movement of $Y(1, r_k; t)$ for $k \gg 1$.

1.2.3 Spectral Theory and Identifiability

The material in Sect. 1.2.1 is due to [11], where a closed relation between unknown coefficients and the given data is derived. The study in Sect. 1.2.2, on the other hand, is due to [4], where a closed form between the input output data and unknown coefficients is not expected and efforts are made to reach a best approximation by reducing output error. This approach is called identification, taken in applied sciences. Several theoretical problems, however, admit the uniqueness of the solution. This question is called identifiability from the viewpoint of inverse problems.

Take the initial-boundary value problem of the parabolic equation

$$
\begin{aligned}
&\frac{\partial u}{\partial t} - \frac{\partial^2 u}{\partial x^2} + p(x)u = 0, \quad 0 < x < 1, && 0 < t < T \\
&-\frac{\partial u}{\partial x} + hu\Big|_{x=0} = 0, \ \frac{\partial u}{\partial x} + Hu\Big|_{x=1} = f(t), && 0 < t < T \\
&u|_{t=0} = 0, && 0 < x < 1,
\end{aligned} \tag{1.40}
$$

where $p \in C^1[0, 1]$, $h \in \mathbf{R}$, and $H \in \mathbf{R}$. Assume that the coefficients (p, h, H) are unknown, while the boundary values of u, that is, $f = f(t)$ in (1.40) and $g(t) = u|_{x=1}$, $0 < t < T$, are known. The method of Sect. 1.2.1 is to derive a closed relation between (p, h, H) and (f, g). It may not be valid for this setting, but the approach in Sect. 1.2.2 still works.

We take a reference system

$$
\begin{aligned}
&\frac{\partial v}{\partial t} - \frac{\partial^2 v}{\partial x^2} + q(x)v = 0, && 0 < x < 1, \ 0 < t < T \\
&-\frac{\partial v}{\partial x} + jv\Big|_{x=0} = 0, \ \frac{\partial v}{\partial x} + Jv\Big|_{x=1} = f(t), && 0 < t < T \\
&v|_{t=0} = 0, && 0 < x < 1
\end{aligned} \tag{1.41}
$$

with the prescribed coefficients $(q, j, J) \in C^1[0, 1] \times \mathbf{R} \times \mathbf{R}$, and try to reduce the output error

$$
\varepsilon(t) = (v - u)|_{x=1}, \quad 0 < t < T
$$

in some norm. So, can we conclude $(q, j, J) = (p, h, H)$ by $\varepsilon \equiv 0$? Or, can we conclude that (q, j, J) is close to (p, h, H) if ε is small? This is a problem different from identification, called identifiability composed of uniqueness and stability.

A result of [15] says the following. Note that $f \equiv 0$ implies $u \equiv 0$ for any (q, j, J) in (1.41), and therefore, concluding $(q, j, J) = (p, h, H)$ by $\varepsilon \equiv 0$ is impossible.

Theorem 1.7 (Pierce) *The condition $\varepsilon(t) \equiv 0$ implies*

$$(q, j, J) = (p, h, H),$$

provided that $f \not\equiv 0$.

The other uniqueness arises in a hyperbolic inverse problem studied by [13]. It is concerned with the determination of the air pressure from the sound caused by thunder. Let

$$\mathbf{R}^3_+ = \{(x, y, z) \mid x, y \in \mathbf{R},\ z > 0\}, \quad p, q \in C^0[0, \infty) \cap L^\infty(0, \infty).$$

Given $f = f(t) > 0$ with $f(0) = 1$, we take

$$\frac{\partial^2 u}{\partial t^2} - \Delta u = p(z)u + f(t)\delta(x, y, z), \quad (x, y, z) \in \mathbf{R}^3_+,\ t > 0$$

$$\left.\frac{\partial u}{\partial z}\right|_{z=0} = 0,\ t > 0, \quad u|_{t=0} = \left.\frac{\partial u}{\partial t}\right|_{t=0} = 0, \quad (x, y, z) \in \mathbf{R}^3_+.$$

Let

$$\frac{\partial^2 v}{\partial t^2} - \Delta v = q(z)v + f(t)\delta(x, y, z), \quad (x, y, z) \in \mathbf{R}^3_+,\ t > 0$$

$$\left.\frac{\partial v}{\partial z}\right|_{z=0} = 0,\ t > 0, \quad v|_{t=0} = \left.\frac{\partial v}{\partial t}\right|_{t=0} = 0, \quad (x, y, z) \in \mathbf{R}^3_+$$

be the reference system, and take two points on the surface of the earth with different distances from the origin denoted by

$$P_1 = (x_1, y_1, 0),\ P_2 = (x_2, y_2, 0), \quad 0 < x_2^2 + y_2^2 < x_1^2 + y_1^2.$$

Theorem 1.8 (Lavrentiev–Reznickaja) *The equalities*

$$v(x_1, y_1, 0; t) = u(x_1, y_1, 0, t),\ v(x_2, y_2, 0; t) = u(x_2, y_2, 0; t), \quad 0 < t < \infty,$$

imply

$$q(z) = p(z), \quad 0 < z < \infty,$$

provided that $q(0) = p(0)$.

Most inverse problems on partial differential equations are ill-posed, with lack of stability in standard metrics. So far, it has been known that the method of *Carleman's estimate* has a generality for a class of these problems. Theorems 1.7 and 1.8, however, follow from the *Gel'fand– Levitan theory* [7]. It is an inverse theory of the general expansion theorem of Titchmarsh–Kodaira, constructing spectral density function $\rho = \rho(\lambda)$ from the differential operator

$$-\frac{d^2}{dx^2} + p(x) \text{ on } [0, \infty), \qquad (-\frac{d}{dx} + h)\cdot\Big|_{x=0} = 0.$$

Hence the Gel'fand–Levitan theory reconstructs (p, h) from ρ. These theories are constructive and concerned with the singular boundary value problem.

We thus meet the first aspect of Liouville's theory, in direct and inverse Sturm–Liouville problems. The following subsection is a summary on the Gel'fand–Levitan theory and its operator theoretical structure.

1.2.4 Summary of the Gel'fand–Levitan Theory

To describe regular versions of the above theories, take

$$P = (p, h, X) \in X \equiv C^1[0, 1] \times \mathbf{R} \times \mathbf{R},$$

and let A_P be the Sturm-Liouville operator

$$-\frac{d^2}{dx^2} + p(x)$$

with the boundary condition

$$(-\frac{d}{dx} + h)\cdot\Big|_{x=0} = (\frac{d}{dx} + H)\cdot\Big|_{x=1} = 0.$$

Its spectrum denoted by $\sigma(A_P)$ consists of simple eigenvalues denoted by

$$\sigma(A_P) = \{\lambda_n(P)\}_{n=0}^{\infty}, \qquad -\infty < \lambda_0(P) < \lambda_1(P) < \cdots \to +\infty.$$

Let, furthermore, $\Phi_n = \Phi_n(\cdot; P)$ be the eigenfunction of A_P corresponding to $\lambda_n = \lambda_n(P)$, normalized by $\Phi_n|_{x=0} = 1$. The normalizing constants $\{\rho_n(P)\}_{n=0}^{\infty}$ are defined by

$$\rho_n(P) = \int_0^1 \Phi_n(x; P)^2 dx,$$

and then the set of two sequences

$$C(P) = \{\lambda_n(P), \rho_n(P)\}$$

is called the *spectral characteristics* of A_P. In this regular case, the Gel'fand–Levitan theory asserts that P is reconstructed by $C(P)$. In particular, for $P, Q \in X$ the relation $C(P) = C(Q)$ implies $P \equiv Q$. In the problem of A. Pierce, for example, the data (f, g) determine the spectral characteristics $C(P)$ and hence the uniqueness $Q \equiv P$ follows.

Actual reconstruction of P by $C(P)$ is as follows. Given $p \in C^1[0, 1]$, $h \in \mathbf{R}$, and $\lambda \in \mathbf{R}$, let $\Phi = \Phi(x; p, h, \lambda)$ be the solution to

$$(-\frac{d^2}{dx^2} + p(x))\Phi(x) = \lambda\Phi(x), \ 0 < x < 1, \quad \Phi|_{x=0} = 1, \ \Phi'\big|_{x=0} = h,$$

which satisfies $\Phi_n(\cdot; P) = \Phi(\cdot; p, h, \lambda_n(P))$ by the definition. Then there arises

$$\Phi(x; p, h, \lambda) = \cos\sqrt{\lambda}x + \int_0^x K(x, y)\cos\sqrt{\lambda}y dy, \quad 0 \le x \le 1 \tag{1.42}$$

with $K = K(x, y) \in C^2(\bar{D})$ for

$$D = \{(x, y) \mid 0 < y < x < 1.$$

This $K = K(x, y)$ is independent of $\lambda \in \mathbf{R}$ and solves

$$\begin{aligned} &K_{xx} - K_{yy} = p(x)K, && (x, y) \in D \\ &K(x, x) = h + \frac{1}{2}\int_0^x p(s)ds, \ K_y(x, 0) = 0, \quad 0 \le x \le 1. \end{aligned}$$

By means of Parseval's relation

$$\int_0^1 g(x)^2 dx = \sum_{n=0}^{\infty} \frac{1}{\rho_n} E(\lambda_n)^2, \quad E(\lambda_n) = \int_0^1 g(x)\Phi_n(x; P)dx,$$

valid for $g \in L^2(0, 1)$, we next arrive at the equality

$$F(x, y) + \int_0^x K(x, z)F(z, y)dz + K(x, y) = 0 \tag{1.43}$$

via (1.42), where $F = (\partial^2 f/\partial x \partial y)$ for

$$f(x,y)=\sum_{n=0}^{\infty}\{\frac{1}{\rho_n}\frac{\sin\sqrt{\lambda_n}x\cdot\sin\sqrt{\lambda_n}y}{\lambda_n}-\frac{2}{\pi^2}\frac{\sin n\pi x\cdot\sin n\pi y}{n^2}\}$$

$$+\frac{1}{\rho_0}\frac{\sin\sqrt{\lambda_0}x\cdot\sin\sqrt{\lambda_0}y}{\lambda_0}-xy. \tag{1.44}$$

Given a pair of sequences $\{\lambda_n,\rho_n\}_{n=0}^{\infty}$ with

$$-\infty<\lambda_0<\cdots\to\infty,\qquad \rho_n>0,$$

conversely, let $f = f(x,y)$ be the right-hand side of (1.44). Under suitable assumptions on the asymptotic behaviors of $\{\lambda_n,\rho_n\}$, say,

$$\lambda_n^{\frac{1}{2}}=n\pi+\frac{b_1}{n}+\frac{b_3}{n^3}+O(\frac{1}{n^3}),\quad \rho_n=\frac{1}{2}+\frac{a_1}{n^2}+O(\frac{1}{n^4}),\quad n\to\infty \tag{1.45}$$

for some $b_1,b_3,a_1\in\mathbf{R}$, the function

$$F=(\partial^2 f/\partial x\partial y)$$

is C^2, and furthermore, the integral equation (1.43) has a unique solution $K = K(x,y)\in C^2(\bar{D})$. Then we define $(p,h,H)\in C^1[0,1]\times\mathbf{R}\times\mathbf{R}$ by

$$p=2\frac{d}{dx}K(x,x),\qquad h=K(0,0),\qquad H=-\frac{\Phi'(1;p,h,\lambda_n)}{\Phi(1;p,h,\lambda_n)}.$$

This H is actually independent of n, and the spectral characteristics of the Sturm–Liouville operator A_P, $P=P(p,h,H)\in X$, coincide with $\{\lambda_n,\rho_n\}$.

Asymptotic formula (1.45) actually holds if $p\in C^1[0,1]$ and p' is absolutely continuous on $[0,1]$, but does not hold for general $p\in C^1[0,1]$, in which case we have only

$$\lambda_n^{\frac{1}{2}}=n\pi+\frac{b_1}{n}+O(\frac{1}{n^3}),\ \rho_n=\frac{1}{2}+O(\frac{1}{n^2}),\quad n\to\infty. \tag{1.46}$$

Hence a regularity gap arises between the direct mapping $(p,h,H)\mapsto\{\lambda_n,\rho_n\}$ and the inverse mapping $\{\lambda_n,\ \rho_n\}\mapsto(p,h,H)$.

Equality (1.42) is generalized as

$$\Phi(x;q,j,\lambda)=\Phi(x;p,h,\lambda)+\int_0^x K(x,y)\Phi(y;p,h,\lambda)dy,\quad 0\le x\le 1, \tag{1.47}$$

with $K(x,y)=K(x,y;q,j;p,h)\in C^2(\bar{D})$ defined by

$$K_{xx}-K_{yy}+p(y)K=q(x)K,\qquad (x,y)\in D$$

$$
\begin{aligned}
&K(x,x) = (j-h) + \frac{1}{2}\int_0^x q(s) - p(s)\,ds, && 0 \le x \le 1 \\
&K_y(x,0) = hK(x,0), && 0 \le x \le 1. \qquad (1.48)
\end{aligned}
$$

These relations are applicable to establish identifiability concerning the parabolic problem

$$
\begin{aligned}
&\frac{\partial u}{\partial t} - \frac{\partial^2 u}{\partial x^2} + p(x)u = 0, && 0 < x < 1,\ 0 < t < T \\
&-\frac{\partial u}{\partial x} + hu\Big|_{x=0} = \frac{\partial u}{\partial x} + Hu\Big|_{x=1} = 0, && 0 < t < T \\
&u|_{t=0} = a(x), && 0 < x < 1, \qquad (1.49)
\end{aligned}
$$

where $P = P(p, h, H) \in X$ and $a \in L^2(0,1)$ are unknown, to be determined by, for instance,

$$
f_j = u|_{x=j}, \quad j = 0, 1,\ 0 < t < T.
$$

For the problem to determine P and a by

$$
g_1 = u|_{x=x_0},\ g_2 = \frac{\partial u}{\partial x}\Big|_{x=x_0}, \quad 0 < t < T,
$$

for instance, another modification of (1.42) is necessary, where a sensitive dependence on the observation point $x_0 \in (0,1)$ arises for the identifiability [22].

There are direct equalities to connect the solutions to parabolic equations [20]. For instance, if $u = u(x,t)$ solves the first equality of (1.49) and

$$
-\frac{\partial u}{\partial x} + hu\Big|_{x=0} = 0, \quad 0 < t < T
$$

then for $K = K(x,y)$ defined by (1.48), the function

$$
v(x,t) = u(x,t) + \int_0^x K(x,y)v(y,t)\,dy
$$

satisfies

$$
\begin{aligned}
&\frac{\partial v}{\partial t} - \frac{\partial^2 v}{\partial x^2} = q(x)v, \quad 0 < x < 1,\ 0 < t < T \\
&-\frac{\partial v}{\partial x} + jv\Big|_{x=0} = 0,\ 0 < t < T, \quad v|_{x=0} = u|_{x=0},\ 0 < t < T.
\end{aligned}
$$

A similar formula holds for solutions of hyperbolic equations, which is applied in the study of the stability for hyperbolic inverse problems [23].

So far, we have discussed the Robin boundary condition. To develop spectral theory for the Dirichlet boundary condition, let $\Phi = \Phi(x; p, \infty, \lambda)$ be the solution to

$$(-\frac{d^2}{dx^2} + p(x))\Phi = \lambda\Phi, \ 0 \le x \le 1, \quad \Phi|_{x=0} = 0, \ \Phi'\big|_{x=0} = 1$$

for $p \in C^1[0, 1]$ and $\lambda \in \mathbf{R}$. Then we obtain

$$\Phi(x; q, \infty, \lambda) = \Phi(x; p, \infty, \lambda) + \int_0^x K(x, y)\Phi(y; p, \infty, \lambda)\, dy, \quad 0 \le x \le 1,$$

where

$$K = K(x, \ y; q, \ \infty; \ p, \ \infty) \in C^2(\overline{D})$$

is independent of λ and is defined by the first equation of (1.48) with

$$K(x, x) = \frac{1}{2}\int_0^x q(s) - p(s)\, ds, \ K(x, 0) = 0, \quad 0 \le x \le 1.$$

In comparison with the above formula on the Robin condition, we note that the above second equality gives $k(0, 0) = 0$ as a compatibility condition. This slight difference induces a duality between the spectrum of A_P and that of A_P^* , the differential operator $-\frac{d^2}{dx^2} + p(x)$ with the boundary condition $\cdot|_{x=0} = \cdot|_{x=1} = 0$. This duality induces an answer to a classical question in the spectral inverse problem.

To detect the other symmetry, let $\Phi^*(x; p, H, \lambda)$ be the solution to

$$(-\frac{d^2}{dx^2} + p(x))\Phi^* = \lambda\Phi^*, \ 0 \le x \le 1, \quad \Phi^*\big|_{x=1} = 1, \ \Phi^*\big|_{x=1} = -H.$$

Then in the same way as in (1.47) the relation

$$\Phi^*(x; q, J, \lambda) = \Phi^*(x; p, H, \lambda) + \int_x^1 K^*(x, y)\Phi^*(y; p, H, \lambda)dy, \quad 0 \le x \le 1$$

follows with an appropriate kernel $K^* = K^*(x, y; q, J; p, H) \in C^2(\overline{D^*})$, where

$$D^* = \{(x, y) \mid 0 < x < y < 1\}.$$

Let $\mathscr{K}$ and $\mathscr{K}^*$ be the integral operator on $[0, x]$ and $[x, 1]$ with kernels $K(x, \cdot)$ and $K^*(x, \cdot)$, respectively:

$$(\mathcal{K}g)(x) = \int_0^x K(x,y)g(y)dy, \quad (\mathcal{K}^*g)(x) = \int_x^1 K(x,y)g(y)dy.$$

An integral operator $\mathcal{F}$ on [0, 1] is defined by

$$I_d + \mathcal{F} = (I_d + \mathcal{K})^{-1}(I_d + \mathcal{K}^*).$$

It takes a kernel denoted by $F = F(x, y)$, piecewise C^2 on $\overline{D}$ and $\overline{D^*}$: $(\mathcal{F}g)(x) = \int_0^1 F(x, y)g(y)dy$. Taking this kernel is called the dressing method [16]. A direct calculation assures

$$F(x-0,x) - F(x+0,x) = C_0(Q) - C_0(P)$$

$$C_0(P) = h + H + \frac{1}{2}\int_0^1 p(s)\,ds$$

$$C_0(Q) = j + J + \frac{1}{2}\int_0^1 q(s)\,ds$$

for $Q = (q, j, J)$ and $P = (p, h, H)$.

The value $C_0(P)$ appears in the asymptotic formula (1.46), that is,

$$\lambda_n(P)^{\frac{1}{2}} = n\pi + \frac{C_0(P)}{n} + O(\frac{1}{n^3}), \qquad n \to \infty,$$

if $p \in C^1[0, 1]$. If $C_0(Q) = C_0(P)$ the kernel $F(x, \cdot)$ is continuous on [0, 1], where the solution $K = K(x, \cdot)$ to the Gel'fand–Levitan equation (1.43) is given by the Fredholm determinant

$$\tau(x) = \det(I_d + \mathcal{F}_x),$$

where $\mathcal{F}_x$ denotes the integral operator on $[0, x]$ associates with the kernel $F(x, \cdot)$. Then we obtain $K(x, x) = -\frac{d}{dx}\log\tau(x)$ and hence

$$q(x) = p(x) - 2\frac{d^2}{dx^2}\log\tau(x).$$

1.3 Boundary Identification of Parabolic Equations

1.3.1 Summary of the Result

Here we take the initial-boundary problem of the parabolic equation

$$
\begin{aligned}
&\frac{\partial u}{\partial t} - \frac{\partial^2 u}{\partial x^2} + p(x)u = 0 \qquad\qquad \text{in } 0 < x < 1, 0 < t < \infty \\
&-\frac{\partial u}{\partial x} + hu\Big|_{x=0} = \frac{\partial u}{\partial x} + Hu\Big|_{x=1} = 0, \quad u|_{t=0} = a(x),
\end{aligned} \tag{1.50}
$$

where $(p, h, H) \in X \equiv C^1[0, 1] \times \mathbf{R} \times \mathbf{R}$ and $a \in L^2(0, 1)$. If the coefficients

$$
P = P(p, h, H) \in X \equiv C^1[0, 1] \times \mathbf{R} \times \mathbf{R}
$$

and the initial value $a \in L^2(0, 1)$ are given, there exists a unique solution $u = u(x, t)$, and hence its boundary values $f_0(t) = u|_{x=0}$ and $f_1(t) = u|_{x=1}$ are determined. Suppose, conversely, that the coefficients $P \in X$ and the initial value $a \in L^2(0, 1)$ are unknown, and instead that the boundary values $f_0 = f_0(t)$ and $f_1 = f_1(t)$ of the solution $u = u(x, t)$ are observed for $t \in [T_1, T_2]$ with $0 \le T_1 < T_2 < \infty$. Then we want to determine $(P, a) \in X \times L^2(0, 1)$ by (f_0, f_1).

As is noted in the previous section, one approach to this problem is to give an explicit formula to recover (P, a) from (f_0, f_1), which is hopeless in this case. Here we take a model of (1.50) as in

$$
\begin{aligned}
&\frac{\partial v}{\partial t} + (-\frac{\partial^2}{\partial x^2} + q(x))v = 0 \qquad\qquad \text{in } 0 < x < 1, 0 < t < \infty \\
&(-\frac{\partial}{\partial x} + j)v\Big|_{x=0} = (\frac{\partial}{\partial x} + J)v\Big|_{x=1} = 0, \quad v|_{t=0} = b(x)
\end{aligned}
$$

where $Q = (q, j, L) \in X$ and $b \in L^2(0, 1)$. Computing boundary values of this $v = v(x, t)$, we have the errors

$$
\varepsilon_0(t) = v|_{x=0} - f_0(t), \quad \varepsilon_1(t) = v|_{x=1} - f_1(t), \quad T_1 \le t \le T_2.
$$

Then the question of identifiability is divided into two parts.

1. (Uniqueness) Does the relation

$$
\varepsilon_0(t) = \varepsilon_1(t) = 0, \quad T_1 \le t \le T_2 \tag{1.51}
$$

   imply $(Q, b) = (P, a)$?
2. (Stability) Is (Q, b) near from (P, a) if the errors $(\varepsilon_0, \varepsilon_1)$ are small?

Stability with respect to the initial does not hold in the standard norm because of the smoothing property of the parabolic equation. Here we take the uniqueness, which, however, does not hold without any assumption on (P, a). For instance, $u \equiv 0$ follows from $a = 0$, so that $\varepsilon_0(t) = \varepsilon_1(t) = 0$ arises for $0 \le t \le \infty$ for any $P, Q \in X$, provided that $a = b = 0$. This property violates the uniqueness of the above inverse problem. By this observation we realize that the unknown equation

$E(P, a)$ must provide good data

$$f_0 = u|_{x=0}, \quad f_1 = u|_{x=1}, \qquad T_1 \le t \le T_2 \tag{1.52}$$

to confirm the uniqueness a.

Problem 1.1 What conditions of the data f_0, f_1 in (1.52) are efficient to ensure $(Q, b) = (P, a)$ by (1.51)?

We confirm several notations in the previous section. First, given

$$P = (p, h, H) \in X \equiv C^1[0, 1] \times \mathbf{R} \times \mathbf{R},$$

A_P denotes the differential operator

$$-\frac{d^2}{dx^2} + p(x) \quad \text{in } L^2(0, 1)$$

with the boundary condition

$$(-\frac{d}{dx} + h)\cdot\Big|_{x=0} = (\frac{d}{dx} + H)\cdot\Big|_{x=1} = 0.$$

Its eigenvalues and eigenfunctions are denoted by $\{\lambda_n(P)\}_{n=0}^{\infty}$ and

$$\{\varphi_n(\cdot; P)\}_{n=0}^{\infty},$$

respectively, the latter being normalized by

$$\|\varphi_n(\cdot; P)\|_{L^2(0,1)} = 1, \quad \varphi_n(0, P) > 0.$$

Here, each $\lambda_n(P)$ is simple: $-\infty < \lambda_0(P) < \lambda_1(P) < \cdots \to \infty$. Second, given $P \in X$ and $a \in L^2(0, 1)$, $E(P, a)$ denotes the evolution equation

$$\frac{du}{dt} + A_P u = 0 \text{ in } 0 < t < \infty, \quad u|_{t=0} = a$$

in $L^2(0, 1)$, that is, the parabolic initial-boundary value problem (1.50).

Definition 1.1 The initial value $a \in L^2(0, 1)$ is called *generating* with respect to A_P if it holds that $(a, \varphi_n(\cdot; P)) \neq 0$ for any $n \in \mathbf{N} \equiv \{0, 1, 2, \dots\}$, where $(\ ,\ )$ denotes the L^2-inner product. We write $E(P, a) \in \mathcal{G}$ if (P, a) satisfies this condition.

The condition $E(P, a) \in \mathcal{G}$ is examined by the data $f_0 = u|_{x=0}$ or $f_1 = u|_{x=1}$ for $T_1 \le t \le T_2$, where $u = u(x, t)$ is the solution of $E(P, a)$.

Theorem 1.9 *Let $E(Q, b)$ and $v = v(x, t)$ be a reference equation and its solution, respectively. Suppose*

$$v|_{x=0} = f_0(t), \quad T_1 \leq t \leq T_2,$$

for

$$f_0 = u|_{x=0}, \quad 0 \leq T_1 < T_2 < \infty.$$

Then $E(P, a) \in \mathscr{G}$ is equivalent to $E(Q, b) \in \mathscr{G}$.

The answer to Problem 1.1, therefore, is given by the following theorems. Hence $E(P, a) \in \mathscr{G}$ is necessary and sufficient for the uniqueness in Problem 1.1 to hold.

Theorem 1.10 ([19]) *If $E(P, a) \in \mathscr{G}$ the conditions*

$$\varepsilon_0(t) \equiv v|_{x=0} - f_0(t) = 0, \quad \varepsilon_1(t) \equiv v|_{x=1} - f_1(t) = 0, \quad T_1 \leq t \leq T_2$$

imply $(Q, b) = (P, a)$.

Theorem 1.11 ([19]) *If $E(P, a) \notin \mathscr{G}$, conversely, there is $(Q, b) \in X \times L^2(0, 1)$ such that*

$$\varepsilon_0(t) = \varepsilon_1(t) = 0, \quad 0 \leq t < \infty$$

in spite of $Q \neq P$.

1.3.2 Study on Dirichlet Series

Given $a \in L^2(0, 1)$ and $P \in X$, henceforth, we call

$$N = \sharp\{\varphi_n(\cdot; P) \mid (a, \varphi_n(\cdot; P)) = 0\}$$

the *degenerate index* of a with respect to A_P. By the definition, $E(P, a) \in \mathscr{G}$ if and only if $N = 0$. For $(P, a) \in X \times L^2(0, 1)$ and $(Q, b) \in X \times L^2(0, 1)$, let $u = u(x, t)$ and $v = v(x, t)$ be the solutions of $E(P, a)$ and $E(Q, b)$, respectively. Let N be the degenerate index of a with respect to A_P, which may be finite or infinite. Theorem 1.9 is refined as follows.

Theorem 1.12 *If*

$$v|_{x=0} = u|_{x=0}, \quad T_1 \leq t \leq T_2 \tag{1.53}$$

with $0 \leq T_1 < T_2 < \infty$, the degenerate index of b with respect to A_Q is also N.

Proof Put

$$\lambda_n = \lambda_n(\cdot; P),\ \mu_m = \mu_m(\cdot; Q),\ \varphi_n = \varphi_n(\cdot; P),\ \psi_m = \psi_m(\cdot; Q) \tag{1.54}$$

and confirm

$$u(\cdot, t) = \sum_{n=0}^{\infty} e^{-t\lambda_n}(a, \varphi_n)\varphi_n, \quad v(\cdot, t) = \sum_{m=0}^{\infty} e^{-t\mu_m}(b, \psi_m)\psi_m. \tag{1.55}$$

Since u and v are analytic in $t \in (0, \infty)$, equality (1.53) extends to

$$\sum_{n=0}^{\infty} e^{-t\lambda_n}(a, \varphi_n)\varphi_n(0) = \sum_{m=0}^{\infty} e^{-t\mu_m}(b, \psi_m)\psi_m(0), \quad 0 < t < \infty. \tag{1.56}$$

Divide $\mathbf{N} \equiv \{0, 1, 2, \dots\}$ into

$$\mathbf{N}_1 = \{n \mid (a, \varphi_n) = 0\}, \quad \mathbf{N}_2 = \{n \mid (a, \varphi_n) \neq 0\},$$

to obtain $N = \sharp\mathbf{N}_1$. Let $\mathbf{N}_2 = \{n_1', n_2', \dots\}$. Then (1.53) implies

$$\sum_{n=n_1'}^{\infty} e^{-t\lambda_n}(a, \varphi_n)\varphi_n(0) = \sum_{m=0}^{\infty} e^{-t\mu_m}(b, \psi_m)\psi_m(0), \quad 0 < t < \infty. \tag{1.57}$$

Let m_1' be the first integer satisfying $(b, \psi_m) \neq 0$. By (1.57) it holds that

$$e^{-t\mu_{m_1'}}(b, \psi_{m_1'})\psi_{m_1'}(0) = \sum_{n=n_1'}^{\infty} e^{-t\lambda_n}(a, \varphi_n)\varphi_n(0) - \sum_{m=m_1'+1}^{\infty} e^{-t\mu_m}(b, \psi_m)\psi_m(0),$$

and hence

$$e^{-t\mu_{m_1'}}|(b, \psi_{m_1'}(0))| \le \sum_{n=n_1'}^{\infty} |e^{-t\lambda_n}(a, \varphi_n)| \cdot |\varphi_n(0)| - \sum_{m=m_1'+1}^{\infty} |e^{-t\mu_m}(b, \psi_m)| \cdot |\psi_m(0)|. \tag{1.58}$$

Using the asymptotics as $n, m \to \infty$,

$$\varphi_n(x) = \frac{1}{\sqrt{2}} \cos n\pi x + O(\frac{1}{n}), \quad \psi_m(x) = \frac{1}{\sqrt{2}} \cos m\pi x + O(\frac{1}{m}) \tag{1.59}$$

uniform in $x \in [0, 1]$, we deduce

$$|\varphi_n(0)|,\ |\psi_m(0)| \le C_1, \quad n, m = 0, 1, 2, \dots$$

with $C_1 > 0$. Hence the right-hand side of (1.58) is estimated by

$$C_1\Big(\sum_{n=n_1'}^{\infty} e^{-t\lambda_n}|(a, \varphi_n)| + \sum_{m=m_1'+1}^{\infty} e^{-t\mu_m}|(b, \psi_m)|\Big)$$

$$\le C_1\Big\{e^{-t\lambda_{n_1'}}\Big(\sum_{n=n_1'}^{\infty} e^{-2t(\lambda_n-\lambda_{n_1'})}\Big)^{\frac{1}{2}}\Big(\sum_{n=n_1'}^{\infty}|(a, \varphi_n)|^2\Big)^{\frac{1}{2}}$$

$$+ e^{-t\mu_{m_1'}}\Big(\sum_{m=m_1'+1}^{\infty} e^{-2t(\mu_m-\mu_{m_1'+1})}\Big)^{\frac{1}{2}}\Big(\sum_{m=m_1'+1}^{\infty}|(b, \psi_m)|^2\Big)^{\frac{1}{2}}\Big\}$$

$$\le C_1\Big\{e^{-t\lambda_{n_1'}}\|a\|_{L^2(0,1)}\Big(\sum_{n=n_1'}^{\infty} e^{-2t(\lambda_n-\lambda_{n_1'})}\Big)^{\frac{1}{2}}$$

$$+ e^{-t\mu_{m_1'+1}}\|b\|_{L^2(0,1)}\Big(\sum_{m=m_1'+1}^{\infty} e^{-2t(\mu_m-\mu_{m_1'+1})}\Big)^{\frac{1}{2}}\Big\}$$

by Parseval's relation.

Since the asymptotics

$$\lambda_n^{1/2} = n\pi + O\Big(\frac{1}{n}\Big), \quad \mu_m^{1/2} = m\pi + O\Big(\frac{1}{m}\Big) \tag{1.60}$$

as $n, m \to \infty$ are also valid, the inequality

$$\sum_{n=n_1'}^{\infty} e^{-2t(\lambda_n-\lambda_{n_1'})} + \sum_{m=m_1'+1}^{\infty} e^{-2t(\mu_m-\mu_{m_1'+1})} \le C_T', \quad T \le t < \infty,$$

follows for each $T > 0$ with $C_T' > 0$, and hence

$$e^{-t\mu_{m_1'}}|(b, \psi_{m_1})\psi_{m_1}(0)| \le C_T\{e^{-t\lambda_{n_1'}} + e^{-t\mu_{m_1'+1}}\}, \quad T \le t < \infty,$$

or,

$$|(b, \psi_{m_1})\psi_{m_1}(0)| \le C_T\{e^{-t(\lambda_{n_1'}-\mu_{m_1'})} + e^{-t(\mu_{m_1'+1}-\mu_{m_1'})}\}, \quad T \le t < \infty,$$

with $C_T > 0$. If $\lambda_{n'_1} > \mu_{m'_1}$, therefore, the right-hand side of the above inequality tends to zero as $t \to \infty$. Hence $(b, \psi_{m'_1}) = 0$ follows from $\psi_{m'_1}(0) \neq 0$, which contradicts the definition of m'_1.

By (1.57), similarly, we obtain

$$\begin{aligned}
&e^{-t\lambda_{n'_1}}|(a, \varphi_{n'_1})| \cdot |\varphi_{n'_1}(0)| \\
&\leq \sum_{n=n'_1+1}^{\infty} e^{-t\lambda_n}|(a, \varphi_n)| \cdot |\varphi_n(0)| + \sum_{m=m'_1}^{\infty} e^{-t\mu_m}|(b, \psi_m)| \cdot |\psi_m(0)| \\
&\leq C_1\Big\{e^{-t\lambda_{n'_1+1}}\Big(\sum_{n=n'_1+1}^{\infty} e^{-2t(\lambda_n-\lambda_{n'_1+1})}\Big)^{\frac{1}{2}}\Big(\sum_{n=n'_1+1}^{\infty}|(a, \varphi_n)|\Big)^{\frac{1}{2}} \\
&\quad + \Big(\sum_{m'_1+1}^{\infty} e^{-2t(\mu_m-\mu_{m'_1})}\Big)^{\frac{1}{2}}\Big(\sum_{m=m'_1}^{\infty}|(b, \psi_m)|^2\Big)^{\frac{1}{2}}\Big\} \\
&\leq C_T\{e^{-t\lambda_{n'_1+1}} + e^{-t\mu_{m_1}}\}, \quad 0 < T \leq t < \infty,
\end{aligned}$$

and therefore, $(a, \varphi_{n_1}) = 0$ by $\varphi_{n_1}(0) \neq 0$ in the case of $\mu'_1 > \lambda'_{n_1}$, a contradiction. We thus end up with $\lambda_{n'_1} = \mu_{m'_1}$, and hence (1.57) is reduced to

$$\begin{aligned}
&e^{-t\lambda_{n'_1}}\{(a, \varphi_{n'_1})\varphi_{n'_1}(0) - (b, \psi_{m'_1})\psi_{m'_1}(0)\} \\
&= -\sum_{n=n'_1+1}^{\infty} e^{-t\lambda_n}(a, \varphi_n)\varphi_n(0) + \sum_{m=m'_1+1}^{\infty} e^{-t\mu_m}(b, \psi_m)\psi_m(0).
\end{aligned}$$

For $t \geq T > 0$, similarly, the right-hand side of this inequality is estimated above by

$$C_T\{e^{-t\lambda_{n'_1+1}} + e^{-t\mu_{m'_1+1}}\}$$

with C_T, which implies $(a, \varphi_{n'_1})\varphi_{n'_1}(0) = (b, \psi_{m'_1})\psi_{m'_1}(0)$, and then, (1.57) means

$$\sum_{n=n'_2}^{\infty} e^{-t\lambda_n}(a, \varphi_n)\varphi_n(0) = \sum_{m=m'_2}^{\infty} e^{-t\mu_m}(b, \psi_m)\psi_m(0).$$

Continuing this procedure, we obtain the "only if" part of the following lemma, while its "if" part is obvious from the eigenfunction expansions (1.55). □

Lemma 1.3 *Under the notations of (1.54), equality (1.53) holds if and only if each* $n \in \mathbf{N}_2$ *admits* $m(n) \in \mathbf{N}$ *such that*

$$\lambda_n = \mu_{m(n)}, \ (a, \varphi_n)\varphi_n(0) = (b, \psi_{m(n)})\psi_{m(n)}(0)(\neq 0), \quad \forall n \in \mathbf{N}_2$$
$$(b, \psi_m)\psi_m(0) = 0, \quad \forall m \in \mathbf{N} \setminus \{m(n) \mid n \in \mathbf{N}_2\}. \tag{1.61}$$

To complete the proof, assume, first, $N = \sharp\, \mathbf{N}_1 < \infty$, and let

$$\mathbf{N}_1 = \{n_1, \dots, \dots n_N\}.$$

By the above assertion each $n \neq n_\ell$, $1 \leq \ell \leq N$, admits $m(n) \in \mathbf{N}$ satisfying

$$\lambda_n = \mu_{m(n)}.$$

Then the asymptotics (1.60) imply $m(n) = n$, $n \geq n_0$, for sufficiently large n_0. Since the mapping $n \mapsto m(n)$ is an order homeomorphism, then there arises

$$\sharp\, (\mathbf{N} \setminus \{m(n) \mid n \neq n_\ell, \ \leq \ell \leq N\}) = N.$$

Hence the degenerate index of b with respect to A_Q is also N by (1.61).

Assume, second, $N = \sharp\mathbf{N}_1 = \infty$. If the degenerate index of b with respect to A_Q denoted by M is finite, we obtain $N = M < \infty$ by the above argument, replacing $E(P, a)$ to $E(Q, b)$, a contradiction.

1.3.3 *Goursat Problems for Hyperbolic Equations*

A fundamental ingredient of the Gel'fand–Levitan theory is the construction of transformation kernel of associated differential operators, denoted by $K = K(x, y)$. If the space dimension is one, this K is subject to a hyperbolic equation, while the observed data are reflected to its boundary values. We thus reach various *Goursat problems*, of which well-posedness clarifies uniqueness and stability of the inverse problem.

Let $\Omega \subset \mathbf{R}^2$ be the interior of a triangle $\triangle ABC$ with $\overline{AC} = \overline{BC}$, $\angle ACB = \frac{\pi}{2}$, AB being parallel to either the x-axis or the y-axis. Let, furthermore, $V \in C^1(\overline{\Omega})$ be given. Our proof of Theorems 1.10–1.11 relies on the study of the Goursat problem for the hyperbolic equation

$$K_{xx} - K_{yy} = r(x, y)K \quad \text{in } \Omega. \tag{1.62}$$

We employ the classical d'Alembert formula and Picard's iteration method to approach this problem. Below, ν and $\tau : \mathbf{R}_+ = [0, \infty] \to \mathbf{R}_+$ denote the outer unit normal vector on $\partial\Omega$ and the monotone increasing continuous function satisfying $\tau(0) = 0$, respectively.

Proposition 1.3 *Given $f \in C^2(\overline{AC})$ and $g \in C^2(\overline{BC})$ with $f|_C = g|_C$, there exists a unique $K = K(x, y; r, f, g) \in C^2(\overline{\Omega})$ satisfying (1.62) and*

$$K|_{AC} = f, \quad K|_{BC} = g. \tag{1.63}$$

It holds that

$$\|K(\cdot,\cdot;r,f,g)\|_{C^2(\overline{\Omega})} \le \tau(\|r\|_{C^1(\overline{\Omega})}+1)\cdot\left\{\|f\|_{C^2(\overline{AC})}+\|g\|_{C^2(\overline{BC})}\right\} \tag{1.64}$$

and

$$\begin{aligned}\|K(\cdot,\cdot;r_1,f,g)-K(\cdot,\cdot;r_2,f,g)\|_{C^2(\overline{\Omega})} &\le \tau(\max\{\|r_1\|_{C^1(\overline{\omega})},\|r_2\|_{C^1(\overline{\omega})}\}+1)\\ &\cdot\|r_1-r_2\|_{C^1(\overline{\Omega})}\left\{\|f\|_{C^2(\overline{AC})}+\|g\|_{C^2(\overline{BC})}\right\}.\end{aligned} \tag{1.65}$$

Proposition 1.4 *Given $f \in C^2(\overline{AB})$ and $g \in C^1(\overline{AB})$, there exists a unique $K = K(x,y) = K(x,y;r,f,g) \in C^2(\overline{\Omega})$ satisfying (1.62) and*

$$K|_{AB} = f, \quad \left.\frac{\partial K}{\partial \nu}\right|_{AB} = g \tag{1.66}$$

It holds that

$$\|K(\cdot,\cdot;r,f,g)\|_{C^2(\overline{\Omega})} \le \tau(\|r\|_{C^1(\overline{\Omega})}+1)\cdot\left\{\|f\|_{C^2(\overline{AB})}+\|g\|_{C^1(\overline{AB})}\right\} \tag{1.67}$$

and

$$\begin{aligned}\|K(\cdot,\cdot;r_1,f,g)-K(\cdot,\cdot;r_2,f,g)\|_{C^2(\overline{\Omega})} &\le \tau(\max\{\|r_1\|_{C^1(\overline{\Omega})},\|r_2\|_{C^1(\overline{\Omega})}\}+1)\\ &\cdot\|r_1-r_2\|_{C^1(\overline{\Omega})}\left\{\|f\|_{C^2(\overline{AB})}+\|g\|_{C^1(\overline{AB})}\right\}.\end{aligned} \tag{1.68}$$

Proposition 1.5 *Given $f \in C^2(\overline{AC})$ and $g \in C^2(\overline{AB})$ with $f|_A = g|_A$, there exists a unique $K = K(x,y) = K(x,y;r,f,g) \in C^2(\overline{\Omega})$ satisfying (1.62) and*

$$K|_{AC} = f, \quad K|_{AB} = g. \tag{1.69}$$

It holds that

$$\|K(\cdot,\cdot;r,f,g)\|_{C^2(\overline{\Omega})} \le \tau(\|r\|_{C^1(\overline{\Omega})}+1)\cdot\left\{\|f\|_{C^2(\overline{AC})}+\|g\|_{C^2(\overline{AB})}\right\} \tag{1.70}$$

and

$$\begin{aligned}\|K(\cdot,\cdot;r_1,f,g)-K(\cdot,\cdot;r_2,f,g)\|_{C^2(\overline{\Omega})} &\le \tau(\max\{\|r_1\|_{C^1(\overline{\Omega})},\|r_2\|_{C^1(\overline{\Omega})}\}+1)\\ &\cdot\|r_1-r_2\|_{C^1(\overline{\Omega})}\{\|f\|_{C^2(\overline{AC})}+\|g\|_{C^2(\overline{AB})}\}.\end{aligned} \tag{1.71}$$

Proposition 1.6 *Given $f \in C^2(\overline{AC})$, $g \in C^1(\overline{AB})$, and $h \in \mathbf{R}$, there exists a unique $K = K(x,y) = K(x,y;r,h,f,g) \in C^2(\overline{\Omega})$ satisfying (1.62) and*

$$K|_{AC} = f, \quad \left.\frac{\partial K}{\partial \nu} + hK\right|_{AB} = g. \tag{1.72}$$

It holds that

$$\|K(\cdot, \cdot; r, h, f, g)\|_{C^2(\overline{\Omega})} \le \tau(\|r\|_{C^1(\overline{\Omega})} + |h| + 1) \cdot \{\|f\|_{C^2(\overline{AC})} + \|g\|_{C^1(\overline{AB})}\} \tag{1.73}$$

and

$$\begin{aligned} &\|K(\cdot, \cdot; r_1, h_1, f, g) - K(\cdot, \cdot; r_2, h_2, f, g)\|_{C^2(\overline{\Omega})} \\ &\quad \le \tau(\max\{\|r_1\|_{C^1(\overline{\Omega})} + |h_1|, \|r_2\|_{C^1(\overline{\Omega})} + |h_2|\} + 1) \\ &\quad\quad \cdot\{\|r_1 - r_2\|_{C^1(\overline{\Omega})} + |h_1 - h_2|\} \cdot \{\|f\|_{C^2(\overline{AC})} + \|g\|_{C^1(\overline{AB})}\}. \end{aligned} \tag{1.74}$$

Assume $A = (1, 1)$, $B = (1, -1)$, and $C(0, 0)$, without loss of generality. To apply the d'Ambert formula, we use the variables

$$X = \frac{1}{2}(x + y), \quad Y = \frac{1}{2}(x - y),$$

to reach

$$k_{XY} = Rk \quad \text{in } \omega \tag{1.75}$$

by (1.62), where

$$k(X, Y) = K(X + Y, X - Y), \quad R(X, Y) = r(X + Y, X - Y),$$

and $\omega = \{(X, Y) \mid 0 < X,\ 0 < Y,\ X + Y < 1\}$.

Proof *(Proposition 1.3)* It holds that

$$k|_{Y=0} = F(X),\ 0 \le X \le 1, \quad k|_{X=0} = G(Y),\ 0 \le Y \le 1, \tag{1.76}$$

by (1.63), where $F(X) = f(X)$ and $G(Y) = g(Y)$. The solution $k = k_0(X, Y) \in C^2(\overline{\omega})$ to $k_{XY} = 0$ in ω with (1.76) is unique, and is given by

$$k_0(X, Y) = F(X) + G(Y) - c,$$

where $c = F(0)(= G(0))$. Each $\Gamma \in C^0(\overline{\omega})$, on the other hand, admits the solution $k = k(X, Y) \in C^2(\overline{\omega})$ to

$$k_{XY} = \Gamma \text{ in } \omega, \quad k|_{Y=0} = 0,\ 0 \le X \le 1, \quad k|_{X=0} = 0,\ 0 \le Y \le 1,$$

is unique, and is given by

$$k(X, Y) = \int_0^X d\xi \int_0^Y d\eta\, \Gamma(\xi, \eta) = \iint_{\omega_1(X,Y)} \Gamma(\xi, \eta) d\xi d\eta,$$

where

$$\omega_1 = \{(\xi, \eta) \mid 0 < \xi < X, 0 < \eta < Y\}.$$

Therefore, (1.75) with (1.66) is equivalent to the integral equation of Volterra type,

$$k(X, Y) = k_0(X, Y) + \iint_{\omega_1(X,Y)} R(\xi, \eta)k(\xi, \eta)\, d\xi d\eta \tag{1.77}$$

which admits a unique solution $k \in C^0(\overline{\omega})$, and then, $R \in C^1(\overline{\omega})$ implies $k \in C^2(\omega)$.

This $k = k(X, Y)$, furthermore, satisfies the estimate

$$\|k(\cdot, \cdot)\|_{C^0(\overline{\omega})} \leq \tau_1(\|R\|_{C^0(\overline{\omega})} + 1)\|k_0(\cdot, \cdot)\|_{C^0(\overline{\omega})} \tag{1.78}$$

where $\tau_1 : \mathbf{R}_+ \to \mathbf{R}_+$ is a monotone increasing function. By (1.77) and (1.78), this $k(X, Y) = k(X, Y; R)$ satisfies

$$\|k(\cdot, \cdot; R)\|_{C^2(\overline{\omega})} \leq \tau_2(\|R\|_{C^1(\overline{\omega})} + 1)\|k_0\|_{C^2(\overline{\omega})} \tag{1.79}$$

and

$$\begin{aligned} &\|k(\cdot, \cdot; R_1) - k(\cdot, \cdot; R_2)\| \leq \tau_2(\max\{\|R_1\|_{C^1(\overline{\omega})}, \|R_2\|_{C^1(\overline{\omega})}\} + 1) \\ &\cdot \|R_1 - R_2\|_{C^1(\overline{\omega})} \cdot \|k_0\|_{C^2(\overline{\omega})} \end{aligned} \tag{1.80}$$

where $\tau_2 : \mathbf{R} \to \mathbf{R}$ is a monotone increasing function. Inequalities (1.64) and (1.65) follow from (1.79) and (1.80), respectively. □

Proof *(Propositions 1.4 and 1.5)* The proof is similar to that of Proposition 1.3. By the same transformation of the variables, equalities (1.66) and (1.69) are reduced to

$$\begin{aligned} &k(X, 1 - Y) = F(X) \\ &\frac{1}{\sqrt{2}}(k_X(X, 1 - X) + k_Y(X, 1 - X)) = G(X), \quad 0 \leq X \leq 1 \end{aligned} \tag{1.81}$$

for $F(X) \equiv f(2X - 1)$ and $G(X) \equiv \sqrt{2}g(2X - 1)$, and

$$k(X, 0) = F(X),\ k(X, 1 - X) = G(X), \quad 0 \leq X \leq 1, \tag{1.82}$$

for $F(X) \equiv f(X)$ and $G(X) \equiv g(2X - 1)$, respectively.

Equation (1.75) with (1.81) is equivalent to

$$
\begin{aligned}
k(X, Y) &= k_0(X, Y) + \iint_{\omega_2(X,Y)} R(\xi, \eta)k(\xi, \eta)\, d\xi d\eta \\
k_0(X, Y) &= \frac{1}{2}(F(X) + F(1 - Y)) + \frac{\sqrt{2}}{2}\int_{1-Y}^{X} G(\zeta)\, d\zeta
\end{aligned}
\tag{1.83}
$$

for

$$\omega_2(X, Y) = \{(\xi, \eta) \mid X < \xi, Y < \eta, \eta + \xi < 1\},$$

while equation (1.75) with (1.82) is equivalent to

$$
\begin{aligned}
k(X, Y) &= k_0(X, Y) - \iint_{\omega_3(X,Y)} R(\xi, \eta)k(\xi, \eta)\, d\xi d\eta \\
k_0(X, Y) &= F(X) - F(1 - Y) + G(1 - Y)
\end{aligned}
\tag{1.84}
$$

for

$$\omega_3(X, Y) = \{(\xi, \eta) \mid X < \xi < 1 - Y, 0 < \eta < Y\}.$$

Both integral equations (1.83) and (1.84) are of Volterra type, and from this stage the proof is similar to that to Proposition 1.3. □

Proof *(Proposition 1.6)* By the same transformation, condition (1.72) is reduced to

$$
\begin{aligned}
&k|_{Y=0} = F(X) \\
&\frac{1}{\sqrt{2}}(k_X(1 - Y, Y) + k_Y(Y, 1 - Y)) + \sqrt{2}hk(1 - Y, Y) \\
&\quad = G(Y), \quad 0 \le Y \le 1
\end{aligned}
\tag{1.85}
$$

for $F(X) \equiv f(X)$ and $G(Y) \equiv \sqrt{2}g(1 - 2Y)$. The solution $k = k_0(X, Y) \in C^2(\overline{\omega})$ to $k_{XY} = 0$ in ω with (1.85) is unique, and is given by

$$
\begin{aligned}
k_0(X, Y) = {} & F(X) + F(1 - Y) - e^{-2hY}F(1) \\
& + \int_0^Y e^{-2h(Y-\eta)}\{\sqrt{2}G(\eta) - 4hF(1 - \eta)\}\, d\eta.
\end{aligned}
$$

Given $\Gamma \in C^0(\overline{\omega})$, on the other hand, there is a unique solution $k = k(X, Y) \in C^2(\overline{\omega})$ to $k_{XY} = \Gamma$ in ω with $k|_{Y=0} = 0$, $0 \le X \le 1$, and

$$\frac{1}{\sqrt{2}}(k_X(1 - Y, Y) + k_Y(1 - Y, Y)) + \sqrt{2}hk(1 - Y, Y) = 0, \ 0 \le Y \le 1.$$

It is given by

$$k(X, Y) = -\int\int_{\omega_4} \chi_h(\xi, \eta; X, Y)\Gamma(\xi, \eta)\, d\xi d\eta$$

for $\omega_4(X, Y) = \omega_3(X, Y) \cup \omega_2(1 - Y, 0)$ and

$$\chi_h(\xi, \eta; X, Y) = \begin{cases} 1 & \text{in } \omega_3(X, Y) \\ 2e^{-2h(Y+\eta-1)} & \text{in } \omega_2(1 - Y, 0). \end{cases}$$

Equality (1.75) with (1.85) is, therefore, equivalent to the integral equation

$$k(X, Y) = k_0(X, Y) - \int\int_{\omega_4(X,Y)} \chi_h(\xi, \eta; X, Y)R(\xi, \eta)k(\xi, \eta)\, d\xi d\eta, \tag{1.86}$$

and hence the proposition is verified similarly. □

1.3.4 Deformation Formulae

Given

$$p \in C^1[0, 1], \quad h \in \mathbf{R}, \quad \lambda \in \mathbf{R},$$

let $\Phi = \Phi(\cdot; p, h; \lambda) \in C^2[0, 1]$ be the solution to

$$(-\frac{d^2}{dx^2} + p(x))\Phi = \lambda\Phi \text{ in } 0 < x < 1, \quad \Phi|_{x=0} = 1, \ \Phi_x|_{x=0} = h.$$

If $\lambda = \lambda_n(P)$ for $P = (p, h, H) \in X$, this $\Phi = \Phi(\cdot; p, h; \lambda)$ is an eigenfunction of A_P satisfying

$$\Phi' + H\Phi\big|_{x=1} = 0.$$

There is a formula connecting $\Phi(\cdot; p, h : \lambda)$ and $\Phi(\cdot; q, j; \lambda)$ through an integral transformation for which the kernel K is independent of λ. Recall

$$D = \{(x, y) \mid 0 < y < x < 1\}.$$

Lemma 1.4 *Given $p, q \in C^1[0, 1]$ and $h, j \in \mathbf{R}$, there is a unique solution $K = K(x, y) = K(x, y; q, j; p, h) \in C^2(\overline{D})$ to*

$$K_{xx} - K_{yy} + p(y)K = q(x)K \qquad \text{in } D$$

$$K(x,x) = (j-h) + \frac{1}{2}\int_0^x q(x) - p(s)\,ds, \quad 0 \le x \le 1$$
$$K_y(x,0) = hK(x,0), \qquad 0 \le x \le 1 \tag{1.87}$$

Proof To show the existence of $K = K(\cdot,\cdot; q, j; p, h)$, we extend $q \in C^1[0,1]$ to $\hat{q} \in C^1[0,2]$. Put

$$\Omega = \{(x,y) \mid 0 < y < x < 2 - y\}.$$

By Proposition 1.6, there exists $\hat{K} = \hat{K}(x,y) \in C^2(\overline{\Omega})$ such that

$$\hat{K}_{xx} - \hat{K}_{yy} + p(y)\hat{K} = \hat{q}(x)\hat{K} \qquad \text{in } D$$
$$\hat{K}(x,x) = (j-h) + \frac{1}{2}\int_0^x q(x) - p(s)\,ds, \quad 0 \le x \le 1$$
$$\hat{K}_y(x,0) = h\hat{K}(x,0), \qquad 0 \le x \le 1.$$

Then the restriction $K = \hat{K}\big|_{\overline{D}} \in C^2(\overline{D})$ satisfies (1.87).

To verify the uniqueness, we divide D into

$$\Omega_1 = \{(x,y) \mid 0 < x < y < 1 - y\}, \quad \Omega_2 = D/\overline{\Omega_1}.$$

We prove that the first and the third equations of (1.87) and

$$K(x,x) = 0, \quad 0 \le x \le 1$$

for $K \in C^2(\overline{D})$ imply $K \equiv 0$. In fact, we have $K = 0$ on Ω_1 by Proposition 1.6 and then $K = 0$ on Ω_2 by Proposition 1.3. □

Lemma 1.5 (First Deformation Formula) *For $K = K(\cdot,\cdot; q, j; p, h)$ in Lemma 1.4, it holds that*

$$\Phi(x; q, j; \lambda) = \Phi(x; p, h; \lambda) + \int_0^x K(x, y; q, j; p, h)\Phi(y; p, h; \lambda)\,dy \tag{1.88}$$

for each $\lambda \in \mathbf{R}$.

Proof We show that the right-hand side of (1.88) denoted by $\Psi = \Psi(\cdot,\lambda)$ satisfies $\Psi|_{x=0} = 1$, $\Psi_x|_{x=0} = j$, and

$$\left(-\frac{d^2}{dx^2} + q(x)\right)\Psi = \lambda\Psi \text{ in } 0 < x < 1. \tag{1.89}$$

Let

$$\Phi(x) = \Phi(x; p, h; \lambda), \quad K(x, y) = K(x, y; q, j; p, h)$$

for simplicity. It is obvious that $\Psi|_{x=0} = 1$, while $\Psi_x|_{x=0} = j$ follows from

$$\Psi_x(x, \lambda) = \Phi_x(x, \lambda) + K(x, x)\Phi(x, \lambda) + \int_0^x K_x(x, y)\Phi(y, \lambda)\, dy.$$

Then we obtain

$$\begin{aligned} I \equiv (\frac{d^2}{dx^2} + \lambda)\Psi(x, \lambda) &= (\frac{d^2}{dx^2} + \lambda)\Phi(x, \lambda) + (\frac{d}{dx}K(x, x) + K_x(x, x))\Phi(x, \lambda) \\ &+ K(x, x)\Phi_x(x, \lambda) + \int_0^x \{\lambda K(x, y) + K_{xx}(x, y)\}\Phi(y, \lambda)\, dy \\ &= p(x)\Phi(x, \lambda) + (\frac{d}{dx}K(x, x) + K_x(x, x))\Phi(x, \lambda) + K(x, x)\Phi_x(x, \lambda) \\ &+ \int_0^x \{\lambda K(x, y) + K_{yy}(x, y) - p(y)K(x, y) + q(x)K(x, y)\}\Phi(y, \lambda)\, dy, \end{aligned}$$

using the first equality of (1.87), and hence

$$\begin{aligned} I = p(x)\Phi(x, \lambda) &+ (\frac{d}{dx}K(x, x) + K_x(x, x))\Phi(x, \lambda) + K(x, x)\Phi_x(x, \lambda) \\ &+ [K_y(x, y)\Phi(y, \lambda) - K(x, y)\Phi_y(y, \lambda)]_{y=0}^{y=x} \\ &+ \int_0^x K(x, y)\{(\lambda + q(x) - p(y))\Phi(y, \lambda) + \Phi_{yy}(y, \lambda)\}dy. \end{aligned}$$

The last equality of (1.87) implies

$$\begin{aligned} I &= p(x)\Phi(x, \lambda) + (\frac{d}{dx}K(x, x) + K_x(x, x) + K_y(x, x))\Phi(x, \lambda) \\ &\quad + \int_0^x K(x, y)q(x)\Phi(y, \lambda)\, dy \\ &= \left(p(x) + 2\frac{d}{dx}K(x, x)\right)\Phi(x, \lambda) + q(x)\int_0^x K(x, y)\Phi(y, \lambda)dy, \end{aligned}$$

and hence

$$I = q(x)\Psi(x, \lambda)$$

by the second equality of (1.87). Thus (1.89) follows. □

In the first deformation formula, we connect two functions φ and ψ satisfying

$$(-\frac{d^2}{dx^2} + p(x))\varphi = \lambda\varphi,\ (-\frac{d^2}{dx^2} + q(x))\psi = \lambda\psi \quad \text{in } 0 < x < 1,$$

requiring φ and ψ to satisfy boundary conditions at $x = 0$. There is another formula connecting the above φ and ψ, without imposing any boundary conditions. An example of its application is the interior identification or periodic problems [22].

Let

$$D_+ = \{(x, y) \mid 1 - x < y < x, \frac{1}{2} < x < 1\},$$

$$D_- = \{x < y < 1 - x, 0 < x < \frac{1}{2}\},$$

$I_+ = (\frac{1}{2}, 1)$, and $I_- = (0, \frac{1}{2})$. We have the following lemma by Proposition 1.3.

Lemma 1.6 *Given $p \in C^1[0,1]$ and $q_\pm \in C^1(\overline{I}_\pm)$, there is a unique solution to $K = K^\pm(x, y) \in C^2(\overline{D}_\pm)$ such that*

$$\begin{aligned} &K^\pm_{xx} - K^\pm_{yy} + p(y)K^\pm = q_\pm(x)K^\pm && \text{in } D_\pm \\ &K^\pm(x, x) = \frac{1}{2}\int_{\frac{1}{2}}^{x} q_\pm(x) - p(s)\,ds, && x \in \overline{I}_\pm \\ &K^\pm(x, 1 - x) = 0, && x \in \overline{I}_\pm \end{aligned} \tag{1.90}$$

Then we obtain the following lemma.

Lemma 1.7 (Second Deformation Formula) *Let $\varphi = \varphi(x) \in C^2[0, 1]$ satisfy*

$$(-\frac{d^2}{dx^2} + p(x))\varphi = \lambda\varphi \quad \text{in } 0 < x < 1$$

with $\lambda \in \mathbf{R}$. Take $K^\pm = K^\pm(x, y)$ in Lemma 1.6 and put

$$\psi_\pm(x) = \varphi(x) + \int_{1-x}^{x} K^\pm(x, y)\varphi_\pm(y)dy, \quad x \in \overline{I}_\pm.$$

Then, $\psi_\pm \in C^2(\overline{I}_\pm)$ satisfies

$$(-\frac{d^2}{dx^2} + q_\pm(x))\psi_\pm = \lambda\psi_\pm, \quad x \in \overline{I}_\pm \tag{1.91}$$

with $\psi_\pm(\frac{1}{2}) = \varphi(\frac{1}{2})$ and $\psi'_\pm(\frac{1}{2}) = \varphi'(\frac{1}{2})$.

Proof The equality $\psi_\pm(\frac{1}{2}) = \varphi(\frac{1}{2})$ is immediate, while $\psi'_\pm(\frac{1}{2}) = \varphi'(\frac{1}{2})$ follows from

$$\psi'_\pm(x) = \varphi'(x) + K^\pm(x,x)\varphi(x) + \int_{1-x}^{x} K'_x(x,y)\varphi(y)\,dy, \quad x \in \overline{I}_\pm$$

and $K^\pm(\frac{1}{2},\frac{1}{2}) = 0$. Then the first equality of (1.90) implies

$$\begin{aligned}
I &\equiv (\frac{d^2}{dx^2} + \lambda)\psi_\pm(x) = (\frac{d^2}{dx^2} + \lambda)\varphi(x) + (\frac{d}{dx}K^\pm(x,x) + K_x^\pm(x,x))\varphi(x) \\
&+ K^\pm(x,x)\varphi'(x) + K_x^\pm(x,1-x)\varphi(1-x) + \int_{1-x}^{x} \{K_{xx}^\pm(x,y) + \lambda K^\pm(x,y)\}\varphi(y)\,dy \\
&= p(x)\varphi(x) + (\frac{d}{dx}K^\pm(x,x) + K_x^\pm(x,x))\varphi(x) + K^\pm(x,x)\varphi'(x) \\
&+ K_x^\pm(x,1-x)\varphi(1-x) + \int_{1-x}^{x} \{K_{yy}^\pm(x,y) + (q_\pm(x) - p(y) + \lambda)K^\pm(x,y)\}\varphi(y)\,dy,
\end{aligned}$$

which means

$$\begin{aligned}
I &= p(x)\varphi(x) + (\frac{d}{dx}K^\pm(x,x) + K_x^\pm(x,x))\varphi(x) + K^\pm(x,x)\varphi'(x) \\
&+ K_x^\pm(x,1-x)\varphi(1-x) + [K_y^\pm(x,y)\varphi(y) - K^\pm(x,y)\varphi'(y)]_{y=1-x}^{y=x} \\
&+ \int_{1-x}^{x} K^\pm(x,y)\{\varphi''(y) + (q_\pm(x) - p(y) + \lambda)\varphi(y)\}\,dy \\
&= p(x)\varphi(x) + (\frac{d}{dx}K^\pm(x,x) + K_x^\pm(x,x) + K_y^\pm(x,x))\varphi(x) \\
&+ (K_x^\pm(x,1-x) - K_y^\pm(x,1-x))\varphi(1-x) + q_\pm(x)\int_{1-x}^{x} K^\pm(x,y)\varphi(y)\,dy.
\end{aligned}$$

Then the second and the third equalities of (1.90) imply

$$\begin{aligned}
I &= (p(x) + 2\frac{d}{dx}K^\pm(x,x))\varphi(x) + (\frac{d}{dx}K^\pm(x,1-x))\varphi(1-x) \\
&+ q_\pm(x)\int_{1-x}^{x} K^\pm(x,y)\varphi(y)\,dy = q_\pm(x)\{\varphi(x) + \int_{1-x}^{x} K^\pm(x,y)\varphi(y)\,dy\} \\
&= q(x)\psi_\pm(x),
\end{aligned}$$

and hence (1.91). □

1.3.5 Hochstadt Equation

Here we prove Theorems 1.10 and 1.11. For the unknown equation $E(P, a)$, we suppose

$$(a, \varphi_{n_l}(\cdot; P)) = 0, \quad 1 \le l \le N$$
$$(a, \varphi_n(\cdot; P)) \ne 0, \quad n \ne n_\ell,\ 1 \le \ell \le N. \tag{1.92}$$

By the definition, the above N (finite or infinite) is the degenerate index of $a \in L^2(0, 1)$ with respect to A_P. Let $E(Q, b)$ be a model equation and $u = u(\cdot, t)$ and $v = v(\cdot, t)$ be the solutions to $E(P, a)$ and $E(Q, b)$, respectively. By Lemma 1.3, the condition

$$\varepsilon_0(t) \equiv (v - u)|_{x=0} = 0, \quad \varepsilon_1(t) \equiv (v - u)|_{x=1} = 0 \tag{1.93}$$

is equivalent to the following property (Z) on the eigenvalues and the eigenfunctions

$$\{\lambda_n = \lambda_n(P)\}_{n=0}^{\infty}, \quad \{\varphi_n = \varphi_n(\cdot, P)\}_{n=0}^{\infty},$$

and

$$\{\mu_m = \lambda_m(Q)\}_{m=0}^{\infty}, \quad \{\psi_m = \varphi_m(\cdot, Q)\}_{m=0}^{\infty}$$

of A_P and A_Q, respectively:

(Z) Each $n \ne n_\ell$, $1 \le \ell \le N$), admits $m(n) \in \mathbf{N}$ such that

$$\lambda_n = \mu_{m(n)}, \quad (b, \psi_m) = 0,\ m \in \mathbf{N} \setminus \{m(n) \mid n \ne n_\ell,\ 1 \le \ell \le N\},$$

and

$$(a, \varphi_n)\varphi(0) = (b, \psi_{m(n)})\psi_{m(n)}(0)(\ne 0)$$
$$(a, \varphi_n)\varphi(1) = (b, \psi_{m(n)}\psi_{m(n)}(1)(\ne 0). \tag{1.94}$$

For the moment we take the other normalization of eigenfunctions, using $\Phi = \Phi(\cdot; p, h; \lambda)$ defined in the preceding subsection, that is,

$$\Phi_n = \Phi(\cdot; p, h; \lambda_n(P)), \quad \Psi_m = \Phi(\cdot; q, j; \lambda_m(Q))$$

for $P = (p, h, H)$ and $Q = (q, j, J)$. Then, Φ_n and Ψ_m are eigenfunctions of A_P and A_Q corresponding to λ_n and μ_m normalized by $\Phi_n(0) = 1$ and $\Psi_m(0) = 1$, respectively.

Lemma 1.8 *Under the assumption of (1.92), the coefficient $Q \in X$ satisfies (1.93) for some $b \in L^2(0, 1)$ if and only if each $n \ne n_\ell$, $1 \le \ell \le N$, admits $m(n) \in \mathbf{N}$ satisfying*

$$\lambda_n = \mu_{m(n)}, \quad \Phi_n(1) = \Psi_{m(n)}(1). \tag{1.95}$$

Then, such $Q \in X$ admits a unique $b \in L^2(0, 1)$ satisfying (1.93).

Proof The "only if" part follows immediately from the above argument, as (1.94) implies (1.95).

Suppose, conversely, that $Q \in X$ satisfies (6.11) and (6.12). Here we show the unique existence of $b \in L^2(0, 1)$ satisfying

$$\begin{aligned} (b, \psi_m) &= \frac{\varphi_n(0)}{\psi_m(0)}(a, \varphi_n), && m = m(n),\ n \neq n_\ell,\ 1 \le \ell \le N \\ (b, \psi_m) &= 0, && m \notin \{m(n) \mid n \neq n_\ell,\ 1 \le \ell \le N\}. \end{aligned} \tag{1.96}$$

In fact, the assertion is obvious if

$$\sharp\{\varphi_n \mid (a, \varphi_n) \neq 0\} < \infty,$$

because then it follows that

$$\sharp(\mathbf{N} \setminus \{m(n) \mid n \neq n_\ell,\ 1 \le \ell \le N\}) < \infty.$$

In the other case of

$$\sharp\{\varphi_n \mid (a, \varphi_n) \neq 0\} = \infty$$

the relation $m(n) = n$, $n \ge n_0$, follows with some $n_0 \in \mathbf{N}$ by the proof of Lemma 1.3. The inequality

$$\left| \frac{\varphi_n(0)}{\psi_{m(n)}(0)} \right| \le C_1, \quad m = m(n),\ n \neq n_\ell,\ 1 \le \ell \le N,$$

therefore, holds by the asymptotics (1.59) with C_1. Hence such $b \in L^2(0, 1)$ is determined by the Fourier coefficients (1.96). The "if" part and the last statement of this lemma have been thus verified, and the proof is complete. □

By Lemma 1.8, Theorems 1.10 and 1.11 are reduced to the following assertions.

Proposition 1.7 *If $N = 0$, condition (1.95) imply $Q = P$.*

Proposition 1.8 *If $N = 1$, there exists $Q \in X \setminus \{P\}$ satisfying (1.95).*

For the moment we take the case that $N \in \mathbf{N}$ is arbitrary.

Lemma 1.9 *In the case of $N < \infty$, $Q = (q, j, J) \in X$ satisfies the conditions (6.11) and (6.12) if and only if there exist $K = K(x, y) \in C^2(\overline{D})$ and $c_\ell, d_\ell \in \mathbf{R}$, $1 \le \ell \le N$, such that*

$$
\begin{aligned}
&K_{xx} - K_{yy} + p(y)K = q(x)K && \text{in } D \\
&K_y(x,0) = hK(x,0), && 0 < x < 1 \\
&K(1,y) = \sum_{l=1}^{N} c_\ell \varphi_{n_\ell}(y),\ K_x(1,y) = \sum_{\ell=1}^{N} d_\ell \varphi_{n_\ell}(y), && 0 \le y \le 1
\end{aligned} \quad (1.97)
$$

and

$$
\begin{aligned}
&K(x,x) = (j-h) + \frac{1}{2}\int_0^x q(s) - p(s)\, ds, \quad 0 < x < 1 \\
&J = H - K(1,1).
\end{aligned} \quad (1.98)
$$

Proof Let

$$K = K(x,y) = K(x,y;q,j;p,h) \in C^2(\overline{D})$$

be the kernel in Lemma 1.5 and put

$$\Phi(\cdot;\lambda) = \Phi(\cdot;p,h;\lambda), \quad \Psi(\cdot;\lambda) = \Psi(\cdot;q,j;\lambda)$$

for $\lambda \in \mathbf{R}$. Then the first deformation formula reads

$$\Psi(x;\lambda) = \Phi(x;\lambda) + \int_0^x K(x,y)\Phi(y;\lambda)\, dy, \quad 0 \le x \le 1.$$

Equalities (1.95) for $Q \in X$ mean

$$
\begin{aligned}
&\Psi_x(1;\lambda) + J\Psi(1;\lambda) = \Phi_x(1;\lambda) + K(1,1)\Phi(1;\lambda) + J\Phi(1;\lambda) \\
&\quad + \int_0^1 \{K_x(1,y) + JK(1,y)\}\Phi(y;\lambda)\, dy = 0, \\
&\Psi(1;\lambda) - \Phi(1;\lambda) = \int_0^1 K(1,y)\Phi(y,\lambda)\, dy = 0
\end{aligned}
$$

for $\lambda = \lambda_n$, $n \ne n_\ell$, $1 \le \ell \le N$, or,

$$
\begin{aligned}
&\int_0^1 K(1,y)\Psi_n(y)\, dy = (J - H + K(1,1))\Phi_n(1) \\
&\quad + \int_0^1 \{K_x(1,y) + JK(1,y)\}\Phi_n(y)\, dy = 0, \quad n \ne n_\ell,\ 1 \le \ell \le N.
\end{aligned}
$$

Hence we obtain

$$\int_0^1 K(1,y)\varphi_n(y)\,dy = (J - H + K(1,1))\varphi_n(1) + \int_0^1 K_x(1,y)\varphi_n(y)\,dy = 0$$

for $n \neq n_\ell$, $1 \leq \ell \leq N$.

The Riemann–Lebesgue theorem assures

$$\lim_{n\to\infty} \int_0^1 K_x(1,y)\varphi_n(y)\,dy = 0,$$

while we have the asymptotics (1.59). Hence the above second equality splits in $J = H - K(1,1)$ and

$$\int_0^1 K_x(1,y)\varphi_n(y)\,dy = 0, \quad n \neq n_\ell,\ 1 \leq \ell \leq N.$$

Thus we obtain the result and the proof is complete. □

Lemma 1.10 *Given $p, q \in C^1[0,1]$, $h \in \mathbf{R}$, and $c_\ell, d_\ell \in \mathbf{R}$, $1 \leq \ell \leq N$, we have a unique solution $K = K(x,y) \in C^2(\overline{D})$ to (1.97), given by*

$$K(x,y) = \sum_{\ell=1}^{N} g_\ell(x)\varphi_{n_\ell}(y), \tag{1.99}$$

where $g_\ell = g_\ell(x)$ is the solution to

$$(-\frac{d^2}{dx^2} + q(x))g_\ell(x) = \lambda_{n_\ell} g_\ell \ \ in\ 0 < x < 1, \quad g_\ell(1) = c_\ell, \quad g_\ell'(1) = d_\ell. \tag{1.100}$$

Proof It is obvious that $K = K(x,y) \in C^2(\overline{D})$ given by (1.99)–(1.100) satisfies the desired relations, recalling

$$\lambda_{n_\ell} = \lambda_{n_\ell}(P), \quad \varphi_{n_\ell} = \varphi_{n_\ell}(\cdot; P).$$

To show the uniqueness of the solution to (1.97), we derive $K \equiv 0$ on $\overline{D}$, assuming $c_\ell = d_\ell = 0$ for $1 \leq \ell \leq N$. As in the proof of Lemma 1.4, we divide D into

$$\Omega_1 = \{(x,y) \mid 0 < y < x < 1 - y\}, \quad \Omega_2 = D \setminus \overline{\Omega_1}.$$

First, Proposition 1.4 implies $K = 0$ on $\overline{\Omega_2}$, and then $K = 0$ on $\overline{\Omega_1}$ follows from Proposition 1.6. The proof is complete. □

Theorem 1.13 (Hochstadt Equation) *Assume (1.92) with $N < \infty$, and let*

$$\varphi = (\varphi_{n_1}(\cdot; P), \ldots, \varphi_{n_N}(\cdot; P))^T, \quad \Lambda = diag(\lambda_{n_1}(P), \ldots, \lambda_{n_N}(P)).$$

Let, furthermore, $\cdot$ *and* I *denote the inner product and the unit matrix in* $\mathbf{R}^N$*, respectively. Then,* $Q = (q, j, J) \in X$ *satisfies (1.93) for some* $b \in L^2(0, 1)$ *if and only if there exists* $g \in C^2([0, 1] \to \mathbf{R}^N)$ *such that*

$$\frac{d^2 g}{dx^2} = [(2\frac{d}{dx}(g \cdot \varphi) + p)I - \Lambda]g, \quad 0 \le x \le 1. \tag{1.101}$$

Furthermore, this $Q = (q, j, J) \in X$ *is given by*

$$q = p + 2\frac{d}{dx}(g \cdot \varphi), \quad j = h + (g \cdot \varphi)|_{x=0}, \quad J = H - (g \cdot \varphi)|_{x=1}, \tag{1.102}$$

and the mapping $g \mapsto Q$ *is injective.*

Proof By Lemmas 1.9 and 1.10, we have

$$\sum_{\ell=1}^{N} g_\ell(x)\varphi_{n_\ell}(x) = (j - h) + \frac{1}{2}\int_0^x q(s) - p(s)\, ds, \quad 0 < x < 1$$

$$J = H - \sum_{\ell=1}^{N} g_\ell(1)\varphi_{n_\ell}(1)$$

and then (1.101)–(1.102) holds by (1.100) for $g = (g_1, \ldots, g_N)^T$.

To conclude the proof, suppose that both g_1 and g_2 satisfy (1.101)–(1.102) for $Q = (q, j, J) \in X$. Then both

$$K_1(x, y) = g_1(x) \cdot \varphi(y)$$

and

$$K_2(x, y) = g_2(x) \cdot \varphi(y)$$

satisfy the first two equalities of (1.97) and the first equality of (1.98). Then Lemma 1.4 implies $K_1 \equiv K_2$ on $\overline{D}$, and hence $g_1 \equiv g_2$ on $\overline{D}$. □

Proof *(Proposition 1.7)* By the proof of Lemma 1.10, there arises $K \equiv 0$ in this case. Then we obtain $Q = P$ by (1.98). □

Proof *(Proposition 1.8)* There is actually a non-trivial solution $g \not\equiv 0$ to (1.101). Then $Q = (q, j, J) \in X$ defined by (1.102) satisfies (1.95), and $Q \neq P$ by Theorem 1.13 and the proof is complete. □

Equation (1.101) arises in the study of several inverse Sturm–Liouville problems [8, 19].

1.4 Eigenvalue Problems

1.4.1 Vibrating Membrane

Here we turn to the classical theory, completeness of eigenfunctions associated with regular symmetric elliptic boundary value problems. Let $\Omega \subset \mathbf{R}^n$ be a bounded domain with smooth boundary $\partial\Omega$ and take the problem

$$-\Delta\psi = \lambda\psi \text{ in } \Omega, \quad \psi|_{\partial\Omega} = 0, \tag{1.103}$$

where λ is a constant. It arises with the separation of variables for the heat equation

$$u_t = \Delta u \text{ in } \Omega \times (0, T), \quad u|_{\partial\Omega} = 0, \quad u|_{t=0} = u_0(x),$$

or the wave equation

$$u_{tt} = \Delta u \text{ in } \Omega \times (0, T), \quad u|_{\partial\Omega} = 0, \quad (u, u_t)|_{t=0} = (u_1(x), u_0(x)). \tag{1.104}$$

The expansion theorem is thus necessary to justify the method of superposition, using eigenfunctions $\psi_k(x)$, $k = 1, 2, \cdots$, that is, their completeness.

More precisely, problem (1.103) always admits the trivial solution $\psi \equiv 0$ and if a non-trivial solution $\psi \not\equiv 0$ exists, such λ and ψ are called eigenvalue and eigenfunction, respectively. The eigenvalue problem for one space dimension with regular coefficients is called the Sturm–Liouville problem, where all eigenvalues are real, countably many, and simple. The expansion theorem of Mercer, furthermore, holds so that the eigenfunction expansion is valid with the uniform convergence for C^2 functions satisfying the boundary condition. Multiple eigenvalues, however, can exist if $n \geq 2$, that is, there may be an eigenvalue which shares plurally linearly independent eigenfunctions.

In the case of (1.104) the solution u is given by

$$u(x, t) = \sum_{k=1}^{\infty} \psi_k(x) \{(u_1, \psi_k) \cos\omega_k t + (u_0, \psi_k) \sin\omega_k t/\omega_k\}$$

with the right-hand side converging in $C^1(\mathbf{R}; L^2(\Omega)) \cap C(\mathbf{R}; H_0^1(\Omega))$ if the initial values are in

$$(u_1, u_0) \in H_0^1(\Omega) \times L^2(\Omega)$$

where $(\ ,\ )$ denotes the L^2 inner product and $\omega_k = \sqrt{\lambda_k}$. In this context, this $\{\omega_k\}_{k=1}^{\infty}$ indicates characteristic vibration numbers or tones if Ω is regarded as a vibrating object. The deformation of this object on the boundary is set to be zero in (1.103), while the Neumann problem arises if it is not bounded,

$$-\Delta\psi = \lambda\psi \text{ in } \Omega, \qquad \left.\frac{\partial\psi}{\partial\nu}\right|_{\partial\Omega} = 0, \tag{1.105}$$

where ν denotes the outer unit normal vector.

In the case that $\Omega = \{(x,y) \mid x^2+y^2<1\} \subset \mathbf{R}^2$ is unit disc we use polar coordinates

$$x = r\cos\theta, \quad y = r\sin\theta$$

which reduce (1.103) to

$$\psi_{rr} + \frac{1}{r}\psi_r + \frac{1}{r^2}\psi_{\theta\theta} + \lambda\psi = 0, \quad \psi(1,\theta) = 0 \tag{1.106}$$

by

$$\Delta = \frac{\partial^2}{\partial r^2} + \frac{1}{r}\frac{\partial}{\partial r} + \frac{1}{r^2}\frac{\partial^2}{\partial\theta^2}.$$

Then we apply the separation of variables $\psi(r,\theta) = f(r)h(\theta)$ to (1.106) which implies

$$\frac{r^2(f_{rr} + \frac{f_r}{r} + \lambda f)}{f} = -\frac{h_{\theta\theta}}{h},$$

where the left- and the right-hand sides are independent of θ and r, respectively. Hence this quantity is a constant, more precisely, k^2 with some integer $k \geq 0$. In fact $h = h(\theta)$ is 2π-periodic which implies

$$\begin{aligned} &h(\theta) = a\cos k\theta + b\sin k\theta, \\ &r^2 f'' + rf' + (r^2\lambda - k^2)f = 0, \ f(1) = 0, \end{aligned} \tag{1.107}$$

where a, b are constants. Any eigenvalue λ is positive in (1.103), and we put $\rho = \lambda^{1/2}r$, $f(r) = J(\rho)$, to obtain *Bessel's equation*

$$\frac{d^2J}{d\rho^2} + \frac{1}{\rho}\frac{dJ}{d\rho} + (1 - \frac{k^2}{\rho^2})J = 0. \tag{1.108}$$

This equation admits a solution in the form

$$J(\rho) = \sum_{m=0}^{\infty} a_m\rho^m = \frac{\rho^k}{2^k k!}\left(1 - \frac{\rho^2}{2(2k+2)} + \frac{\rho^4}{2\cdot 4(2k+2)(2k+4)} - \cdots\right)$$

denoted by $J = J_k(\rho)$. It is called the Bessel function of the k-th order. Writing $\lambda = \omega^2$, we have $f_k(r) = J_k(\omega r)$ and hence (1.107) is equivalent to $J_k(\omega) = 0$. More precisely, if $\{\omega_{k\ell}\}_{\ell=1,2,\cdots}$ denotes the set of zeros of the k-th Bessel function, $J_k = J_k(r)$ in $r > 0$, then the eigenvalues and the eigenfunctions of (1.103) on the unit disc are given by $\left\{\omega_{k\ell}^2\right\}_{\ell=1,2,\cdots,k=0,1,\cdots}$ and

$$J_k(\omega_{k\ell}r)\cos k\theta, \quad J_k(\omega_{k\ell}r)\sin k\theta,$$

respectively. The eigenvalues are thus double except for $k = 0$.

In the case that $\Omega = \left\{(x, y, z) \mid x^2 + y^2 + z^2 < 1\right\}$ is a three-dimensional unit ball, the polar coordinates are given by

$$x = r\sin\theta\cos\varphi, \ \ y = r\sin\theta\sin\varphi, \ \ z = r\cos\theta$$

with $(r, \theta, \varphi) \in (0, 1) \times [0, \pi] \times [0, 2\pi)$. Since

$$\Delta\psi = \frac{1}{r^2\sin\theta}\left[\frac{\partial}{\partial r}\left(r^2\psi_r\sin\theta\right) + \frac{\partial}{\partial\varphi}\left(\frac{\psi_\varphi}{\sin\theta}\right) + \frac{\partial}{\partial\theta}\left(\psi_\theta\sin\theta\right)\right],$$

the separation of variables $\psi = Y(\theta, \varphi)f(r)$ induces

$$\frac{(r^2f')' + \lambda r^2 f}{f} = -\frac{1}{Y\sin\theta}\left[\frac{\partial}{\partial\varphi}\left(\frac{Y_\varphi}{\sin\theta}\right) + \frac{\partial}{\partial\theta}\left(Y_\theta\sin\theta\right)\right]. \tag{1.109}$$

This quantity is again a constant denoted by μ and in this way we get

$$\Delta^*Y + \mu Y = 0, \quad 0 \le \theta \le \pi, \ 0 \le \varphi < 2\pi, \tag{1.110}$$

where Δ^* denotes the *Laplace–Beltrami operator* on the unit sphere

$$\Delta^*Y = \frac{1}{\sin\theta}\left[\frac{\partial}{\partial\varphi}\left(\frac{Y_\varphi}{\sin\theta}\right) + \frac{\partial}{\partial\theta}\left(Y_\theta\sin\theta\right)\right].$$

This $Y = Y(\theta, \varphi)$ is regular at $\theta = 0, \pi$, and 2π-periodic with respect to φ. Then (1.109) implies

$$f'' + \frac{2}{r}f' + (\lambda - \frac{\mu}{r^2})f = 0.$$

Taking the ansatz $f = r^t$, we obtain

$$t^2 + t - \mu = 0 \tag{1.111}$$

besides $\lambda = 0$, and hence $\mu = \ell(\ell + 1)$ where $t = \ell, -\ell - 1$ in (1.111).

The solution to (1.110) with $\mu = \ell(\ell+1)$, generally, is called the ℓ-th *spherical harmonic function* denoted by $Y_\ell(\theta, \varphi)$. Using

$$(r^2 f')' - \ell(\ell+1) f + \lambda r^2 f = 0,$$

we obtain

$$f(r) = S_\ell(\lambda^{1/2} r), \quad S(\rho) = J_{\ell+\frac{1}{2}}(\rho)/\rho^{1/2}$$

and the eigenvalues are determined by the boundary condition

$$J_{\ell+\frac{1}{2}}(\lambda^{1/2}) = 0, \quad \lambda > 0, \tag{1.112}$$

where $J_k(\rho)$ is the solution to (1.108).

We take further separation of variables

$$Y(\theta, \varphi) = p(\varphi) q(\theta)$$

for (1.110), $\mu = \ell(\ell+1)$, to obtain

$$-\frac{p_{\varphi\varphi}}{p} = \frac{(q_\theta \sin\theta)_\theta \sin\theta}{q} + \ell(\ell+1)\sin^2\theta.$$

This quantity is again k^2, $k = 0, 1, 2, \cdots$, by the 2π periodicity in φ of p. Using $z = \cos\theta$, on the other hand, the equation for q is transformed into

$$((1-z^2)q_z)_z + (\ell(\ell+1) - \frac{k^2}{1-z^2})q = 0, \quad -1 < z < 1, \tag{1.113}$$

which is called the *associated Legendre equation*. If $k = 0$, (1.113) stands for the Legendre equation

$$(1-z^2)q_{zz} - 2zq_z + \ell(\ell+1)q = 0$$

which has the power series solution if and only if $\ell = 0, 1, 2, \cdots$. Then (1.113) has a solution

$$q_{k,\ell}(z) = (1-z^2)^{k/2} \left(\frac{d}{dz}\right)^\ell P_\ell(z),$$

regular at $z = \pm 1$, where $P_\ell(z)$ is the Legendre polynomial defined by

$$P_\ell(z) = \frac{1}{2^\ell \ell!} \left(\frac{d}{dz}\right)^\ell (z^2-1)^\ell.$$

If $\{\lambda_{\ell m}\}_{m=1,2,\cdots}$ denotes the set of solutions to (1.112) the eigenfunctions are thus given by

$$\left\{Y^{\pm}_{k,\ell}(\theta,\varphi)S_\ell(\lambda^{1/2}_{\ell m}r) \mid 0 \le k \le \ell,\ \ell = 0, 1, \cdots, m = 1, 2, \cdots\right\}$$

for $Y^{\pm}_{k,\ell}(\theta,\varphi) = q_{k,\ell}(\cos\theta)e^{\pm \imath k\varphi}$.

Even in the general domain, eigenvalues in (1.103) or (1.105) are real, countably many, and bounded from below, labeled as

$$\lambda_1 < \lambda_2 \le \cdots$$

according to their multiplicities. Thus the first eigenvalue λ_1 is always simple, and is positive and zero according to (1.103) and (1.105), respectively. The L^2-normalized eigenfunctions form a complete orthonormal system in $L^2(\Omega)$. The number of the nodal domains of the k-th eigenfunction is fewer than or equal to k. The nodal domain here denotes the subdomain of Ω where ψ_k has a definite sign.

Weyl's formula says

$$A(\lambda) = \begin{cases} \frac{\lambda}{4\pi}\,|\Omega|\,(1+o(1))\,, & n=2, \\ \frac{\lambda^{3/2}}{16\pi^2}\,|\Omega|\,(1+o(1))\,, & n=3, \end{cases} \qquad \lambda \to +\infty,$$

where $A(\lambda)$ denotes the number of eigenvalues of (1.103) less than or equal to λ. If $\lambda_1(\Omega) > 0$ denotes the first eigenvalue of (1.103) then the *Faber–Krahn isoperimetric inequality* ensures

$$\lambda_1(\Omega) \ge \lambda_1(\Omega^*), \tag{1.114}$$

where Ω^* denotes the ball (or disc) with the same volume as Ω. The *Pólya–Szegö–Weinberger isoperimetric inequality*, on the other hand, indicates

$$\mu_2(\Omega) \le \mu_2(\Omega^*), \tag{1.115}$$

where $\mu_2(\Omega)$ denotes the second eigenvalue of (1.105), of which the first eigenvalue is always zero. In the cases of (1.114) and (1.115) the equality holds if and only if $\Omega = \Omega^*$. Kac's problem, on the other hand, is to determine Ω by eigenvalues.

1.4.2 Gel'fand Triple

The above Dirichlet and Neumann problems have the *weak form*

$$u \in V,\ \ a(u,v) = \lambda b(u,v), \quad \forall v \in V,$$

defined for

$$a(u, v) = \int_{\Omega} \nabla u \cdot \nabla v \, dx, \ u, v \in V$$

$$b(u, v) = \int_{\Omega} uv \, dx, \qquad u, v \in H, \tag{1.116}$$

where $V = H_0^1(\Omega)$ and $V = H^1(\Omega)$ for (1.103) and (1.105), respectively, and $H = L^2(\Omega)$. The spaces $H_0^1(\Omega)$ and $H^1(\Omega)$ are Hilbert spaces where $a : V \times V \to \mathbf{R}$ is a bounded symmetric bilinear form.

More precisely, $H^1(\Omega)$ is composed of functions on Ω, square-integrable, including their first derivatives; and $H_0^1(\Omega)$ is the closure of $C_0^\infty(\Omega)$, the set of infinitely differentiable functions on Ω with compact supports. It also holds that $b = b(\cdot, \cdot)$ is the inner product in $H = L^2(\Omega)$, and, therefore, the abstract theory below guarantees a complete orthonormal system in $L^2(\Omega)$ derived from the eigenfunctions of (1.103) or (1.105).

Henceforth H denotes an infinite-dimensional Hilbert space with the inner product $(\ , \ )$ and the norm $| \ |$. If it has a countable dense subset, it is said to be separable. In this case the space H admits a complete orthonormal system (composed of countable members), and henceforth this condition is assumed unless otherwise stated. Here $\{\varphi\}_{i=1}^\infty \subset H$ is an orthonormal system if and only if

$$\left(\varphi_i, \varphi_j\right) = \delta_{ij} = \begin{cases} 1, \ i = j \\ 0, \ i \neq j. \end{cases}$$

This system is said to be complete if its linear hull

$$\{\sum_{i=1}^n \alpha_i \varphi_i \mid \alpha_i \in \mathbf{R}, \ n = 1, 2, \cdots\}$$

is dense in H. The completeness of the orthonormal system $\{\varphi_i\}_{i=1}^\infty$ is equivalent to the convergence

$$\lim_{n\to\infty} \left| f - \sum_{i=1}^n (f, \varphi_i) \varphi_i \right| = 0, \quad f \in H,$$

which is also equivalent to Parseval's relation

$$\sum_{i=1}^\infty |(f, \varphi_i)|^2 = |f|^2, \quad f \in H. \tag{1.117}$$

Even if $\{\varphi_i\}_{i=1}^\infty$ is not necessarily complete, the left-hand side is less than or equal to the right-hand side in (1.117), in which case it is called Bessel's inequality. In the present section we show that the complete orthonormal system is obtained by the eigenvalue problem with symmetry and compactness.

The mapping $b = b(\ ,\ ) : H \times H \to \mathbf{R}$ is bilinear if it satisfies

$$\begin{aligned} b(\alpha u + \beta v, w) &= \alpha b(u, w) + \beta b(v, w) \\ b(u, \alpha v + \beta w) &= \alpha b(u, v) + \beta b(u, w), \quad u, v, w \in H,\ \alpha, \beta \in \mathbf{R}. \end{aligned}$$

It is symmetric if

$$b(u, v) = b(v, u), \quad u, v \in H.$$

It is positive definite if there is $\delta > 0$ satisfying

$$b(u, u) \geq \delta\, |u|^2\,, \quad u \in H.$$

Finally, it is bounded if there is $M > 0$ satisfying

$$|b(u, v)| \leq M\, |u| \cdot |v|\,, \quad u, v \in H.$$

Such a $b(\ ,\ )$ provides the inner product to the vector space H, and then H becomes a Hilbert space with its topology equivalent to the original norm $|\cdot|$.

Let V be another Hilbert space with the inner product $((\ ,\ ))$ and the norm $\|\cdot\|$ satisfying $V \subset H$ as a vector space. We suppose that this inclusion involves the topology, that is, there is $K > 0$ such that

$$|u| \leq K\, \|u\|\,, \quad u \in V,$$

and, furthermore, that V is dense in H. Throughout this section these relations are written as $V \hookrightarrow H$ for short. In this case, identifying the dual space H' with H by Riesz's representation theorem and restricting the domain of each $T : H \to \mathbf{R}$ to V, we obtain

$$V \hookrightarrow H \cong H' \subset V',$$

where V' is the dual space of V, that is, the Gel'fand triple.

Given a symmetric, positive definite, and bounded bilinear form

$$a = a(\ ,\ ) : V \times V \to \mathbf{R},$$

we can introduce the bounded linear operator $\tilde{A} : V \to V'$ satisfying

$$a(u, v) = \left\langle \tilde{A}u, v \right\rangle_{V', V}, \quad u, v \in V, \tag{1.118}$$

where $\langle\ ,\ \rangle_{V',V}$ denotes the pairing between V' and V, that is,

$$\langle T, v\rangle_{V',V} = T(v), \quad T \in V',\ v \in V.$$

Then the (unbounded) linear operator A on H is defined by

$$D(A) = \left\{u \in V \mid \tilde{A}u \in H\right\}, \quad Au = \tilde{A}u,\ u \in D(A).$$

We show that this A is *self-adjoint* in H with the domain $D(A)$. Its range and kernel are given by

$$R(A) = \{Au \mid u \in D(A)\}, \quad N(A) = \{u \in D(A) \mid Au = 0\}.$$

Theorem 1.14 *The linear operator $\tilde{A} : V \to V'$ defined by (1.118) is an isomorphism, that is, one-to-one, onto, and bounded with its inverse.*

Proof The linearity of $\tilde{A}$ is obvious. We recall that the norm in V is denoted by $\|\ \|$. From the boundedness of $a(\ ,\ )$ we have

$$|\langle \tilde{A}u, v\rangle_{V',V}| = |a(u,v)| \leq M\|u\| \cdot \|v\|, \quad u, v \in V,$$

and hence it follows that

$$\|\tilde{A}u\|_{V'} = \sup\left\{|\langle \tilde{A}u, v\rangle_{V',V}| \mid v \in V,\ \|v\| = 1\right\} \leq M\|u\|.$$

Thus $\tilde{A} : V \to V'$ is a bounded linear operator. The relation

$$\left\langle \tilde{A}u, u\right\rangle_{V',V} = a(u,u) \geq \delta\|u\|^2, \quad u \in V,$$

on the other hand, implies

$$\left\|\tilde{A}u\right\|_{V'} = \sup\{\frac{|\langle \tilde{A}u, v\rangle_{V',V}|}{\|v\|} \mid v \in V \setminus \{0\}\} \geq \delta\|u\|.$$

The operator $\tilde{A}$ is thus one-to-one, and, therefore, if

$$R(\tilde{A}) = V' \tag{1.119}$$

is proven then $T = \tilde{A}^{-1} : V' \to V$ satisfies

$$\|Tf\| \leq \delta^{-1}\|f\|_{V'}, \quad f \in V',$$

and hence is bounded. To show (1.119) we replace the inner product in V with $a(\ ,\ )$. Given $f \in V'$, we apply Riesz's representation theorem to obtain $u \in V$ such that

$$\left\langle \tilde{A}u, v \right\rangle_{V',V} = a(u,v) = \langle f, v \rangle_{V',V}, \quad v \in V,$$

which means $\tilde{A}u = f$ as desired. □

For the Dirichlet problem (1.103) we provide the inner products (1.116) for $H = L^2(\Omega)$ and $V = H_0^1(\Omega)$. This $a(u,v)$ is called the Dirichlet form, which is actually an inner product in $V = H_0^1(\Omega)$ by the Poincaré inequality. If $\partial\Omega$ is C^2 it holds that

$$D(A) = \left\{ u \in V \mid \tilde{A}u \in H \right\} = H_0^1(\Omega) \cap H^2(\Omega)$$

and $Au = -\Delta u$ for $u \in D(A)$, where $H^2(\Omega)$ is the Hilbert space composed of square-integrable functions in Ω, including their second derivatives.

Concluding this subsection, we confirm several fundamental facts on the self-adjoint operator. Let A be a linear operator with the domain $D(A)$ and the range $R(A)$ in the Hilbert space H with inner product $(\ ,\ )$ and norm $|\cdot|$. Hence $D(A) \subset H$ is a linear subspace and the mapping $A : D(A) \subset H \to H$ is provided with the property

$$A(\alpha u + \beta v) = \alpha Au + \beta Av, \quad \alpha, \beta \in \mathbf{R},\ u, v \in D(A).$$

In the case that $D(A)$ is dense in H, $\overline{D(A)} = H$, the adjoint operator A' is defined by

$$A'v = v'$$

if and only if $(Au, v) = (u, v')$ for any $u \in D(A)$. The densely defined linear operator A is called self-adjoint if $A = A'$ holds, including their domains.

The linear operator A is said to be closed if it satisfies the condition that $\{u_j\} \subset D(A)$, $u_j \to u$, and $Au_j \to v$ implies $u \in D(A)$ and $Au = v$. If the linear operator A is densely defined then its adjoint A' is closed, and, therefore, a self-adjoint operator is always closed. For a densely defined closed operator between Banach spaces we have the closed range theorem, of which the special form is stated as follows.

Henceforth we put

$$L^\perp = \{u \in H \mid (u, v) = 0 \text{ for any } v \in L\},$$

where $L \subset H$. If L is a linear subspace, $\overline{L} = H$ if and only if $L^\perp = \{0\}$. We have $R(A)^\perp = N(A')$ and $R(A')^\perp = N(A)$ easily, but there arises the following theorem.

Theorem 1.15 *If H is a Hilbert space and $A : D(A) \subset H \to H$ is a closed linear operator satisfying $\overline{D(A)} = H$, then it holds that $\overline{R(A)} = N(A')^{\perp}$ and $\overline{R(A')} = N(A)^{\perp}$. Furthermore, $R(A)$ is closed if and only if $R(A')$ is so.*

A self-adjoint operator A is said to be positive definite if there is $\tilde{\delta} > 0$ such that $(Au, u) \geq \tilde{\delta}\,|u|^2$ for any $u \in D(A)$. In this case we have

$$|Au| \geq \tilde{\delta}\,|u|\,, \quad u \in D(A), \tag{1.120}$$

and hence $N(A) = \{0\}$ and the closedness of $R(A)$. Theorem 1.15 then guarantees $R(A) = H$, and, therefore, $A^{-1} : H \to H$ is well-defined as a closed linear operator. Since Banach's open mapping theorem guarantees that a closed operator with the domain of full space is bounded, any positive definite self-adjoint operator on a Hilbert space has a bounded inverse. The self-adjoint operator takes the spectral decomposition from which the non-negative self-adjoint operator admits its fractional powers.

1.4.3 Symmetric Forms

Following Sect. 1.4.2, let $V \hookrightarrow H \hookrightarrow V'$ be a Gel'fand triple where H and V are Hilbert spaces provided with the inner products $(\ ,\ )$ and $((\ ,\ ))$ and with the norms $|\cdot|$ and $\|\cdot\|$, respectively. The bilinear form

$$a = a(\ ,\ ) : V \times V \to \mathbf{R}$$

is symmetric, positive definite, and bounded. Then the bounded linear operator $\tilde{A} : V \to V'$ and the linear operator $A : D(A) \subset H \to H$ are defined by

$$\left\langle \tilde{A}u, v \right\rangle_{V',V} = a(u, v), \quad u, v \in V, \tag{1.121}$$

and $A = \tilde{A}\Big|_{D(A)}$ with $D(A) = \left\{ u \in V \mid \tilde{A}u \in H \right\}$, respectively.

Theorem 1.16 *The above A is self-adjoint in H.*

Proof We prove, first, that $D(A)$ is dense in V, and hence is also dense in H. It suffices to show that

$$f \in V', \ \langle f, v\rangle_{V',V} = 0, \ \forall v \in D(A) \quad \Rightarrow \quad f = 0.$$

To this end, given $f \in V'$, we take $u \in V$ satisfying $\tilde{A}u = f$ by Theorem 1.14. Then we obtain

$$a(u, v) = \langle f, v\rangle_{V',V} = 0, \quad \forall v \in D(A)$$

by the above assumption. Given $g \in H$, now we take $v = \tilde{A}^{-1}g \in D(A)$, to reach

$$0 = a(u, v) = \langle u, g\rangle = (u, g),$$

and hence $u = 0$ because $g \in H$ is arbitrary. Then $f = \tilde{A}u = 0$ follows.

We show, next, $R(A) = H$ by $R(\tilde{A}) = V'$ and

$$D(A) = \{u \in V \mid \tilde{A}u \in H\}.$$

Since $Au = 0$ for $u \in D(A)$ implies

$$(Au, u) \geq \delta \|u\|_V^2 = 0,$$

the operator $A : D(A) \to H$ is one-to-one. It is also closed, and hence the operator $T = A^{-1} : H \to H$ is bounded by the open mapping theorem. Furthermore, since

$$a = a(\ ,\ ) : V \times V \to \mathbf{R}$$

is symmetric, we have

$$\begin{aligned}(Tu, v) &= (Tu, Aw) = (Aw, Tu) = a(w, Tu) = a(Tu, w)\\ &= (A \cdot Tu, w) = (u, w) = (u, Tv)\end{aligned}$$

for $v = Aw \in R(A) = H$ and $u \in H$.

Thus, T is bounded and symmetric, and, hence, is self-adjoint. Since $T^{-1} = A$, including their domains, A is self-adjoint in H by the following theorem, where T may not be bounded. □

Theorem 1.17 *If T is a one-to-one self-adjoint operator in a Hilbert space H, then T^{-1} is also self-adjoint.*

Proof We show, first, $R(T)^{\perp} = \{0\}$, which implies that $D(T^{-1}) = R(T)$ is dense in H. In fact, given $u \in R(T)^{\perp}$, we have

$$(Tv, u) = 0, \quad \forall v \in D(T).$$

Since T is self-adjoint, this property implies $u \in D(T)$ and $(v, Tu) = 0$. Using

$$\overline{D(T)} = H, \quad N(T) = \{0\},$$

we obtain $Tu = 0$ and $u = 0$ in turn. Thus $R(T)^{\perp} = \{0\}$ or $\overline{R(T)} = H$ is proven, and, therefore, $\left(T^{-1}\right)'$ is well-defined.

Given $v \in D((T^{-1})')$, we have

$$(u, v) = (T^{-1}Tu, v) = (Tu, (T^{-1})'v), \quad u \in D(T),$$

which implies

$$(T^{-1})'v \in D(T), \quad T(T^{-1})'v = v$$

by $T' = T$. Thus we obtain

$$v \in D(T^{-1}), \quad T^{-1}v = (T^{-1})'v.$$

Given $v \in D(T^{-1})$ and $w = T^{-1}v$, conversely, we have

$$w \in D(T), \quad v = Tw,$$

and therefore

$$(u, w) = (TT^{-1}u, w) = (T^{-1}u, Tw) = (T^{-1}u, v), \quad \forall u \in D(T^{-1}).$$

This property means $v \in D((T^{-1})')$ and $w = \left(T^{-1}\right)' v$. We have thus

$$(T^{-1})' = T^{-1},$$

including the domains. □

It is obvious that the self-adjoint operator A in Theorem 1.16 is positive definite. In connection with its fractional powers it holds that $D(A^{1/2}) = V$ and

$$a(u, v) = (A^{1/2}u, A^{1/2}v), \quad u, v \in V.$$

1.4.4 Compact Operators

We recall that H is a separable Hilbert space with the inner product $(\ ,\ )$ and the norm $|\cdot|$. A bounded linear operator $T : H \to H$ is said to be compact if it maps bounded sets to relatively compact sets, that is, $\{Tu_n\}$ contains a converging subsequence if $\{u_n\}$ is bounded in H. By the fact that any bounded sequence in H admits a weakly converging subsequence, the bounded linear operator $T : H \to H$ is compact if and only if $u_n \rightharpoonup u$ implies $Tu_n \to Tu$. Here and henceforth, $u_n \rightharpoonup u$ and $u_n \to u$ indicate the weak and the strong convergences so that $\lim_{n\to\infty}(u_n, v) = (u, v)$ for any $v \in H$ and $\lim_{n\to\infty} |u_n - u| = 0$, respectively. This property, referred to as the complete continuity, however, is not equivalent to the compactness for the nonlinear operator.

In the Gel'fand triple $V \hookrightarrow H \hookrightarrow V'$ we say that $V \hookrightarrow H$ is compact if the embedding (or, identity) mapping

$$i : V \to H$$

is compact. In the case that the symmetric, positive definite, and bounded bilinear form

$$a = a(\ ,\ ) : V \times V \to \mathbf{R}$$

exists, the inner product $((\ ,\ ))$ of V is replaced by $a(\ ,\ )$, and then, this compactness means

$$a(u_n - u, v) \to 0, \ \forall v \in V \ \Rightarrow \ |u_n - u| \to 0,$$

where $u_n, u \in V, n = 1, 2, \cdots$.

We recall that $((\ ,\ ))$ and $\|\cdot\|$ indicate the inner product and the norm in V, respectively.

Theorem 1.18 *Given the Gel'fand triple $V \hookrightarrow H \hookrightarrow V'$ with compact inclusion $V \hookrightarrow H$ and with symmetric, positive definite, and bounded bilinear form*

$$a = a(\ ,\ ) : V \times V \to \mathbf{R},$$

the bounded linear operator $T = A^{-1} : H \to H$ is compact, where A is the self-adjoint operator introduced in Theorem 1.17.

Proof Let $\{u_n\} \subset H$ be a bounded sequence. Then it holds that

$$\begin{aligned} \|Tu_n\|^2 &= a(Tu_n, Tu_n) = (u_n, Tu_n) \\ &\le |u_n| \cdot |Tu_n| \le K\,|u_n| \cdot \|Tu_n\| \end{aligned}$$

and hence

$$\|Tu_n\| \le K\,|u_n|.$$

Since $\{Tu_n\} \subset V$ is bounded in V, it has a subsequence converging in H. Thus the proof is complete. □

For the moment, T denotes a general linear operator on H not necessarily bounded. Given $\lambda \in \mathbf{R}$, the set of solutions to

$$Tv = \lambda v \tag{1.122}$$

forms a linear subspace of H called the eigenspace associated with the eigenvalue λ. If the dimension of this space is greater than or equal to 1, that is, (1.122) admits a non-trivial solution $0 \neq v \in D(T)$, we say that λ and v are an eigenvalue and an eigenfunction of T, respectively. We see that if T is self-adjoint then the eigenfunctions u, v associated with different eigenvalues are orthogonal so that $(u, v) = 0$ holds in H.

Theorem 1.19 *If $T : H \to H$ is a compact self-adjoint bounded linear operator then $\|T\|$ or $-\|T\|$ is an eigenvalue of T, where $\|T\|$ denotes the operator norm of T.*

Proof We may suppose that $\lambda = \|T\| \neq 0$. The linear operator

$$S = \lambda^2 - T^2 : H \to H$$

is bounded, self-adjoint, and non-negative so that

$$(Sv, v) \geq 0, \quad \forall v \in H.$$

Since $[u, v] = (Su, v)$ provides a non-negative and bounded bilinear form on $H \times H$, the Schwarz inequality is valid as

$$|[u, v]|^2 \leq [u, u] \cdot [v, v], \quad u, v \in H.$$

Letting $v = Su$, we get

$$\begin{aligned} |Su|^4 &\leq (Su, u) \cdot (S^2u, Su) \leq (Su, u) \cdot \left|S^2u\right| \cdot |Su| \\ &\leq (Su, u) \cdot |Su|^2 \cdot \|S\|, \end{aligned}$$

or, equivalently,

$$|Su| \leq \|S\|^{1/2} (Su, u)^{1/2}, \quad u \in H. \tag{1.123}$$

Since

$$\|T\| = \sup \{|Tu| \mid u \in H, \ |u| = 1\}$$

we have $\{u_n\} \subset H$ satisfying $|u_n| = 1$ and $|Tu_n| \to \lambda$, and then it follows that

$$((\lambda^2 - T^2)u_n, u_n) = \lambda^2 - |Tu_n|^2 \to 0.$$

Hence we obtain

$$\left(\lambda^2 - T^2\right) u_n \to 0 \tag{1.124}$$

by (1.123).

Since T is compact, $\{u_n\}$ admits a subsequence denoted by the same symbol and $w \in H$ such that

$$Tu_n \to w \tag{1.125}$$

which implies $\lambda = |w| \neq 0$ and $T^2 u_n \to Tw$. Therefore, $\lambda^2 u_n \to Tw$ follows from (1.124), and hence

$$\lambda^2 T u_n \to T^2 w.$$

We thus obtain $\lambda^2 w = T^2 w$ by (1.125).

Then $v = (\lambda + T)w$ satisfies $(\lambda - T)v = 0$ so that if $v \neq 0$ then $\lambda = \|T\|$ is an eigenvalue of T. The other case of $v = 0$, $-\|T\|$ is an eigenvalue of T by $|w| = \lambda$. Thus the proof is complete. □

1.4.5 Eigenfunction Expansion

Henceforth, the eigenfunction v is normalized as $|v| = 1$ and an orthonormal basis is taken for each eigenspace.

Theorem 1.20 *If H is a Hilbert space and $T : H \to H$ is a compact, self-adjoint, and bounded linear operator then its eigenvalues are discrete and accumulate to 0, and each eigenspace is of finite dimension. The family composed of orthonormal basis of these spaces forms a basis of $R(T)$, and in particular, $R(T)$ is closed.*

Proof In the first step, we show that linearly independent $\{u_n\}_{n=1}^{\infty} \subset H$ in $|u_n| = 1$ cannot satisfy $Tu_n = \lambda_n u_n$ and $\lambda_n \to \lambda \neq 0$, and therefore, eigenvalues of T are discrete and can accumulate only to 0. In fact, assume such $\{u_n\}_{n=1}^{\infty}$, and let M_n be the linear hull of $\{u_1, \cdots, u_n\}$. Then there is $v_n \in M_n$ satisfying

$$|v_n| = 1, \qquad \mathrm{dist}\,(v_n, M_{n-1}) \equiv \inf_{v \in M_{n-1}} |v_n - v| = 1.$$

Then we shall show that $\left\{\lambda_n^{-1} T v_n\right\}$ cannot accumulate in H, which, however, contradicts the compactness of T.

For this purpose we take $m < n$ so that

$$\lambda_n^{-1} T v_n - \lambda_m^{-1} T v_m = v_n - \left(\lambda_m^{-1} T v_m - \lambda_n^{-1} (T - \lambda_n) v_n\right). \tag{1.126}$$

We have $Tv_m \in M_{n-1}$ by $v_m \in M_{n-1}$, while $(T - \lambda_n)u_n = 0$ implies

$$(T - \lambda_n)v_n \in M_{n-1}.$$

The second term on the right-hand side of (1.126) thus belongs to M_{n-1}. We obtain

$$\left|\lambda_n^{-1} T v_n - \lambda_m^{-1} T v_m\right| \geq \mathrm{dist}\,(v_n, M_{n-1}) = 1,$$

and, therefore, $\left\{\lambda_n^{-1} T v_n\right\}$ has no subsequence converging in H.

We proceed to the second step of the proof. First, we have $v_1 \in H$ and $\lambda_1 \in \mathbf{R}$ satisfying

$$|v_1| = 1, \quad |\lambda_1| = \|T\|, \quad Tv_1 = \lambda_1 v_1$$

by Theorem 1.19. Then we take $H_1 = \{v_1\}^\perp$, to note that

$$T_1 = T|_{H_1} : H_1 \to H_1$$

is a compact, self-adjoint, and bounded linear operator. There are thus $v_2 \in H_1$ and $\lambda_2 \in \mathbf{R}$ satisfying

$$|v_2| = 1, \quad |\lambda_2| = \|T_1\| \leq \|T\|, \quad Tv_2 = \lambda_2 v_2.$$

We continue the argument, taking $H_2 = \{v_1, v_2\}^\perp$, and in this way obtain a sequence of eigenvalues denoted by $\{\lambda_j\}$.

In the case that this λ_j reaches 0, there is n such that $\lambda_n = 0$ and $\lambda_j \neq 0$ for $1 \leq j \leq n-1$. Then it holds that $H_{n-1} \equiv \{v_1, \cdots, v_{n-1}\}^\perp \subset N(T)$ and $Tv_j = 0$ for $j \geq n$. We have, on the other hand, $v_j \in R(T)$ for $j < n$ and hence $R(T)^\perp \subset H_{n-1}$. By $N(T) = R(T)^\perp$, therefore, we obtain $H_{n-1} = N(T)$ and $R(T)$ is a linear hull of

$$\{v_1, \cdots, v_{n-1}\}.$$

In the other case of $\lambda_n \neq 0$ for any n and it holds that

$$|\lambda_1| \geq |\lambda_2| \geq \cdots \to 0$$

by the first part of this theorem. Now we shall show that any $w \in R(T)$ has the property

$$\lim_{n\to\infty} \left| w - \sum_{i=1}^{n} (w, v_j)v_j \right| = 0. \tag{1.127}$$

In fact, since $w \in R(T)$ there is $u \in H$ such that $w = Tu$. Letting $b_j = (w, u_j)$ and $c_j = (u, v_j)$, we have

$$b_j = (w, v_j) = (Tu, v_j) = (u, Tv_j) = \lambda_j(u, v_j) = \lambda_j c_j.$$

Hence we obtain $T(c_j v_j) = b_j v_j$, which implies

$$w - \sum_{j=1}^{n} b_j v_j = T(u - \sum_{j=1}^{n} c_j v_j) = T_n(u - \sum_{j=1}^{n} c_j v_j).$$

Since $\|T_n\| = |\lambda_{n+1}|$ and

$$\left| u - \sum_{j=1}^{n} c_j v_j \right| \leq |u| ,$$

therefore, we obtain

$$\left| w - \sum_{j=1}^{n} b_j v_j \right| \leq |\lambda_{n+1}| \cdot |u| \to 0,$$

which means that (1.127) holds. □

1.4.6 Minimax Principle

Henceforth, the dimension of eigenspace associated with an eigenvalue is called its multiplicity.

Theorem 1.21 *Given the Gel'fand triple of Hilbert spaces $V \hookrightarrow H \hookrightarrow V'$, suppose that H is separable and of infinite dimension and that the inclusion $V \hookrightarrow H$ is compact. Suppose, furthermore, that $b = b(\ ,\ ) : H \times H \to \mathbf{R}$ and $a = a(\ ,\ ) : V \times V \to \mathbf{R}$ are symmetric, positive definite, and bounded bilinear forms. Then the eigenvalue problem*

$$v \in V, \quad a(v, w) = \mu b(v, w), \qquad \forall w \in V,$$

provides countably many eigenvalues, each of which has finite multiplicity denoted by

$$0 < \mu_1 \leq \mu_2 \leq \cdots \to +\infty.$$

The associated normalized system of eigenvectors $\{v_i\}_{i=1}^{\infty}$ forms a complete orthonormal system in H in the sense that $b(v_i, v_j) = \delta_{ij}$ and

$$\lim_{n\to\infty} \left| v - \sum_{i=1}^{n} \alpha_i v_i \right| = 0, \quad \alpha_i = b(v, v_i), \tag{1.128}$$

holds for any $v \in H$.

Proof Replacing the inner product in H to $b(\ ,\ )$, we take the self-adjoint operator A associated with $a(\ ,\ )$ so that $\tilde{A} : V \to V'$ is defined by (1.121), and $A : D(A) \subset H \to H$ is the restriction of $\tilde{A}$ to

$$D(A) = \left\{ u \in V \mid \tilde{A}u \in H \right\}.$$

Then $T = A^{-1} : H \to H$ is a compact, self-adjoint, bounded linear operator, and the eigenfunctions associated with non-zero eigenvalues to T provides an orthonormal basis $\{v_i\}$ of $R(T) = D(A)$ satisfying

$$b(v_i, v_j) = \delta_{ij}.$$

Any $v \in D(A)$ thus admits (1.128) which is valid for any $v \in H$ by $\overline{D(A)} = H$. □

In the above theorem the eigenvalues are labeled according to their multiplicities and each eigenvalue takes one normalized eigenvector. These eigenvectors with the same eigenvalue, furthermore, are arranged to be orthogonal to each other. Under this agreement it holds that

$$\begin{aligned} a(v_i, v_j) &= \mu_j \delta_{ij} \\ b(v, v) &= \sum_{k=1}^{\infty} c_k^2, \quad v \in H \\ a(v, v) &= \sum_{k=1}^{\infty} \mu_k c_k^2, \quad v \in V, \end{aligned} \tag{1.129}$$

for $c_k = b(v, v_k)$, which implies

$$\mu_1 = \min \left\{ a(v, v)/b(v, v) \mid v \in V \setminus \{0\} \right\}. \tag{1.130}$$

Henceforth, $R[v] = a(v, v)/b(v, v)$ is called the Rayleigh quotient and (1.130) the Rayleigh principle. A generalization of the Rayleigh principle is described by the following theorem, referred to as the minimax principle.

Theorem 1.22 *It holds that*

$$\mu_k = \min_{L_k} \max_{v \in L_k \setminus \{0\}} R[v] = \max_{V_k} \min_{v \in V_k \setminus \{0\}} R[v],$$

where $\{L_k\}$ and $\{V_k\}$ denote the families of all subspaces of H with the dimension k and the co-dimension $k - 1$, respectively.

Proof To prove the first equality we set

$$\Lambda_k = \inf_{L_k} \sup_{v \in L_k \setminus \{0\}} R[v]$$

and

$$H_n = \{v_1, \cdots, v_n\}^\perp = \{v \in H \mid b(v, v_i) = 0, \ \ 1 \le i \le n\}.$$

Given L_k, a k-dimensional subspace of H, we have $\tilde{v} \in L_k \setminus \{0\}$ in $\tilde{v} \in H_{k-1}$ and hence

$$\sup_{v \in L_k \setminus \{0\}} R[v] \ge R[\tilde{v}] \ge \mu_k$$

by (1.129), which implies $\Lambda_k \ge \mu_k$. If L_k is the linear hull of $\{v_1, \cdots, v_k\}$, on the other hand, we obtain

$$\max_{v \in L_k \setminus \{0\}} R[v] = \mu_k$$

again by (1.129), and, therefore, Λ_k is attained by this L_k and is equal to μ_k.

To prove the second equality we put

$$\Sigma_k = \sup_{V_k} \inf_{v \in V_k \setminus \{0\}} R[v].$$

If W_n denotes the linear hull of $\{v_1, \cdots, v_n\}$, then any V_k satisfies $V_k \cap W_k \ne \{0\}$. Here (1.129) guarantees

$$\inf_{v \in V_k \setminus \{0\}} R[v] = \min_{v \in V_k \setminus \{0\}} R[v] \le \mu_k$$

and hence $\Sigma_k \le \mu_k$. Given a $(k-1)$-codimensional subspace $V_k = H_{k-1}$, on the other hand, we have

$$\inf_{v \in V_k \setminus \{0\}} R[v] = \min_{v \in V_k \setminus \{0\}} R[v] = \mu_k.$$

Thus Σ_k is attained and is equal to μ_k, and the proof is complete. □

1.4.7 Hilbert–Schmidt and Nuclear Operators

Let H be a separable Hilbert space over $\mathbf{R}$, and let $\{\varphi_i\}$ and $\{\psi_i\}$ be complete orthonormal systems. Given the bounded linear operator $T : H \to H$, we obtain

$$\begin{aligned}
\sum_{i=1}^{\infty} \|T\varphi_i\|^2 &= \sum_{i,j=1}^{\infty} |(T\varphi_i, \psi_j)|^2 = \sum_{i,j=1}^{\infty} |(\varphi_i, T'\psi_j)|^2 \\
&= \sum_{i,j=1}^{\infty} |(T'\psi_j, \varphi_i)|^2 = \sum_{j=1}^{\infty} \|T'\psi_i\|^2
\end{aligned}$$

which implies

$$\sum_{i=1}^{\infty} \|T\varphi_i\|^2 = \sum_{i=1}^{\infty} \|T'\varphi_i\|^2 = \sum_{i=1}^{\infty} \|T'\psi_i\|^2 = \sum_{i=1}^{\infty} \|T\psi_i\|^2.$$

If

$$|||T||| \equiv (\sum_{i=1}^{\infty} \|T\varphi_i\|^2)^{1/2} < +\infty$$

is the case, we call this $T : H \to H$ a Hilbert–Schmidt operator.

Theorem 1.23 *The Hilbert–Schmidt operator $T : H \to H$ is compact.*

Proof Using $v = \sum_{i=1}^{\infty}(v, \varphi_i)\varphi_i$, we obtain

$$\begin{aligned}\|Tv\| &= \|\sum_{i=1}^{\infty}(v, \varphi_i)T\varphi_i\| \leq \sum_{i=1}^{\infty} |(v, \varphi_i)|\|T\varphi_i\| \\ &\leq \left(\sum_{i=1}^{\infty} |(v, \varphi_i)|^2\right)^{1/2} \left(\sum_{i=1}^{\infty} \|T\varphi_i\|^2\right)^{1/2} = |||T||| \cdot \|v\|\end{aligned}$$

and hence $\|T\| \leq |||T|||$. It holds that

$$\|T - T_n\|^2 \leq |||T - T_n|||^2, \quad T_n v = \sum_{i,j=1}^{n-1} (T\varphi_i, \varphi_j)(v, \varphi_i)\varphi_j$$

by this inequality, while

$$Tv = \sum_{i=1}^{\infty}(v, \varphi_i)T\varphi_i = \sum_{i,j=1}^{\infty} (v, \varphi_j)(T\varphi_i, \varphi_j)\varphi_j$$

implies

$$(T - T_n)\varphi_i = \sum_{k,j=n}^{\infty} (T\varphi_k, \varphi_j)(\varphi_i, \varphi_k)\varphi_j = \sum_{j=n}^{\infty}(T\varphi_i, \varphi_j)\varphi_j,$$

and hence

$$\|(T - T_n)\varphi_i\|^2 = \sum_{j=n}^{\infty} |(T\varphi_i, \varphi_j)|^2$$

$$|||T - T_n|||^2 = \sum_{i=1}^{\infty} \|(T - T_n)\varphi_i\|^2 = \sum_{j=n}^{\infty} \|T^*\varphi_j\|^2.$$

Thus it holds that

$$\lim_{n\to\infty} |||T - T_n||| = 0.$$

Since $\dim R(T_n) < +\infty$, the Hilbert–Schmidt operator $T : H \to H$ is compact, because the norm limit of a sequence of compact operators in a Banach space is compact. □

Given a compact bounded linear operator $T : H \to H$, conversely, we have that

$$T^*T : H \to H$$

is compact, self-adjoint, and bounded. Hence there arises $\lambda_i \downarrow 0$ and a set of orthonormal difencutions $\{e_i\}$ such that $T^*Te_i = \lambda_i^2 e_i$ and

$$T^*Tv = \sum_{i=1}^{\infty} \lambda_i^2 (v, e_i)e_i, \quad v \in H.$$

If $v \in H_1^\perp$ for $H_1 = \overline{\langle e_i \mid i = 1, 2, \cdots \rangle}$, there arises $T^*Tv = 0$, and hence $Tv = 0$ by $\|Tv\|^2 = (T^*Tv, v) = 0$. Thus we obtain

$$Tv = \sum_{i=1}^{\infty} (v, e_i) f_i, \quad f_i = Te_i.$$

Letting

$$\|T\|_p = \left(\sum_i \lambda_i^p\right)^{1/p}, \quad 1 \le p < \infty,$$

we have $\|T\|_2 < +\infty$ if and only if T is Hilbert–Schmidt. We say that T is nuclear if $\|T\|_1 < \infty$. Then the trace of this operator,

$$\mathrm{tr}\,(T) = \sum_{i=1}^{\infty} (Te_i, e_i),$$

and also the Fredholm determinant

$$\det(I - T) = \lim_{n\to\infty} \det(I - T_n), \quad T_n v = \sum_{i=1}^{n} (v, e_i) f_i,$$

are well-defined. If $I - zT$ for $z \in \mathbf{C}$, is invertible, it holds that

$$\log\det(I - zT) = \mathrm{tr}\,(\log(I - zT)) = \sum_{n=1}^{\infty} \frac{z^n}{n} \mathrm{tr}\,(T)^n$$

by [18].

1.5 Nodal Properties

1.5.1 The First Eigenfunction

The fundamental result of the Strum–Liouville theory is on the number of zeros of eigenfunction for one-dimensional regular second order elliptic boundary value operators. We begin with the general case.

Theorem 1.21 is applicable to (1.103), or its weak form,

$$\varphi \in H_0^1(\Omega), \quad (\nabla\varphi, \nabla v) = \lambda(\varphi, v), \ \forall v \in H_0^1(\Omega),$$

defined for (1.116), $V = H_0^1(\Omega)$, and $H = L^2(\Omega)$. In fact, the Rellich–Kondrachov theorem guarantees the compact embedding of $V \hookrightarrow H$ if Ω is bounded. There arises also the positivity (1.120) because of the Poincaré inequality indicated by

$$a(u, u) \geq \delta b(u, u), \quad u \in V,$$

with $\delta > 0$. There arises countably many eigenvalues to (1.103) with finite multiplicities

$$0 < \lambda_1 \leq \lambda_2 \leq \cdots \to +\infty,$$

associated with the eigenfunctions $\{\phi_i\}$ taken to be complete orthonormal eigenfunctions of $H = L^2(\Omega)$.

The first eigenvalue to (1.103) denoted by $\lambda_1 = \lambda_1(\Omega)$ is simple, which means that its eigenspace is one-dimensional. This simplicity is a consequence of the definite sign of an eigenfunction $\varphi_1 = \varphi_1(x)$, described by

$$-\Delta\phi_1 = \lambda_1\phi_1, \ \phi_1 > 0 \quad \text{in } \Omega, \quad \phi_1|_{\partial\Omega} = 0. \tag{1.131}$$

To confirm this fact, we use the minimax principle in Sect. 1.4.6, which is reduced to the Rayleigh principle

$$\lambda_1 = \inf\{\|\nabla v\|_2^2 \mid v \in H_0^1, \ \|v\|_2^2 = 1\}. \tag{1.132}$$

The following facts are be proven in turn:

1. The infimum λ_1 in (1.132) is attained by a function $\phi_1(x) \geq 0$.
2. This function $\phi_1(x)$ is a classical solution to (1.131) and $\phi_1(x) > 0$ holds for any $x \in \Omega$.
3. The constant λ_1 is the minimum eigenvalue, i.e.,

$$-\Delta\phi = \lambda\phi, \ \varphi \not\equiv 0 \text{ in } \Omega, \quad \phi|_{\partial\Omega} = 0 \tag{1.133}$$

implies $\lambda \geq \lambda_1$. Furthermore, any solution ϕ to (1.133) with $\lambda = \lambda_1$ is a constant multiplication of ϕ_1.

The first fact is already contained in Theorem 1.22. To confirm directly, let $v_k \in H_0^1$, $k = 1, 2, \ldots$, be a minimizing sequence:

$$\|v_k\|_2^2 = 1, \quad \|\nabla v_k\|_2^2 \to \lambda_1.$$

This sequence is bounded in H_0^1 so that has a weak limit $v \in H_0^1$, passing to a subsequence. This property means

$$\lim_{k\to\infty} \int_\Omega \nabla v_k \cdot \nabla \psi \, dx = \int_\Omega \nabla v \cdot \nabla \psi \, dx$$

for any $\psi \in H_0^1$, which implies

$$\|\nabla v\|_2^2 \leq \liminf_{k\to\infty} \|\nabla v_k\|_2^2 = \lambda_1.$$

The problem left, is therefore, whether $\|v\|_2^2 = 1$ holds or not. This property actually follows from the Rellich–Kondrachov theorem which assures $v_k \to v$ strongly in L^2 by $v_k \rightharpoonup v$ weakly in H_0^1 so that

$$\|v_k\|_2^2 = 1 \to \|v\|_2^2 = 1.$$

The second fact is proven by the Lagrangian multiplier principle, which guarantees that the minimizer $\phi_1(x)$ of λ_1 is a weak solution to (1.131). Here we suppose $\phi_1(x) \geq 0$ by replacing $|\phi|$ for $\phi \in H_0^1(\Omega)$ in the minimizing process. In fact, there arises

$$|\phi| \in H_0^1(\Omega)$$

in this case, and then the regularity theorem and the strong maximum principle imply the smoothness and the positivity of $\phi_1(x)$, respectively, that is, (1.133).

The third fact may be proven by Jacobi's method. We take the above $\varphi_1(x) > 0$, $x \in \Omega$ and an arbitrary $\varphi(x)$ satisfying (1.133) for $\lambda = \lambda_1$. Then the Hopf lemma to $\varphi_1(x)$ assures

$$\psi = \varphi/\varphi_1 \in C^2(\bar{\Omega}).$$

Applying (1.133) to $\varphi = \psi \cdot \varphi_1$, we obtain

$$\begin{aligned} \varphi_1 \Delta\psi + 2\nabla\varphi_1 \cdot \nabla\psi &= \varphi_1 \left(\Delta\psi + \nabla \log \varphi_1^2 \cdot \nabla\psi \right) \\ &= \varphi_1 [\varphi_1^{-2} \nabla \cdot (\varphi_1^2 \nabla\psi)] = 0 \end{aligned}$$

and hence $\nabla \cdot (\varphi_1^2 \nabla\psi) = 0$ in Ω. Then it follows that

$$\int_\Omega \varphi_1^2 |\nabla\psi|^2 \, dx = 0$$

and therefore, $\psi(x)$ is a constant.

The Neumann problem (1.105) is equivalent to

$$(-\Delta + 1)\phi = (\lambda + 1)\phi \text{ in } \Omega, \qquad \left. \frac{\partial\phi}{\partial\nu} \right|_{\partial\Omega} = 0,$$

which takes the weak form

$$\phi \in H^1(\Omega), \quad \int_\Omega \nabla\phi \cdot \nabla v + \phi v \, dx = (\lambda + 1) \int_\Omega \phi v \, dx, \ \forall v \in H^1(\Omega),$$

where Theorem 1.21 is applicable. Henceforth, eigenvalues are so ordered as

$$\lambda_0 = 0 < \lambda_1 \leq \lambda_2 \leq \cdots.$$

1.5.2 Sturm–Liouville Problems

The simplest case of (1.103) is $\Omega = (0, 1)$, that is,

$$-\phi'' = \lambda\phi \quad \text{in } 0 < x < 1, \qquad \phi|_{x=0,1} = 0.$$

The eigenvalues and eigenfunctions of this problem are

$$\lambda_j = (j\pi)^2, \ \phi_j(x) = \sqrt{2} \sin jx, \quad j = 1, 2, \cdots.$$

The corresponding problem to (1.105) is

$$-\phi'' = \lambda\phi \quad \text{in } 0 < x < 1, \qquad \phi'\big|_{x=0,1} = 0$$

where the eigenvalues and eigenfunctions are given by

$$\lambda_j = (j\pi)^2,\ \phi_j(x) = c_j \cos jx, \quad j = 0, 1, 2, \cdots,$$

where $c_0 = 1$ and $c_j = \sqrt{2}$, $j \geq 1$.

These results are applicable even to higher-dimensional problems. For example, if $\Omega = (0, a) \times (0, b)$ then the eigenvalues and the eigenfunctions of (1.103) are given by

$$\phi_{jk} = c_{jk} \sin(j\pi \frac{x_1}{a}) \cdot \sin(k\pi \frac{x_2}{b}),\ \lambda_{jk} = \{(\frac{j}{a})^2 + (\frac{k}{b})^2\}\pi^2, \quad j, k = 1, 2, \cdots,$$

by the principle of separation of variables.

The Liouville transformation reduces one-dimensional problem on finite interval with regular coefficients to the normal form A_P for $P = (p, h, H)$ or A_p^* in Sect. 1.2.3. To see this property, let $y = y(x)$, $a \leq x \leq b$ is a solution to

$$\frac{d}{dx}(p(x)\frac{dy}{dx}) + \{\mu r(x) - q(x)\}y = 0 \quad \text{in } a < x < b$$

with continuous $p(x) > 0$, $r(x) > 0$, and $q(x)$ defined for $x \in [a, b]$.

Assume

$$p \in C^1[a, b], \quad pr \in C^2[a, b],$$

and put

$$z = \frac{1}{K}\int_a^x (\frac{r(x)}{p(x)})^{1/2} dx, \quad K = \int_a^b (\frac{r(x)}{p(x)})^{1/2} dx.$$

Then u and λ defined by

$$u = (p(x)r(x))^{1/4} y, \quad \lambda = K^2 \mu$$

satisfy

$$\frac{d^2u}{dz^2} + (\lambda - g(z))u = 0 \quad \text{in } 0 < z < 1$$

for

$$g(z) = \frac{f''(z)}{f(z)} - k^2 k(z), \quad f(z) = (p(x)r(x))^{1/4}, \quad k(z) = q(x)/r(x).$$

If $y(x)$ is provided with the boundary condition

$$y'(a) - \alpha y(a) = y'(b) + \beta y(b) = 0, \quad \alpha, \beta \in \mathbf{R},$$

then there arises

$$u'(0) - hu(0) = u'(1) + Hu'(1) = 0$$

for suitable $h, H \in \mathbf{R}$. If this condition is replaced by

$$y(a) = y(b) = 0,$$

on the other hand, it follows that

$$u(0) = u(1) = 0.$$

We thus obtain the Sturm–Louville problem in normal form,

$$\begin{aligned} -\frac{d^2\phi}{dx^2} + p(x)\phi = \lambda\phi \ \text{ in } 0 < x < 1 \\ -\frac{d\phi}{dx} + h\phi\Big|_{x=0} = \frac{d\phi}{dx} + H\phi\Big|_{x=1} = 0 \end{aligned} \tag{1.134}$$

and

$$-\frac{d^2\phi}{dx^2} + p(x)\phi = \lambda\phi \ \text{ in } 0 < x < 1, \quad \phi|_{x=0} = \phi|_{x=1} = 0. \tag{1.135}$$

These problems are two-point boundary value problems of the second order, and consequently, any eigenvalue is simple, denoted by

$$\lambda_0 < \lambda_1 < \lambda_2 < \cdots \to +\infty, \quad \lambda_1^* < \lambda_2^* < \cdots \to +\infty$$

for (1.134) and (1.135), respectively. Under this notation, there comes the asymptotic formulas

$$\lambda_j^{1/2} = j\pi + O(\frac{1}{j}), \ \lambda_j^{*1/2} = j\pi + \frac{1}{j}, \quad j \to \infty.$$

Furthermore, the eigenfunctions $\phi_j(x)$ and $\phi_j^*(x)$ associated with λ_j and λ_j^* can take the forms

$$\phi_j(x) = \sqrt{2}\cos j\pi x + O(\frac{1}{j}), \ \phi_j^*(x) = \sqrt{2}\sin j\pi x + O(\frac{1}{j}), \quad j \to \infty$$

uniformly in $x \in [0, 1]$. One of the main assertions of the Sturm–Liouville theory is that each $\phi_j(x)$ and $\phi_j^*(x)$ has exactly j-zero's on $[0, 1]$. Here, each zero of ϕ_j^* on the boundary $x = 0, 1$ is counted by $1/2$.

Given $P = (p, h, H) \in C^1[0, 1] \times \mathbf{R} \times \mathbf{R}$, let A_P and A_p^* denote the operators

$$-\frac{d^2}{dx^2} + p(x), \quad (-\frac{d}{dx} + h)\cdot\Big|_{x=0} = (\frac{d}{dx} + H)\cdot\Big|_{x=1} = 0$$

and

$$-\frac{d^2}{dx^2} + p(x), \quad \cdot|_{x=0} = \cdot|_{x=1} = 0,$$

respectively. The eigenvalues of these operators are denoted by

$$\lambda_0(P) < \lambda_1(P) < \cdots \to +\infty, \quad \lambda_1^*(p) < \lambda_2^*(p) <\to +\infty,$$

respectively. We say that A_P is symmetric if $p(1-x) = p(x)$ for $0 < x < 1$ and $h = H$, while A_p is so if $p(1-x) = p(x)$ for $0 < x < 1$. A classical result by G. Borg asserts that the symmetric Sturm Liouville operator is determined by its spectrum [5].

1.5.3 Comparison Theorems

Given

$$0 < p = p(t) \in C^1(a, b), \quad g = g(t) \in C(a, b),$$

we clarify the property of $x = x(t)$ satisfying

$$Lx \equiv (p(t)x')' + g(t)x = 0 \quad \text{in } a < t < b. \tag{1.136}$$

Theorem 1.24 *Let* $g_2(t) > g_1(t)$,

$$(px_1')' + g_1x_1 = 0, \ (px_2')' + g_2x_2 = 0 \quad \text{in } (a, b), \tag{1.137}$$

and $a < t_1 < t_2 < b$ *be successive zeros of* $x_1 = x_1(t)$. *Then there is* $t_3 \in (t_1, t_2)$ *such that* $x_2(t_3) = 0$.

Proof Assume the contrary, and

$$x_2(t), \ x_1(t) > 0 \quad \text{in } (t_1, t_2)$$

without loss of generality. Since

$$(px_1')'x_2 - (px_2')'x_1 = (g_2 - g_1)x_1x_2$$

by (1.137), there arises

$$\begin{aligned}0 &< \int_{t_1}^{t_2} (px_1')' - (px_2')'x_1 \, dx \\ &= \left[px_1'x_2 - px_2'x_1\right]_{t=x_1}^{t=x_2} \\ &= p(t_2)x_1'(t_2)x_2(t_2) - p(t_1)x_1'(t_1)x_2(t_1) \le 0\end{aligned}$$

because $x_1'(t_1) \le 0 \le x_1'(t_1)$ follows from the assumption, a contradiction. □

Theorem 1.25 *The zeros of two linearly independent solutions to (1.136) separate one another.*

Proof Assume the contrary, and let $g_1 \equiv g_2$, $\{x_1, x_2\}$ be linearly independent, and $t_1 < t_2$ be successive zeros of x_1 in the previous proof.

Then it follows that

$$\begin{aligned}0 &\le \int_{t_1}^{t_2} (px_1')' - (px_2')'x_1 \, dt \\ &= p(t_2)x_1'(t_2)x_2(t_2) - p(t_1)x_1'(t_1)x_2(t_1) \le 0,\end{aligned}$$

which implies

$$x_1'(t_2)x_2(t_2) = x_1'(t_1)x_2(t_1) = 0,$$

and therefore, $x_2(t_2) = x_2(t_1) = 0$. Hence $\{x_1, x_2\}$ are not linearly independent, a contradiction. □

Here we apply the *Prüfer transformation*, using $y = p(t)x'$, which reduces (1.136) to

$$x' = \frac{y}{p(t)}, \quad y' = -g(t)x \qquad \text{in } a < t < b.$$

Let

$$x = r \sin\theta, \ y = r\cos\theta$$

be polar-like coordinates. Since

$$x' = r'\sin\theta + r\theta'\cos\theta, \quad y' = r'\cos\theta - r\theta'\sin\theta,$$

there arises

$$r' = x' \sin\theta + y' \cos\theta = \frac{y}{p} \sin\theta - gx \cos\theta = (\frac{1}{p} - g) r \sin\theta \cos\theta$$

$$\theta' = \frac{x' \cos\theta - y' \sin\theta}{r} = \frac{\frac{y}{p} \cos\theta + gx \sin\theta}{r} = \frac{\cos^2\theta}{p} + g \sin^2\theta. \quad (1.138)$$

If $x \not\equiv 0$, the functions x and x' do not vanish simultaneously. Hence

$$r^2 = x^2 + y^2 = (px')^2 + x^2 > 0,$$

and therefore,

$$\theta = \tan^{-1}(\frac{x}{y}) = \tan^{-1}(\frac{x}{px'})$$

is well-defined. It holds that $x(t) = 0$ if and only if $\theta(t) \in \pi\mathbf{Z}$.

The boundary condition at $t = a$ for (1.136) takes the form

$$x(a) \cos\alpha - p(a) x'(a) \sin\alpha = 0 \quad (1.139)$$

for some α. Since $x = r \sin\theta$ and $y = x'p = r \cos\theta$, there arises

$$x \cos\theta - px' \sin\theta = 0.$$

Hence (1.139) is equivalent to $\theta(a) \equiv \alpha \pmod{\pi}$.

Theorem 1.26 *Let* $0 < p_i = p_i(t) \in C^1[a, b]$ *and* $g_i = g_i(t) \in C[a, b]$ *for* $i = 1, 2$ *satisfy*

$$0 < p_2(t) \le p_1(t), \qquad g_2(t) \ge g_1(t)$$
$$L_1 x_1 \equiv (p_1 x_1')' + g_1 x_1 = 0, \quad L_2 x_2 \equiv (p_2 x_2')' + g_2 x_2 = 0$$
$$\theta_2(a) \ge \theta_1(a).$$

Then it follows that

$$\theta_2(t) \ge \theta_1(t), \quad a \le t \le b.$$

Moreover, if $g_1 > g_2$ *in* (a, b) *then there arises*

$$\theta_2(t) > \theta_1(t) \quad a < t \le b.$$

Proof By (1.138) it holds that

$$\theta_i' = \frac{1}{p_i}\cos^2\theta_i + g_i\sin^2\theta_i, \quad i = 1, 2,$$

and hence

$$\begin{aligned}(\theta_2 - \theta_1)' &= \frac{1}{p_2}\cos^2\theta_2 + g_2\sin^2\theta_2 - \frac{1}{p_1}\cos^2\theta_2 - g_1\sin^2\theta_1 \\ &= (g_1 - \frac{1}{p_1})(\sin^2\theta_2 - \sin^2\theta_1) + h \qquad (1.140)\end{aligned}$$

for

$$\begin{aligned}h &= \frac{1}{p_2}\cos^2\theta_2 + (g_2 - g_1 + \frac{1}{p_1})\sin^2\theta_2 - \frac{1}{p_1}\cos^2\theta_1 - \frac{1}{p_1}\sin^2\theta_1 \\ &= (g_2 - g_1)\sin^2\theta_1 + (\frac{1}{p_2} - \frac{1}{p_1})\cos^2\theta_2 \\ &\quad + \frac{1}{p_1}(\cos^2\theta_2 + \sin^2\theta_2) - \frac{1}{p_1}(\cos^2\theta_1 + \sin^2\theta_1) \\ &= (g_2 - g_1)\sin^2\theta_1 + (\frac{1}{p_2} - \frac{1}{p_1})\cos^2\theta_2 \geq 0. \qquad (1.141)\end{aligned}$$

Putting $u = \theta_2 - \theta_1$, therefore, we obtain $u' \geq fu$ for

$$f = (g_1 - \frac{1}{p_1})(\sin\theta_2 + \sin\theta_1)\cdot\frac{\sin\theta_2 - \sin\theta_1}{\theta_2 - \theta_1} \in C[a, b],$$

which implies

$$(e^F u)' \geq 0, \quad F(t) = \int_t^b f(s)\,ds.$$

and hence $e^{F(t)}u(t) \geq e^{F(a)}u(a) \geq 0$, or,

$$\theta_2(t) \geq \theta_1(t), \quad t \in [a, b].$$

Let $g_2 > g_1$ in (a, b), and assume $\theta_2 = \theta_1$ on $[a, c]$ for some $c \in (a, b]$. Then the above proof using (1.140)–(1.141) implies

$$\theta_1(=\theta_2) \equiv 0 \mod \pi \text{ on } [a, c].$$

Then we obtain

$$\theta_1' = \frac{1}{p_1}\cos^2\theta_1 + g_1\sin^2\theta = \frac{1}{p_1} > 0 \quad \text{on } [a,c],$$

and hence $\theta_1 \equiv 0 \pmod{\pi}$ is impossible.

Thus there arises $t_j \downarrow a$ such that $\theta_2(t_j) > \theta_1(t_j)$. Then we obtain

$$e^{F(t)}u(t) \geq e^{F(t_j)}u(t_j) > 0, \quad t \in [t_j, b]$$

again, and hence $\theta_2(t) > \theta_1(t)$ for $t \geq t_j$, which implies

$$\theta_2 > \theta_1 \quad \text{in } (a,b]$$

as desired. □

Given

$$0 < p = p(t) \in C^1[a,b], \quad 0 < r = r(t) \in C[a,b], \quad q = q(t) \in C[a,b],$$

we take the eigenvalue problem

$$\begin{aligned} &(px')' + (\lambda r - q)x = 0 \ \text{ on } [a,b] \\ &x(a)\cos\alpha - p(a)x'(a)\sin\alpha = x(b)\cos\beta - p(b)x'(b)\sin\beta = 0. \end{aligned} \tag{1.142}$$

All of the eigenvalues are simple, denoted by

$$-\infty < \lambda_0 < \lambda_1 < \cdots \to +\infty.$$

Let $x_j = x_j(t) \not\equiv 0$ be the eigenfunction corresponding to λ_j.

Theorem 1.27 *The above $x_j(t)$ has exactly j zeros in (a,b).*

Proof We may assume $0 \leq \alpha < \pi$ and $0 < \beta \leq \pi$. The cases that $(\alpha, \beta) = (0, \pi)$ and $\alpha > 0$ are sufficient to examine, changing the end points if necessary. Given $\lambda \in \mathbf{R}$, let $\phi = \phi(t,\lambda)$ be the solution to

$$(p\phi')' + (\lambda r - q)\phi = 0 \ \text{ in } (a,b), \quad \phi(a,\lambda) = \sin\alpha, \quad p(a)\phi'(a,\lambda) = \cos\alpha.$$

Then, λ is an eigenvalue of (1.142) if and only if

$$\phi(b,\lambda)\cos\beta - p(b)\phi'(b,\lambda)\sin\beta = 0.$$

Following the Prüfer transformation, we put

$$x = \phi(t,\lambda) = r\sin\theta, \quad y = p(t)x' = r\cos\theta.$$

Note $r = r(t,\lambda) > 0$ and $\theta = \theta(t,\lambda)$. Then we put $\theta(a,\lambda) = \alpha$ without loss of generality. For each $t \in (a,b]$, the mapping

$$\lambda \mapsto \theta(t,\lambda) \tag{1.143}$$

is strictly increasing by Theorem 1.26. There arises $\phi = 0$ if and only if $\theta \equiv 0 \pmod{\pi}$, and then $\theta' > 0$ by (1.138), that is,

$$\theta' = \frac{1}{p}\cos^2\theta + (\lambda r - q)\sin^2\theta. \tag{1.144}$$

Consequently, if $t_k \in (a,b)$ satisfies

$$\theta(t_k,\lambda) = k\pi,$$

then it holds that $\theta(t,\lambda) > k\pi$ and $\theta(t,\lambda) < k\pi$ for $t > t_k$ and $t < t_k$, respectively, and therefore, each zero of ϕ moves left as λ increases from the strict monotonicity of the mapping (1.143) for $t \in (a,b]$. This property, called (P), guarantees that the k-th zero of $\phi = \phi(t,\lambda)$, denoted by $t_k = t_k(\lambda)$, is continuous and strictly decreasing in λ, because

$$\theta'(t_k,\lambda)\frac{dt_k}{dt} + \frac{\partial\theta}{\partial\lambda}(t_k,\lambda) = 0$$

follows from the implicit function theorem. Since property (P) implies the theorem for the case of $(\alpha,\beta) = (0,\pi)$, henceforth we take the case $\alpha > 0$.

We show

$$\lim_{\lambda\uparrow+\infty}\theta(c,\lambda) = +\infty, \quad \lim_{\lambda\downarrow-\infty}\theta(c,\lambda) \le 0, \quad \forall c \in (a,b]. \tag{1.145}$$

In fact, first, we have $\theta(t,\lambda) \ge 0$ by (1.138) and $\theta(a,\lambda) = \alpha \ge 0$. Second, there hold

$$p(t) \le \exists P, \ r(t) \ge \exists R > 0, \ q(t) \le \exists Q, \quad t \in [a,b].$$

Then we take the solution $x = \tilde{\phi}(t,\lambda)$ to

$$Px'' + (\lambda R - Q)x = 0, \quad x|_{t=a} = \phi(a,\lambda), \quad Px'\big|_{t=a} = p(a)\phi'(a,\lambda), \tag{1.146}$$

and introduce the Prüfer transformation

$$\tilde{\phi} = \tilde{r}\sin\tilde{\theta}, \quad P\tilde{\phi} = \tilde{r}\cos\tilde{\theta}, \quad \tilde{\theta}(a,\lambda) = \theta(a,\lambda).$$

It holds that $\theta(c,\lambda) \ge \tilde{\theta}(c,\lambda)$ for any $c \in (a,b)$. Then the exact solution to (1.146) guarantees $\lim_{\lambda\uparrow+\infty}\tilde{\theta}(c,\lambda) = +\infty$ and hence

$$\lim_{\lambda \uparrow +\infty} \theta(c, \lambda) = +\infty.$$

To show

$$\lim_{\lambda \downarrow -\infty} \theta(c, \lambda) = 0,$$

next, we take $0 < \delta \ll 1$ such that $\delta < \alpha < \pi - \delta$. Assume $\lambda < 0$ and let

$$0 < P \leq p(t), \quad 0 < R \leq r(t), \quad q(t) \geq -Q.$$

We obtain

$$\theta' < \frac{1}{P} + \lambda R \sin^2 \delta + Q \tag{1.147}$$

by (1.144), provided that $\delta \leq \theta \leq \pi - \delta$. The right-hand side of (1.147) is negative for $\lambda \ll -1$, and hence

$$\theta(t, \lambda) < \pi - \delta, \quad \forall t \in [a, b].$$

It holds that

$$I_\lambda = \{t \in (a, b) \mid \theta(t, \lambda) \geq \delta\} = (a, t_\lambda), \quad a < \exists t_\lambda < b,$$

similarly, and furthermore, $\liminf_{\lambda \downarrow -\infty} t_\lambda = a$. Then we obtain

$$\lim_{\lambda \downarrow -\infty} \theta(t, \lambda) \leq \delta, \quad \forall t \in (a, b],$$

recalling the strictly decreasing of $\lambda \mapsto \theta(t, \lambda)$, and hence the second equality of (1.145).

Having (1.145) and the above mentioned monotonicity of $\lambda \mapsto \theta(b, \lambda)$, there is λ_0 such that $\theta(b, \lambda_0) = \beta > 0$ by (1.145). Then we obtain

$$0 < \theta(t, \lambda_0) < \pi, \quad \forall t \in (a, b)$$

by the property (P), recalling $0 \leq \alpha < \pi$ and $0 < \beta \leq \pi$. Hence $x = \phi(t, \lambda_0)$ is a solution to (1.142) which does not vanish in (a, b). Next, there is $\lambda_1 > \lambda_0$ such that

$$\theta(b, \lambda_1) = \beta + \pi.$$

Then $x = \phi(t, \lambda_1)$ is a solution to (1.142), taking one zero in (a, b). Continuing this process, we find all the eigenvalues λ_j, $j = 0, 1, 2, \cdots$, and the proof is complete. □

1.5.4 Parabolic Liouville Property

The Liouville property on the solution to linear parabolic equation arises in the spirit of the Sturm–Liouville theory. It is concerned with the boundary control of creation and annhilation of zeros of a function depending on a parameter. The importance of this property has been recognized again via the study of nonlinear problems [6, 14].

Let $D \in \mathbf{R}$ and $J \in \mathbf{R}$ be open space and time intervals, respectively, and $S = D \times J$. Assume that

$$a = a(x,t) \geq \exists a > 0, \quad b = b(x,t), \quad c = c(x,t)$$

are continuous in $(x,t) \in S$, and $u = u(x,t) \in C(\overline{S})$ is a solution to

$$u_t = a(x,t)u_{xx} + b(x,t)u_x + c(x,t)u. \tag{1.148}$$

Given $t \in \overline{J}$, let $z(t)$ be the number of components of $\{x \in D \mid u(x,t) \neq 0\}$. It counts the number of sign changes of $u(\cdot,t)$ in D, finite or infinite, and is equal to

$$\begin{aligned} \sup\{k \in \mathbf{N} \mid \exists x_i \in D,\ 0 \leq i \leq k;\ x_0 < x_1 < \cdots < x_k, \\ u(x_j,t) \cdot u(x_{j+1},t) < 0,\ 0 \leq \forall j \leq k-1\}. \end{aligned}$$

Given $\tau \in J$, let $S_\tau = S \cap \{t < \tau\}$,

$$\partial_p S_\tau = (D \times \{0\}) \bigcup (\partial D) \times (0,\tau]$$

be its parabolic boundary, and $\hat{z}(\tau)$ be the number of sign changes of u on $\partial_p S_\tau$ defined similarly.

Theorem 1.28 *It holds that $z(\tau) \leq \hat{z}(\tau)$ for each $\tau \in J$.*

Proof First, taking $u = e^{\lambda t}U$ with $\lambda \gg 1$, we may assume $c \leq 0$. Second, it suffices to show that if there are precisely $n_\pm$ disjoint intervals on $\partial_p S_\tau$ such that $\pm u > 0$, then

$$U^\pm = \{(x,t) \in S_\tau \mid \pm u(x,t) > 0\}$$

has at most $n_\pm$ components in S_τ. Since $u \in C(\overline{S})$, furthermore, this property follows if closure of each component of $U^\pm$ intersects $\partial_p S_\tau$ in at least one interval.

Given a component of U^+ denoted by $F \neq \emptyset$, the continuous function u attains a positive maximum on $\overline{F}$. Having $c \leq 0$, on the other hand, we obtain

$$u_t \leq au_{xx} + bu_x \quad \text{in } F,$$

and hence this maximum is not attained in F or on $\overline{F} \cap \{t = \tau\}$ by the strong maximum principle. Hence there is $Q \in \partial_P S_\tau \cap F$ such that $u(Q) > 0$, which implies $u > 0$ on $\partial_P S_\tau$ near Q as is desired. □

If a, b, c and u are regular, and $u_0 = u(x, 0)$ takes 0 of order m at $x = 0$. Then there arises a sharp form of Theorem1.28 [3] as

$$z(t_1) - z(t_2) \geq \begin{cases} m, & m : even \\ m - 1, & m : odd, \end{cases} \quad 0 < -t_1, t_2 \ll 1.$$

Theorem 1.28 is applicable to confirm the Morse–Smale property of a semilinear parabolic equation in one space dimension, for example,

$$u_t = u_{xx} + f(u) \text{ in } I \times (0, +\infty), \quad u|_{\partial I} = 0 \tag{1.149}$$

for $I = (0, 1)$. If (1.149) is provided with uniformly bounded global-in-time dynamics, stable and unstable manifolds of non-degenerate stationary solutions ϕ and ψ, denoted by $W^s(\phi)$ and $W^u(\psi)$, respectively, intersect transversally. If this intersection is empty, it holds that $i(\psi) > i(\phi)$, where $i(\psi)$ and $i(\phi)$ denote the Morse indices of ψ and ϕ, respectively [2].

1.6 Inverse Sturm–Liouville Problems

1.6.1 Uniqueness Theorems

Here we take the spectral inverse problem for the Sturm–Liouville operator

$$-\frac{d^2}{dx^2} + p(x), \quad (-\frac{d}{dx} + h)\cdot\Big|_{x=0} = (\frac{d}{dx} + H)\cdot\Big|_{x=1} = 0,$$

denoted by A_P for

$$P = (p, h, H) \in C^1[0, 1] \times \mathbf{R} \times \mathbf{R}.$$

This operator is a normal form of the general operator through a change of variables, the Liouville transformation. We provide with the boundary condition $\cdot|_{x=0} = 0$ and $\cdot|_{x=1} = 0$ if $h = \infty$ and $H = \infty$, respectively.

The spectrum of A_P denoted by $\sigma(A_P)$ consists of simple eigenvalues

$$-\infty < \lambda_0(P) < \lambda_1(P) < \cdots \to \infty,$$

provided with the asymptotics

$$\lambda_n^{1/2} = n\pi + O(1/n), \qquad h, H \in \mathbf{R}$$

$$\lambda_n^{1/2} = (n + \frac{1}{2})\pi + O(1/n), \qquad h \in \mathbf{R},\ H = \infty \text{ or } h = \infty,\ H \in \mathbf{R}$$

$$\lambda_n^{1/2} = (n+1)\pi + O(1/n), \qquad h = H = \infty \tag{1.150}$$

for $\lambda_n = \lambda_n(P)$ as $n \to \infty$. Its eigenfunctions $\varphi_n = \varphi_n(\cdot, P)$, $n = 0, 1, \cdots$, normalized by

$$\|\varphi_n(\cdot, P)\|_{L^2(0,1)} = 1,$$

furthermore, take the asymptotics

$$\varphi_n(x) = \frac{1}{\sqrt{2}} \cos n\pi x + O(\frac{1}{n}), \quad h, H \in \mathbf{R}$$

$$\varphi_n(x) = \frac{1}{\sqrt{2}} \sin(n + \frac{1}{2})\pi x + O(\frac{1}{n}), \quad h = \infty,\ H \in \mathbf{R}$$

$$\varphi_n(x) = \frac{1}{\sqrt{2}} \cos(n + \frac{1}{2})\pi x + O(\frac{1}{n}), \quad h \in \mathbf{R},\ H = \infty$$

$$\varphi_n(x) = \frac{1}{\sqrt{2}} \sin(n+1)\pi x + O(\frac{1}{n}), \quad h = H = \infty$$

as $n \to \infty$. Each n-th eigenfunction φ_n has exactly n-zeros in (0,1) and each zero point of φ_{n+1} lies between a successive pair of the zeros of φ_n. We recall the following examples.

Example 1.1 If $p \equiv 0$ and $h = H = 0$ it holds that $\lambda_n(P) = (n\pi)^2$, $n = 0, 1, 2, \cdots$, and

$$\varphi_n(x; p) = \begin{cases} 1, & n = 0 \\ \frac{1}{\sqrt{2}} \cos n\pi x, & n = 1, 2, \cdots. \end{cases}$$

Example 1.2 If $p \equiv 0$ and $h = H = \infty$, we have

$$\lambda_n(P) = ((n+1)\pi)^2, \quad \varphi_n(x; p) = \frac{1}{\sqrt{2}} \sin(n+1)\pi x$$

for $n = 0, 1, 2, \cdots$.

V. Ambarzumian found the following theorem [1].

Theorem 1.29 (Ambarzumian) *The condition* $\lambda_n(p, 0, 0) = (n\pi)^2$, $n = 0, 1, 2, \ldots$, *implies* $p \equiv 0$.

A natural extension is the following question:

Problem 1.2 Does the condition $\lambda_n(q, h, H) = \lambda_n(p, h, H)$, $n = 0, 1, 2, \cdots$, imply $q \equiv p$?

G. Borg proved the following theorem later [5].

Theorem 1.30 (Borg) *The conditions*

$$\lambda_n(q, h, H) = \lambda_n(p, h, H), \qquad n = 0, 1, 2, \cdots$$
$$\lambda_n(q, h, H^*) = \lambda_n(p, h, H^*), \qquad n = 1, 2, \cdots \tag{1.151}$$

imply $q \equiv p$, where $h, H, H^* \in \mathbf{R}$ and $H \neq H^*$.

Then many results have been presented.

Here we take the case

$$p \in C_s^1 \equiv \{r \in C^1[0, 1] \mid r(1 - x) = r(x),\ 0 \leq x \leq 1)\}, \quad h = H,$$

where the operator A_p is invariant under the action $x \mapsto 1 - x$ and is called a symmetric operator. The following theorem is also due to [5]:

Theorem 1.31 (Borg) *If $p, q \in C_s^1$, $h \in \mathbf{R}$ and*

$$\lambda_n(q, h, h) = \lambda_n(p, h, h), \quad n = 1, 2, \ldots, \tag{1.152}$$

it holds that $q \equiv p$.

This theorem, together with the following theorem due to [9], illustrates that the spectrum determines a half of the Sturm–Liouville operator:

Theorem 1.32 (Hochstadt–Lieberman) *If $p, q \in C^1[0, 1]$,*

$$p(x) = q(x) \quad \text{in } 1/2 \leq x \leq 1,$$

and $h = H \in \mathbf{R}$, the first condition of (1.151) implies $q \equiv p$.

It seems strange that the condition for $n = 0$ is not assumed in (1.152). This property is actually related to the asymptotic formula (1.150).

Given $p \in C_s^1$ and $h \in \mathbf{R}$, we use slightly different notations:

$$\lambda_n(p, h), \ n = 0, 1, 2, \cdots$$

for $\lambda_n(p, h, h)$, and

$$\lambda_n^*(p), \ n = 1, 2, \cdots$$

for $\lambda_{n-1}(p, \infty, \infty)$, $n = 1, 2, \ldots$. The asymptotic formula (1.150) then reads

$$\lambda_n^{1/2}(p,h) = n\pi + O(\frac{1}{n}),\ \lambda_n^*(p)^{1/2} = n\pi + O(\frac{1}{n}), \quad n \to \infty.$$

Theorem 1.33 ([21]) *Let $p, q \in C_s^1$ and $h, j \in \mathbf{R}$ be given. First, the condition*

$$\lambda_n(q, j) = \lambda_n(p, h), \quad n = 0, 1, 2, \cdots \tag{1.153}$$

implies $q \equiv p$ and $j = h$. Second, the condition

$$\lambda_n^*(q) = \lambda_n^*(p), \quad n = 1, 2, \cdots$$

implies $q \equiv p$.

The second part of this theorem is an extension of Theorem 1.31 for the case of $h = \infty$. Note that the eigenvalue $\lambda_n^*(p)$ for $n = 0$ is missed in our new notations. The first part of Theorem 1.31, on the other hand, means that the condition (1.153) for $n = 0$ is recovered by the assumption $j = h$ for the uniqueness $q \equiv p$. Then there arises a natural question.

Problem 1.3 Given $n_1 \geq 1$, $p, q \in C_s^1$, and $h \in \mathbf{R}$, does the condition

$$\lambda_n(q, h, h) = \lambda_n(p, h, h), \quad n \neq n_1 \tag{1.154}$$

imply $q \equiv p$?

The answer is negative. To state the result, given

$$p \in C_s^1,\ h \in \mathbf{R},\ n_1 \in \mathbf{N} \equiv \{0, 1, 2, \cdots\},$$

let

$$Q(p, h, n_1) \equiv \{q \in C_s^1 \mid \lambda_n(q, h) = \lambda_n(p, h),\ n \in \mathbf{N}\setminus\{n_1\}\}.$$

The n-th eigenfunctions of $A_{p,h,h}$ and A_p^*, furthermore, are denoted by

$$\varphi_n(\cdot; p, h), \quad n = 0, 1, 2, \cdots,$$

and

$$\varphi_n^*(\cdot; p), \quad n = 1, 2, \cdots,$$

respectively, under the normalization of

$$\|\varphi_n\|_{L^2(0,1)} = \|\varphi_n^*\|_{L^2(0,1)} = 1, \quad \varphi_n(0),\ \varphi_n^{*\prime}(0) > 0.$$

Theorem 1.34 ([21]) *Given $(p, h, n_1) \in C_s^1 \times \mathbf{R} \times \mathbf{N}$, there arises*

$$Q(p,h,n_1) = \begin{cases} \{p\}, & n_1 = 0 \\ \{p, p - 2\frac{d^2}{dx^2}\log\tau\}, & n_1 \geq 1 \end{cases}$$

where

$$\tau = W(\varphi_{n_1}^*(\cdot; p), \varphi_{n_1}(\cdot; p, h)) \equiv \varphi_{n_1}^{*\prime}\varphi_{n_1} - \varphi_{n_1}^*\varphi_{n_1}'. \tag{1.155}$$

This result has a connection between the theory of iso-spectral flow described in [17]. Below we use a duality between the spectrums $\sigma(A_{p,h})$ and $\sigma(A_p^*)$ for the proof.

1.6.2 Reconstruction Theorems

Given $(p, h, n_1) \in C_s^1 \times \mathbf{R} \times \mathbf{N}$ and

$$\mu_1 \in (\lambda_{n_1-1}(p,h), \lambda_{n_1+1}(p,h)),$$

we can find $q \in C_s^1$ and $j \in \mathbf{R}$ such that

$$\lambda_n(q,j) = \lambda_n(p,h),\ \forall n \neq n_1, \quad \lambda_{n_1}(q,j) = \mu_1, \tag{1.156}$$

where $\lambda_{-1}(p,h) = -\infty$. This $(q, j) \in C_s^1 \times \mathbf{R}$ is unique by Theorem 1.33, but is given explicitly.

To state the result precisely, we use $\varphi^e = \varphi^e(\cdot; p, \lambda)$ and $\varphi^o = \varphi^o(\cdot; p, \lambda)$ for $(p, \lambda) \in C_s^1 \times \mathbf{R}$, defined by

$$(-\frac{d^2}{dx^2} + p(x))\varphi = \lambda\varphi \text{ in } 0 \leq x \leq 1, \quad \varphi(\frac{1}{2}) = 1,\ \varphi'(\frac{1}{2}) = 0$$
$$(-\frac{d^2}{dx^2} + p(x))\varphi = \lambda\varphi \text{ in } 0 \leq x \leq 1, \quad \varphi(\frac{1}{2}) = 0,\ \varphi'(\frac{1}{2}) = 1.$$

Then it holds that

$$\varphi^e(1-x) = \varphi^e(x),\ \varphi^o(1-x) = -\varphi^o(x) \quad \text{in } 0 \leq x \leq 1. \tag{1.157}$$

We will see that

$$\varphi_n(\cdot; p, h) = \begin{cases} \varphi^e(\cdot; p, \lambda_n(p,h)), & n : even \\ \varphi^o(\cdot; p, \lambda_n(p,h)), & n : odd \end{cases} \tag{1.158}$$

and

$$\varphi_n^*(\cdot; p) = \begin{cases} \varphi^e(\cdot; p, \lambda_n^*(p)), \ n : odd \\ \varphi^o(\cdot; p, \lambda_n^*(p)), \ n : even \end{cases} \tag{1.159}$$

for n-th eigenfunctions $\varphi_n(\cdot, p, h)$ and $\varphi_n^*(\cdot, p)$ of $A_{p,h}$ and A_p^*, respectively, which are not necessarily L^2-normalized. Then we obtain the following theorems.

Theorem 1.35 ([21]) *Given $p \in C_s^1$, $h \in \mathbf{R}$, $n_1 \in \mathbf{N}$, and*

$$\mu_1 \in (\lambda_{n_1-1}(p, h), \lambda_{n_1+1}(p, h)),$$

there exists a unique $q \in C_s^1$ and $j \in \mathbf{R}$ satisfying (1.156), that is,

$$q = p - 2\frac{d^2}{dx^2}\log\tau, \quad j = h - 2\left.\frac{d}{dx}\log\tau\right|_{x=0}$$

for

$$\tau = W(f_1, \varphi_{n_1}(\cdot; p, h)), \quad f_1 = \begin{cases} \varphi^e(\cdot; p, \mu_1), & n_1\text{: even} \\ -\varphi^o(\cdot; p, \mu_1), & n_1\text{: odd.} \end{cases}$$

Theorem 1.36 ([21]) *Given $p \in C_s^1$, $n_1 \in \mathbf{N}^*$ and*

$$\mu_1 \in (\lambda_{n_1-1}^*(p), \lambda_{n_1+1}^*(p)),$$

there exists a unique $q \in C_s^1$ satisfying

$$\lambda_n^*(q) = \lambda_n^*(p), \ \forall n \neq n_1, \quad \lambda_{n_1}^*(q) = \mu_1,$$

where $\lambda_0^*(p) = -\infty$, which is given by

$$q = p - 2\frac{d^2}{dx^2}\log\tau$$

for

$$\tau = W(f_1^*, \varphi_{n_1}^*(\cdot; p)), \quad f_1^* = \begin{cases} \varphi^o(\cdot; p, \mu_1), & n_1\text{: odd} \\ -\varphi^e(\cdot; p, \mu_1), & n_1\text{: even.} \end{cases}$$

Thus, any eigenvalue in $\sigma(A_{p,h})$ or $\sigma(A_p^*)$ moves to an arbitrary value keeping the order with the others fixed in the category of symmetric operators, through the solutions to the linear equation

$$(-\frac{d^2}{dx^2} + p(x))\varphi = \lambda\varphi \text{ in } 0 \leq x \leq 1. \tag{1.160}$$

Example 1.3 The spectrum $\sigma(A_q^*)$ for

$$q(x) = \frac{-2\pi^2\{\cos^2\pi(x-\frac{1}{2}) - \pi^2(x-\frac{1}{2})^2\}}{\pi^2(x-\frac{1}{2})^2\sin^2\pi(x-\frac{1}{2}) + 2\pi\sin\pi(x-\frac{1}{2})\cos\pi(x-\frac{1}{2}) + \cos^2\pi(x-\frac{1}{2})}$$

is given by $\{(n\pi)^2 \mid n = 0, 2, 3\}$.

In fact, for $p = 0$ we have $\sigma(A_p^*) = \{(n\pi)^2 \mid n = 1, 2, 3 \ldots\}$ and the solution φ to (1.160) for $\lambda = 0$ is given by

$$\varphi(x) = a(x - 1/2) + b, \quad a, b \in \mathbf{R}.$$

Hence we obtain $n_1 = 1$, $\mu_1 = 0$, and

$$\tau = \text{constant} \times W(x - \frac{1}{2}, \sin\pi x).$$

In Theorems 1.35–1.36 it is necessary to give explicit solutions to (1.160), more precisely, $\{\varphi^{o,e}(\cdot; p, \lambda)\}$ to (1.160), to move $\lambda_{n_1}(\cdot; p, h)$ to μ_1. To move $\lambda_{n_2}(\cdot; p, h)$ to μ_2, then, we require $\{\varphi^{o,e}(\cdot; q, \lambda)\}$. Since we have an explicit formula to express $\{\varphi^{o,e}(\cdot; q, \lambda)\}$ in terms of $\{\varphi^{o,e}(\cdot; p, \lambda)\}$ as in the following theorem, we can continue the process arbitrary many times, to establish the principle of superposition.

Given $(p, h) \in C_s^1 \times \mathbf{R}$, we thus have an explicit formula of $(q, j) \in C_s^1 \times \mathbf{R}$, the eigenvalues of which, $\{\lambda_n(q, h))\}$, coincide with $\{\lambda_n(p, h)\}$ except for a finite number of them, using $\varphi^{e,o}(\cdot, p, \lambda)$, $\lambda \in \mathbf{R}$.

Theorem 1.37 ([21]) *In Theorem 1.35 it holds that*

$$\varphi^{o,e}(x; q, \lambda) = \varphi^{o,e}(x; p, \lambda) + \int_{1-x}^{x} M(x, y)\varphi^{o,e}(y; p, \lambda)dy, \tag{1.161}$$

where

$$M(x, y) = \frac{1}{2}(\mu_1 - \lambda_{n1}(p, h))\{f_1(x)\varphi_{n1}(y; p, h) + f_1(y)\varphi_{n_1}(x; p, h)\}/\tau(x).$$

The same is true between $\{\lambda_n^*(q)\}$ and $\{\lambda_n^*(p)\}$.

Theorem 1.38 ([21]) *In Theorem 1.36, similarly, there arises (1.161) for*

$$M(x, y) = \frac{1}{2}(\mu_1 - \lambda_{n1}^*(p))\{f_1^*(x)\varphi_{n1}^*(y; p) + f_1^*(y)\varphi_{n_1}^*(x; p)\}/\tau(x).$$

A direct consequence is the following theorem.

Theorem 1.39 *The element $(p, h) \in C_s^1$ satisfying*

$$\lambda_n(p, h) = (n\pi)^2$$

except for a finite number of $n \in \mathbf{N}$ *is an elementary function. Similarly, the function* $p \in C_s^1$ *satisfying*

$$\lambda_n^*(p) = (n\pi)^2$$

except for a finite number of $n \in \mathbf{N}^*$ *is an elementary function.*

1.6.3 Derivation of the Hochstadt Equation

In this section we prove the following theorem, and meanwhile, Theorem 1.33.

Theorem 1.40 *Given* $p \in C_s^1$, $h \in \mathbf{R}$, *and* $n_1 \in \mathbf{N}$, *the set*

$$\hat{Q}(p, h, n_1) = \{(q, j) \in C_s^1 \times \mathbf{R} \mid \lambda_n(q, j) = \lambda_n(p, h),\ n \neq n_1\}$$

is homeomorphic to $\mathcal{G}(p, h, n_1)$, *the set of solutions* $g = g(x) \in C^2[0, 1]$ *to*

$$\begin{aligned} \frac{d^2 g}{dx^2} &= [2\frac{d}{dx}(g\varphi_{n_1}(\cdot; p, h)) + p - \lambda_{n_1}(p, h)]g \\ g(1-x) &= (-1)^{n_1+1} g(x), \quad 0 \leq x \leq 1, \end{aligned} \tag{1.162}$$

through the relation

$$q = p + 2\frac{d}{dx}(g\varphi_{n_1}(\cdot; p, h)), \quad j = h + \left.(g\varphi_{n_1}(\cdot; p, h))\right|_{x=0}.$$

We begin with the symmetry of eigenfunctions of symmetric operators. Put

$$\varphi_n = \varphi_n(\cdot; p, h)$$

for simplicity, given $(p, h) \in C_s^1 \times \mathbf{R}$.

Lemma 1.11 *It holds that*

$$\varphi_n(1-x) = (-1)^n \varphi_n(x),\ 0 \leq x \leq 1$$

for $n \in \mathbf{N}$.

Proof Since $A_{p,h,h}$ is symmetric, the equality

$$\varphi_n(1-x) = c_n \varphi_n(x), \quad 0 \leq x \leq 1$$

holds with $c_n \in \mathbf{R}$. The conditions $c_n \neq 1$ and $c_n \neq -1$ imply $\varphi_n(\frac{1}{2}) = 0$ and $\varphi_n'(\frac{1}{2}) = 0$, respectively.

Since

$$\varphi_n(\frac{1}{2}) = \varphi_n'(\frac{1}{2}) = 0$$

means $\varphi_n(\cdot) \equiv 0$, which is impossible, the equalities $\varphi_n(\frac{1}{2}) = 0$ and $\varphi_n'(\frac{1}{2}) = 0$ do not occur simultaneously. Hence we have

$$c_n = \pm 1.$$

By the Sturm–Liouville theorem, on the other hand, the function $\varphi_n(\cdot)$ has exactly n-zeros in $(0, 1)$. Hence we obtain $c_n = (-1)^n$. □

Lemma 1.12 *Both $\{\varphi_{2n}\}_{n=0}^{\infty}$ and $\{\varphi_{2n+1}\}_{n=0}^{\infty}$ are complete and orthogonal system in $L^2(0, 1/2)$.*

Proof By Lemma 1.11, $\varphi = \varphi_{2n}$ satisfies

$$(-\frac{d^2}{dx^2} + p(x))\varphi = \lambda\varphi \text{ in } 0 < x < 1, \quad \varphi'(0) - h\varphi(0) = \varphi'(\frac{1}{2}) = 0 \tag{1.163}$$

for $\lambda = \lambda_{2n}(p, h)$ and has n-zeros in $(0, 1/2)$. This property means that $\{\varphi_{2n}\}_{n=0}^{\infty}$ coincides with the totality of eigenfunctions of the eigenvalue problem (1.163) by the Sturm–Liouville theory. Therefore, it is complete and orthogonal each other in $L^2(0, \frac{1}{2})$. So is true for $\{\varphi_{2n+1}\}_{n=0}^{\infty}$ by the same reasoning. □

Proof *(Theorem 1.33 for the first case)* Assume (1.153) for $p, q \in C_s^1$ and $h, j \in \mathbf{R}$. Then, Lemma 1.11 implies

$$\Phi_n(1; q, j) = \Phi_n(1; p, h), \quad n = 0, 1, 2, \cdots,$$

where $\Phi_n(\cdot; p, h)$ denotes the n-th eigenfunction of $A_{p,h}$ normalized by

$$\Phi_n(0; p, h) = 1.$$

These conditions imply $q \equiv p$ and $j = h$ by Proposition 1.7, and the proof is complete. □

The second case of Theorem 1.33 is proven similarly. First, we note the following lemmas concerning the eigenfunctions $\{\varphi_n^*(\cdot; p)\}_{n=0}^{\infty}$ of A_p^* for $p \in C_s^1$. Put

$$\varphi_n^* = \varphi_n^*(\cdot; p)$$

for simplicity.

Lemma 1.13 *It holds that*

$$\varphi_n^*(1-x) = (-1)^{n+1}\varphi_n^*(x),\ 0 \le x \le 1$$

for $n \in \mathbf{N}^*$.

Lemma 1.14 *Both* $\{\varphi_{2n}^*\}_{n=1}^\infty$ *and* $\{\varphi_{2n-1}^*\}_{n=1}^\infty$ *are complete and orthogonal systems in* $L^2(0, 1/2)$.

Proof *(Theorem 1.33 for the second case)* The assumption

$$\lambda_n^*(q) = \lambda_n^*(p), \quad n \in \mathbf{N}^* \tag{1.164}$$

with $p, q \in C_s^1$ implies

$$\Phi_n^{*\prime}(1; q) = \Phi_n^{*\prime}(1; p), \quad n \in \mathbf{N}^*, \tag{1.165}$$

where $\Phi_n^*(1; p)$ denotes the n-th eigenfunction of A_p^* normalized by $\Phi_n^{*\prime}\big|_{x=0} = 1$. Hence the result follows from the following fact. □

Theorem 1.41 *Equalities (1.164)–(1.165) imply* $q = p$ *for* $p, q \in C^1[0, 1]$.

For the proof of this theorem we use a variant of the first deformation formula in Sect. 1.3.4. Let $p \in C^1[0, 1]$ and $\lambda \in \mathbf{R}$ be given. Then, $\Phi^* = \Phi^*(\cdot; p, \lambda)$ denotes the solution to

$$(-\frac{d^2}{dx^2} + p(x))\Phi^* = \lambda\Phi^* \text{ in } 0 < x < 1, \quad \Phi^*\big|_{x=0} = 0, \quad \Phi_x^*\big|_{x=0} = 1.$$

We connect $\Phi^*(\cdot; q, \lambda)$ and $\Phi^*(\cdot; p, \lambda)$ by an integral transformation of which kernel is independent of λ.

Lemma 1.15 *Each* $p, q \in C^1[0, 1]$ *admits a unique solution* $K^* = K^*(x, y; q, p) \in C^2(\overline{D})$ *to*

$$\begin{aligned} &K_{xx}^* - K_{yy}^* + p(y)K^* = q(x)K^* && \text{in } D \\ &K^*(x, x) = \frac{1}{2}\int_0^x q(s) - p(s)\, ds, && 0 \le x \le 1 \\ &K^*(x, 0) = 0, && 0 \le x \le 1, \end{aligned} \tag{1.166}$$

where $D = \{(x, y) \mid 0 < y < x < 1\}$.

Lemma 1.16 (Third Deformation Formula) *For* $K^* = K^*(\cdot, \cdot; q, p)$ *in Lemma 1.15, the identity*

$$\Phi^*(x;q,\lambda) = \Phi^*(x;p,\lambda) + \int_0^x K^*(x,y;q,p)\Phi^*(y;p,\lambda)\,dy, \quad 0 \le x \le 1 \tag{1.167}$$

holds for any $\lambda \in \mathbf{R}$.

The proof is similar to that of the first deformation formula in Sect. 1.3.4. We note that the last equality of (1.166) implies

$$K^*(0,0) = 0,$$

and consequently, the boundary condition at $x = 0$ of Φ^* is invariant under this transformation. From this stage the proof of Theorem 1.41 is similar to that of Proposition 1.7.

Having Theorem 1.33, we turn to the proof of Theorem 1.40.

Lemma 1.17 *Given* $(p, h, n_1) \in C_s^1[0,1] \times \mathbf{R} \times \mathbf{N}$*, we have* $(q, j) \in \hat{Q}(p, h, n_1)$ *if and only if there exists* $K \in C^2(\overline{D})$ *and* $c \in \mathbf{R}$ *such that*

$$\begin{aligned}
&K_{xx} - K_{yy} + p(y)K = q(x)K && \text{in } \overline{D}\\
&K(x,x) = (j-h) + \frac{1}{2}\int_0^x q(s) - p(s)\,ds, && 0 \le x \le 1\\
&K_y(x,0) = hK(x,0), && 0 \le x \le 1\\
&j = h - K(1,1)\\
&\int_0^1 \{K_x(1,y) + jK(1,y)\}\varphi_n(y)\,dy = 0, && n \ne n_1
\end{aligned} \tag{1.168}$$

and

$$K(\frac{1}{2},y) = 0,\ K_x(\frac{1}{2},y) = c\varphi_{n_1}(y), \quad 0 \le y \le 1/2 \tag{1.169}$$

if n_1 *is even, and*

$$K(\frac{1}{2},y) = c\varphi_{n_1}(y),\ K_x(\frac{1}{2},y) = 0, \quad 0 \le y \le 1/2 \tag{1.170}$$

if n_1 *is odd, where*

$$\varphi_n = \varphi_n(\cdot; p, h)$$

denotes the n-th eigenfunction of $A_{p,h}$ *normalized by* $\|\varphi_n\|_{L^2(0,1)} = 1$.

Proof Let $K = K(x,y;q,j;p,h)$ be the solution to the first three equalities of (1.168). Then the first deformation formula

$$\Phi(x;q,j,\lambda)=\Phi(x;p,h,\lambda)+\int_0^x K(x,y;q,j;p,h)\Phi(y;p,h,\lambda)dy,\ 1\le x\le 1 \tag{1.171}$$

arises, where $\Phi=\Phi(x;p,h,\lambda)$ denotes the solution to

$$(-\frac{d^2}{dx^2}+p(x))\Phi=\lambda\Phi \text{ in } 0<x<1,\quad \Phi|_{x=0}=1,\quad \Phi'\big|_{x=0}=h.$$

Suppose that $(q,j)\in\hat{Q}_{p,h,n_1}$, which means

$$\lambda_n(q,j)=\lambda_n(p,h),\quad \forall n\neq n_1.$$

Thus we obtain

$$\Phi'(1;q,j;\lambda)+j\Phi(1;q,j;\lambda)=0 \text{ for } \lambda=\lambda_n(p,h), n\neq n_1, \tag{1.172}$$

or equivalently,

$$(j-h+K(1,1))\varphi_n(1)+\int_0^1\{K_x(1,y)+jK(1,y)\}\varphi_n(y)\,dy=0,\ n\neq n_1,$$

which splits into the last two equalities of (1.168) by the argument in Sect. 1.3.4.

Similarly, since

$$\varphi_{2n}'(\frac{1}{2})=\varphi_{2n}'(\frac{1}{2};q,j)=0,\quad \varphi_{2n+1}(\frac{1}{2})=\varphi_{2n+1}(\frac{1}{2};q,j)=0$$

we have

$$K(\frac{1}{2},\frac{1}{2})\varphi_n(\frac{1}{2})+\int_0^{1/2}K_x(\frac{1}{2},y)\varphi_n(y)\,dy=0,\qquad n:even,\ n\neq n_1$$

$$\int_0^{1/2}K(\frac{1}{2},y)\varphi_n(y)\,dy=0,\qquad n:odd,\ n\neq n_1. \tag{1.173}$$

By the same reasoning, the first equality of (1.173) splits into

$$K(\frac{1}{2},\frac{1}{2})=0,\quad \int_0^{1/2}K_x(\frac{1}{2},y)\varphi_n(y)\,dy=0,\ n:even,\ n\neq n_1. \tag{1.174}$$

By Lemma 1.12 these equalities imply the first and the second equalities of (1.169) according to the cases that n_1 is even and odd, respectively. Here,

$$K(\frac{1}{2},\frac{1}{2})=0$$

follows always from (1.168) because $\varphi_{n_1}(\frac{1}{2}) = 0$ if n_1 is odd.

Suppose, conversely, the existence of $K \in C^2(\overline{D})$ and $c \in \mathbf{R}$ satisfying (1.168)–(1.169). By the first three equalities of (1.168) we obtain

$$K = K(\cdot, \cdot; q, j; p, h)$$

and hence the first deformation formula (1.171). The last two equalities of (1.168) then imply (1.172), which means

$$\lambda_n(p, h) \in \sigma(A_{q,j}), \quad \forall n \neq n_1.$$

In other words, each $n \neq n_1$ admits $m(n) \in \mathbf{N}$ such that

$$\lambda_n(p, h) = \lambda_{m(n)}(q, j), \quad n \neq n_1.$$

We show

$$m(n) = n, \quad n \neq n_1 \tag{1.175}$$

to conclude the proof of $(q, j) \in \hat{Q}(p, h, n_1)$.

In fact, the assumption (1.169) implies

$$\Phi'(\frac{1}{2}; q, j; \lambda_n(p, h)) = 0, \quad n : even, \ n \neq n_1$$

$$\Phi(\frac{1}{2}; q, j; \lambda_n(p, h)) = 0, \quad n : odd, \ n \neq n_1$$

by (1.171) and Lemma 1.12, which means

$$m(n) \equiv n \bmod 2, \quad n \neq n_1. \tag{1.176}$$

First, the asymptotics of eigenvalues guarantee

$$m(n) = n, \quad n \geq n_0,$$

for $n_0 \in \mathbf{N}$ sufficiently large. Second, the mapping $n \longmapsto m(n)$ is injective and order-preserving. Hence we obtain (1.175) by (1.176), which completes the proof.

□

The following lemma is a variant of Lemma 1.10.

Lemma 1.18 *Each $p, q \in C_s^1$ and $h, j, c \in \mathbf{R}$ admits the unique solution $K = K(x, y) \in C(\overline{D})$ to the first, the third, and the last equalities of (1.168) and the first equality of (1.169). It is given by*

$$K(x, y) = g(x)\varphi_{n_1}(y),$$

where $g = g(x) \in C^2[0, 1]$ *is the solution to*

$$(-\frac{d^2}{dx^2} + q(x))g = \lambda_{n_1}(p, h)g, \quad g'(1/2) = c, \quad g(1/2) = 0. \tag{1.177}$$

Similarly, the solution $K = K(x, y) \in C^2(\overline{D})$ *to the first and the third equalities of (1.168) and the second equality of (1.169) is unique, and is given by*

$$K(x, y) = g(x)\varphi_{n_1}(y),$$

with $g = g(x) \in C^2[0, 1]$ *satisfying*

$$(-\frac{d^2}{dx^2} + q(x))g = \lambda_{n_1}(p, h)g, \quad g'(\frac{1}{2}) = 0, \quad g(\frac{1}{2}) = c. \tag{1.178}$$

Proof Since it is obvious that

$$K = g(x)\varphi_{n_1}(y) \in C^2(\overline{D})$$

with $g = g(x)$ defined by (1.177) or (1.178) satisfies the required equalities, it suffices to show the uniqueness of such K. Hence we derive $K \equiv 0$ on $\overline{D}$, assuming the first, the third, and the last equalities of (1.168) and

$$K(\frac{1}{2}, y) = K_x(\frac{1}{2}, y) = 0, \quad 0 \le y \le 1/2. \tag{1.179}$$

So far, we have divided $D = \{(x, y) \mid 0 < y < x < 1\}$ into

$$\Omega_1 = \{(x, y) \mid 0 < y < x < 1 - y\}, \quad \Omega_2 = D \setminus \overline{\Omega}_1.$$

Now divide Ω_1 into

$$\begin{aligned}
\Omega_3 &= \{(x, y) \mid 0 < y < x < 1/2 - y\} \\
\Omega_4 &= \{(x, y) \mid 0 < y < x - 1/2 < 1/2 - y\} \\
\Omega_5 &= \Omega_1 \setminus (\overline{\Omega}_3 \cup \overline{\Omega}_4).
\end{aligned}$$

Put $A = (0, 0)$, $B = (1, 0)$, $C = (1, 1)$, $E = (1/2, 0)$, $F = (1/4, 1/4)$, $G(3/4, 1/4)$, and $H = (1/2, 1/2)$.

By Proposition 1.4, the first equality of (1.168) and (1.179) imply $K = 0$ on $\overline{\Omega}_5$. Then Proposition 1.6, the first and the third equalities of (1.168), $K = 0$ on $\overline{EF}$, and $K = 0$ on $\overline{EG}$ imply $K = 0$ on $\overline{\Omega}_3 \cup \overline{\Omega}_4$.

Here, the last equality of (1.168) means

$$K_x(1, y) + jK(1, y) = d\varphi_{n_1}(y), \quad 0 \le y \le 1 \tag{1.180}$$

with some $d \in \mathbf{R}$, while $K = 0$ on $\bar{\Omega}_4$ gives $K_x(1, 0) = K(1, 0) = 0$. Hence there arises $d = 0$. Again Proposition 1.6 ensures $K = 0$ on $\overline{\Omega}_2$ by the first equality of (1.168), (1.180) with $d = 0$, and $K = 0$ on $\overline{BH}$. We thus end up with $K = 0$ on $\overline{D}$. □

Proof *(Theorem 1.40)* This theorem is a consequence of Lemmas 1.17 and 1.18 as in the proof of Theorem 1.24. □

1.6.4 τ-Functions

Here we show Theorem 1.34. We put

$$\lambda_{n_1} = \lambda_{n_1}(p, h), \quad \varphi_{n_1} = \varphi(\cdot; p, h, \lambda_{n_1}(p, h))$$

to study the Hochstadt equation

$$\begin{aligned} \frac{d^2 g}{dx^2} &= (2\frac{d}{dx}(g\varphi_{n_1}) + p - \lambda_{n_1})g \\ g(1-x) &= (-1)^{n_1+1} g(x), \quad 0 \le x \le 1. \end{aligned} \tag{1.181}$$

By Theorem 1.40 we have

$$\begin{aligned} Q_{p,h,n_1} &\equiv \{q \in C_s^1 \mid \lambda_n(q, h) = \lambda_n(p, h),\ n \neq n_1\} \\ &= \{p + 2\frac{d}{dx}(g\varphi_{n_1}) \mid g \in C^2[0, 1], (1.181), g(0) = 0\}. \end{aligned}$$

Lemma 1.19 *For $g \in C^2[0, 2]$ satisfying (1.181) it holds that*

$$\varphi_{n_1} g' - \varphi_{n_1}' g = (\varphi_{n_1} g)^2 + \gamma, \quad 0 \le x \le 1 \tag{1.182}$$

where $\gamma \in \mathbf{R}$.

Proof Using $(-\frac{d^2}{dx^2} + p(x))\varphi_{n_1} = \lambda_{n_1}\varphi_{n_1}$, we obtain

$$(\varphi_{n_1} g' - \varphi_{n_1}' g)' = \varphi_{n_1} g'' - \varphi_{n_1}'' g = 2(\varphi_{n_1} g)' \varphi_{n_1} g = \{(\varphi_{n_1} g)^2\}'$$

by (1.181), and hence (1.182). □

Proof *(Theorem 1.34 for the case of $n_1 = 0$)* We derive $g \equiv 0$ from (1.181), $g(0) = 0$, and $n_1 = 0$. If $\gamma = 0$ in (1.182), then it holds that

$$g(0) = g'(0) = 0$$

and hence $g \equiv 0$.

Assume the contrary, $\gamma \neq 0$, and take $x_0 \in [0, 1]$ with $g(x_0) = 0$. Then the value $g'(x_0)$ has the same sign as that of γ, because $\varphi_{n_1} > 0$ holds on $[0, 1]$ by $n_1 = 0$. Since

$$g(0) = g(1) = 0$$

follows from $g(0) = 0$ and the second equality of (1.181), this property contradicts to $g \in C^2[0, 1]$. Hence we obtain $\gamma = 0$. □

Proceeding to the other case of $n_1 \neq 0$, we note the following theorem proven similarly to Theorem 1.40. Let

$$Q^*(p, n_1) = \{q \in C_s^1 \mid \lambda_n^*(q) = \lambda_n^*(p),\ n \neq n_1\}$$

for $p \in C_s^1$ and $n_1 \in \mathbf{N}^* \equiv \{1, 2, \dots\}$, and put

$$\lambda_{n_1}^* = \lambda_{n_1}^*(p), \quad \varphi_{n_1}^* = \varphi_{n_1}^*(\cdot; p)$$

for simplicity.

Theorem 1.42 *The set $Q^*(p, n_1)$ is homeomorphic to $\mathscr{G}^*(p, n_1)$, the set of solutions $f = f(x) \in C^2[0, 1]$ to*

$$\begin{aligned} \frac{d^2 f}{dx^2} &= (2\frac{d}{dx}(f\varphi_{n_1}^*) + p - \lambda_{n_1}^*)f \\ f(1-x) &= (-1)^{n_1} f(x), \quad 0 \leq x \leq 1 \end{aligned} \tag{1.183}$$

through the relation $q = p + 2\frac{d}{dx}(f\varphi_{n_1}^*)$.

Here is a key theorem for the proof of Theorem 1.34 of $n_1 \geq 1$.

Theorem 1.43 *For each $p, q \in C_s^1$ and $n_1 \in \mathbf{N}^*$, the condition*

$$\lambda_n^*(p) = \lambda_n^*(q), \quad n \in \mathbf{N}^* \setminus \{n_1\} \tag{1.184}$$

is equivalent to the existence of $h \in \mathbf{R}$ such that

$$\lambda_n(p, h) = \lambda_{n_1}(q, h), \quad n \in \mathbf{N} \setminus \{n_1\}. \tag{1.185}$$

If $q \not\equiv p$, furthermore, it holds that

$$\lambda_{n_1}^*(p) = \lambda_{n_1}(q, h), \quad \lambda_{n_1}(p, h) = \lambda_{n_1}^*(q). \tag{1.186}$$

Proof *((1.184) implies (1.185) for some $h \in \mathbf{R}$)* It suffices to take the case $q \not\equiv p$. Since (1.184) means $q \in Q^*(p, n_1)$, there exists $f \in C^2[0, 1]$ satisfying (1.183) and $f \not\equiv 0$ such that

$$q = p + 2\frac{d}{dx}(f\varphi_{n_1}^*). \tag{1.187}$$

Here we show $f(0) \neq 0$.

In fact, if $f(0) = 0$ then $f(1) = 0$ by (1.183) and hence

$$\lambda_{n_1}^* (= \lambda_{n_1}^*(p)) \in \sigma(A_q^*).$$

Combining this property with (1.184), we obtain $\sigma(A_p^*) \subset \sigma(A_q^*)$, which implies

$$\sigma(A_p^*) = \sigma(A_q^*)$$

by the asymptotics of eigenvalues. Hence we obtain $q \equiv p$ by Theorem 1.33, a contradiction.

Having $f(0) \neq 0$, we define $h = f'(0)/f(0) \in \mathbf{R}$ to obtain

$$f'(0) - hf(0) = f'(1) + hf(1) = 0,$$

recalling (1.183). Hence equalities (1.183) and (1.187) imply

$$\lambda_{n_1}^* = \lambda_{m_1}(q, h), \quad f = c\varphi_{m_1}(\cdot; q, h) \tag{1.188}$$

for $m_1 \in \mathbf{N} \equiv \{0, 1, 2, \cdots\}$ and $c \in \mathbf{R} \setminus \{0\}$. Now we show that f has exactly n_1-zeros in $(0, 1)$, which implies $m_1 = n_1$ by the Sturm–Liouville theory.

To this end we recall that

$$\varphi_{n_1}^* = \varphi_{n_1}^*(\cdot; p, \lambda_{n_1}(p))$$

has $(n_1 - 1)$-zeros in (0,1), and therefore, $(n_1 + 1)$-zeros on $[0, 1]$ denoted by

$$x_0^* = 0 < x_1^* < \cdots < x_{n_1-1}^* < x_{n_1}^* = 1.$$

We show:

(a) $f(x_1^*) \neq 0$ for $0 \leq i \leq n_1$.
(b) f has at least one zero in $I_i \equiv (x_i^*, x_{i+1}^*)$ for $0 \leq i \leq n_1 - 1$.
(c) f has at most one zero in I_i for $0 \leq i \leq n_1 - 1$.

Since the functions f and $\varphi_{n_1}^*$ satisfy (1.183) and

$$(-\frac{d^2}{dx^2} + p(x))\varphi^*_{n_1} = \lambda^*_{n_1}\varphi^*_{n_1}, \quad \varphi^*_{n_1}(0) = 0, \quad \varphi^*_{n_1}(0) = 1,$$

respectively, we obtain

$$(\varphi^*_{n_1} f' - \varphi^*_{n_1}{}')' = \varphi^*_{n_1} f'' - \varphi^*_{n_1}{}'' f = 2(\varphi^*_{n_1} f)'\varphi^*_{n_1} f = ((\varphi^*_{n_1} f)^2)'$$

as in Lemma 1.19. Hence there arises

$$\varphi^*_{n_1} f' - \varphi^*_{n_1}{}' f = (\varphi^*_{n_1} f)^2 + \gamma \tag{1.189}$$

with $\gamma \in \mathbf{R}$. □

Proof *((a))* If $f(x^*_i) = 0$, then we obtain $\gamma = 0$. By (1.183) and Lemma 1.13, on the other hand, it holds that

$$\varphi^*_{n_1}(\frac{1}{2}) f(\frac{1}{2}) = 0.$$

Hence $f(\frac{1}{2}) = f'(\frac{1}{2}) = 0$ follows from (1.189) with $\gamma = 0$. This property implies $f \equiv 0$, a contradiction. □

Proof *((b))* Equality (1.189) implies

$$\left[\varphi^*_{n_1} f' - \varphi^*_{n_1}{}' f\right]_{x=x^*_i}^{x=x^*_{i+1}} = \left[(\varphi^*_{n_1} f)^2\right]_{x=x^*_i}^{x=x^*_{i+1}} = 0$$

and hence

$$\varphi^*_{n_1}{}'(x^*_i) f(x^*_i) = \varphi^*_{n_1}{}'(x^*_{i+1}) f(x^*_{i+1}), \quad 0 \le i \le n_1 - 1.$$

If $f \neq 0$ on $\overline{I}_i = [x^*_i, x^*_{i+1}]$, then

$$\varphi^*_{n_1}{}'(x^*_i)\varphi^*_{n_1}{}'(x^*_{i+1}) > 0,$$

which contradicts $\varphi^*_{n_1}(x^*_i) = \varphi^*_{n_1}(x^*_{i+1}) = 0$ and $\varphi(x) \neq 0$ for $x \in I_i$. □

Proof *((c))* Suppose that there exists y_1 and y_2 in $x^*_i < y_1 < y_2 < x^*_{i+1}$ such that

$$f(y_1) = f(y_2) = 0, \quad f(x) \neq 0, \; y_1 < x < y_2.$$

Then equality (1.189) implies

$$[\varphi^*_{n_1} f' - \varphi^*_{n_1}{}' f]_{x=y_1}^{x=y_2} = [(\varphi^*_{n_1} f)^2]_{x=y_1}^{x=y_2} = 0$$

and hence

$$\varphi_{n_1^*}(y_2)f'(y_2) = \varphi_{n_1}^*(y_1)f'(y_1).$$

If $\varphi_{n_1}^*(x) \neq 0$ holds for $x \in I_i$, the we obtain $f'(y_1)f'(y_2) > 0$, which contradicts

$$f(y_1) = f(y_2) = 0, \quad f(x) \neq 0, \ y_1 < x < y_2.$$

□

So far, we have assumed (1.184) and $q \not\equiv p$, to reach $f \not\equiv 0$ and $h \in \mathbf{R}$ satisfying (1.187) and (1.188) with $m_1 = n_1$. Here we put

$$F(x, y) = -\varphi_{n_1}^*(x)f(y).$$

By (1.187) and (1.188) this F satisfies

$$\begin{aligned} &F_{xx} - F_{yy} + q(y)F = p(x)F && \text{in } D \\ &F(x, x) = \frac{1}{2}\int_0^x p(s) - q(s)\, ds, && 0 \le x \le 1 \\ &F_y(x, 0) = hF(x, 0), && 0 \le x \le 1, \end{aligned}$$

and therefore, $F = K(\cdot, \cdot; p, h; q, h)$. We thus obtain

$$\begin{aligned} \Phi(x; p, h, \lambda) &= \Phi(x; q, h, \lambda) + \int_0^x F(x, y)\Phi(y; q, h, \lambda)\, dy \\ &= \Phi(x; q, h, \lambda) - \varphi_{n_1}^*(x)\int_0^x f(y)\Phi(y; q, h, \lambda)\, dy, \quad (1.190) \end{aligned}$$

which implies

$$\begin{aligned} &\Phi'(1; p, h, \lambda) + h\Phi(1; p, h, \lambda) = [\Phi'(1; q, h, \lambda) + h\Phi(1; q, h, \lambda)] \\ &- \varphi_{n_1}^{*\,\prime}(1)\int_0^1 f(y)\Phi(y; q, h, \lambda)\, dy, \quad (1.191) \end{aligned}$$

by $\varphi_{n_1}^*(1) = 0$.

Using (1.188) with $m_1 = n_1$, on the other hand, we obtain

$$\begin{aligned} &\Phi'(1; q, h; \lambda_m(q, h)) + h\Phi(1; q, h, \lambda_m(q, h)) \\ &= \int_0^1 f(y)\Phi(y; q, h, \lambda_m(q, h))\, dy = 0, \quad m \neq n_1, \end{aligned}$$

and therefore,

$$\Phi'(1; p, h, \lambda) + h\Phi(1; p, h, \lambda) = 0, \quad \lambda = \lambda_m(q, h), \ m \in \mathbf{N} \setminus \{n_1\}$$

by (1.191) and (1.188) with $m_1 = n_1$. This property means

$$\lambda_m(q, h) = \lambda_{n(m)}(p, h), \quad m \in \mathbf{N} \setminus \{n_1\} \tag{1.192}$$

for some $n(m) \in \mathbf{N}$. We now show

$$n(m) = m, \quad m \in \mathbf{N} \setminus \{n_1\} \tag{1.193}$$

to conclude the proof of (1.185).

If n_1 is even, there arises $\varphi^*_{n_1}(\frac{1}{2}) = 0$ and hence

$$\Phi(\frac{1}{2}; p, h, \lambda) = \Phi(\frac{1}{2}; q, h, \lambda). \quad \lambda \in \mathbf{R}$$

by (1.190). We thus obtain

$$\Phi(\frac{1}{2}; p, h, \lambda_m(q, h)) = 0, \quad m : odd. \tag{1.194}$$

Equality (1.191), on the other hand, gives

$$\Phi'(\frac{1}{2}; p, h, \lambda) = \Phi'(\frac{1}{2}; q, h, \lambda) - \varphi^{*}_{n_1}{}'(\frac{1}{2}) \int_0^{1/2} f(y)\Phi(y; q, h, \lambda)\, dy.$$

Then we have

$$\Phi'(\frac{1}{2}; q, h, \lambda_m(q, h)) = 0, \quad m : even$$

and

$$\int_0^{1/2} f(y)\varphi(y; q, h, \lambda_m(q, h))\, dy = 0, \quad m : even, \ m \neq n_1$$

by (1.188) with $m_1 = n_1$. Hence, it holds that

$$\Phi'(\frac{1}{2}; p, h, \lambda_m(q, h)) = 0, \quad m : even, \ m \neq n_1. \tag{1.195}$$

Then Lemma 1.13 implies

$$n(m) \equiv m \bmod 2, \quad m \neq n_1, \tag{1.196}$$

by (1.194)–(1.195). By the proof of Lemma 1.17, equalities (1.192) and (1.196) imply (1.193).

The case n_1: *odd* is similarly treated. In fact, since

$$f(\frac{1}{2}) = 0, \quad {\varphi_{n_1}^*}'(\frac{1}{2}) = 0$$

there arises

$$\Phi'(\frac{1}{2}; q, h, \lambda) = \Phi'(\frac{1}{2}; p, h, \lambda), \quad \lambda \in \mathbf{R}$$

by (1.190), and in particular,

$$\Phi'(\frac{1}{2}; p, h, \lambda_m(q, h)) = 0, \quad m : even. \tag{1.197}$$

By (1.188) with $m_1 = n_1$, on the other hand, we have

$$\begin{aligned} &\Phi(\frac{1}{2}; p, h, \lambda_m(q, h)) \\ &= \Phi(\frac{1}{2}; q, h, \lambda_m(q, h)) - \varphi_{n_1}^*(\frac{1}{2}) \int_0^{1/2} f(y)\Phi(y; q, h, \lambda_m(q, h))\, dy \\ &= 0, \quad m : odd,\ m \neq n_1. \end{aligned} \tag{1.198}$$

Equalities (1.197) and (1.198) imply (1.196), and then (1.193) follows.

We have proven that (1.184) implies the existence of $h \in \mathbf{R}$ satisfying (1.185). Meanwhile we have observed (1.188) with $m_1 = n_1$ in the case of $q \not\equiv p$.

Proof *((1.186) for the case of $q \not\equiv p$)* Continue to write

$$\varphi_{n_1} = \varphi_{n_1}(\cdot; p, h), \quad \lambda_{n_1} = \lambda_{n_1}(\cdot; p, h).$$

Since (1.185) means $q \in Q(p, n_1)$, there exists $g \in C^2[0, 1]$ such that (1.181) holds and

$$q = p + 2\frac{d}{dx}(g\varphi_{n_1}), \quad g(0) = 0, \quad g \not\equiv 0 \tag{1.199}$$

by Theorem 1.40.

This property assures the existence of $m_1 \in \mathbf{N}^*$ such that

$$\lambda_{n_1} = \lambda_{m_1}^*(q), \quad g = \text{constant} \cdot \varphi^*(\cdot; q, \lambda_{m_1}^*(q)). \tag{1.200}$$

Then the above proof guarantees $m_1 = n_1$ and hence the second equality of (1.186). The first equality follows similarly. □

We conclude the proof of Theorem 1.42 with the following proof.

Proof *((1.185) implies (1.184) for $q \not\equiv p$)* We take $g \in C^2[0,1]$ in the above proof to put

$$F^*(x,y) = -\varphi_{n_1}(x)g(y).$$

It holds that

$$\begin{aligned} &F^*_{xx} - F^*_{yy} + q(y)F^* = p(x)F^* && \text{in } D \\ &F^*(x,x) = \frac{1}{2}\int_0^x p(s) - q(s)\,ds, && 0 \le x \le 1 \\ &F^*(x,0) = 0, && 0 \le x \le 1 \end{aligned}$$

by $\varphi_{n_1} = \varphi_{n_1}(\cdot, p, h)$, (1.199), and (1.200). This property means

$$F^* = K^*(\cdot,\cdot; p, q)$$

and hence

$$\Phi^*(x; p, \lambda) = \Phi^*(x; q, \lambda) - \varphi_{n_1}(x)\int_0^x g(y)\Phi^*(y; q, \lambda)\,dy. \tag{1.201}$$

Using (1.200) with $m_1 = n_1$ and (1.201), we have

$$\varphi^*(1; p, \lambda^*_m(q)) = 0, \quad m \in \mathbf{N} \setminus \{n_1\}, \tag{1.202}$$

while, as in the proof of (1.194)–(1.195), there arises

$$\begin{aligned} &\Phi^*(\tfrac{1}{2}; q, \lambda^*_m(q)) = 0, && m: even,\ m \ne n_1 \\ &(\Phi^*)'(\tfrac{1}{2}; q, \lambda^*_m(q)) = 0, && m: odd,\ m \ne n_1 \end{aligned} \tag{1.203}$$

by (1.200) with $m_1 = n_1$ and (1.201). Then the proof of (1.193) is valid to ensure (1.184) by (1.202)–(1.203), and the proof is complete. □

Having Theorem 1.42, we conclude the proof of Theorem 1.34.

Proof *(Theorem 1.34 for the case of $n_1 \ge 1$)* We divide the proof into the following four parts.

(a) The number of elements in $Q(p, h, n_1)$ different from p is at most one.
(b) If $n_1 \ge 1$ and $\lambda^*_{n_1}(p) = \lambda_{n_1}(p, h)$, then

$$Q(p, h, n_1) = \{p\}.$$

(c) For $p \in C_s^1[0, 1]$, $h \in \mathbf{R}$, and $n_1 \in \mathbf{N}^*$, the function τ defined by (1.155) satisfies

$$(d^2/dx^2) \log \tau \in C_s^1.$$

Furthermore, $(d^2/dx^2) \log \tau = 0$ if and only if

$$\lambda_{n_1}^*(p) = \lambda_{n_1}(p, h).$$

(d) It holds that

$$p - 2\frac{d^2}{dx^2} \log \tau \in Q(p, h, n_1).$$

□

Proof *((a))* It suffices to take the case $n_1 \geq 1$. Suppose

$$q_1, q_2 \in Q(p, h, n_1), \quad q_1 \not\equiv p, \quad q_2 \not\equiv p.$$

Then, Theorem 1.43 implies

$$\lambda_{n_1}^*(q_1) = \lambda_{n_1}^*(p) = \lambda_{n_1}^*(q_2), \quad \forall n \in \mathbf{N}^* \setminus \{n_1\}$$
$$\lambda_{n_1}^*(q_1) = \lambda_{n_1}^*(p, h) = \lambda_{n_1}^*(q_2).$$

Hence it holds that

$$\sigma(A_{q_1}^*) = \sigma(A_{q_2}^*),$$

and then $q_1 \equiv q_2$ follows from Theorem 1.33. □

Proof *((b))* Let

$$\lambda_{n_1}^*(p) = \lambda_{n_1}(p, h), \quad n_1 \in \mathbf{N}^*,$$

and assume the existence of $q \in Q(p, h, n_1) \setminus \{p\}$. Then there arises

$$\lambda_n(p, h) = \lambda_n(q, h), \ \forall n \neq n_1, \quad q \not\equiv p.$$

Theorem 1.43, however, assures $\lambda_{n_1}^*(p) = \lambda_{n_1}(q, h)$, and therefore,

$$\lambda_{n_1}(p, h) = \lambda_{n_1}(q, h)$$

by from the assumption $\lambda_{n_1}^*(p) = \lambda_{n_1}(p, h)$. Summing up these relations, we get

$$\sigma(A_{p,h,h}) = \sigma(A_{q,h,h}),$$

and hence $q \equiv p$ by Theorem 1.33, a contradiction. □

Proof *((c))* Recall

$$\lambda_{n_1} = \lambda_{n_1}(p, h), \ \varphi_{n_1} = \varphi_{n_1}(\cdot; p, h), \quad \lambda^*_{n_1} = \lambda^*_{n_1}(p), \ \varphi^*_{n_1} = \varphi^*_{n_1}(\cdot; p),$$

and

$$\tau = (\varphi^*_{n_1})'\varphi_{n_1} - \varphi^*_{n_1}\varphi'_{n_1}.$$

By the Sturm–Liouville theorem, φ_{n_1} and $\varphi^*_{n_1}$ have n_1-zeros and $(n_1 + 1)$-zeros on $[0, 1]$, denoted by

$$0 < x_1 < x_2 < \cdots < x_{n_1} < 1$$

and

$$0 = x^*_0 < x^*_1 < \cdots < x^*_{n_1-1} < x^*_{n_1} = 1,$$

respectively. By the comparison theorem, these $\{x_i\}_{i=1}^{n_1}$ and $\{x^*_i\}_{i=0}^{n_1}$ are interlaced as

$$0 = x^*_0 < x_1 < x^*_1 < \cdots < x^*_{n_1-1} < x^*_{n_1} = 1, \tag{1.204}$$

because of

$$\lambda_{n_1} = \lambda^*_{n_1}(q) \in (\lambda^*_{n_1-1}(p), \lambda^*_{n_1+1}(p))$$

and

$$\lambda^*_{n_1} = \lambda_{n_1}(p, h) \in (\lambda_{n_1-1}(p, h), \lambda_{n_1+1}(p, h)).$$

We first show $\tau > 0$ on $[0, 1]$. In fact, since

$$\tau' = \varphi^{*\,\prime\prime}_{n_1}\varphi_{n_1} - \varphi^*_{n_1}\varphi''_{n_1} = (\lambda_{n_1} - \lambda^*_{n_1})\varphi^*_{n_1}\varphi_{n_1} \tag{1.205}$$

we have only to prove

$$\tau(x_i) = -\varphi^*_{n_1}(x_i)\varphi'_{n_1}(x_i) > 0, \ \tau(x^*_i) = (\varphi^*_{n_1})'(x^*_i)\varphi_{n_1}(x^*_i) > 0, \quad 1 \le i \le n_1.$$

These inequalities, however, are obvious by (1.204).

Second, $\tau \in C^3[0, 1]$ follows from (1.205), which implies

$$(d^2/dx^2)\log\tau \in C^1[0,1].$$

Third,

$$\tau(1-x)=\tau(x), \quad 0\le x\le 1 \tag{1.206}$$

holds by Lemmas 1.11–1.12. We obtain

$$\frac{d^2}{dx^2}\log\tau(1-x)=\frac{d^2}{dx^2}\log\tau(x),$$

and hence $(d^2/dx^2)\log\tau \in C_s^1$.

Proceeding to the latter part, we note that $\lambda_{n_1}^*=\lambda_{n_1}$ implies $\tau'=0$ by (1.205). Then, this τ is constant and hence

$$(d^2/dx^2)\log\tau=0. \tag{1.207}$$

Equality (1.207), conversely, implies

$$\tau(x)=\alpha e^{\beta x}, \quad 0\le x\le 1$$

with $\alpha,\beta\in\mathbf{R}$, and hence $\tau'=\alpha\beta e^{\beta x}$. Since (1.205) implies $\tau'(0)=0$, we obtain

$$\alpha\beta=0, \quad \tau'\equiv 0, \quad \lambda_{n_1}^*=\lambda_{n_1},$$

in turn. □

Proof *((d))* We show that

$$q=p-2\frac{d^2}{dx^2}\log\tau \tag{1.208}$$

satisfies

$$\lambda_n(q,h)=\lambda_n(p,h), \quad \forall n\in\mathbf{N}\setminus\{n_1\}.$$

By Theorem 1.40 it suffices to show that

$$g(x)=(\lambda_{n_1}^*-\lambda_{n_1})\varphi_{n_1}^*(x)/\tau(x)\in C^2[0,1]$$

satisfies

$$\frac{d^2g}{dx^2}=(2\frac{d}{dx}(g\varphi_{n_1})+p-\lambda_{n_1})g$$
$$g(1-x)=(-1)^{n_1+1}g(x), \quad 0\le x\le 1 \tag{1.209}$$

with

$$g(0) = 0, \quad q - p = 2\frac{d}{dx}(g\varphi_{n_1}).$$

In fact, the second equality of (1.209) follows from Lemma 1.11 and (1.206). Equality $g(0) = 0$, on the other hand, is obvious, while

$$q - p = 2\frac{d}{dx}(g\varphi_{n_1})$$

holds by (1.208) and

$$\begin{aligned} 2\frac{d}{dx}(g\varphi_{n_1}) &= 2(\lambda^*_{n_1} - \lambda_{n_1})\frac{d}{dx}(\varphi^*_{n_1}\varphi_{n_1}/\tau) \\ &= -2\frac{d}{dx}(\tau'/\tau) = -2\frac{d^2}{dx^2}\log\tau \end{aligned}$$

derived from (1.205). Finally, setting

$$\hat{g}(x) = \varphi^*_{n_1}(x)/\tau(x),$$

we obtain

$$\begin{aligned} \hat{g}'' &= ({\varphi^*_{n_1}}''\tau^2 - 2(\varphi^*_{n_1})'\tau\tau' + 2\varphi^*_{n_1}(\tau')^2 - \varphi^*_{n_1}\tau\tau'')/\tau^3 \\ &= \frac{\varphi^*_{n_1}}{\tau}[(p - \lambda_{n_1}) - 2(\lambda_{n_1} - \lambda^*_{n_1})(\varphi^*_{n_1})'\varphi_{n_1}/\tau + 2(\tau')^2/\tau^2 - \tau''/\tau] \end{aligned}$$

by $\varphi^*_{n_1} = \varphi^*_{n_1}(\cdot, p)$ and (1.206), and hence

$$\begin{aligned} \hat{g}'' &= \hat{g}[p - \lambda_{n_1} + (\lambda_{n_1} - \lambda^*_{n_1})(1 - 2(\varphi^*_{n_1})'\varphi_{n_1}/\tau) + 2(\tau')^2/\tau^2 - \tau''/\tau^2] \\ &= \hat{g}[p - \lambda_{n_1} + (\lambda_{n_1} - \lambda^*_{n_1})(\varphi^*_{n_1}\varphi_{n_1})'/\tau + 2(\tau')^2/\tau^2 - \tau''/\tau], \end{aligned}$$

recalling

$$\tau = (\varphi^*_{n_1})'\varphi_{n_1} - \varphi^*_{n_1}\varphi'_{n_1}.$$

We thus reach

$$\begin{aligned} \hat{g}'' &= \hat{g}[p - \lambda_{n_1} + 2((\tau')^2/\tau^2 - \tau''/\tau)] \\ &= \hat{g}[p - \lambda_{n_1} - 2\frac{d^2}{dx^2}\log\tau] \end{aligned}$$

again by (1.205). Then there arises

$$g'' = g(p - \lambda_{n_1} - 2\frac{d^2}{dx^2}\log\tau) = (2\frac{d}{dx}(g\varphi_{n_1}) + p - \lambda_{n_1})g$$

by

$$q - p = 2\frac{d}{dx}(g\varphi_{n_1}),$$

and the proof is complete. □

1.6.5 Rearrangement of Eigenvalues

Here we prove Theorem 1.35. The proof of Theorem 1.36 is similar and omitted. Given $(p, \lambda) \in C^1_s \times \mathbf{R}$, recal that $\varphi = \varphi^e(\cdot; p, \lambda)$ denotes the solution φ to

$$(-\frac{d^2}{dx^2} + p(x))\varphi = \lambda\varphi, \quad 0 \le x \le 1 \tag{1.210}$$

with

$$\varphi(1/2) = 1, \quad \varphi'(1/2) = 0,$$

while $\varphi = \varphi^o(\cdot; p, \lambda)$ denotes the solution φ to (1.210) with

$$\varphi(1/2) = 0, \quad \varphi'(1/2) = 1.$$

It holds that

$$\begin{aligned} &\varphi^e(1-x; p, h) = \varphi^e(x; p, h) \\ &\varphi^o(1-x; p, \lambda) = -\varphi^o(x; p, \lambda), \quad x \in [0, 1]. \end{aligned} \tag{1.211}$$

Regarding Lemmas 1.11 and 1.13, we put

$$\varphi_n(\cdot; p, h) = \begin{cases} \varphi^e(\cdot; p, \lambda_n(p, h)), \ n : even \\ \varphi^o(\cdot; p, \lambda_n(p, h)), \ n : odd \end{cases}$$

$$\varphi_n^*(\cdot; p) = \begin{cases} \varphi^e(\cdot; p, \lambda_n^*(p)), \ n : odd \\ \varphi^o(\cdot; p, \lambda_n^*(p)), \ n : even. \end{cases}$$

Let

$$\varphi_n = \varphi_n(\cdot; p, h), \quad \lambda_n = \lambda_n(p, h), \quad n_1 \in \mathbf{N}$$

and take

$$\mu_1 \in (\lambda_{n_1-1}, \lambda_{n_1+1})$$

It holds that

$$\varphi_n(1-x) = (-1)^n \varphi_n(x), \quad \varphi_n^*(1-x) = (-1)^{n+1}\varphi_n(x) \tag{1.212}$$

for $x \in [0, 1]$. Let

$$f_1 = \begin{cases} \varphi^o(\cdot; p, \mu_1), & n_1 : even \\ -\varphi^e(\cdot; p, \mu_1), & n_1 : odd. \end{cases} \tag{1.213}$$

Since the uniqueness of such (q, j) follows from Theorem 1.33, we have only to prove the following facts.

(a) $\tau = \tau(x)$ given by

$$\tau = W(f_1, \varphi_{n_1}) \equiv f_1' \varphi_{n_1} - f_1 \varphi_{n_1}'$$

for f_1 in (1.213) satisfies $\tau \in C^3[0, 1]$ and

$$\tau(1-x) = \tau(x), \ \tau(x) > 0, \quad 0 \le x \le 1.$$

(b) $(q, j) \in C_s^1[0, 1] \times \mathbf{R}$ defined by

$$q = p - 2\frac{d^2}{dx^2} \log \tau(x), \quad j = h - \frac{d}{dx} \log \tau(x)\Big|_{x=0} \tag{1.214}$$

satisfies $\lambda_n(q, j) = \lambda_n$ for $n \neq n_1$.

(c) $(q, j) \in C_s^1[0, 1] \times \mathbf{R}$ defined by (1.214) satisfies $\lambda_{n_1}(q, j) = \mu_1$.

Proof *((a))* Since

$$\tau' = (\lambda_{n_1} - \mu_1) f_1 \varphi_{n_1} \in C^2[0, 1] \tag{1.215}$$

we obtain $\tau \in C^3[0, 1]$. Since equality (1.211) implies

$$f_1(1-x) = (-1)^{n_1+1} f_1(x), \quad \varphi_{n_1}(1-x) = (-1)^{n_1} \varphi_{n_1}(x), \tag{1.216}$$

it holds that

$$\tau(1-x) = \tau(x), \quad 0 \le x \le 1.$$

The proof of

$$\tau(x) > 0, \quad 0 \le x \le 1,$$

is similar to the previous subsection. Regarding (1.215), we show the following facts:

(d) τ is positive at each zero of f_1.
(e) τ is positive at each zero of φ_{n_1}.
(f) $\tau(0) = \tau(1) > 0$.

□

Proof *((f))* It suffices to show

$$f_1'(0) - hf_1(0) > 0.$$

Note that $f_1 = f_1(x)$, and hence $\tau = \tau(x)$ depends on μ_1, and put

$$t(\mu_1) = \tau(0)/\varphi_{n_1}(0) = f_1'(0) - hf_1(0).$$

It is obvious that $\mu_1 \mapsto t(\mu_1)$ is continuous. Since

$$f_1(1-x) = (-1)^{n_1+1} f_1(x), \quad \lambda_{n_1-1} < \mu_1 < \lambda_{n_1+1},$$

this f_1 cannot be an eigenfunction of $A_{p,h}$ by (1.212). Hence

$$t(\mu_1) \equiv f_1'(0) - hf_1(0) \neq 0, \quad \lambda_{n_1-1} < \mu_1 < \lambda_{n_1+1}.$$

If $\mu_1 = \lambda_{n_1}$, on the other hand, it holds that

$$\tau' = (\mu_1 - \lambda_{n_1}) f_1 \varphi_{n_1} \equiv 0,$$

so that

$$t(\lambda_{n_1}) = \frac{\tau(0)}{\varphi_{n_1}(0)} = \frac{\tau(\frac{1}{2})}{\varphi_{n_1}(0)} = \frac{f_1'(\frac{1}{2})\varphi_{n_1}'(\frac{1}{2}) - f_1'(\frac{1}{2})\varphi_{n_1}(\frac{1}{2})}{\varphi_{n_1}(0)} > 0.$$

Hence $\tau(0) = \tau(1) > 0$ follows. □

Proof *((d) and (e))* Here we use Theorem 1.24. We assume that n_1 is odd and put

$$m = \frac{1}{2}(n_1 + 1).$$

The proof for the other case that n_1 is even is similar.

Regarding (1.216) and that φ_n has n zeros in (0, 1) by the Sturm–Liouville theory, let the zeros of φ_{n_1+1} be $\{x_i\}_{i=1}^m \bigcup \{x_i'\}_{i=1}^m$ as in

$$0 < x_m' < x_{m-1}' < \cdots < x_1' < \frac{1}{2} < x_1 < \cdots < x_{m-1} < x_m < 1$$

with

$$x_i' + x_i = 1, \quad 1 \leq i \leq m.$$

Similarly, let the zeros of φ_{n_1-1} be $\{y_i\}_{i=1}^{m-1} \bigcup \{y_i'\}_{i=1}^{m-1}$ as in

$$0 < y_{m-1}' < y_{m-2}' < \cdots < y_1' < \frac{1}{2} < y_1 < \cdots < y_{m-2} < y_{m-1} < 1$$

with

$$y_i' + y_i = 1, \quad 1 \leq i \leq m.$$

By the proof of Lemma 1.11, φ_{n_1+1} is the next eigenfunction of φ_{n_1-1} of the eigenvalue problem

$$(-\frac{d^2}{dx^2} + p(x))\varphi = \lambda\varphi \text{ in } \frac{1}{2} \leq x < 1, \quad \varphi'(\frac{1}{2}) = \varphi'(1) + h\varphi(1) = 0.$$

Hence $\{x_i\}_{i=1}^m \bigcup \{x_i'\}_{i=1}^m$ and $\{y_i\}_{i=1}^{m-1} \bigcup \{y_i'\}_{i=1}^{m-1}$ are interlaced as

$$\begin{aligned} 0 < x_m' < y_{m-1}' < x_{m-1}' < \cdots < y_2' < x_2' < y_1' < x_1' < \frac{1}{2} \\ < x_1 < y_1 < x_2 < y_2 < \cdots < x_{m-1} < y_{m-1} < x_m < 1. \end{aligned} \tag{1.217}$$

Since

$$\begin{aligned} &(-\frac{d^2}{dx^2} + p(x))f_1 = \mu_1 f_1, && 0 \leq x \leq 1 \\ &(-\frac{d^2}{dx^2} + p(x))\varphi_{n_1-1} = \lambda_{n_1-1}\varphi_{n_1-1}, && 0 \leq x \leq 1 \\ &(-\frac{d^2}{dx^2} + p(x))\varphi_{n_1+1} = \lambda_{n_1+1}\varphi_{n_1+1}, && 0 \leq x \leq 1 \end{aligned} \tag{1.218}$$

and $\lambda_{n_1-1} < \mu_1 < \lambda_{n_1+1}$, Theorem 1.24 ensures the following properties.

(α) There exists at least one zero of f_1 in each interval

$$(y_{m-1}', y_{m-2}'), \ldots, (y_2', y_1'), (y_1', y_1), (y_1, y_2), \ldots, (y_{m-2}, y_{m-1}).$$

(β) There exists at most one zero of f_1 in each interval

$$(x'_m, x'_{m-1}), \dots, (x'_2, x'_1), (x'_1, x_1), (x_1, x_2), \dots, (x_{m-1}, x_m).$$

We now recall $f_1(\frac{1}{2}) = -1$ and

$$f_1(1-x) = f_1(x), \quad 0 \le x \le 1.$$

By (1.217), (α), and (β), we have the following properties.

(γ) f_1 has just one zero in each interval (y'_i, x'_i) and (x_i, y_i) for $1 \le i \le m$.
(δ) f_1 has just one zero in each interval (x'_i, y'_{i-1}) and (y_{i-1}, x_i) for $1 \le i \le m$, where $y'_0 = y_0 = \frac{1}{2}$.

Consequently, the number of zeros of f_1 on [0, 1], denoted by

$$\mathscr{Z}(f_1),$$

is either $2m$ or $2(m-1)$, and these zeros $\{z_i\}_{i=1}^s \bigcup \{z'_i\}_{i=1}^m$ for $s = m$ or $s = m-1$ of f_1 satisfy

$$\begin{aligned} 0 \le z'_m < x'_m < y'_{m-1} < z'_{m-1} < x'_{m-1} < \cdots < y'_1 < z'_1 < x'_1 < \frac{1}{2} \\ < x_1 < z_1 < y_1 < \cdots < x_{m-1} < z_{m-1} < y_{m-1} < x_m < z_m \le 1 \end{aligned}$$

in the first case, and

$$\begin{aligned} 0 < x'_m < y'_{m-1} < z'_{m-1} < x'_{m-1} < \cdots < y'_1 < z'_1 < x'_1 < \frac{1}{2} \\ < x_1 < z_1 < y_1 < \cdots < x_{m-1} < z_{m-1} < y_{m-1} < x_m < 1 \end{aligned}$$

in the second case. It holds also that

$$z'_i + z_i = 1, \quad 1 \le i \le s.$$

The eigenfunction φ_{n_1}, on the other hand, satisfies

$$\varphi_{n_1}(1-x) = -\varphi_{n_1}(x), \quad 0 \le x \le 1$$

and has n_1 zeros on [0, 1], denoted by $\{w_i\}_{i=0}^{m+1} \bigcup \{w'_i\}_{i=0}^{m+1}$ as

$$0 < w'_{m-1} < \cdots < w'_1 < w_0 = \frac{1}{2} < w_1 < \cdots < w_{m-1} < 1$$

with $w_i' + w_i = 1$ for $1 \le i \le m-1$. Here, φ_{n_1-1}, φ_{n_1}, and φ_{n_1+1} are successive eigenfunctions of $A_{p,h,h}$, and hence there arises

$$y_i < w_i < x_{i+1},\ x_{i+1}' < w_i' < y_i', \quad 1 \le i \le m-1.$$

These relations imply

$$\begin{aligned} 0 \le z_m' &< w_{m-1}' < z_{m-1}' < \cdots < w_1' < z_1' < w_0 = \frac{1}{2} \\ &< z_1 < w_1 < \cdots < z_{m-1} < w_{m-1} < z_m \le 1 \end{aligned} \tag{1.219}$$

and

$$\begin{aligned} 0 < w_{m-1}' &< z_{m-1}' < \cdots < w_1' < z_1' < w_0 = \frac{1}{2} \\ &< z_1 < w_1 < \cdots < z_{m-1} < w_{m-1} < 1 \end{aligned} \tag{1.220}$$

in the case of $\mathscr{Z}(f_1) = 2m$ and $\mathscr{Z}(f_1) = 2(m-1)$, respectively.

Since $\varphi_{n_1}'(\frac{1}{2}) = 1$ and $f_1(\frac{1}{2}) = -1$, relations (1.219) and (1.220) imply

$$\begin{aligned} \tau(z_i) &= f_1'(z_i)\varphi_{n_1}(z_i) > 0, & 1 \le i \le s \\ \tau(w_i) &= -f_1(w_i)\varphi_{n_1}'(w_i) > 0, & 1 \le i \le m-1 \end{aligned}$$

for $s = m$ and $s = m-1$, respectively, which guarantees $\tau > 0$ on $[0, 1]$ as in the previous subsection. □

Proof *((b))* To apply Theorem 1.40, we show that

$$g = (\mu_1 \lambda_{n_1}) f_1/\tau = (\mu_1 - \lambda_{n_1}) f_1 / W(f_1, \varphi_{n_1}) \in C^2[0, 1]$$

satisfies (1.162) and

$$\begin{aligned} p + 2\frac{d}{dx}(g\varphi_{n_1}) &= p - 2\frac{d^2}{dx^2}\log\tau \\ h + (g\varphi_{n_1})(0) &= h - \frac{d}{dx}\log\tau\bigg|_{x=0} . \end{aligned} \tag{1.221}$$

Equalities (1.221) follow from $g\varphi_{n_1} = -(d/dx)\log\tau$, which is obtained by

$$g\varphi_{n_1} = (\mu_1 - \lambda_{n_1}) f_1 \varphi_{n_1} / W(f_1, \varphi_{n_1}) = -\tau'/\tau = -\frac{d}{dx}\log\tau.$$

The second equality of (1.162) is a consequence of (1.213) and Lemma 1.11. Finally, the first equality of (1.162) is shown by the identity

$$I \equiv \lambda_{n_1} + \frac{\tau}{f_1} \cdot \frac{d^2}{dx^2}(f_1/\tau) = p - 2\frac{d^2}{dx^2}\log\tau. \tag{1.222}$$

This proof is essentially the same as in the previous subsection.

In fact we have

$$\begin{aligned} I &\equiv \lambda_{n_1} + \frac{\tau}{f_1} \cdot \frac{d^2}{dx^2}(f_1/\tau) \\ &= \lambda_{n_1} + \frac{\tau}{f_1} \cdot \{f_1''/\tau - 2f_1'\tau'/\tau^2 + f_1(-\tau''/\tau^2 + 2(\tau')^2/\tau^3)\} \\ &= \lambda_{n_1} + f_1''/f_1 - 2(f_1'\tau')/(f_1\tau) - \tau''/\tau + 2(\tau')^2/\tau^2. \end{aligned}$$

From (1.215) and (1.218) it follows that

$$I = p + (\lambda_{n_1} - \mu_1) - 2(\lambda_{n_1} - \mu_1) f_1 \varphi_{n_1}/\tau - \tau''/\tau + 2(\tau')^2/\tau^2$$

and then (1.213) implies

$$\begin{aligned} I &= p - (\lambda_{n_1} - \mu_1)\{f_1'\varphi_{n_1} + f_1\varphi_{n_1}'\}/\tau - \tau''/\tau + 2(\tau')^2/\tau^2 \\ &= p - 2(\tau''/\tau + 2(\tau')^2/\tau^2) = p - 2\frac{d^2}{dx^2}\log\tau. \end{aligned}$$

The proof is thus complete. □

Proof *((c))* We show that

$$\varphi = \varphi_{n_1}/\tau (= \varphi_{n_1}/W(f_1, \varphi_{n_1}) \not\equiv 0) \in C^2[0,1]$$

has n_1 zeros in (0, 1) and satisfies

$$\begin{aligned} &(-\frac{d^2}{dx^2} + q(x))\psi = \mu_1\psi \quad \text{in } 0 < x < 1 \\ &(-\frac{d}{dx} + j)\psi\Big|_{x=0} = (\frac{d}{dx} + j)\psi\Big|_{x=1} = 0 \end{aligned} \tag{1.223}$$

for $q = p - 2(d^2/dx^2)\log\tau$ and $j = h - (d/dx)\log\tau\big|_{x=0}$.

First, ψ obviously has n_1-zeros in (0, 1), because so does φ_{n_1}. Second, using

$$\psi' = \varphi_{n_1}'/\tau - \varphi_{n_1}\tau'/\tau^2,$$

we have

$$\psi'(0)/\psi(0) = \varphi'_{n_1}(0)/\varphi_{n_1}(0) - \tau'(0)/\tau(0) = h - \frac{d}{dx}\log\tau\Big|_{x=0} = j$$

which implies the boundary condition in (1.223) by

$$\varphi_{n_1}(1-x) = (-1)^{n_1}\varphi_{n_1}(x),\ \ \tau(1-x) = \tau(x), \quad 0 \le x \le 1.$$

Finally, the first equality of (1.223) follows from

$$\begin{aligned}
&\mu_1 + \frac{\tau}{\varphi_{n_1}} \cdot \frac{d^2}{dx^2}(\varphi_{n_1}/\tau) \\
&\quad = \mu_1 + \varphi''_{n_1}/\varphi_{n_1} - 2(\varphi'_{n_1}\tau')/(\varphi_{n_1}\tau) - \tau''/\tau + 2(\tau')^2/\tau^2 \\
&\quad = p + (\mu_1 - \lambda_{n_1}) - 2(\lambda_{n_1} - \mu_1) f_1 \varphi'_{n_1}/\tau - \tau''/\tau + 2(\tau')^2/\tau^2 \\
&\quad = p - (\lambda_{n_1} - \mu_1)(f_1'\varphi_{n_1} + f_1\varphi'_{n_1})/\tau - \tau''/\tau + 2(\tau')^2/\tau^2 \\
&= p - 2(\tau''/\tau - (\tau')^2/\tau^2) = p - 2\frac{d^2}{dx^2}\log\tau.
\end{aligned}$$

The proof is complete. □

1.6.6 *Principle of Superposition*

Here we show the following proof, using the second deformation formula.

Proof *(Theorem 1.38)* Put

$$K(x, y) = (\mu_1 - \lambda_{n_1}) \cdot f_1(x)\varphi_{n_1}(y)/W(f_1, \varphi_{n_1})(x).$$

From the proof of the property (b) in the preceding subsection,

$$g \equiv (\mu_1 - \lambda_{n_1}) f_1/W(f_1, \varphi_{n_1})$$

satisfies

$$(-\frac{d^2}{dx^2} + q(x))g_1 = \lambda_{n_1} g_1, \quad g_1 \cdot \varphi_{n_1} = -\frac{d}{dx}\log\tau.$$

This

$$K = K(x, y) = g_1(x) \cdot \varphi_{n_1}(x),$$

therefore, satisfies

$$\begin{aligned} &K_{xx} - K_{yy} + p(y)K = q(x)K \\ &2\frac{d}{dx}K(x,x) = -2\frac{d^2}{dx^2}\log\tau(x) = q(x) - p(x) \\ &K(x, 1-x) = (-1)^{n_1}K(x,x) \end{aligned}$$

From the proof of the property (*c*) in the preceding subsection, similarly,

$$\psi_1 = (\mu_1 - \lambda_{n_1})\varphi_{n_1}/W(f_1, \varphi_{n_1})$$

satisfies

$$(-\frac{d^2}{dx^2} + q(x))\psi_1 = \mu_1\psi_1, \quad \psi_1 \cdot f_1 = -\frac{d}{dx}\log\tau.$$

Hence

$$L(x,y) = (\mu_1 - \lambda_{n_1}) \cdot \varphi_{n_1}(x) f_1(y)/W(f_1, \varphi_{n_1})(x)$$

satisfies

$$\begin{aligned} &L_{xx} - L_{yy} + p(y)L = q(x)L \\ &2\frac{d}{dx}L(x,x) = q(x) - p(x) \\ &L(x, 1-x) = (-1)^{n_1+1}L(x,x) = (-1)^{n_1+1}K(x,x). \end{aligned}$$

Then

$$\begin{aligned} M(x,y) &= \frac{1}{2}\{K(x,y) + L(x,y)\} \\ &= \frac{1}{2} \cdot (\mu_1 - \lambda_{n_1})\{f_1(x)\varphi_{n_1}(y) + f_1(y)\varphi_{n_1}(x)\}/W(f_1, \varphi_{n_1})(x) \end{aligned}$$

satisfies the requirement in Lemma 1.6,

$$\begin{aligned} &M_{xx} - M_{yy} + p(y)M = q(x)M \qquad \text{in } D_+ \bigcup D_2 \\ &M(x,x) = \frac{1}{2}\int_{1/2}^{x} q(s) - p(s)\,ds, \quad 0 \le x \le 1 \\ &M(x, 1-x) = 0, \quad 0 \le x \le 1, \end{aligned}$$

and then, the second deformation formula, Lemma 1.7 ensures the theorem. □

1.7 Linearization of the Hochstadt Equation

1.7.1 *Introduction*

Having completed the proof of Theorems 1.33, 1.34, 1.35, 1.36, and 1.38, we turn to an algebraic structure hidden by them. Here we note that

$$p - 2(d^2/dx^2)\log\tau$$

in Theorem 1.34 is derived from the conclusion of Theorem 1.43.

In fact, given

$$q \in Q(p, h, n_1) \setminus \{p\}, \quad n_1 \geq 1,$$

let

$$\varphi_{n_1} = \varphi_{n_1}(\cdot; p, h)), \quad \varphi^*_{n_1} = \varphi^*_{n_1}(\cdot; p, \lambda^*_{n_1}(p))).$$

Then there is $g \in C^2[0,1]$ satisfying (1.162), $g(0) = 0$, and

$$q - p = 2\frac{d}{dx}(g\varphi_{n_1}).$$

by Theorem 1.40. Also, by Theorems 1.42–1.43, there exists $f \in C^2[0,1]$ satisfying (1.183) and

$$q - p = 2\frac{d}{dx}(f\varphi^*_{n_1}).$$

In the proof of Theorem 1.43, we have shown

$$g = \text{constant} \times \varphi^*_{n_1}(\cdot; q) \not\equiv 0, \quad f = \text{constant} \times \varphi_{n_1}(\cdot; q, h) \not\equiv 0.$$

Let

$$\lambda_{n_1} = \lambda_{n_1}(p, h), \quad \lambda^*_{n_1} = \lambda^*_{n_1}(p).$$

Put, furthermore,

$$\psi_{n_1} = \varphi_{n_1}(\cdot; q, h), \quad \mu_{n_1} = \lambda_{n_1}(q, h)$$
$$\psi^*_{n_1} = \varphi^*_{n_1}(\cdot; q), \quad \mu^*_{n_1} = \lambda^*_{n_1}(q).$$

We have, finally,

$$\lambda_{n_1} = \mu_{n_1}^*, \quad \lambda_{n_1}^* = \mu_{n_1} \tag{1.224}$$

by Theorem 1.43.

These relations ensure

$$(\psi_{n_1}\varphi_{n_1}^*)' = \text{constant} \times (\psi_{n_1}^*\varphi_{n_1})',$$

which is reduced to

$$(\psi_{n_1}\varphi_{n_1}^*)' = (\psi_{n_1}^*\varphi_{n_1})'$$

by setting $x = 0$. It holds that

$$\psi_{n_1}\varphi_{n_1}^* = \psi_{n_1}^*\varphi_{n_1} + \text{constant}$$

and hence

$$\psi_{n_1}\varphi_{n_1}^* = \psi_{n_1}^*\varphi_{n_1},$$

again by setting $x = 0$. Noting (1.204), we reach

$$\psi_{n_1}^* = c\varphi_{n_1}^*, \quad \psi_{n_1} = c\varphi_{n_1}$$

for some $c \in C^2[0, 1]$.

We substitute these equalities into

$$(\psi_{n_1}^{*\,\prime}\psi_{n_1} - \psi_{n_1}^*\psi_{n_1}')' = (\lambda_{n_1}^* - \lambda_{n_1})\psi_{n_1}^*\psi_{n_1}$$

derived from (1.224), to reach

$$\begin{aligned} &2cc'((\varphi_{n_1}^*)'\varphi_{n_1} - \varphi_{n_1}^*\varphi_{n_1}^{*\,\prime}) + c^2((\varphi_{n_1}^*)'\varphi_{n_1} - \varphi_{n_1}^*\varphi_{n_1}')' \\ &\quad = (\lambda_{n_1}^* - \lambda_{n_1})c^2\varphi_{n_1}^*\varphi_{n_1}. \end{aligned}$$

Since

$$\tau = (\varphi_{n_1}^*)'\varphi_{n_1} - \varphi_{n_1}^*\varphi_{n_1}', \quad \tau' = (\lambda_{n_1} - \lambda_{n_1}^*)\varphi_{n_1}^*\varphi_{n_1},$$

we thus obtain $(c\tau)' = 0$, and hence $c\tau = \text{constant}$.

Putting $x = 0$, we end up with

$$c\tau \equiv 1,$$

which implies

$$\psi_{n_1}^* = \varphi_{n_1}^*/\tau, \quad \psi_{n_1} = \varphi_{n_1}/\tau.$$

Then the equality $q = p - 2\frac{d^2}{dx^2}\log\tau$ follows from

$$\psi_{n_1} = \varphi_{n_1}/\tau, \quad q = \mu_{n_1} + \psi_{n_1}''/\psi_{n_1}$$

through a calculation similar to the previous subsection.

The Hochstadt equation

$$\frac{d^2g}{dx^2} = [(2\frac{d}{dx}(g\cdot\varphi) + p)I - \Lambda]g, \quad 0 \le x \le 1$$

has arisen in connection with the boundary identification for coefficients of the parabolic equation in Sect. 1.3.5, where

$$\varphi = {}^T(\varphi_{n_1}(\cdot; P), \ldots, \varphi_{n_N}(\cdot; P)), \quad \Lambda = \mathrm{diag}(\lambda_{n_1}(P), \ldots, \lambda_{n_N}(P))$$

for $P = (p, h, H) \in C^1[0, 1] \times \mathbf{R} \times \mathbf{R}$. The above method reduces this equation to the linear equation

$$(-\frac{d^2}{dx^2} + p(x))f = \mu f, \quad 0 \le x \le 1$$

for the case of $N = 1$, and then the principle of superposition to the general case.

The dressing method, however, gives an explicit form of τ-function after these superpositions of N-times, and then its convergence as $N \to \infty$ is examined [10]. This section is devoted to the description of these methods for the linearization of the Hochstadt equation.

1.7.2 Duality Method

Noting the duality observed in Theorem 1.43, here we study the Hochstadt equation in one component,

$$\frac{d^2g}{dx^2} = (2\frac{d}{dx}(g\varphi) + p - \lambda)g, \quad g \not\equiv 0, \quad 0 \le x \le 1 \tag{1.225}$$

for $\varphi = \varphi_{n_1}(\cdot; P)$ and $\lambda = \lambda_{n_1}(P)$, where

$$P = (p, h, H) \in C^1[0, 1] \times \mathbf{R} \times \mathbf{R}, \quad n_1 \in \mathbf{N}.$$

It holds that

$$(-\frac{d^2}{dx^2} + p(x))\varphi = \lambda\varphi \text{ in } 0 < x < 1, \quad \varphi(0) \cdot \varphi(1) \neq 0. \tag{1.226}$$

We show that the solution to equation (1.225) is represented by that of the linear equation

$$(-\frac{d^2}{dx^2} + p(x))\varphi^* = \lambda^*\varphi^*, \quad 0 \leq x \leq 1 \tag{1.227}$$

for $\lambda^* \in \mathbf{R}$.

Theorem 1.44 *Each $g = g(x) \in C^2[0, 1]$ in (1.225) admits $\lambda^* \in \mathbf{R}$ and $\varphi^* \in C^2[0, 1]$ satisfying (1.227) and*

$$\tau(x) \equiv 1 + (\lambda - \lambda^*) \int_0^x \varphi^*(y)\varphi(y)\, dy > 0, \quad 0 \leq x \leq 1. \tag{1.228}$$

It holds that

$$g = \begin{cases} (\lambda^* - \lambda)\varphi^*/\tau, & \lambda \neq \lambda^* \\ \varphi/\{\int_0^x \varphi(y)^2\, dy + \alpha\}, & \lambda = \lambda^* \end{cases} \tag{1.229}$$

with $\alpha > 0$. The function g given by (1.229), conversely, satisfies (1.225).

Proof It holds that

$$g''\varphi - g\varphi'' = 2(g\varphi)'(g\varphi) = [(g\varphi)^2]'$$

by (1.225) and (1.226), and hence

$$g'\varphi - g\varphi' = (g\varphi)^2 + \beta, \quad \beta \in \mathbf{R}. \tag{1.230}$$

If $\beta = 0$, we have $-(1/u)' = \varphi^2$ for $u = g/\varphi$ and hence

$$g(x) = -\frac{\varphi(x)}{\int_0^x \varphi(y)^2\, dy + \alpha}$$

holds with $\alpha > 0$.

Let $\beta \neq 0$, and put

$$K(x, y) = g(x)\varphi(y).$$

Recall that $\Phi = \Phi(\cdot; p, h; \lambda)$ denotes the solution to

$$(-\frac{d^2}{dx^2} + p(x))\Phi = \lambda\Phi \text{ in } 0 < x < 1, \quad \Phi(0) = 1, \quad \Phi'(0) = h,$$

and define $Q = (q, j, H)$ by

$$q = p + 2\frac{d}{dx}(g\varphi), \quad j = h + (g \cdot \varphi)|_{x=0}, \quad J = H - (g\varphi)|_{x=1}. \tag{1.231}$$

Set

$$\lambda_n = \lambda_n(P), \ \mu_m = \lambda_m(Q), \ \Phi_n = \Phi(\cdot; p, h, \lambda_n(p)), \ \Psi_m = \Phi(\cdot; q, h; \lambda_m(Q)).$$

The argument in Sect. 1.3.5 ensures

$$\lambda_n = \mu_{m(n)}, \ \Phi_n(1) = \Psi_{m(n)}(1), \quad n \neq n_1$$

with $m(n) \in \mathbf{N}$ through

$$\begin{aligned} &K_{xx} - K_{yy} + p(y)K = q(x)K && \text{in } D \\ &K(x, x) = (j - h) + \frac{1}{2}\int_0^x q(s) - p(s)\,ds, && 0 \le x \le 1 \\ &K_y(x, 0) = hK(x, 0), && 0 \le x \le 1. \end{aligned}$$

Here we take

$$F(x, y) = -g(x)\varphi(x)$$

to obtain

$$F_{xx} - F_{yy} + q(y)F = p(x)F, \quad \frac{d}{dx}F(x, x) = \frac{1}{2}(p(x) - q(x)).$$

Put

$$j^* = \begin{cases} g'(0)/g(0), & \text{if } g(0) \neq 0 \\ \infty, & \text{if } g(0) = 0, \end{cases} \qquad h^* = \begin{cases} j^* - (j - h), & \text{if } j^* \in \mathbf{R} \\ \infty, & \text{if } j^* = \infty. \end{cases} \tag{1.232}$$

In the case of $j^* \in \mathbf{R}$, we obtain

$$F(x, x) = (h^* - j^*) + \int_0^x p(s) - q(s)\,ds, \ F_y(x, 0) = j^*F(x, 0), \quad 0 \le x \le 1.$$

Then the first deformation formula assures

$$\begin{aligned} \Phi(x; p, h^*, \lambda) &= \Phi(x; q, j^*, \lambda) + \int_0^x F(x, y)\Phi(y; q, j^*, \lambda)\,dy \\ &= \Phi(x; q, j^*, \lambda) - \varphi(x)\int_0^x g(y)\Phi(y; q, j^*, \lambda)\,dy, \quad 0 \le x \le 1 \end{aligned} \tag{1.233}$$

for any $\lambda \in \mathbf{R}$. If $j = \infty$, similarly, we obtain

$$F(x, x) = \frac{1}{2}\int_0^x p(s) - q(s)\, ds, \; F(x, 0) = 0, \quad 0 \le x \le 1,$$

which ensures

$$\begin{aligned}\Phi^*(x; p, \lambda) &= \Phi^*(x; q, \lambda) + \int_0^x F(x, y)\Phi^*(y; q, \lambda)\, dy \\ &= \Phi^*(x; q, \lambda) - \varphi(x)\int_0^x g(y)\Phi^*(y; q, \lambda)\, dy\end{aligned}$$

where $\Phi^* = \Phi^*(\cdot; p, \lambda)$ denotes the solution to

$$(-\frac{d^2}{dx^2} + p(x))\Phi^* = \lambda\Phi^* \text{ in } 0 < x < 1, \quad \Phi^*(0) = 1, \quad \Phi^{*\prime}(0) = 1.$$

Henceforth, we write

$$\Phi(\cdot; p, \infty, \lambda) = \Phi^*(\cdot; p, \lambda).$$

Put $Q^* = (q, j^*, J^*)$ for

$$J^* = \begin{cases} -g'(1)/g(1), & \text{if } g(1) \neq 0 \\ \infty, & \text{if } g(1) = 0. \end{cases}$$

Then, A_{Q^*} denotes the differential operator $-\frac{d^2}{dx^2} + q(x)$ in $L^2(0, 1)$ with the boundary condition $(-\frac{d}{dx} + j^*)\cdot\big|_{x=0} = (\frac{d}{dx} + J^*)\cdot\big|_{x=1} = 0$. Henceforth,

$$(-\frac{d}{dx} + j^*)\cdot\Big|_{x=0} = 0 \quad \text{and} \quad (\frac{d}{dx} + J^*)\cdot\Big|_{x=1} = 0$$

stand for $\cdot|_{x=0} = 0$ and $\cdot|_{x=1} = 0$ if $j^* = \infty$ and $J^* = \infty$, respectively. The eigenvalues and the eigenfunctions of A_{Q^*} are denoted by $\{\lambda_n(Q^*)\}$ and $\{\varphi_n(\cdot; Q^*)\}$, respectively, with the normalization

$$\|\varphi_n(\cdot; Q^*)\|_{L^2(0,1)} = 1, \quad \varphi_n(0; Q^*) > 0.$$

Since $g \not\equiv 0$, the equality (1.225) with (1.231), and (1.232) means

$$\lambda(= \lambda_{n_1}(P)) = \lambda_{m_1}(Q^*), \quad g = \text{constant} \times \varphi_{m_1}(\cdot; Q^*)$$

for some m_1, while (1.233) implies

$$\Phi(1; p, h^*; \lambda) = \Phi(1; q, j^*; \lambda) - \varphi(1) \int_0^1 g(y)\Phi(y; q, j^*; \lambda)\, dy$$

$$\Phi_x(1; p, h^*; \lambda) = \Phi_x(1; q, j^*; \lambda) - \varphi'(1) \int_0^1 g(y)\Phi(y; q, j^*; \lambda)\, dy$$
$$- \varphi(1)g(1)\Phi(1; q, j^*; \lambda).$$

Thus we obtain

$$\Phi(1; p, h^*; \lambda) = \Phi(1; q, j^*; \lambda)$$
$$\Phi_x(1; p, h^*; \lambda) = \Phi_x(1; q, j^*; \lambda) + (J - H)\Phi(1; q, j^*, \lambda)$$

for $\lambda = \lambda_m(Q^*)$ with $m \neq m_1$.

To summarize the result, let

$$P^* = (p, h^*, H^*) \tag{1.234}$$

for

$$H^* = \begin{cases} J^* - (J - H), & \text{if } J^* \in \mathbf{R} \\ \infty, & \text{if } J^* = \infty, \end{cases}$$

and put

$$\Psi_m^* = \Phi(\cdot; q, j^*, \lambda_m(Q^*)), \qquad \Phi_n^* = \Phi(\cdot; p, h^*, \lambda_n(P^*)).$$

Then each $m \neq m_1$ admits $n(m)$ such that

$$\lambda_m(Q^*) = \lambda_{n(m)}(P^*), \qquad \begin{cases} \Psi_m^*(1) = \Phi_{n(m)}^*(1), & \text{if } J^* \in \mathbf{R} \\ (\Psi_m^*)'(1) = (\Phi_{n(m)}^*)'(1), & \text{if } J^* = \infty. \end{cases} \tag{1.235}$$

The argument in Sect. 1.3.5 ensures $\sharp(\mathbf{N} \setminus \{n(m) \mid m \neq m_1\}) = 1$, and now we put

$$\{n_1^*\} = \mathbf{N} \setminus \{n(m) \mid m \neq n_1\}. \tag{1.236}$$

Then (1.235) is equivalent to the existence of $m^*(n)$ for $n \neq n_1^*$ such that

$$\lambda_n(P^*) = \lambda_{m^*(n)}(Q^*), \qquad \begin{cases} \Phi_n^*(1) = \Psi_{m^*(n)}^*(1), & \text{if } H^* \in \mathbf{R} \\ (\Phi_n^*)'(1) = (\Psi_{m^*(n)}^*)'(1), & \text{if } H^* = \infty. \end{cases}$$

By the argument developed so far, therefore, there is $f = f(x) \in C^2[0, 1]$ satisfying

$$\frac{d^2f}{dx^2} = (2\frac{d}{dx}(f\varphi^*) + p - \lambda^*)f, \quad q = p + 2\frac{d}{dx}(f\varphi^*) \tag{1.237}$$

for $\varphi^* = \varphi_{n_1^*}(\cdot; P^*)$ and $\lambda^* = \lambda_{n_1^*}(P^*)$. If $j^* \in \mathbf{R}$ and $J^* \in \mathbf{R}$, furthermore, we have

$$j^* = h^* + (f\varphi^*)\big|_{x=0}, \quad \text{and} \quad J^* = H^* + (f\varphi^*)\big|_{x=1}, \tag{1.238}$$

respectively.

Combining (2.48) with (2.13), we get

$$q - p = 2\frac{d}{dx}(g\varphi) = 2\frac{d}{dx}(f\varphi^*).$$

In the case of $j^* = \infty$, we have $g|_{x=0} = \varphi^*|_{x=0} = 0$, while $j^* \in \mathbf{R}$ implies

$$(g\varphi)|_{x=0} = j - h = j^* - h^* = (f\varphi^*)\big|_{x=0}$$

by (1.231), (1.232), and (1.238). Consequently, we arrive at

$$g\varphi = f\varphi^*. \tag{1.239}$$

Equality (1.239) produces nothing in the case $P^* = P$, or equivalently $Q^* = Q$. The assumption $\beta \neq 0$ in (1.230), however, implies $P^* \neq P$. This fact is obvious if $j^* = \infty$. In the case of $j^* \in \mathbf{R}$, we have

$$\begin{aligned}\beta &= g'(0)\varphi(0) - g(0)\varphi'(0) - (g(0)\varphi(0))^2 \\ &= (j^* - h - g(0)\varphi(0))g(0)\varphi(0) = (j^* - j)(j - h)\end{aligned}$$

by (1.231). Hence $\beta \neq 0$ implies $j^* \neq j$, and therefore,

$$\varphi^* \not\equiv \varphi. \tag{1.240}$$

Using (1.239), we define a piecewise C^2-function $c = c(x)$ by

$$c = g/\varphi^* = f/\varphi.$$

We substitute $g = c\varphi^*$ and $f = c\varphi$ into

$$f''g - fg'' = (\lambda - \lambda^*)fg$$

derived from

$$-f'' + qf = \lambda^* f, \quad -g'' + qg = \lambda g,$$

to obtain $(c\tau)' = 0$ as in the preceding section, where

$$\tau = (\varphi^*)'\varphi - \varphi^*\varphi'.$$

Note that c has only a finite number of zeros by $g \not\equiv 0$. Thus

$$c\tau|_{I_\ell} = \alpha_\ell \in \mathbf{R}, \quad 1 \le l \le s,$$

holds, where $I_\ell = (y_\ell, y_{\ell+1})$ with

$$0 = y_0 < y_1 < \cdots < y_s < y_{s+1} = 1.$$

If $\alpha_\ell = 0$ for some ℓ, then $\tau = 0$ on $\overline{I}_\ell$ follows, and hence $\lambda = \lambda^*$ by

$$\tau' = (\lambda - \lambda^*)\varphi^*\varphi. \tag{1.241}$$

Then φ^* depends linearly on φ, and therefore, $\varphi^* = \varphi$ holds by the normalization, which contradicts (1.240). Therefore, we have $\tau \neq 0$ in I_ℓ for any ℓ, and hence there arises $c = \alpha_\ell/\tau$ in I_l, or

$$g = \alpha_\ell\, \varphi^*/\tau \quad \text{in } I_\ell. \tag{1.242}$$

By

$$q = \lambda + g''/g,$$

therefore, we have

$$\begin{aligned} q &= \lambda + \frac{\tau}{\varphi^*}\frac{d^2}{dx^2}(\varphi^*/\tau) = \lambda + \frac{\tau}{\varphi^*}\{(\varphi^*)''/\tau - 2(\varphi^*)'\tau'/\tau^2 + \varphi^*(\frac{1}{\tau})''\} \\ &= \lambda + \frac{\tau}{\varphi^*}\{(\varphi^*)''/\tau - 2(\varphi^*)'\tau'/\tau^2 + \varphi^*(-\tau''/\tau^2 + 2(\tau')^2 + 2(\tau')^2/\tau^3)\} \\ &= \lambda + (\varphi^*)''/\varphi^* - 2\varphi^*\tau'/\varphi^*\tau - \tau''/\tau + 2(\tau')^2/\tau^2. \end{aligned} \tag{1.243}$$

Since

$$\lambda^* + (\varphi^*)''/\varphi^* = p,$$

furthermore we get

$$\begin{aligned} q &= p + (\lambda - \lambda^*) - 2(\lambda - \lambda^*)\varphi(\varphi^*)'/\tau - \tau''/\tau + 2(\tau')^2/\tau^2 \\ &= p - (\lambda - \lambda^*)\{(\varphi^*)'\varphi - \varphi^*\varphi'\}/\tau - \tau''/\tau + 2(\tau')^2/\tau^2 \end{aligned}$$

$$= p - 2(\tau''/\tau - (\tau')^2/\tau^2) = p - 2\frac{d^2}{dx^2}\log\tau \tag{1.244}$$

by (1.241), in each I_ℓ, $1 \le \ell \le s$.

Since $q - p \in C^2[0,1]$, we have

$$\tau > 0 \quad \text{on } [0,1].$$

By

$$q - p = -2\frac{d^2}{dx^2}\log\tau = 2\frac{d}{dx}(g\varphi)$$

with (1.242), we obtain $\alpha_\ell = \lambda^* - \lambda$, $1 \le \ell \le s$, and hence

$$g = (\lambda^* - \lambda)\varphi^*/\tau. \tag{1.245}$$

There arises, furthermore, (1.228) by (1.241). The first part of the theorem is complete with these properties.

We have

$$g\varphi = -\frac{d}{dx}\log\tau$$

by (1.245), and hence

$$q = p - 2\frac{d^2}{dx^2}\log\tau, \quad j = h - \frac{d}{dx}\log\tau\Big|_{x=0}, \quad J = H + \frac{d}{dx}\log\tau\Big|_{x=1} \tag{1.246}$$

by (1.231).

The second part of this theorem follows from a direct calculation. If $g = g(x)$ is given by the second formula of (1.229), it follows that

$$\frac{\varphi}{g} = \int_0^x \varphi(y)^2 dy + \alpha,$$

and hence

$$g'\varphi - g\varphi' = (g\varphi)^2$$

Then it follows that (1.225) from

$$g'' = [2(g\varphi)' + \varphi''/\varphi]g.$$

If $g = g(x)$ is given by the first formula of (1.229), we have (1.245). Since

$$\lambda + \frac{\tau}{\varphi^*}\frac{d^2}{dx^2}(\varphi^*/\tau) = p - 2\frac{d^2}{dx^2}\log\tau = q$$

holds by (1.243)–(1.244), we obtain

$$\frac{d^2 g}{dx^2} = (q-\lambda)g$$

and the proof is complete. □

We conclude this subsection with a theorem on the rearrangement of eigenvalues and the principle of superposition. Given

$$P = (p, h, H) \in C^1[0,1] \times \mathbf{R} \times \mathbf{R} \equiv X, \quad n_1 \in \mathbf{N},$$

put $\Phi_n = \Phi(\cdot; P, \lambda_n(P))$ and

$$W(f, g) = f'g - fg', \quad f, g \in C^1[0,1].$$

Theorem 1.45 *If $Q = (q, j, J) \in X$ satisfies*

$$\lambda_n(P) = \lambda_{m(n)}(Q), \ \Phi_n(1; P) = \Psi_{m(n)}(1; Q), \quad n \neq n_1,$$

there arises (1.246) with $\tau = \tau(x) > 0$ defined by (1.228).

In the first case of $\tau(x)$, the eigenvalues and the eigenfunctions of A_Q are given by

$$\lambda_n(Q) = \lambda_n(P) \ \mathit{for}\ n \neq n_1, \qquad \lambda_{n_1}(Q) = \lambda^* \tag{1.247}$$

and

$$\Psi_n = \begin{cases} \Phi_n + \frac{\lambda-\lambda^*}{\lambda-\lambda_n} \cdot \frac{\varphi^*}{\tau} W(\varphi, \Phi_n), & n \neq n_1 \\ \phi/\tau, & n = n_1, \end{cases} \tag{1.248}$$

where

$$\Psi_n = \Phi(\cdot; Q, \lambda_n(Q)), \quad \varphi = \Phi_{n_1}(\cdot; P), \quad \varphi^* = \Phi_{n_1^*}(\cdot, P^*).$$

Here,

$$P^* = (p, H^*, J^*), \quad n_1 \in \mathbf{N}$$

are defined by (1.234) and (1.236) in the proof of the previous theorem. It holds that

$$m(n) = n, \ n \neq n_1,$$

and hence λ^* must be in

$$\lambda_{n_1-1}(P) < \lambda^* < \lambda_{n_1+1}(P)$$

in this case of $\tau(x)$.

In the second case of $\tau(x)$, on the other hand, the eigenvalues and the eigenfunctions of A_Q are given by

$$\lambda_n(Q) = \lambda_n(P), \quad n = 0, 1, 2, \ldots, \tag{1.249}$$

and

$$\Psi_n = \begin{cases} \Phi_n + \frac{1}{\lambda - \lambda_n} \cdot \frac{\varphi}{\tau} W(\varphi, \Phi_n), & n \neq n_1 \\ \varphi/\tau, & n = n_1, \end{cases} \tag{1.250}$$

respectively.

Proof The first case of $\tau(x)$ arises with $\beta \neq 0$ in the previous proof. The function

$$K(x, y) = g(x)\varphi(y) = (\lambda^* - \lambda)\varphi^*(x)\varphi(y)/\tau(x)$$

satisfies

$$\begin{aligned} &K_{xx} - K_{yy} + p(y)K = q(x)K && \text{in } D \\ &K(x, x) = (j - h) + \frac{1}{2}\int_0^x q(s) - p(s)\, ds, && 0 \leq x \leq 1 \\ &K_y(x, 0) = hK(x, 0), && 0 \leq x \leq 1, \end{aligned}$$

which guarantees

$$\Phi(x; q, j; \lambda) = \Phi(x; p, h; \lambda) + (\lambda^* - \lambda)\frac{\varphi^*(x)}{\tau(x)}\int_0^x \varphi(y)\Phi(y; p, h; \lambda)\, dy$$

For

$$\Phi_n = \Phi(\cdot; p, h; \lambda_n(P)), \quad \lambda_n = \lambda_n(P)$$

we have

$$-(\lambda - \lambda_n)\int_0^x \varphi(y)\Phi_n(y)\, dy = [W(\varphi, \Phi_n)]_{y=0}^{y=x} = W(\varphi, \Phi_n)(x)$$

and hence

$$\Psi_n = \Phi_n + \frac{\lambda - \lambda^*}{\lambda - \lambda_n} \cdot \frac{\varphi^*}{\tau} W(\varphi, \Phi_n), \quad n \neq n_1$$

satisfies

$$
\begin{aligned}
&(-\frac{d^2}{dx^2} + q(x))\Phi_n = \lambda_n \Phi_n \\
&\Psi_n(0) = 1, \quad \Psi_n'(0) = j, \quad \Psi_n'(0) + J\Psi_n(0) = 0.
\end{aligned}
$$

Here, $f = f(x)$ in the proof of Theorem 1.44 satisfies (1.239), which implies

$$
f = (\lambda^* - \lambda)\varphi/\tau.
$$

Using (1.237), we obtain

$$
(-\frac{d^2}{dx^2} + q(x))\Psi = \lambda^* \Psi, \quad \Psi = \varphi/\tau \not\equiv 0.
$$

Then (1.231) implies

$$
\begin{aligned}
\Psi'(0)/\Psi(0) &= \frac{d}{dx}\log\Psi\Big|_{x=0} = \frac{d}{dx}(\log\varphi - \log\tau)\Big|_{x=0} = j \\
-\Psi'(1)/\Psi(1) &= -\frac{d}{dx}\log\Psi\Big|_{x=1} = \frac{d}{dx}(-\log\varphi + \log\tau)\Big|_{x=1} = J. \quad (1.251)
\end{aligned}
$$

We reach (1.247) because $\Psi = \varphi/\tau$ has exactly n_1 zeros on $[0, 1]$, and hence

$$
\Psi = \text{constant} \times \varphi_{n_1}(\cdot; Q).
$$

Thus we obtain $m(n) = n$ for $n \neq n_1$, which requires

$$
\lambda_{n_1 - 1}(P) < \lambda^* < \lambda_{n_1+1}(P).
$$

In the case of $\beta = 0$, the function

$$
K(x, y) = g(x)\varphi(y) = -\varphi(x)\varphi(y)/\tau(x)
$$

forms the kernel of the first deformation formula, similarly, which guarantees

$$
\Phi(x; q, j; \lambda) = \Phi(x; p, h; \lambda) - \frac{\varphi(x)}{\tau(x)} \int_0^x \varphi(y)\Phi(y; p, h; \lambda)\, dy
$$

for any $\lambda \in \mathbf{R}$. Hence

$$
\Psi_n = \Phi_n + \frac{1}{\lambda - \lambda_n} \cdot \frac{\varphi}{\tau} W(\varphi, \Phi_n), \quad n \neq n_1
$$

satisfies

$$(-\frac{d^2}{dx^2} + q(x))\Psi_n = \lambda_n \Psi_n$$
$$\Psi_n(0) = 1, \quad \Psi_n'(0) = j, \quad \Psi_n'(1) + J\Psi_n'(1) = 0.$$

To assure

$$(-\frac{d^2}{dx^2} + q(x))\Psi = \lambda\Psi, \quad \Psi = \frac{\varphi}{\tau} \not\equiv 0, \quad \lambda = \lambda_{n_1}$$

we note

$$\lambda + \frac{\tau}{\varphi}\frac{d^2}{dx^2}(\frac{\varphi}{\tau}) = \lambda + \frac{\tau}{\varphi}\{\varphi''/\tau - 2\varphi'\tau'/\tau^2 - \varphi(\tau''/\tau^2 - 2(\tau')^2/\tau^3)\}$$
$$= \lambda + \frac{\varphi''}{\varphi} - 2\varphi'\tau'/\varphi\tau - (\tau''/\tau - 2(\tau')^2/\tau^2).$$

Using

$$\lambda + \frac{\varphi''}{\varphi} = p, \quad -2\varphi'\tau'/\varphi\tau = -2\varphi'\varphi^2/\varphi\tau = -2\varphi'\varphi/\tau = -\tau''/\tau$$

derived from $\beta = 0$, we obtain

$$\lambda + \frac{\tau}{\varphi}\frac{d^2}{dx^2}(\frac{\varphi}{\tau}) = p - 2(\tau''/\tau - (\tau')^2/\tau^2)$$
$$= p - 2\frac{d^2}{dx^2}\log\tau = q.$$

We have, finally,

$$\Psi'(0)/\Psi(0) = j, \quad -\Psi'(1)/\Psi(1) = J$$

as in (1.251), and hence (1.249) in this case. The proof is complete. □

1.7.3 Dressing Method

Given

$$P = (p, h, H) \in X = C^1[0, 1] \times \mathbf{R} \times \mathbf{R}$$

and a finite set $\Sigma \subset \mathbf{N}$, we can classify $Q = (q, j, J) \in X$ satisfying

$$\lambda_n(P) = \lambda_{m(n)}(Q), \ \Phi_n(1; P) = \Phi_{m(n)}(1; Q), \quad \forall n \in \mathbf{N} \setminus \Sigma, \ \exists m(n) \in \mathbf{N} \tag{1.252}$$

by the superposition of Theorem 1.45. It holds that

$$m(n) = n, \quad n \in \mathbf{N}$$

and there is $0 < \tau = \tau(x) \in C^3[0, 1]$ such that (1.246) holds. Here we show that this τ is represented by a Fredholm determinant of an integral operator derived from the method of dressing.

Continue to use $\Phi = \Phi(\cdot; p, h; \lambda)$ for the solution to

$$(-\frac{d^2}{dx^2} + q(x))\Phi = \lambda\Phi \text{ in } 0 < x < 1, \quad \Phi|_{x=0} = 1, \quad \Phi_x|_{x=0} = h,$$

while $\Phi^* = \Phi^*(\cdot; p, H, ; \lambda)$ denotes the solution to

$$(-\frac{d^2}{dx^2} + q(x))\Phi^* = \lambda^*\Phi^* \text{ in } 0 < x < 1, \quad \Phi^*\big|_{x=1} = 1, \quad \Phi^*_x\big|_{x=1} = -H.$$

We obtain

$$\Phi^*(x; q, J; \lambda) = \Phi(x; p, H; \lambda) + \int_x^1 K^*(x, y; q, J; p, H)\Phi^*(y; p, H; \lambda)\, dy$$

similarly, where $K = K^*(x, y) = K^*(x, y; q, J; p, H) \in C^2(\overline{D^*})$ is the solution to

$$\begin{aligned} &K^*_{xx} - K^*_{yy} + p(y)K^* = q(x)K^* && \text{in } D^* \\ &K^*(x, x) = (J - H) + \frac{1}{2}\int_x^1 q(s) - p(s)\, ds, && 0 \le x \le 1 \\ &-K^*_y(x, 1) = HK^*(x, 1), && 0 \le x \le 1 \end{aligned}$$

for $D^* = \{(x, y) \mid 0 < x < y < 1\})$.

These kernels K and K^* induce the integral operators on $[0, 1]$,

$$(Kf)(x) = \int_0^x K(x, y)f(y)\, dy, \quad (K^*f)(x) = \int_x^1 K^*(x, y)f(y)\, dy.$$

We define the other operator F by $1 + F = (1 + K)^{-1}(1 + K^*)$. It is an integral operator on $[0, 1]$ with the kernel $F = F(x, y)$, C^2 on each $\overline{D}$ and $\overline{D}^*$, respectively, and satisfies

$$\chi_{[0,x]}(y)K(x,y) + F(x,y) + \int_0^x K(x,z)F(z,y)\,dz = \chi_{[x,1]}(y)K^*(x,y), \tag{1.253}$$

using the indicator function χ_I defined by $\chi_I(y) = 1$ if $y \in I$ and $\chi_I(y) = 0$ if $y \notin I$.

Then we put $(1+K)^{-1} = 1 + L$, where L is provided with the kernel $L = L(x,y)$:

$$L(f)(x) = \int_0^x L(x,y)f(y)dy.$$

It satisfies

$$\begin{aligned}
&L_{xx} - L_{yy} + q(y)L = p(x)L && \text{in } D\\
&L(x,x) = (h-j) + \frac{1}{2}\int_0^x p(s) - q(s)\,ds, && 0 \le x \le 1\\
&L_y(x,0) = jL(x,0), && 0 \le x \le 1.
\end{aligned}$$

Then we obtain $F = L + K^* + LK^*$ and hence

$$F(x,y) = \chi_{[0,x]}(y)L(x,y) + \chi_{[x,1]}(y)K^*(x,y) + \int_0^{x\wedge y} L(x,z)K^*(z,y)\,dz. \tag{1.254}$$

Let

$$\Lambda_p = -\frac{\partial^2}{\partial x^2} + \frac{\partial^2}{\partial y^2} + p(x) - p(y).$$

Lemma 1.20 *The kernel* $F = F(x,y) \in C^2(\overline{D}) \cup C^2(\overline{D}^*)$ *satisfies*

$$\begin{aligned}
&\Lambda_p F = -L(x,0)(K_x^*(0,y) - jK^*(0,y)) && \text{in } D \bigcup D^*\\
&F_y(x,0) = (j + J - H + \frac{1}{2}\int_0^1 q(s) - p(s)\,ds)F(x,0), && 0 \le x \le 1\\
&F(x+0,x) - F(x-0,x) &&\\
&\quad = j + J - h - H + \frac{1}{2}\int_0^1 q(s) - p(s)\,ds, && 0 \le x \le 1.
\end{aligned} \tag{1.255}$$

Proof By (1.254) we have

$$F(x,y) = L(x,y) + \int_0^y L(x,z)K^*(z,y)\,dz \quad \text{in } D, \tag{1.256}$$

and hence $F(x,0) = L(x,0)$ for $0 \le x \le 1$. Furthermore, we have

$$F_y(x,y) = L_y(x,y) + L(x,y)K^*(y,y) + \int_0^y L(x,z)K_y^*(z,y)\,dz \quad \text{in } D,$$

which implies

$$\begin{aligned}&F_y(x,0) + L_y(x,0) + L(x,0)K^*(0,0)\\&= (j + J - H + \frac{1}{2}\int_0^1 q(s) - p(s)\,ds)L(x,0), \quad 0 \le x \le 1.\end{aligned}$$

The second equality of (1.255) thus follows.

By (1.254), next, we have

$$\begin{aligned}F(x-0,x) &= L(x,x) + \int_0^x L(x,z)K^*(z,y)\,dz\\F(x+0,x) &= K^*(x,x) + \int_0^x L(x,z)K^*(z,y)\,dz, \quad 0 \le x \le 1,\end{aligned}$$

and therefore,

$$\begin{aligned}&F(x+0,x) - F(x-0,x) = K^*(x,x) - L(x,x)\\&= (J - H + \frac{1}{2}\int_x^1 q(s) - p(s)\,ds) - (h - j + \frac{1}{2}\int_0^x p(s) - q(s)\,ds),\end{aligned}$$

which implies the third equality of (1.255).

To verify the first equality in D, we use (1.256) to deduce

$$\begin{aligned}\Lambda_p F(x,y) &= -L_{xx}(x,y) + L_{yy}(x,y) + (p(x) - p(y))L(x,y)\\&\quad - \int_0^y L_{xx}(x,z)K^*(z,y)\,dz\\&\quad + L_y(x,y)K^*(y,y) + L(x,x)\frac{d}{dy}K^*(y,y) + L(x,y)K_y^*(y,y)\\&\quad + \int_0^y L(x,z)K_{yy}^*(z,y)\,dz + (p(x) - p(y))\int_0^y L(x,z)K^*(z,y)\,dz\\&= (q(y) - p(x))L(x,y) + (p(x) - p(y))L(x,y) + L_y(x,y)K^*(y,y)\\&\quad + L(y,y)\frac{d}{dy}K^*(y,y) + L(x,y)K_y^*(y,y) + (p(x) - p(y))\\&\qquad \int_0^y L(x,z)K^*(z,y)\,dz\end{aligned}$$

$$
\begin{aligned}
&+\int_0^y (q(z)-p(x))L(x,z)K^*(z,y)\,dz\\
&+\int_0^y (p(y)-q(z))L(x,z)K^*(z,y)\,dz\\
&-\int_0^y (L_{zz}(x,z)K^*(z,y)-L(x,z)K^*_{zz}(z,y))\,dz\\
&=(q(y)-p(y))L(x,y)+L_y(x,y)K^*(y,y)+L(y,y)\frac{d}{dy}K^*(y,y)\\
&+L(x,y)K^*_y(x,y)-\left[L_z(x,z)K^*(z,y)-L(x,z)K^*_z(z,y)\right]_{z=0}^{z=y}\\
&=(q(y)-p(y))L(x,y)+L(y,y)2\frac{d}{dy}K^*(y,y)+L_y(x,0)K^*(0,y)\\
&-L(x,0)K^*_x(0,y)\\
&=-L(x,0)(K^*_x(0,y)-jK^*(0,y)),
\end{aligned}
$$

which means the first equality of (1.255) in D.

Finally, (1.254) in D^* reads

$$
F(x,y)=K^*(x,y)+\int_0^x L(x,z)K^*(z,y)\,dz \tag{1.257}
$$

and therefore, we obtain

$$
\begin{aligned}
\Lambda_p F=&-K^*_{xx}(x,y)+K^*_{yy}(x,y)+(p(x)-p(y))K^*(x,y)-(\frac{d}{dx}L(x,x))K^*(x,y)\\
&-L(x,x)K^*_x(x,y)-L_x(x,x)K^*(x,y)-\int_0^x L_{xx}(x,z)K^*(z,y)\,dz\\
&+\int_0^x L(x,z)K^*_{yy}(z,y)\,dz+(p(x)-p(y))\int_0^x L(x,z)K^*(z,y)\,dz\\
=&(p(x)-p(y))K^*(x,y)+(p(x)-p(y))K^*(x,y)-(\frac{d}{dx}L(x,x))K^*(x,y)\\
&-L(x,x)K^*_x(x,y)-L_x(x,x)K^*(x,y)\\
&+\int_0^x (q(z)-p(x))L(x,z)K^*(z,y)\,dz\\
&+\int_0^x (p(y)-q(z))L(x,z)K^*(z,y)\,dz+(p(x)-p(y))\\
&\int_0^x L(x,z)K^*(z,y)\,dz
\end{aligned}
$$

$$
\begin{aligned}
&- \int_0^x L_{zz}(x,z)K^*(z,y) - L(x,z)K^*_{zz}(z,y)\,dz \\
&= (p(x)-q(x))K^*(x,y) - (\frac{d}{dx}K(x,x))K^*(x,y) - L(x,x)K^*_x(x,y) \\
&\quad - L_x(x,x)K^*(x,y) - \left[L_z(x,z)K^*(z,y) - L(x,z)K^*_z(z,y)\right]_{z=0}^{z=x} \\
&= (p(x)-q(x))K^*(x,x) - 2(\frac{d}{dx}L(x,x))K^*(x,y) + L_y(x,0)K^*(0,y) \\
&\quad - L(x,0)K^*_x(0,y) = -L(x,0)(K^*_x(0,y) - jK^*(0,y)).
\end{aligned}
$$

Hence we obtain the first equality of (1.255) in D^*, and the proof is complete. □

Equality (1.253) in D means

$$
K(x,y) + F(x,y) + \int_0^x K(x,z)F(z,y)\,dz = 0, \quad 0 < y < x < 1 \tag{1.258}
$$

called the Gel'fand–Levitan–Marchenko equation. Given $F = F(x,y)$, this (1.258) is resolved by the Fredholm theory. We assume

$$
j + J + \frac{1}{2}\int_0^1 q(s)\,ds = h + H + \frac{1}{2}\int_0^1 p(s)\,ds \tag{1.259}
$$

which makes $F = F(x,y)$ continuous on $[0,1]\times[0,1]$ by (1.254). Let F_x be the integral operator on $[0,x]$ with the kernel $F = F(z,y)$:

$$
(F_x f)(y) = \int_0^x f(z)F(z,y)\,dz.
$$

Lemma 1.21 *Under the assumption of (1.259), the integral equation (1.258) is uniquely solvable with respect to $K = K(x,y)$. There arises the trace formula*

$$
K(x,x) = -\frac{d}{dx}\log\det(I + F_x). \tag{1.260}
$$

Proof We introduce a parameter $\lambda \in \mathbf{C}$ to define the operator F^λ by

$$
I + F^\lambda = (1+\lambda K)^{-1}(I + \lambda K^*).
$$

Under (1.259), the kernel $F^\lambda = F^\lambda(x,y)$ of this operator is also continuous on $[0,1]\times[0,1]$. Then we obtain

$$
\|F^\lambda\| < 1, \quad |\lambda| \ll 1,
$$

which ensures $\det(I + F^\lambda_x) \neq 0$.

It holds that

$$\log\det(I+F_x^\lambda) = \mathrm{tr}\,(\log(I+F_x^\lambda)) = \sum_{n=1}^{\infty}\frac{(-1)^n}{n}\mathrm{tr}\,(F_x^\lambda)^n$$

$$= \sum_{n=1}^{\infty}\frac{(-1)^n}{n}\int_0^x dt_1\ldots\int_0^x dt_n F^\lambda(x,t_1)F^\lambda(t_1,t_2)\ldots F^\lambda(t_n,t_1),$$

and hence

$$\frac{d}{dx}\log\det(I+F_x^\lambda)$$
$$=\sum_{n=1}^{\infty}\frac{(-1)^n}{n}\int_0^x dt_1\ldots\int_0^x dt_n F^\lambda(x,t_1)F^\lambda(t_1,t_2)\ldots F^\lambda(t_{n-1},x). \quad (1.261)$$

If $|\lambda|\ll 1$, the integral equation of $K^\lambda = K^\lambda(x,y)$,

$$K^\lambda(x,y)+F^\lambda(x,y)+\int_0^x K^\lambda(x,z)F^\lambda(z,y)\,dz = 0, \quad 0<y<x<1, \quad (1.262)$$

is uniquely solvable by the Neumann series. With this K^λ, the right-hand side of (1.261) is equal to

$$-K^\lambda(x,x).$$

Since $K^\lambda = \lambda K$ is a solution to (1.262), there arises

$$\det(I+F_x^\lambda) = \exp(-\lambda\int_0^x K(y,y)\,dy).$$

Since both sides of the above equality are holomorphic in $\lambda\in\mathbf{C}$, we obtain

$$\det(I+F_x) = \exp(-\lambda\int_0^x K(y,y)\,dy)\neq 0.$$

Equation (1.258) is, in particular, uniquely solvable, and it follows that (1.260). □

Given $P\in X$ and finite set $\Sigma\subset\mathbf{N}$, we have the following theorem [10].

Theorem 1.46 (Iwasaki) *The τ-function to (1.252) takes the form*

$$\tau(x) = \det\left(\delta_{\ell\ell'}+\beta_\ell\int_0^x\sum_k e_k(y)\varphi_{n_{\ell'}}(y)\,dy\right)_{1\le\ell,\ell'\le N} \quad (1.263)$$

where $\beta_\ell \in \mathbf{R}$, $1 \le \ell \le N$, $e_k = e_k(x)$, $1 \le k \le N$, *is a solution to*

$$(-\frac{d^2}{dx^2} + p(x))e_k = \mu_k e_k, \quad \mu_k \in \mathbf{R},$$

and $\phi_{n_\ell} = \Phi(\cdot, P, \lambda_{n_\ell}(P))$, $1 \le \ell \le N$.

A technical proposition is necessary for this purpose.

Proposition 1.9 *Let* $J \subset [0, 1]$ *be an open interval and* $\Sigma \subset \mathbf{N}$ *be finite. Then* $\{\varphi_n \mid n \in \mathbf{N} \setminus \Sigma\}$ *is complete in* $L^2(J)$, *which means that the equalities*

$$\int_J f(x)\varphi_n(x)\, dx = 0, \quad n \in \mathbf{N} \setminus \Sigma \tag{1.264}$$

imply $f = 0$, *where* $f \in L^2(J)$.

Proof Using a zero-extension of $f \in L^2(J)$ defined by

$$\tilde{f} = \begin{cases} f \text{ in } J \\ 0 \text{ in } (0, 1) \setminus J, \end{cases}$$

we obtain

$$\int_0^1 \tilde{f}\varphi_n\, dx = 0, \quad n \in \mathbf{N} \setminus \Sigma,$$

and hence

$$\tilde{f}(x) = \sum_{\ell=1}^{N} c_\ell \varphi_{n_\ell}(x), \quad x \in (0, 1),$$

where $c_\ell \in \mathbf{R}$ and $\Sigma = \{n_\ell \mid 1 \le \ell \le N\}$. We operate

$$(-\frac{d^2}{dx^2} + p(x))^s, \quad s = 0, 1, 2, \ldots,$$

on both sides of this equality to obtain

$$\sum_{\ell=1}^{N} \lambda_{n_\ell}^s c_\ell \varphi_{n_\ell}(x) = 0, \quad x \in (0, 1) \setminus J$$

Since $\lambda_n \neq \lambda_m$ for $n \neq m$ and $\varphi_{n_\ell} \not\equiv 0$ in $(0, 1) \setminus J$, assured by the unique continuation theorem, this equality implies $c_\ell = 0$ for $1 \le \ell \le N$, and hence $f = \tilde{f}\big|_J = 0$. □

Proof *(Theorem 1.46)* We use the argument in Sect. 1.3.5. First, the assumption (1.252) implies

$$J = H - K(1,1)$$

for $K(x,y) = K(x,y;q,j;p,h)$, which means (1.259). Hence we obtain

$$K(x,x) = -\frac{d}{dx}\log\det(1+F_x)$$

by Lemma 1.21, and it holds that $\tau(x) = \det(1+F_x)$.

Second, equality

$$K^*(x,y) = -g(x)\cdot\varphi(y), \quad \varphi = {}^T(\varphi_{n_1},\cdots,\varphi_{n_N}), \quad g = {}^T(g_1,\cdots,g_N) \tag{1.265}$$

holds with $g_\ell = g_\ell(x)$, $1 \leq \ell \leq N$, satisfying

$$(-\frac{d^2}{dx^2} + q(x))g_\ell = \nu_\ell g_\ell, \quad 0 \leq x \leq 1$$

for some $\nu_\ell \in \mathbf{R}$.

Third, (1.252) is equivalent the existence of m_ℓ, $1 \leq \ell \leq N$, and $n(m)$, $m \neq m_\ell$, $1 \leq \ell \leq N$, such that

$$\lambda_m(Q) = \lambda_{n(m)}(P), \ \Phi_m(1;Q) = \Phi_{n(m)}(1;P), \quad m \in \mathbf{N}\setminus\{m_\ell \mid 1 \leq \ell \leq N\}.$$

This property implies

$$L(x,y) = e(x)\cdot\psi(y), \quad \psi = {}^T(\psi_{m_1},\ldots,\psi_{m_N}), \quad e = {}^T(e_1,\ldots,e_N) \tag{1.266}$$

for

$$\psi_m = \Phi(\cdot, Q, \lambda_m(Q)),$$

where $e = e_\ell(x)$, $1 \leq \ell \leq N$, is a solution to

$$(-\frac{d^2}{dx^2} + p(x))e_\ell = \beta_\ell e_\ell, \quad 0 \leq x \leq 1$$

for some $\beta_\ell \in \mathbf{R}$.

Using (1.256) valid in D, we obtain

$$F(x,y) = F_1(x,y) \equiv \sum_{\ell=1}^N e_\ell(x)u_\ell(y), \quad 0 < y < x < 1$$

with $u_\ell \in C^2[0, 1]$. We have, similarly, (1.257) in D^*, and hence

$$F(x, y) = F_2(x, y) \equiv \sum_{\ell=1}^{N} v_\ell(x)\varphi_{n_\ell}(y), \quad 0 < x < y < 1 \tag{1.267}$$

by (1.257) and (1.265), where $v_\ell \in C^2[0, 1]$.

From the first equality of (1.255), we have

$$\begin{aligned} \Lambda_p F_2 &= \sum_{\ell=1}^{N}(-\frac{d^2}{dx^2} + p(x) - \lambda_{n_\ell})v_\ell(x)\varphi_{n_\ell}(y) \\ &= -L(x, 0)(K_x^*(0, y) - jK^*(0, y)) \quad x < y < 1 \end{aligned}$$

recalling $\lambda_{n_\ell} = \lambda_{n_\ell}(P)$. This relation extends to D by (1.265) because

$$\{\varphi_{n_\ell}(y) \mid 1 \le \ell \le N\}$$

is linearly independent on $[x, 1]$ from the proof of Proposition 1.9, which ensures

$$\Lambda_p F_2 = \Lambda_p F_1 \quad \text{in } D.$$

It holds that

$$F_{2y}(x, 0) = hF_2(x, 0), \quad 0 \le x \le 1$$

by $\varphi_{n_1} = \Phi(\cdot; P, \lambda_n(P))$, while $F_{1y}(x, 0) = hF_1(x, 0)$, $0 \le x \le 1$, is examined by the second equality of (1.255) and (1.259). Finally, we have

$$F_1(x, x) = F_2(x, x), \quad 0 \le x \le 1$$

by the third equality of (1.255) and (1.259), which guarantees $F_1 \equiv F_2$ in D. Hence it holds that

$$\sum_{\ell=1}^{N} v_\ell(x)\varphi_{n_1}(y) = \sum_{\ell=1}^{N} e_\ell(x)u_\ell(y), \quad 0 < x < y.$$

By the proof of Proposition 1.9 again, we see that $u_\ell(y)$ is differentiable on $[0, x]$ arbitrarily many times, and it holds that

$$v_\ell(x) = \sum_k \beta_{\ell k} e_k(x), \quad 1 \le \ell \le N$$

with $\beta_{\ell k} \in \mathbf{R}$, and hence

$$F(x,y) = \sum \beta_{k\ell} e_k(x) \varphi_{n_\ell}(y).$$

Replacing e_ℓ by its constant times, we put $\beta_{k\ell} = \beta_\ell$ to be independent of k, to reach

$$F(x,y) = \sum_{k,\ell=1}^{N} \beta_\ell e_\ell(x) \varphi_{n_\ell}(y).$$

Then we have

$$\tau(x) = \det(1 + F_x) = \det \left(\delta_{\ell\ell'} + \beta_\ell \int_0^x \sum_k e_k(y) \varphi_{n_{\ell'}}(y) dy \right)_{1 \le \ell, \ell' \le N}$$

for this degenerate kernel. □

References

1. V. Ambarzumian, Über eine Frage der Eigenwert theories. Zeitschr. Phys. **53**, 690–695 (1929)
2. S. Angenent, The Morse-Smale property for a semi-linear parabolic equation. J. Differ. Equations **62**, 427–442 (1986)
3. S. Angenent, The zero set of a solution of a parabolic equation. J. Reine Angew. Math. **390**, 79–68 (1988)
4. A. Bamberger, G. Chavent, P. Lailly, About the stability of the inverse problems in $1 - D$ wave equation – an application of seismographic profiles. Appl. Math. Optim. **5**, 1–47 (1979)
5. G. Borg, Eine Umkehrung der Sturm-Liouvilleschen Eigenwertaufgabe. Acta Math. **78**, 1–46 (1946)
6. V.A. Galaktionov, *Geometric Sturmian Theory of Nonlinear Parabolic Equations and Applications* (Chapman and Hall/CRC, Boca Raton, 2004)
7. I.M. Gel'fand, B.M. Levitan, On the determination of a differential equation from its spectral function. Amer. Math. Soc. Transl. (2) **1**, 253–304 (1955)
8. H. Hochstadt, The inverse Sturm-Liouville problem. Commun. Pure Appl. Math. **26**, 715–729 (1973)
9. H. Hochstadt, B. Lieberman, An inverse Sturm-Liouville problem with mixed given data. SIAM J. Appl. Math. **34**, 676–680 (1978)
10. K. Iwasaki, On the inverse Sturm-Liouville problem with spatial symmetry. Funkc. Ekvac. **31**, 25–74 (1988)
11. B.F. Jones Jr., The determination of a coefficient in a parabolic differential equation, Part I, existence and uniqueness. J. Math. Mech. **11**, 907–918 (1962)
12. M. Kac, Can one hear the shape of a drum? Amer. Math. Monthly **73**, 1–23 (1966)
13. M.M. Lavrentiev, M.M. Reznickaja, Soviet Math. Dokl. **20**, 1416–1418 (1979)
14. H. Matano, Nonincrease of the lap-number of a solution for a one-dimensional semilinear parabolic equation. J. Fac. Sci. Univ. Tokyo, Sect. IA Math. **29**, 401–441 (1982)
15. A. Pierce, Unique identification of eigenvaules and coefficients in a parabolic problem. SIAM J. Control Optim. **17**, 494–499 (1979)
16. Ch. Pöppe, D. Sattinger, Fredholm determinants and the τ-function for the Kadomtsev, Petviashivili hierarchy. Publ. RIMS Kyoto Univ. **24**, 505–538 (1988)

17. J. Pöschel, E. Trubowitz, *Inverse Spectral Theory* (Academic Press, New York, 1987)
18. B. Simon, Notes on infinite determinants of Hilbert space operators. Adv. Math. **24**, 244–273 (1977)
19. T. Suzuki, Uniqueness and nonuniqueness in an inverse problem for parabolic equations. J. Differ. Equations **47**, 296–316 (1983)
20. T. Suzuki, Gel'fand-Levitan's theory, deformation formulas, and inverse problems. J. Fac. Sci. Univ. Tokyo Sec. IA **32**, 223–271 (1985)
21. T. Suzuki, On the inverse Sturm-Liouville problem for spatially symmetric operators, I. J. Differ. Equations **56**, 165–194 (1985); III, ibid, **58**, 257–281 (1985)
22. T. Suzuki, Inverse problems for heat equations on comact intervals and on circles, I. J. Math. Soc. Jpn. **38**, 39–65 (1986)
23. T. Suzuki, Stability of an inverse hyperbolic problem. Inverse Probl. **4**, 273–299 (1988)

Chapter 2
Surface Theory

Abstract The Liouville integral is a parametric representation of s surface with constant Gaussian curvature. We start with fundamental properties of surfaces, and then derive this Liouville integral via complex variables. The representation formula, surprisingly, induces a quantized blowup mechanism of the Boltzmann–Poisson (BP) equation because of its geometric feature of nonlinearity, associated with a flat plane and a sphere. The complex analysis used, furthermore, leads to a new insight into the statistical mechanics, that is, recursive hierarchy in the context of Onsager's theory on many point vortices, as the location of blowup points of the family of solutions to the BP equation is controlled by the Hamiltonian concerning the system of point vortices. We thus reach a fusion of the theories of surface, complex variables, elliptic equations, and statistical mechanics through BP, or Gel'fand equations in two-space dimensions. These phenomena encountered in the interaction of analysis, geometry, and physics, spread into a wide area in the theory of nonlinear partial differential equations, as is described in later chapters. In this chapter, after a summary in the first section, we confirm the classical surface theory in the second section. The Liouville integral is induced in the third section, which leads to a motivation of the study of BP equation in the final section, that is, the quantized blowup mechanism and recursive hierarchy.

2.1 Introduction

2.1.1 Representation of Surfaces

There is a category of surfaces provided with the first integral, called *Liouville's integral*. It is usually presented in the context of complex geometry. Although this chapter is concerned with surfaces immersed in $\mathbf{R}^3$, the complex variable takes a role in their mathematical analysis.

More precisely, a surface with positive constant Gaussian curvature is locally represented by the solution to the Gel'fand equation

T. Suzuki, *Liouville's Theory in Linear and Nonlinear Partial Differential Equations*, Springer Asia Pacific Mathematics Series 8,
https://doi.org/10.1007/978-981-95-1975-0_2

$$-\Delta u = \lambda e^{u} \quad \text{in } \Omega, \tag{2.1}$$

where $\Omega \subset \mathbf{R}^2$ is a bounded domain and $\lambda > 0$ is a constant. There is a Liouville integral for (2.1), which induces the meromorphic function $F = F(z)$ on Ω satisfying

$$\rho(F) \equiv \frac{|F'|}{1+|F|^2} = \left(\frac{\lambda}{8}\right)^{1/2} e^{u/2}. \tag{2.2}$$

Let S^2 be the sphere with radius 1, and regard the above F as a conformal mapping $\overline{F} : \Omega \to S^2$ through the inverse stereographic transformation. The above $\rho(F)$ in (2.2), called the spherical derivative, ensures the connection of standard line elements on S^2 and Ω, denoted by $d\sigma$ and ds, respectively, as

$$\frac{d\sigma}{ds} = \rho(F)$$

under $\overline{F}$. Then the immersed volume and length of Ω and $\partial\Omega$ in S^2 under $\overline{F}$ are given by

$$\overline{F}(\Omega) = \frac{1}{8}\int_{\Omega} \lambda e^{u}\, dx, \quad |\overline{F}(\partial\Omega)| = \left(\frac{\lambda}{8}\right)^{1/2} |\partial\Omega|, \tag{2.3}$$

respectively.

2.1.2 Quantization and Hamiltonian Control

Elliptic problems are usually accompanied with boundary conditions. If we impose

$$u|_{\partial\Omega} = 0 \tag{2.4}$$

on (2.1), this problem is equivalent to realizing conformal immersion of Ω into S^2 satisfying

$$\left.\frac{d\sigma}{ds}\right|_{\partial\Omega} = \left(\frac{\lambda}{8}\right)^{1/2}, \tag{2.5}$$

where $d\sigma$ and ds denote the standard metrics on S^2 and Ω, respectively, associated with this immersion.

Problem (2.1) with (2.4), on the other hand,

$$-\Delta u = \lambda e^{u} \text{ in } \Omega, \quad u|_{\partial\Omega} = 0, \tag{2.6}$$

is regarded as a nonlinear eigenvalue problem in the context of elliptic theory, to find u and λ, simultaneously. Then we are concerned with the family of solutions $(\lambda, u) = (\lambda_k, u_k)$, $k = 1, 2, \cdots$, to (2.6) satisfying

$$\lambda_k \to 0. \tag{2.7}$$

Passing to a subsequence, we then obtain

$$\Sigma_k = \int_\Omega \lambda_k e^{u_k} \to 8\pi m \tag{2.8}$$

for some $m = 0, 1, \cdots, +\infty$ [14]. If $m = 0$ and $m = +\infty$, it holds that $\|u_k\|_\infty \to 0$ and $u_k \to +\infty$ locally uniformly in Ω, respectively. If $0 < m < +\infty$, there arises

$$\|u_k\|_\infty \to +\infty.$$

It holds also that $\mathscr{S} \subset \Omega$ and $\sharp \mathscr{S} = m$, where

$$\mathscr{S} = \{x_0 \in \overline{\Omega} \mid \exists x_k \to x_0, \ u_k(x_k) \to +\infty\} \tag{2.9}$$

denotes the blowup set.

Regarding (2.3), we see that the case (2.7)–(2.8) corresponds to the formation of m-covering of conformally immersed domains of Ω into S^2 satisfying (2.5) for $\lambda = \lambda_k \downarrow 0$.

The proof of this theorem, Theorem 2.6 of this chapter, is executed by the theory of complex functions, that is, *residue analysis* in accordance with the Liouville integral. The convergence (2.8) comes from vanishing of poles of the first order of an analytic function around the blowup point, while that for the second order induces Theorem 2.7. Thus, if $0 < m < \infty$ and $\mathscr{S} = \{x_1^*, \cdots, x_m^*\}$, then $x_* = (x_1^*, \cdots, x_m^*)$ is a critical point of

$$H(x_1, \cdots, x_m) = \frac{1}{2}\sum_{j=1}^{m} R(x_j) + \sum_{1 \le i < j \le m} G(x_i, x_j), \tag{2.10}$$

where $G = G(x, x')$ is Green's function for $-\Delta$ under the Dirichlet boundary condition, and

$$R(x) = \left[G(x, x') + \frac{1}{2\pi} \log |x - x'| \right]_{x'=x}$$

is the Robin function.

This profile of the solution is interpreted as a *recursive hierarchy* in the context of statistical mechanics, that is, *Onsager's theory* on point vortices [17]. This terminology of recursive hierarchy is used in [20] because (2.6), or its equivalent form (2.92) below, arises as a mean field limit of many point vortices in canonical statistical mechanics, where the motion of vortices is subject to the Kirchhoff equation, that is, a Hamilton system derived from the *Hamiltonian* (2.10).

Theorem 2.5 in this chapter is an alternative form of Theorem 2.6 mentioned above. This theorem and Theorem 2.7 thus have been recognized as a recursive hierarchy in mathematical physics [20].

2.1.3 Uniqueness

The above geometric structure of (2.6) induces a question, parametrization of the solution (λ, u) to (2.6) by

$$\Sigma = \int_{\Omega} \lambda e^{u}. \tag{2.11}$$

More precisely, do the surfaces immersed in S^2 satisfying (2.5) increase their areas from 0 to π monotonically, to create one covering of S^2 eventually? If the answer is affirmative, one gets the connectivity of the trivial solution $(\lambda, u) = (0, 0)$ to the one-point blowup solution (λ_*, u_*), called the singular limit described by

$$\lambda_* = 0, \quad u_* = 8\pi G(\cdot, x_0), \; \nabla R(x_0) = 0,$$

in λu space. This property is actually the case if this singular limit is formed as the immersed area denoted by Σ_k converges to 8π below. The result follows from the uniqueness theorem [5, 19] that any $\Sigma \in (0, 8\pi)$ admits the solution (λ, u) to (2.1) with (2.4) satisfying (2.11).

Here we assume the simply-connectedness of Ω, to execute a symmetrization in the context of equi-measurable rearrangement induced by the isometric inequality on S^2. Then there arises an associated Legendre equation under the stereographic transformation. This structure actually comes from the separation of variables for the Laplace–Beltrami operator on S^2. Then we reach a uniqueness theorem via the spectral analysis, that is, Theorem 2.9, which is used to justify the above described mean field theory of many point vortices mathematically [6], that is, the *propagation of chaos*.

2.1.4 Organization of Chap. 2

The quantized blowup mechanism and recursive hierarchy have been a driving force to understand real principles hidden behind the nonlinearity. Among them are the quantized blowup mechanism in both finite and infinite time, and the Hamilton dynamics of collapses formed in infinite time of the *Smoluchowski–*

Poisson equation described in Chap. 5. Others are the higher-dimensional version in both elliptic and parabolic equations [20] and normalized Ricci flow in accordance with the Trudinger–Moser inequality [21]. This chapter thus offers the first step to access these remarkable events in theoretical sciences.

We begin with a confirmation of the theory of surfaces and curves in Sect. 2.2. We define scalar and mean curvatures for surfaces immersed in three-dimensional Euclidean space, and turn to the notions of differential forms, tangent and cotangent spaces, frames and covariant derivatives, and connections to reach the Theorem Egregium of Gauss. In Sect. 2.3 we describe the conformal geometry, Liouville integral for surfaces of positive constant Gaussian curvature, and its analytic profile. Section 2.4 is devoted to the proof of the above mentioned results on (2.6): the quantized blowup mechanism, recursive hierarchy, and parametrization of solution by the immersed area in S^2 of Ω under $\overline{F}$.

2.2 Riemann Surfaces

2.2.1 *Surfaces*

A surface is a set of points in $\mathbf{R}^3$ represented by two parameters,

$$\boldsymbol{x}(u, v) = \begin{pmatrix} x_1(u, v) \\ x_2(u, v) \\ x_3(u, v) \end{pmatrix}, \quad (u, v) \in \Omega.$$

This $\boldsymbol{x} = \boldsymbol{x}(u, v)$ is hence regarded as a mapping $\Omega \to \mathbf{R}^3$, where $\Omega \subset \mathbf{R}^2$ is the parameter region and $\boldsymbol{x}(\Omega) = \mathscr{M}$. Let the distance between two points on $\mathscr{M}$, $\boldsymbol{x}(u, v)$ and $\boldsymbol{x}(u + \Delta u, v + \Delta v)$ be Δs, the area of the parallelogram made by

$$\boldsymbol{a} = \boldsymbol{x}(u + \Delta u, v) - \boldsymbol{x}(u, v), \quad \boldsymbol{b} = \boldsymbol{x}(u, v + \Delta v) - \boldsymbol{x}(u, v)$$

be ΔS, and the normal unit vector on $\mathscr{M}$ at $\boldsymbol{x}(u, v)$ be $\boldsymbol{n}$. Using

$$\boldsymbol{x}(u + \Delta u, v + \Delta v) - \boldsymbol{x}(u, v) = \boldsymbol{x}_u(u, v)\Delta u + \boldsymbol{x}_v(u, v)\Delta v + o\left(\sqrt{\Delta u^2 + \Delta v^2}\right),$$

we have

$$\begin{aligned}(\Delta s)^2 &= |\boldsymbol{x}(u + \Delta u, v + \Delta v) - \boldsymbol{x}(u, v)|^2 \\ &= |\boldsymbol{x}_u|^2 \Delta u^2 + 2\boldsymbol{x}_u \cdot \boldsymbol{x}_v \Delta u \Delta v + |\boldsymbol{x}_v|^2 \Delta v^2 + o\left(\Delta u^2 + \Delta v^2\right), \end{aligned} \tag{2.12}$$

where $\cdot$ denotes the inner product, $\boldsymbol{x}_u = \boldsymbol{x}_u(u, v) = \frac{\partial \boldsymbol{x}}{\partial u}(u, v)$, and $\boldsymbol{x}_v = \boldsymbol{x}_v(u, v) = \frac{\partial \boldsymbol{x}}{\partial v}(u, v)$. Equality (2.12) is represented by the first fundamental form,

$$ds^2 = Edu^2 + 2Fdudv + Gdv^2,$$

with

$$E = |\boldsymbol{x}_u|^2, \quad F = \boldsymbol{x}_u \cdot \boldsymbol{x}_v, \quad G = |\boldsymbol{x}_v|^2$$

standing for the first fundamental quantities. Similarly, we obtain

$$\Delta S = |\boldsymbol{a} \times \boldsymbol{b}| = |\boldsymbol{x}_u(u, v) \times \boldsymbol{x}_v(u, v)| \, \Delta u \Delta v + o\left(\Delta u^2 + \Delta v^2\right) \tag{2.13}$$

by

$$\boldsymbol{a} = \boldsymbol{x}(u + \Delta u, v) - \boldsymbol{x}(u, v) = \boldsymbol{x}_u(u, v)\Delta u + o(\Delta u)$$
$$\boldsymbol{b} = \boldsymbol{x}(u, v + \Delta v) - \boldsymbol{x}(u, v) = \boldsymbol{x}_v(u, v)\Delta v + o(\Delta v),$$

where $\times$ denotes the outer product in $\mathbf{R}^3$. Equality (2.13) then induces the area element $dS = |\boldsymbol{x}_u \times \boldsymbol{x}_v| \, dudv$.

Since $\boldsymbol{x}_u(u, v)$ and $\boldsymbol{x}_v(u, v)$ are tangent to the curves on $\mathscr{M}$ where v and u are constants, respectively, the unit vector $\boldsymbol{n}$ defined below is perpendicular to them:

$$\boldsymbol{n} = \frac{\boldsymbol{x}_u \times \boldsymbol{x}_v}{|\boldsymbol{x}_u \times \boldsymbol{x}_v|}.$$

Henceforth, we use $d\boldsymbol{x} = \boldsymbol{x}_u du + \boldsymbol{x}_v dv$ to write the first fundamental form as

$$ds^2 = d\boldsymbol{x} \cdot d\boldsymbol{x} = |\boldsymbol{x}_u|^2 \, du^2 + 2\boldsymbol{x}_u \cdot \boldsymbol{x}_v dudv + |\boldsymbol{x}_v|^2 \, dv^2.$$

We have, on the other hand,

$$\begin{aligned} |\boldsymbol{x}_u \times \boldsymbol{x}_v|^2 &= (\boldsymbol{x}_u \times \boldsymbol{x}_v) \cdot (\boldsymbol{x}_u \times \boldsymbol{x}_v) \\ &= (\boldsymbol{x}_u \cdot \boldsymbol{x}_u)(\boldsymbol{x}_v \cdot \boldsymbol{x}_v) - (\boldsymbol{x}_u \cdot \boldsymbol{x}_v)^2 \\ &= EG - F^2 \end{aligned}$$

and hence

$$dS = \sqrt{EG - F^2} dudv, \quad \boldsymbol{n} = \frac{\boldsymbol{x}_u \times \boldsymbol{x}_v}{\sqrt{EG - F^2}}.$$

2.2.2 Curves

A curve in three-dimensional space is parmetrized by the length parameter: $\mathscr{C} = \{\boldsymbol{x}(s)\}$. Then $\boldsymbol{t} = d\boldsymbol{x}/ds$ is the unit tangent vector, and the plane containing $\boldsymbol{P} = \boldsymbol{x}(s)$ and perpendicular to $\boldsymbol{t}$ is called the normal plane. We take $\boldsymbol{P}_1$, $\boldsymbol{P}_2$ on $\mathscr{C}$ near $\boldsymbol{P}$ and let π be the plane made by these three points, $\boldsymbol{P}_1$, $\boldsymbol{P}_2$, and $\boldsymbol{P}$. As $\boldsymbol{P}_1, \boldsymbol{P}_2 \to \boldsymbol{P}$, this π converges to a plane orthogonal to the normal plane, called the osculating plane. The intersection of normal and osculating planes forms a line on which we take a unit vector $\boldsymbol{n}$, called the principal normal vector.

Near $\boldsymbol{P}$ the curve $\mathscr{C}$ is approximated by a circle on the osculating plane. If its radius is denoted by ρ and the direction of $\boldsymbol{n}$ is taken toward its center, it holds that

$$\frac{d\boldsymbol{t}}{ds} = \frac{1}{\rho}\boldsymbol{n}$$

with a scalar $1/\rho$, the curvature of $\mathscr{C}$ at $\boldsymbol{P}$, by $\frac{d\boldsymbol{t}}{ds} \cdot \boldsymbol{t} = 0$ derived from $|\boldsymbol{t}|^2 = 1$. This ρ stands for the radius of the approximate circle mentioned above. The tangent and the principal normal vectors are perpendicular to each other. Then the bi-principal normal vector is defined by $\boldsymbol{b} = \boldsymbol{t} \times \boldsymbol{n}$.

This $\boldsymbol{b}$ changes its direction as $\mathscr{C}$ twists, and the rate is detected by torsion, denoted by τ. To clarify this role, note that $\boldsymbol{b} = \boldsymbol{t} \times \boldsymbol{n}$ implies

$$\frac{d\boldsymbol{b}}{ds} = \frac{d\boldsymbol{t}}{ds} \times \boldsymbol{n} + \boldsymbol{t} \times \frac{d\boldsymbol{n}}{ds} = \frac{1}{\rho}\boldsymbol{n} \times \boldsymbol{n} + \boldsymbol{t} \times \frac{d\boldsymbol{n}}{ds} = \boldsymbol{t} \times \frac{d\boldsymbol{n}}{ds}.$$

We have also

$$\boldsymbol{b} \cdot \frac{d\boldsymbol{b}}{ds} = 0$$

by $\boldsymbol{b} \cdot \boldsymbol{b} = 1$, and hence $d\boldsymbol{b}/ds$ is perpendicular to both $\boldsymbol{t}$ and $\boldsymbol{b}$. Regarding these facts, we define τ by

$$\frac{d\boldsymbol{b}}{ds} = -\tau\boldsymbol{n}$$

including its sign. It holds also that $\boldsymbol{n} = \boldsymbol{b} \times \boldsymbol{t}$, which implies

$$\frac{d\boldsymbol{n}}{ds} = \frac{d\boldsymbol{b}}{ds} \times \boldsymbol{t} + \boldsymbol{b} \times \frac{d\boldsymbol{t}}{ds} = -\tau\boldsymbol{n} \times \boldsymbol{t} + \boldsymbol{b} \times \frac{1}{\rho}\boldsymbol{n} = \tau\boldsymbol{b} - \frac{1}{\rho}\boldsymbol{t}$$

because $\boldsymbol{n} \times \boldsymbol{t} = -\boldsymbol{b}$ and $\boldsymbol{b} \times \boldsymbol{n} = -\boldsymbol{t}$. We thus end up with the Frenet–Serret formula,

$$\frac{d\boldsymbol{t}}{ds} = \frac{1}{\rho}\boldsymbol{n}, \quad \frac{d\boldsymbol{n}}{ds} = -\frac{1}{\rho}\boldsymbol{t} + \tau\boldsymbol{b}, \quad \frac{d\boldsymbol{b}}{ds} = -\tau\boldsymbol{n},$$

where ρ, τ, and s are curvature radius, torsion, and length parameter, respectively.

Recall $\boldsymbol{x} = \boldsymbol{x}(u, v)$, a parameter representation of the surface $\mathscr{M}$, and let $\mathscr{C}$ be a curve on $\mathscr{M}$. This curve $\mathscr{C}$ is represented by $(u, v) = (u(s), v(s))$, or

$$\boldsymbol{x} = \boldsymbol{x}(u(s), v(s)),$$

using the length parameter s of $\mathscr{C}$. The unit tangential vector of $\mathscr{C}$ is given by

$$\boldsymbol{t} = \frac{d\boldsymbol{x}}{ds} = \boldsymbol{x}_u \frac{du}{ds} + \boldsymbol{x}_v \frac{dv}{ds} \tag{2.14}$$

and it holds that

$$\frac{d\boldsymbol{t}}{ds} = \frac{1}{\rho_{\mathscr{C}}} \boldsymbol{n}_{\mathscr{C}},$$

where $\rho_{\mathscr{C}}$ and $\boldsymbol{n}_{\mathscr{C}}$ denote the curvature radius and the principal normal unit vector of $\mathscr{C}$, respectively. If $\psi_{\mathscr{C}}$ denotes the angle between $\boldsymbol{n}$ and $\boldsymbol{n}_{\mathscr{C}}$, then $\cos \psi_{\mathscr{C}} = \boldsymbol{n} \cdot \boldsymbol{n}_{\mathscr{C}}$, or

$$\frac{\cos \psi_{\mathscr{C}}}{\rho_{\mathscr{C}}} = \boldsymbol{n} \cdot \frac{d\boldsymbol{t}}{ds}$$

arises with the unit normal vector of $\mathscr{M}$ denoted by

$$\boldsymbol{n} = \frac{\boldsymbol{x}_u \times \boldsymbol{x}_v}{|\boldsymbol{x}_u \times \boldsymbol{x}_v|}.$$

Since (2.14) implies

$$\frac{d\boldsymbol{t}}{ds} = \boldsymbol{x}_{uu} \left(\frac{du}{ds}\right)^2 + 2\boldsymbol{x}_{uv} \left(\frac{du}{ds}\right)\left(\frac{dv}{ds}\right) + \boldsymbol{x}_{vv} \left(\frac{dv}{ds}\right)^2 + \boldsymbol{x}_u \frac{d^2u}{ds^2} + \boldsymbol{x}_v \frac{d^2v}{ds^2}$$

it holds that

$$\boldsymbol{n} \cdot \frac{d\boldsymbol{t}}{ds} = \boldsymbol{n} \cdot \left\{ \boldsymbol{x}_{uu} \left(\frac{du}{ds}\right)^2 + 2\boldsymbol{x}_{uv} \left(\frac{du}{ds}\right)\left(\frac{dv}{ds}\right) + \boldsymbol{x}_{vv} \left(\frac{dv}{ds}\right)^2 \right\}$$

by

$$\boldsymbol{n} \cdot \boldsymbol{x}_u = \boldsymbol{n} \cdot \boldsymbol{x}_v = 0.$$

Then the second fundamental quantities are defined by

$$L = \boldsymbol{x}_{uu} \cdot \boldsymbol{n}, \quad M = \boldsymbol{x}_{uv} \cdot \boldsymbol{n}, \quad N = \boldsymbol{x}_{vv} \cdot \boldsymbol{n}.$$

2.2.3 Curvatures

Writing

$$\frac{\cos\psi_{\mathscr{C}}}{\rho_{\mathscr{C}}} = L\left(\frac{du}{ds}\right)^2 + 2M\frac{du}{ds}\frac{dv}{ds} + N\left(\frac{dv}{ds}\right)^2 = \frac{Ldu^2 + 2Mdudv + Ndv^2}{Edu^2 + 2Fdudv + Gdv^2},$$

furthermore, we call

$$Ldu^2 + 2Mdudv + Ndv^2$$

the second fundamental form. Using

$$k = dv/du,$$

the direction of the infinitesimal vector $d\boldsymbol{x}$, we have

$$\frac{\cos\psi_{\mathscr{C}}}{\rho_{\mathscr{C}}} = \frac{L + 2Mk + Nk^2}{E + 2Fk + Gk^2}. \tag{2.15}$$

Here, L, M, N, E, F, G are determined by $\boldsymbol{P} \in \mathscr{M}$, while

$$k = \frac{dv}{du} = \frac{dv/ds}{du/ds}$$

is determined by the tangential vector $\boldsymbol{t}$ of $\mathscr{C}$, as well as $\boldsymbol{P}$, because of

$$\boldsymbol{t} = \frac{du}{ds}\boldsymbol{x}_u + \frac{dv}{ds}\boldsymbol{x}_v.$$

Let $\prod$ be the plane containing $\boldsymbol{P}$ made by the tangential vector $\boldsymbol{t}$ of $\mathscr{C}$ and the normal vector $\boldsymbol{n}$ of $\mathscr{M}$, and let $\mathscr{C}'$ be the curve on $\mathscr{M}$ cut by $\prod$. Since the center of curvature $\mathscr{C}'$ at $\boldsymbol{P}$ is on $\prod$ it holds that $(\psi_{\mathscr{C}'}, \rho_{\mathscr{C}'}) = (0, R)$ or $(\psi_{\mathscr{C}'}, \rho_{\mathscr{C}'}) = (\pi, -R)$ where R denotes the curvature of $\mathscr{C}'$ at $\boldsymbol{P} \in \mathscr{M}$. The right-hand side of (2.15), on the other hand, is determined by $\boldsymbol{P}$ and $\boldsymbol{t}$ so that we have

$$\frac{\cos\psi_{\mathscr{C}}}{\rho_{\mathscr{C}}} = \frac{\cos\psi_{\mathscr{C}'}}{\rho_{\mathscr{C}'}} = \frac{1}{R} \tag{2.16}$$

at $\boldsymbol{P}$. This $1/R$ is called the normal curvature of $\mathscr{M}$ at $\boldsymbol{P}$ with the direction $\boldsymbol{t}$.

Fixing $\boldsymbol{P}$, we seek $\boldsymbol{t}$ such that the normal curvature attains the minimum or the maximum. In these extremal cases, $\boldsymbol{t}$ and $1/R$ are called the principal direction and the principal curvature, respectively. They are actually obtained by

$$\frac{d}{dk}\left(\frac{1}{R}\right) = \frac{d}{dk} \cdot \frac{L + 2Mk + Nk^2}{E + 2Fk + Gk^2} = 0,$$

or, equivalently,

$$\frac{1}{R}(F + Gk) = M + Nk. \tag{2.17}$$

Then, since

$$\frac{1}{R}(E + 2Fk + Gk^2) = L + 2Mk + Nk^2$$

it holds that

$$\frac{1}{R}(E + Fk) = L + Mk. \tag{2.18}$$

We have, on the other hand,

$$k = -\frac{F/R - M}{G/R - N}$$

again by (2.17) and then

$$\left(\frac{F}{R} - M\right)\left(\frac{F}{R} - M\right) - \left(\frac{E}{R} - L\right)\left(\frac{G}{R} - N\right) = 0 \tag{2.19}$$

follows from (2.18). Thus, the principal curvatures $1/R_1$ and $1/R_2$ are the solutions to (2.19).

Writing (2.19) as

$$(EG - F^2)\frac{1}{R^2} - (GL + EN - 2FM)\frac{1}{R} + LN - M^2 = 0, \tag{2.20}$$

we have

$$2H \equiv \frac{1}{R_1} + \frac{1}{R_2} = \frac{GL + EN - 2FM}{EG - F^2}, \quad K \equiv \frac{1}{R_1} \cdot \frac{1}{R_2} = \frac{LN - M^2}{EG - F^2}. \tag{2.21}$$

The quantities

$$2H = \frac{1}{R_1} + \frac{1}{R_2}, \quad K = \frac{1}{R_1} \cdot \frac{1}{R_2}. \tag{2.22}$$

are called the *mean curvature* and the *Gaussian curvature* of $\mathscr{M}$ at $\boldsymbol{P}$, respectively. If the Gaussian curvature Gaussian curvature is positive, then the surface is convex to one side there. Similarly, if it is negative then it looks like a saddle. It is known that the Gaussian curvature is determined by the first fundamental quantities.

Equation (2.19) on $1/R$ takes an equivalent form on k by (2.17) and (2.18), that is,

$$FL - EM + (GL - EN)k + (GM - FN)k^2 = 0, \tag{2.23}$$

which provides the principal directions k_1 and k_2 as the solution. These directions are associated with the infinitesimal vectors

$$d\boldsymbol{x}_1 = \boldsymbol{x}_u du_1 + \boldsymbol{x}_v dv_1, \quad d\boldsymbol{x}_2 = \boldsymbol{x}_u du_2 + \boldsymbol{x}_v dv_2$$

through $k_1 = dv_1/du_1$ and $k_2 = dv_2/du_2$, which implies

$$\begin{aligned} d\boldsymbol{x}_1 \cdot d\boldsymbol{x}_2 &= Edu_1du_2 + F(du_2dv_1 + du_1dv_2) + Gdv_1dv_2 \\ &= (E + F(k_1 + k_2) + Gk_1k_2)\, du_1du_2. \end{aligned}$$

We obtain, on the other hand,

$$k_1 + k_2 = -\frac{GL - EN}{GM - FN}, \quad k_1k_2 = \frac{FL - EM}{GM - FN}$$

by (2.23) and hence

$$\begin{aligned} E + F(k_1 + k_2) + Gk_1k_2 &= \frac{1}{GM - FN} \\ &\quad \cdot \{(GM - FN)E - F(GL - EN) + G(FL - EM)\} = 0. \end{aligned}$$

Thus it holds that $d\boldsymbol{x}_1 \cdot d\boldsymbol{x}_2 = 0$, and, therefore, the principal directions are perpendicular to each other.

2.2.4 Differential Forms

Each scalar function

$$f = f(x_1, x_2, x_3) \tag{2.24}$$

is called a zero form. Then a 1-, 2-, and 3-forms are given by

$$fdx_1 + gdx_2 + hdx_3, \quad fdx_2 \wedge dx_3 + gdx_3 \wedge dx_1 + hdx_1 \wedge dx_2, \quad fdx_1 \wedge dx_2 \wedge dx_3,$$

respectively. Here, the wedge product $\wedge$ is an operation satisfying

$$dx_i \wedge dx_j = -dx_j \wedge dx_i, \tag{2.25}$$

which results in

$$dx_i \wedge dx_i = 0.$$

Given 1-forms

$$\omega_1 = \sum_{i=1}^{3} P_i dx_i, \quad \omega_2 = \sum_{i=1}^{3} Q_i dx_i,$$

we define

$$\omega_1 \wedge \omega_2 = \sum_{i=1}^{3} P_i Q_i \, dx_i \wedge dx_j = (P_2 Q_3 - P_3 Q_2) \, dx_2 \wedge dx_3$$
$$+ (P_3 Q_1 - P_1 Q_3) \, dx_3 \wedge dx_1 + (P_1 Q_2 - P_2 Q_1) \, dx_1 \wedge dx_2,$$

regarding (2.25). For 1- and 2-forms denoted by

$$\omega = \sum_{i=1}^{3} P_i dx_i \tag{2.26}$$

and

$$\theta = Q_1 dx_2 \wedge dx_3 + Q_2 dx_3 \wedge dx_1 + Q_3 dx_1 \wedge dx_2, \tag{2.27}$$

respectively, we have

$$\omega \wedge \theta = (P_1 dx_1 + P_2 dx_2 + P_3 dx_3) \wedge (Q_1 dx_2 \wedge dx_3 + Q_2 dx_3 \wedge dx_1 + Q_3 dx_1 \wedge dx_2)$$
$$= (P_1 Q_1 + P_2 Q_2 + P_3 Q_3) \, dx_1 \wedge dx_2 \wedge dx_3$$

similarly. Then it follows that $d^2 = 0$.

The outer differentiation is denoted by d. It acts on 0-, 1-, and 2-forms denoted by f, ω, and θ in (2.24), (2.26), and (2.27), respectively, as

$$df = \frac{\partial f}{\partial x_1} dx_1 + \frac{\partial f}{\partial x_2} dx_2 + \frac{\partial f}{\partial x_3} dx_3,$$

$$d\omega = dP \wedge dx_1 + dQ \wedge dx_2 + dR \wedge dx_3 = (\frac{\partial R}{\partial x_2} - \frac{\partial Q}{\partial x_3}) \, dx_2 \wedge dx_3$$
$$+ (\frac{\partial P}{\partial x_3} - \frac{\partial R}{\partial x_1}) \, dx_3 \wedge dx_1 + (\frac{\partial Q}{\partial x_1} - \frac{\partial P}{\partial x_2}) \, dx_1 \wedge dx_2$$

$$\begin{aligned} d\theta &= dP \wedge dx_2 \wedge dx_3 + dQ \wedge dx_3 \wedge dx_1 + dR \wedge dx_1 \wedge dx_2 \\ &= (\frac{\partial P}{\partial x_1} + \frac{\partial Q}{\partial x_2} + \frac{\partial R}{\partial x_3})\, dx_1 \wedge dx_2 \wedge dx_3. \end{aligned}$$

Given 1-forms ω_1 and ω_2, we obtain

$$d(\omega_1 \wedge \omega_2) = d\omega_1 \wedge \omega_2 - \omega_1 \wedge d\omega_2.$$

The *divergence formula* of Gauss, furthermore, is represented by

$$\int\int\int_V d\theta = \int\int_{\partial V} \theta,$$

where $V \subset \mathbf{R}^3$ is a bounded domain with smooth boundary ∂V, θ is a (smooth) 2-form, and

$$dx_i \wedge dx_j, \quad i \neq j,$$

is a projection of dS to the plane $x = x_k$, $k \neq i, j$, with (i, j, k) oriented in a right-hand system.

The Hodge operator $*$ is defined for (2.24), (2.26), and (2.27) by

$$\begin{aligned} *f &= f\, dx_1 \wedge dx_2 \wedge dx_3, \\ *\omega &= P\, dx_2 \wedge dx_3 + Q\, dx_3 \wedge dx_1 + R\, dx_1 \wedge dx_2, \\ *\theta &= P\, dx_1 + Q\, dx_2 + R\, dx_3, \end{aligned}$$

respectively. We put also

$$*\,(f\, dx_1 \wedge dx_2 \wedge dx_3) = f,$$

which ensures $** = Id$.

Let $\varphi : D \subset \mathbf{R}^2 \to \mathscr{M} \subset \mathbf{R}^3$ be a parametric representation of surface, denoted by

$$\varphi(u, v) = x \equiv \begin{pmatrix} x_1(u, v) \\ x_2(u, v) \\ x_3(u, v) \end{pmatrix}.$$

Given a smooth function $f : \mathscr{M} \to \mathbf{R}$, its pull-back is defined by

$$\varphi^* f = f \circ \varphi : D \to \mathbf{R}.$$

The pull-backs of 1- and 2-forms on $\mathscr{M}$ in the forms of (2.26) and (2.27), respectively, are given by

$$\varphi^*\omega = \varphi^* P \cdot d(\varphi^* x_1) + \varphi^* Q \cdot d(\varphi^* x_2) + \varphi^* R \cdot d(\varphi^* x_3),$$
$$\varphi^*\theta = \varphi^* P \cdot d(\varphi^* x_2) \wedge d(\varphi^* x_3) + \varphi^* Q \cdot d(\varphi^* x_3) \wedge d(\varphi^* x_1)$$
$$+ \varphi^* R \cdot d(\varphi^* x_1) \wedge d(\varphi^* x_2).$$

The following theorems are fundamental.

Theorem 2.1 *Let f, ω, and θ be* 0-, 1-, *and* 2-*forms, respectively. Then it holds that*

$$\varphi^*(df) = d(\varphi^* f),\ \varphi^*(dw) = d(\varphi^* w),\ \varphi^*(d\theta) = d(\varphi^*\theta).$$

Theorem 2.2 *If ω, θ are* 1, 2-*forms, respectively, it holds that*

$$\int\int_{\mathscr{M}} \theta = \int\int_D \varphi^*\theta,\quad \int_{\partial\mathscr{M}} \omega = \int_{\partial D} \varphi^*\theta.$$

Green's formula indicates

$$\int\int_D d(\varphi^*\omega) = \int_{\partial D} \varphi^*\omega,$$

where ∂D is oriented counter-clockwise. Then the Stokes formula

$$\int\int_{\mathscr{M}} d\omega = \int_{\partial\mathscr{M}} \omega$$

arises by Theorems 2.1 and 2.2.

2.2.5 *Tangent and Cotangent Spaces*

Let $\mathscr{M} = \bigcup_{d=1}^{s} U_\alpha$ be a global representation of a surface, and

$$\varphi_\alpha : D_\alpha \subset \mathbf{R}^2 \to U_\alpha \subset \mathscr{M}$$

be a local chart. Hence

$$\varphi_\beta^{-1} \circ \varphi_\alpha : \widetilde{D}_\alpha = \varphi_\alpha^{-1}(U_\beta \cap U_\alpha) \subset D_\alpha \to \varphi_\beta^{-1} \circ \varphi_\alpha(\widetilde{D}_\alpha) \subset D_\beta$$

is a diffeomorphism, provided that

$$U_\beta \cap U_\alpha \neq \emptyset. \tag{2.28}$$

If (2.28) arises, and $(u_\alpha, u_\alpha) \in D_\alpha$ and $(u_\beta, u_\beta) \in D_\beta$ are local coordinates, we have

$$u_\beta = u_\beta(u_\alpha, v_\alpha), \quad v_\beta = v_\beta(u_\alpha, v_\alpha),$$

which induces

$$du_\beta = \frac{\partial u_\beta}{\partial u_\alpha} du_\alpha + \frac{\partial u_\beta}{\partial v_\alpha} dv_\alpha, \quad dv_\beta = \frac{\partial v_\beta}{\partial u_\alpha} du_\alpha + \frac{\partial v_\beta}{\partial v_\beta} dv_\beta.$$

The vector spaces are identified with this relation,

$$\langle du_\alpha, dv_\alpha \rangle \sim \langle du_\alpha, dv_\alpha \rangle,$$

and the cotangent space of $\mathscr{M}$ at $x \in U_\alpha$ is defined by

$$T_x^* \mathscr{M} = \langle du_\alpha, du_\beta \rangle / \sim .$$

Using the dual basis

$$\langle \frac{\partial}{\partial u_\alpha}, \frac{\partial}{\partial v_\alpha} \rangle$$

of $\langle du_\alpha, dv_\alpha \rangle$ and a naturally induced transformation of variables, we define the tangent space of $\mathscr{M}$ at $x \in U_\alpha$ by

$$T_x \mathscr{M} = \langle \frac{\partial}{\partial u_\alpha}, \frac{\partial}{\partial v_\alpha} \rangle / \sim,$$

and then

$$\mathscr{X}(M) = U_{x \in \mathscr{M}} T_x \mathscr{M}$$

stands for the set of tangential vector fields on $\mathscr{M}$. Under the local chart

$$\varphi : (u, v) \in D \subset \mathbf{R}^2 \mapsto x = \begin{pmatrix} x_1(u, v) \\ x_2(u, v) \\ x_3(u, v) \end{pmatrix} \in \mathscr{M} \subset \mathbf{R}^3,$$

the tangential vector field

$$X = \xi x_u + \eta x_v \quad \text{on } \mathscr{M},$$

that is, $X \in \mathscr{X}(\mathscr{M})$, is defined by

$$Xf = \xi \frac{\partial}{\partial u}(\varphi^* f) + \eta \frac{\partial}{\partial v}(\varphi^* f),$$

where $f, \xi, \eta : \mathscr{M} \to \mathbf{R}$ are 0-forms.

2.2.6 Frames and Covariant Derivatives

Given a surface $\mathscr{M}$ and a point $x \in \mathscr{M}$, let $e_1, e_2 \in T_x\mathscr{M}$, and e_3 be a right-handed orthonormal system in $\mathbf{R}^3$. We call

$$\{e_1, e_2, e_3\} = \{e_1(x), e_2(x), e_3(x)\}$$

an orthonormal *frame* if it is smooth in $x \in \mathscr{M}$. Then it holds that

$$dx = \omega^1 e_1 + \omega^2 e_2$$

for some 1-forms ω^1 and ω^2, and this $\{\omega^1, \omega^2\}$ is called the *dual frame*.

The vector field de_j on $\mathscr{M}$ takes the form

$$de_j = \sum_{i=1}^{3} \omega^i_j e_i$$

for $j = 1, 2, 3$, and each ω^i_j is called a *connection form*. This connection form is a 1-form on $\mathscr{M}$. Since $e_i \cdot e_j = \delta_{ij}$ implies

$$de_i \cdot e_j + e_i \cdot de_j = 0,$$

it holds that

$$\omega^i_j + \omega^j_i = 0. \tag{2.29}$$

We thus end up with

$$d\begin{pmatrix} e_1 \\ e_2 \\ e_3 \end{pmatrix} = \begin{pmatrix} 0 & \omega^2_1 & \omega^3_1 \\ -\omega^2_1 & 0 & \omega^3_2 \\ -\omega^3_1 & -\omega^3_2 & 0 \end{pmatrix} \begin{pmatrix} e_1 \\ e_2 \\ e_3 \end{pmatrix}. \tag{2.30}$$

The first two and the last equations are called the *Gauss equation* and the *Weingarten equation*, respectively.

Since

$$0 = d^2x = d(\omega^1 e_1 + \omega^2 e_2), \quad d(\omega e) = (d\omega)e - \omega \wedge (de),$$

there arises

$$d\omega^1 + \omega_2^1 \wedge \omega^2 = 0, \quad d\omega^2 + \omega_1^2 \wedge \omega^1 = 0, \quad \omega_1^3 \wedge \omega^1 + \omega_2^3 \wedge \omega^2 = 0.$$

It follows also that

$$d\omega_j^i - \sum_{k=1}^{3} \omega_j^k \wedge \omega_k^i = 0$$

from $0 = d^2 e$.

Let

$$\mathscr{M} : x = x(u, v) \in \mathbf{R}^3, \quad (u, v) \in D \subset \mathbf{R}^2$$

be a parametric representation, and $e_3 = n$ be the unit normal vector on $\mathscr{M}$. Then we take a frame on $\mathscr{M}$ denoted by $\{e_1, e_2, e_3\}$ and regard $\{e_1, e_2\}$ as tangential vector fields, $e_1, e_2 \in \mathscr{X}(\mathscr{M})$.

Given 0-forms $\xi^i : \mathscr{M} \to \mathbf{R}$, $i = 1, 2$, we put

$$X = \sum_{i=1}^{2} \xi^i e_i \in \mathscr{X}(\mathscr{M}).$$

Then it holds that

$$dX = \sum_{i=1}^{2} (d\xi^i e_i + \xi^i de_i) = \sum_{i=1}^{2} \{d\xi^i e_i + \xi^i \sum_{k=1}^{3} \omega_i^k e_k\},$$

using the connection forms ω_j^i defined by (2.29). We have

$$dX = \sum_{i=1}^{2} (d\xi^i + \sum_{k=1}^{2} \xi^k \omega_k^i) e_i + \sum_{i=1}^{2} \xi^i \omega_i^3 e_3$$

and call the first term on the right-hand side, denoted by ∇X, the *covariant derivative* of X.

2.2.7 Connections

Given $Y \in \mathscr{X}(\mathscr{M})$, now we put

$$\nabla_X Y = \sum_{i=1}^{2} \{\langle Y, d\xi^i \rangle + \sum_{k=1}^{2} \xi^k \langle Y, \omega_k^i \rangle\} e_i \in \mathscr{X}(\mathscr{M}),$$

where $\langle\ ,\ \rangle$ denotes the pairing, to obtain the bilinear form

$$\nabla : \mathscr{X}(\mathscr{M}) \times \mathscr{X}(\mathscr{M}) \to \mathscr{X}(\mathscr{M}). \tag{2.31}$$

If $X, Y \in \mathscr{X}(\mathscr{M})$ and f is a 0-form on $\mathscr{M}$, there arises

$$\nabla_{fX} Y = f\nabla_X Y, \ \nabla_X(fY) = (Xf)Y + f\nabla_X Y$$
$$\nabla_X Y - \nabla_Y X = [X, Y] \equiv XY - YX.$$

Such a bilinear form ∇ in (2.31) is called a *connection*.

Using the embedding $T_x\mathscr{M} \hookrightarrow \mathbf{R}^3$, we put

$$g(X, Y) = X \cdot Y, \quad X, Y \in \mathscr{X}(\mathscr{M}).$$

Recall that $\cdot$ denotes the inner product in $\mathbf{R}^3$. Then this g induces a metric on $\mathscr{M}$, and $(\mathscr{M}, g)$ is called a *Riemann surface*. For each $X, Y \in \mathscr{X}(\mathscr{M})$, this $g(X, Y)$ is a 0-form on $\mathscr{M}$. Then it holds that

$$X(g(Y, Z)) = g(\nabla_X Y, Z) + g(Y, \nabla_X Z), \quad X, Y, Z \in \mathscr{X}(\mathscr{M}).$$

In other words, the above ∇ is a *Levi–Chivita connection* with respect to g.

To define the *Cristoffel symbol* Γ_{ij}^k, let $x_1 = u$ and $x_2 = v$. Put

$$\nabla_{\frac{\partial}{\partial x_j}} \frac{\partial}{\partial x_i} = \sum_k \Gamma_{ij}^k \frac{\partial}{\partial x_k}.$$

Then it holds that

$$\Gamma_{ij}^k = \frac{1}{2} \sum_\ell g^{kl} (\frac{\partial g_{i\ell}}{\partial x_j} + \frac{\partial g_{j\ell}}{\partial x_i} - \frac{\partial g_{ij}}{\partial x_\ell}) \tag{2.32}$$

for

$$g^{ij} = (g_{ij})^{-1}, \quad g_{ij} = g(\frac{\partial}{\partial x_i}, \frac{\partial}{\partial x_j}).$$

Since $\{g_{ij}\}$ are the first fundamental quantities,

$$g_{11} = E, \quad g_{12} = g_{21} = F, \quad g_{22} = G,$$

it follows that

$$\begin{aligned} \Gamma_{11}^1 &= \frac{GE_u - 2FF_u + FE_v}{2(EG - F^2)}, & \Gamma_{12}^1 = \Gamma_{21}^1 &= \frac{GE_v - 2FG_u}{2(EG - F^2)} \\ \Gamma_{22}^1 &= \frac{2GF_v - GG_u - FG_v}{2(EG - F^2)}, & \Gamma_{11}^2 &= \frac{2EF_u - FF_v - FE_u}{2(EG - F^2)} \\ \Gamma_{12}^2 = \Gamma_{21}^2 &= \frac{2EG_u - FF_v}{2(EG - F^2)}, & \Gamma_{22}^2 &= \frac{EG_v - 2FF_u + FG_v}{2(EG - F^2)}, \end{aligned} \tag{2.33}$$

recalling $x_1 = u, x_2 = v$.

2.3 Liouville Integral

2.3.1 *Review of the Surface Theory*

Summarizing Sect. 2.2, let

$$\mathscr{M} : x = \begin{pmatrix} x_1(u, v) \\ x_2(u, v) \\ x_3(u, v) \end{pmatrix} \in \mathbf{R}^3, \quad (u, v) \in D \subset \mathbf{R}^2 \tag{2.34}$$

be a parametric surface. Writing

$$dx = x_u du + x_v dv,$$

we define the first fundamental form $I = ds^2$ by

$$I = dx \cdot dx = E du^2 + 2F du dv + G dv^2,$$

where

$$E = |x_u|^2, \quad F = x_u \cdot x_v, \quad G = |x_v|^2$$

stand for the first fundamental quantities. Fix $P \in \mathscr{M}$, and n be the unit normal vector on $\mathscr{M}$ at P. Then it holds that

$$n = \frac{x_u \times x_v}{\sqrt{EG - F^2}},$$

where $\times$ denotes the outer product in $\mathbf{R}^3$.

Given a unit tangential vector $\boldsymbol{t}$ on $\mathcal{M}$ at P, we take the plane $\prod$ made by n and $\boldsymbol{t}$, and curve $\mathcal{C}$ on $\mathcal{M}$ cut by $\prod$. Parametrize this $\mathcal{C}$ by the arc-length as in $x = x(s) \in \mathcal{C}$. If ρ_C and n_C denote the curvature radius and the principal unit normal vector of $\mathcal{C}$, it holds that

$$\frac{d\boldsymbol{t}}{ds} = \frac{1}{\rho_C} n_C,$$

and the normal curvature of $\mathcal{M}$ at P in the direction $\boldsymbol{t}$ is defined by (2.16):

$$\frac{1}{R} = \frac{\cos\psi_C}{\rho_C},$$

where ψ_C denotes the angle made by n and n_C: $(\psi_C, \rho_C) = (0, R)$ or $(\pi, -R)$. The principal curvatures stand for the minimum and maximum of this $\frac{1}{R}$ concerning $\boldsymbol{t} \in S^2$, denoted by $\frac{1}{R_1}$ and $\frac{1}{R_2}$, respectively. Then the mean and Gauss curvatures of $\mathcal{M}$ at P are defined by (2.22).

Since $\mathcal{C} : x = x(u(s), v(s)) \in \mathcal{M}$ is parametrized by the arc-length parameter s, it holds that

$$\begin{aligned}\frac{\cos\psi_C}{\rho_C} &= n \cdot \frac{d\boldsymbol{t}}{ds} = L(\frac{du}{ds})^2 + 2M\frac{du}{ds}\frac{dv}{ds} + N(\frac{dv}{ds})^2 \\ &= \frac{II}{I} = \frac{Ldu^2 + 2Mdudv + Ndv^2}{Edu^2 + 2Fdudv + Gdv^2},\end{aligned}$$

where

$$\begin{aligned}L &= x_{uu} \cdot n = -x_u \cdot n_u \\ M &= x_{uv} \cdot n = -x_u \cdot n_v = -x_v \cdot n_u \\ N &= x_{vv} \cdot n = -x_v \cdot n_v\end{aligned} \tag{2.35}$$

stand for the second fundamental quantities, and

$$II = Ldu^2 + 2Mdudv + Ndv^2$$

denotes the second fundamental form.

Then $\frac{1}{R_1}$ and $\frac{1}{R_2}$ are two solutions to (2.20), which guarantees (2.21):

$$2H = \frac{GL + EN - 2FM}{EG - F^2} = \mathrm{tr}\, A, \quad K = \frac{LN - M^2}{EG - F^2} = \det A \tag{2.36}$$

for the *Weingarten matrix*

$$A = \begin{pmatrix} E & F \\ F & G \end{pmatrix}^{-1} \begin{pmatrix} L & M \\ M & N \end{pmatrix} \equiv \begin{pmatrix} A_1^1 & A_2^1 \\ A_1^2 & A_2^2 \end{pmatrix}. \tag{2.37}$$

Then (2.35) implies the Weingarten formula

$$n_u = -A_1^1 x_u - A_1^2 x_v, \quad n_v = -A_2^1 x_u - A_2^2 x_v. \tag{2.38}$$

Given a Riemann metric $ds^2 = g_{ij} dx_i dx_j$, we recall the Christoffel symbol (2.32),

$$\Gamma_{k\ell}^i = \frac{1}{2} g^{im} (g_{mk,\ell} + g_{m\ell,k} - g_{k\ell,m}),$$

where $(g^{jk}) = (g_{jk})^{-1}$ and $g_{mk,\ell} = \frac{\partial g_{mx}}{\partial x_\ell}$. In the parametric surface we obtain

$$\begin{aligned}
\Gamma_{11}^1 E + \Gamma_{11}^2 F &= x_{uu} \cdot x_u, \quad \Gamma_{11}^1 F + \Gamma_{11}^2 G = x_{uu} \cdot x_v, \\
\Gamma_{12}^1 E + \Gamma_{12}^2 F &= x_{uv} \cdot x_u, \quad \Gamma_{12}^1 F + \Gamma_{12}^2 G = x_{uv} \cdot x_v \\
\Gamma_{11}^1 E + \Gamma_{11}^2 F &= x_{vv} \cdot x_u, \quad \Gamma_{22}^1 F + \Gamma_{22}^2 G = x_{uv} \cdot x_u
\end{aligned}$$

by this definition, which ensures the Gauss formula

$$\begin{aligned}
x_{uu} &= \Gamma_{11}^1 x_u + \Gamma_{11}^2 x_v + Ln \\
x_{uv} &= \Gamma_{12}^1 x_u + \Gamma_{12}^2 x_v + Mn \\
x_{vv} &= \Gamma_{22}^1 x_u + \Gamma_{22}^2 x_v + Nn.
\end{aligned} \tag{2.39}$$

A direct calculation, on the other hand, induces (2.33).

Equalities (2.38) and (2.39) are summarized as the fundamental equation in surface theory,

$$F_u = F\Omega, \quad F_v = F\wedge, \tag{2.40}$$

for $F = [x_u \; x_v \; n]$ and

$$\Omega = \begin{pmatrix} \Gamma_{11}^1 & \Gamma_{11}^2 & -A_1^1 \\ \Gamma_{11}^2 & \Gamma_{12}^2 & -A_1^2 \\ L & M & 0 \end{pmatrix}, \quad \wedge = \begin{pmatrix} \Gamma_{21}^1 & \Gamma_{22}^1 & -A_2^1 \\ \Gamma_{21}^2 & \Gamma_{22}^2 & -A_1^2 \\ M & N & 0 \end{pmatrix}. \tag{2.41}$$

Theorem 2.3 *The coefficient matrices Ω and $\wedge$ in (2.41) satisfy the Lie structure*

$$\Omega_v - \wedge_u = \Omega \wedge - \wedge \Omega. \tag{2.42}$$

Proof Equality (2.42) arises as a compatibility condition to (2.40). In fact, we have

$$F_{uv} = (F\Omega)_v = F_v\Omega + F\Omega_v = F(\wedge\Omega + \Omega_v)$$

and similarly,

$$F_{vu} = F(\Omega \wedge + \wedge_u).$$

Then we the obtain the result because the matrix F is non-singular. □

Equality (2.42) is composed of the Theorem Egregium,

$$\begin{aligned} K = & \frac{E(E_vG_v - 2F_uG_v + G_u)^2}{4(EG - F^2)^2} \\ & + \frac{F(E_uG_v - E_vG_u - 2E_vF_v - 2F_uG_u + 4F_uF_v)}{4(EG - F^2)} \\ & + \frac{G(E_uG_u - 2E_uF_v + E_v^2)^2}{4(EG - F^2)} - \frac{E_{vv} - 2F_{uv} + G_{uu}}{2(EG - F^2)} \end{aligned}$$

by (2.21), and the *Codazzi–Mainardi equation*

$$\begin{aligned} L_v - M_u &= L\Gamma_{12}^1 + M(\Gamma_{12}^2 - \Gamma_{11}^1) - N\Gamma_{11}^2 \\ M_v - N_u &= L\Gamma_{22}^1 + M(\Gamma_{22}^2 - \Gamma_{12}^1) - N\Gamma_{12}^2. \end{aligned}$$

2.3.2 Soap Films and Soap Bubbles

Given a local parametrization $x = x(u, v) : D \subset \mathbf{R}^2 \to \mathscr{M} \subset \mathbf{R}^3$, the first and the second fundamental forms

$$I = Edu^2 + 2Fdudv + Gdv^2, \quad II = Ldu^2 + 2Mdudv + Ndv^2$$

define the bilinear forms

$$(X, Y)_I = \sum I_{ij}X_iY_j, \quad I_{11} = E,\ I_{12} = I_{21} = F,\ I_{22} = G$$

$$(X, Y)_{II} = \sum J_{ij}X_iY_j, \quad J_{11} = L,\ J_{12} = J_{21} = M,\ J_{22} = N$$

for

$$X = X_1x_u + X_2x_v, \quad Y = Y_1x_u + Y_2x_v \ \in \mathscr{X}(\mathscr{M}).$$

It holds that

$$(X, Y)_I = \begin{pmatrix} E & F \\ F & G \end{pmatrix} X \cdot Y, \quad (X, Y)_{II} = \begin{pmatrix} L & M \\ M & N \end{pmatrix} X \cdot Y$$

and hence the Weingarten matrix $A = \begin{pmatrix} E & F \\ F & G \end{pmatrix}^{-1} \begin{pmatrix} L & M \\ M & N \end{pmatrix}$ in (2.37) satisfies

$$(AX, Y)_I = (X, Y)_{II}.$$

Then we obtain (2.36) and hence equation (2.20) is equivalent to

$$\det |\lambda I - A| = 0, \quad \lambda = \frac{1}{R}.$$

If

$$|x_u| = |x_v| = 1, \quad x_u \cdot x_v = 0 \quad \text{at } P \in \mathscr{M}, \tag{2.43}$$

these x_u and x_v are principal directions under an orthogonal transformation if necessary, and then it holds that

$$Ax_u = k_1 x_u, \quad Ax_v = k_2 x_v, \tag{2.44}$$

where $k_1 = \frac{1}{R_1}$ and $k_2 = \frac{1}{R_2}$ are the principal curvatures. We thus obtain

$$S^{-1}AS = \begin{pmatrix} k_1 & 0 \\ 0 & k_2 \end{pmatrix}, \quad S = [x_u \; x_v]$$

which implies $(E, F, G) = (1, 0, 1)$ and hence $(L, M, N) = (k_1, 0, k_2)$ by (2.37). Then it holds that

$$n_u = -k_1 x_u, \quad n_v = -k_2 x_v$$

by (2.38).

The second fundamental form II is identified with a smooth function valued in symmetric matrices. Then it produces a bilinear symmetric form on $\mathscr{X}(\mathscr{M})$, the set of vector fields on $\mathscr{M}$, denoted by $\mathscr{B}$. There arises

$$\mathscr{B}(\xi, \eta) = -\frac{\partial n}{\partial \xi} \cdot \eta, \quad \xi, \eta \in \mathscr{X}(\mathscr{M}). \tag{2.45}$$

In fact, it suffices to take the local chart satisfying (2.43) to confirm (2.45). Then we obtain

$$\frac{\partial n}{\partial \xi} = \frac{d}{ds} n(\cdot + s\xi)\Big|_{s=0} = (\xi \cdot \nabla) n = (\xi_1 \frac{\partial}{\partial u} + \xi_2 \frac{\partial}{\partial v}) n, \quad \xi = \xi_1 x_u + \xi_2 x_v,$$

and therefore,

$$\begin{aligned}(\xi, \eta)_{II} &= \xi_1 \eta_2 L + \xi_1 \eta_2 M + \xi_2 \eta_1 M + \xi_2 \eta_2 N \\ &= -\xi_1 \eta_2 n_u \cdot x_u - \xi_1 \eta_2 n_u \cdot x_v - \xi_2 \eta_1 n_v \cdot x_u - \xi_2 \eta_2 n_v \cdot x_v \\ &= -\frac{\partial n}{\partial \xi} \cdot \eta\end{aligned}$$

for $\eta = \eta_1 x_u + \eta_2 x_v$ by (2.35).

Here we take a family of surfaces $\mathscr{M}_\varepsilon$ parametrized by

$$\boldsymbol{x}^\varepsilon = \boldsymbol{x} + \varepsilon f \boldsymbol{n},$$

where f is a smooth function on $\mathscr{M}$. If it holds that

$$\delta A(0) = \frac{d}{d\varepsilon} A(\varepsilon)\Big|_{\varepsilon=0} = 0 \tag{2.46}$$

for any f, this $\mathscr{M}$ is called a minimal surface, where $A(\varepsilon)$ denotes the area of $\mathscr{M}_\varepsilon$. Since

$$A(\varepsilon) = \int_{\mathscr{M}_\varepsilon} dS_\varepsilon, \; dS_\varepsilon = \left| \boldsymbol{x}_u^\varepsilon \times \boldsymbol{x}_v^\varepsilon \right| du dv$$

it holds that

$$\begin{aligned}\boldsymbol{x}_u^\varepsilon &= \boldsymbol{x}_u + \varepsilon f_u \boldsymbol{n} + \varepsilon f \boldsymbol{n}_u = (1 - \varepsilon f k_1) \boldsymbol{x}_u + \varepsilon f_u \boldsymbol{n} \\ \boldsymbol{x}_v^\varepsilon &= \boldsymbol{x}_v + \varepsilon f_v \boldsymbol{n} + \varepsilon f \boldsymbol{n}_v = (1 - \varepsilon f k_2) \boldsymbol{x}_v + \varepsilon f_v \boldsymbol{n}.\end{aligned}$$

Since (2.43) implies

$$\boldsymbol{x}_u \times \boldsymbol{x}_v = \boldsymbol{n}, \quad \boldsymbol{x}_v \times \boldsymbol{n} = \boldsymbol{x}_u, \quad \boldsymbol{n} \times \boldsymbol{x}_u = \boldsymbol{x}_v$$

it holds that

$$\begin{aligned}\boldsymbol{x}_u^\varepsilon \times \boldsymbol{x}_v^\varepsilon &= (1 - \varepsilon f k_1)(1 - \varepsilon f k_2) \boldsymbol{n} - \varepsilon (1 - \varepsilon f k_1) f_v \boldsymbol{x}_v - \varepsilon (1 - \varepsilon f k_2) f_u \boldsymbol{x}_u \\ &= (1 - 2\varepsilon f H) \boldsymbol{n} - \varepsilon (f_v \boldsymbol{x}_v + f_u \boldsymbol{x}_u) + \varepsilon^2 f^2 K \boldsymbol{n} + \varepsilon^2 f (k_1 f_v \boldsymbol{x}_v + k_2 f_u \boldsymbol{x}_u)\end{aligned}$$

by $2H = k_1 + k_2$ and $K = k_1 k_2$. We thus obtain

$$\left| \boldsymbol{x}_u^\varepsilon \times \boldsymbol{x}_v^\varepsilon \right|^2 = 1 - 4\varepsilon f H + \varepsilon^2 (4 f^2 H^2 + 2 f^2 K + f_u^2 + f_v^2) + O(\varepsilon^3),$$

and therefore,

$$|\boldsymbol{x}_u^\varepsilon \times \boldsymbol{x}_v^\varepsilon| = 1 - 2\varepsilon f H + \varepsilon^2 (f^2 K + \frac{f_u^2 + f_v^2}{2}) + O(\varepsilon^3).$$

Equality (2.43) implies

$$f_u^2 + f_v^2 = |df|^2, \quad df = f_u \boldsymbol{x}_u + f_v \boldsymbol{x}_v \quad \text{at } \boldsymbol{P} \in \mathscr{M}.$$

Since f^2 and $|df|^2$ are functions on $\mathscr{M}$, it holds that

$$A(\varepsilon) = A(0) - 2\varepsilon \int_{\mathscr{M}} f H \, dS + \varepsilon^2 \int_{\mathscr{M}} f^2 K + \frac{1}{2} |df|^2 \, dS + O(\varepsilon^3).$$

Hence equality (2.46) holds for any f if and only if $H = 0$ so that the minimal surface is a surface where the mean curvature vanishes. This minimal surface is physically realized as a soap film. It is stable in the direction of f if

$$2f^2 K + |df|^2 \geq 0.$$

Hence if $K \geq 0$ it is stable under any deformation.

A soap bubble is realized as an extremal closed surface of the area functional under the constraint that the volume of the enclosed body is a constant. If $\mathscr{O}$ denotes this body its volume is given by

$$V = \int_{\mathscr{O}} dx = \frac{1}{3} \int_{\partial \mathscr{O}} x \cdot n \, dS,$$

and then the Lagrange multiplier principle $\delta A(0) + \lambda \delta V(0) = 0$ realizes

$$H = \frac{\lambda}{6} = \text{constant}$$

for such a surface.

2.3.3 Conformal Geometry

Parametrization (2.34) is called conformal if

$$|x_u|^2 = |x_v|^2 \equiv E = E(u, v), \quad x_u \cdot x_v = 0, \tag{2.47}$$

everywhere in D. Then it holds that

$$A = \frac{1}{E}\begin{pmatrix} L & M \\ M & N \end{pmatrix}, \quad 2H = \operatorname{tr} A = \frac{L+N}{E}, \quad K = \det A = \frac{LN - M^2}{E^2}$$

and $E = G = e^{2\sigma}$, $F = 0$ for $\sigma = \sigma(u, v)$. Then we obtain

$$\begin{aligned}
\Gamma^1_{11} &= \frac{2\sigma_u e^{4\sigma}}{2e^{4\sigma}} = \sigma_u, & \Gamma^1_{12} &= \Gamma^1_{21} = \frac{2\sigma_v e^{4\sigma}}{2e^{4\sigma}} = \sigma_v, \\
\Gamma^1_{22} &= \frac{-2\sigma_u e^{4\sigma}}{2e^{4\sigma}} = -\sigma_u, & \Gamma^2_{11} &= \frac{-2\sigma_v e^{4\sigma}}{2e^{4\sigma}} = -\sigma_v \\
\Gamma^2_{12} &= \Gamma^2_{21} = \frac{2\sigma_u e^{4\sigma}}{2e^{4\sigma}} = \sigma_u, & \Gamma^2_{22} &= \frac{2\sigma_v e^{4\sigma}}{2e^{4\sigma}} = \sigma_v
\end{aligned}$$

by (2.33).

It holds also that

$$A = e^{-2\sigma}\begin{pmatrix} L & M \\ M & N \end{pmatrix} = \begin{pmatrix} A^1_1 & A^1_2 \\ A^2_1 & A^2_2 \end{pmatrix}$$

and hence

$$\Omega = \begin{pmatrix} \sigma_u & \sigma_v & -e^{-2\sigma} L \\ -\sigma_v & \sigma_u & -e^{-2\sigma} L \\ L & M & 0 \end{pmatrix}, \quad \Lambda = \begin{pmatrix} \sigma_v & -\sigma_u & -e^{-2\sigma} M \\ \sigma_u & \sigma_v & -e^{-2\sigma} N \\ M & N & 0 \end{pmatrix}$$

by (2.41). Equality (2.40) now implies

$$\begin{aligned}
&\sigma_{uu} + \sigma_{vv} + e^{-2\sigma}(LN - M^2) = 0 \\
&L_v - M_u = \sigma_v(L + N) \\
&N_u - M_v = \sigma_u(L + N),
\end{aligned} \tag{2.48}$$

and in particular, the Gauss equation ensures

$$K = \frac{LN - M^2}{E^2} = -e^{-2\sigma}(\sigma_{uu} + \sigma_{vv}) = -\frac{\Delta \log E}{2E} \tag{2.49}$$

by $E = e^{2\sigma}$.

From the Codazzi–Mainardi equation, on the other hand, there arises

$$\begin{aligned}
E_v H &= 2e^{2\sigma}\sigma_v H = 2E\sigma_v H = (L + N)\sigma_v = L_v - M_u \\
E_u H &= 2e^{2\sigma}\sigma_u H = 2E\sigma_u H = (L + N)\sigma_u = N_u - M_v.
\end{aligned}$$

Then we obtain

$$M_v + \frac{1}{2}(L - N)_u = N_u - E_u H + \frac{1}{2}(L - N)_u$$

$$= -E_u H + \frac{1}{2}(L+N)_u$$
$$= -E_u H + (EH)_u = EH_u \qquad (2.50)$$

and similarly,

$$M_u - \frac{1}{2}(L-N)_v = L_v - E_v H - \frac{1}{2}(L-N)_v$$
$$= -E_v H + \frac{1}{2}(L+N)_v$$
$$= -E_v H + (EH)_v = EH_v. \qquad (2.51)$$

Now we use the complex variable $z = u + \sqrt{-1}v$. Since

$$\frac{\partial}{\partial z} = \frac{1}{2}(\frac{\partial}{\partial u} - \sqrt{-1}\frac{\partial}{\partial v}), \quad \frac{\partial}{\partial \overline{z}} = \frac{1}{2}(\frac{\partial}{\partial u} + \sqrt{-1}\frac{\partial}{\partial v})$$

it holds that

$$EH_z = \frac{1}{2}E(H_u - \imath H_v)$$
$$= \frac{1}{2}\{M_v + \frac{1}{2}(L-N)_u - \sqrt{-1}(M_u - \frac{1}{2}(L-N)_v)\}$$
$$= \frac{1}{4}\left\{(L-N-2\imath M)_u + \sqrt{-1}(L-N-2\imath M)_v\right\} = \frac{1}{2}\phi_{\overline{z}},$$

where $\phi = L - N - 2\sqrt{-1}M$.

A closed surface $\mathscr{M}$ of genous 1 with constant mean curvature H is realized by a doubly periodic conformal mapping $x = x(u, v) : \mathbf{R}^2 \to \mathbf{R}^3$. Then above

$$\phi = L - N - 2\sqrt{-1}M$$

is holomorphic by $\phi_{\overline{z}} = 2EH_z = 0$, and therefore, a constant by Liouville's theorem on bounded entire holomorphic functions in Sect. 3.2.7.

Since

$$|\phi|^2 = (L-N)^2 + 4M^2 = (L+N)^2 + 4(M^2 - LN) = 4E^2(H^2 - K)$$

there arises

$$\frac{\lambda^2}{4E^2} = H^2 - K = H^2 + \frac{\Delta \log E}{2E}, \quad \lambda = |\phi|,$$

so that

$$\Delta \log E + 2EH^2 - \frac{\lambda^2}{2E} = 0.$$

Therefore, putting

$$E = \frac{\lambda}{2H} e^u,$$

we obtain

$$-\Delta u = 2EH^2 - \frac{\lambda^2}{2E} = \lambda H e - v - \lambda H e^{-2} = 2\lambda H \sinh u$$

or

$$-\Delta u = \lambda \sinh u \text{ in } \mathbf{R}^2, \text{ doubly periodic} \tag{2.52}$$

if $H = \frac{1}{2}$.

What has been done in differential geometry is to create a doubly periodic conformal immersion $x : \mathbf{R}^2 \to \mathbf{R}^3$, conversely, from the solution u to (2.52). The equation

$$-\Delta u = \lambda \sinh u, \ u > 0 \text{ in } \Omega, \quad u|_{\partial\Omega} = 0, \tag{2.53}$$

is used for this purpose, where $\Omega \subset \mathbf{R}^2$ is a rectangle. Then the theory of partial differential equations is applied to solve this question in the affirmative [24].

2.3.4 Liouville Integral and Spherical Derivatives

Gauss curvature is given by (2.49) under the conformal local chart (2.47). If this K is a positive constant there arises

$$-\Delta u = e^u \quad \text{in } \Omega = D \subset \mathbf{R}^2 \tag{2.54}$$

for $u = \log E + \log(2K)$. There is an integral of this equation found by J. Liouville. Here we take the complex variable $z = x_1 + \sqrt{-1}x_2$ and $\overline{z} = x_1 - \sqrt{-1}x_2$ for $x = (x_1, x_2) \in \Omega$, to introduce the function

$$s = u_{zz} - \frac{1}{2}u_z^2. \tag{2.55}$$

From (2.54) follows $u_{z\overline{z}} = -\frac{1}{4}e^u$ and hence

$$s_{\overline{z}} = u_{zz\overline{z}} - u_z u_{z\overline{z}} = -\frac{1}{4}e^u u_z + \frac{1}{4}e^u u_z = 0,$$

which means that $s = s(z)$ is a holomorphic function of $z \in \Omega \subset \mathbf{C}$.

Regarding (2.55) as a Riccati equation, we see that $\phi = e^{-u/2}$ satisfies

$$\phi_{zz} + \frac{1}{2}s\phi = 0. \tag{2.56}$$

Taking a point $x^* = (x_1^*, x_2^*) \in \Omega$, we introduce the fundamental system of solutions $\{\phi_1, \phi_2\}$ to the linear ordinary equation (2.56) in z by

$$\left(\phi_1, \frac{\partial \phi_1}{\partial z}\right)\Bigg|_{z=z^*} = (1, 0), \quad \left(\phi_2, \frac{\partial \phi_2}{\partial z}\right)\Bigg|_{z=z^*} = (0, 1) \tag{2.57}$$

for $z^* = x_1^* + \imath x_2^*$. These $\phi_1 = \phi_1(z)$ and $\phi_2 = \phi_2(z)$ are analytic functions in $z \in \Omega$ and the relation

$$\phi \equiv e^{-u/2} = \overline{f}_1(\overline{z})\phi_1(z) + \overline{f}_2(\overline{z})\phi_2(z)$$

holds for some functions $\overline{f}_1$ and $\overline{f}_2$ of $\overline{z}$. There arises

$$f_1(z) = C_1\phi_1(z), \quad f_2(z) = C_2\phi_2(z) \tag{2.58}$$

for

$$C_1 = e^{-u/2}\Big|_{x=x^*}, \quad C_2 = \frac{\lambda}{8}e^{u/2}\Big|_{x=x^*} \tag{2.59}$$

if $x^* = (x_1^*, x_2^*) \in \Omega$ is a critical point of $u = u(x)$, where

$$f_1(z) = \overline{\overline{f}_1(\overline{z})}, \quad f_2(z) = \overline{\overline{f}_2(\overline{z})}.$$

In fact we have

$$W(\phi_1, \phi_2) \equiv \phi_1\phi_{2z} - \phi_{1z}\phi_2 = 1$$

and hence

$$\overline{f}_1(\overline{z}) = W(\phi, \phi_2) = \phi\phi_{2z} - \phi_z\phi_2, \quad \overline{f}_2(\overline{z}) = W(\phi_1, \phi) = \phi_1\phi_z - \phi_{1z}\phi.$$

Here we confirm that these functions are independent of z. Then, putting $z = z^*$, we get

$$\overline{f}_1(\overline{z}) = \phi(z^*, \overline{z}), \quad \overline{f}_2(\overline{z}) = \phi_z(z^*, \overline{z}).$$

Since $\phi = e^{-u/2}$ is real-valued it solves also

$$\phi_{\overline{z}\overline{z}} + \frac{1}{2}\overline{s}\phi = 0 \tag{2.60}$$

for $\overline{s}(\overline{z}) = \overline{s(z)}$, and so do $\overline{f}_1(\overline{z}) = \phi(z^*, \overline{z})$ and $\overline{f}_2(\overline{z}) = \phi_z(z^*, \overline{z})$.

A fundamental system of solutions to (2.60) is given by $\{\overline{\phi}_1, \overline{\phi}_2\}$ for

$$\overline{\phi}_1(\overline{z}) = \overline{\phi_1(z)}, \quad \overline{\phi}_2(\overline{z}) = \overline{\phi_1(z)},$$

which satisfies

$$\left(\overline{\phi}_1, \frac{\partial \overline{\phi}_1}{\partial \overline{z}}\right)\Bigg|_{\overline{z}=\overline{z}^*} = (1, 0), \quad \left(\overline{\phi}_2, \frac{\partial \overline{\phi}_2}{\partial \overline{z}}\right)\Bigg|_{\overline{z}=\overline{z}^*} = (0, 1).$$

Then we obtain

$$\begin{aligned}
\overline{f}_1(\overline{z^*}) &= \phi(z^*, \overline{z^*}) = e^{-u/2}\Big|_{x=x^*} = C_1, \\
\frac{\partial}{\partial \overline{z}}\overline{f}_1(\overline{z^*}) &= \phi_{\overline{z}}(z^*, \overline{z^*}) = \frac{\partial}{\partial \overline{z}} e^{-u/2}\Big|_{x=x^*} = 0 \\
\overline{f}_2(\overline{z^*}) &= \phi_z(z^*, \overline{z^*}) = \frac{\partial}{\partial \overline{z}} e^{-u/2}\Big|_{x=x^*} = 0
\end{aligned}$$

and

$$\begin{aligned}
\frac{\partial}{\partial \overline{z}}\overline{f}_2(\overline{z^*}) &= \phi_{z\overline{z}}(z^*, \overline{z^*}) = \frac{1}{4}\Delta e^{-u/2}\Big|_{x=x^*} \\
&= -\frac{1}{8}e^{-u/2}\Delta u\Big|_{x=x^*} = \frac{\lambda}{8}e^{u/2}\Big|_{x=x^*} = C_2,
\end{aligned}$$

using $\nabla u(x^*) = 0$. These relations imply (2.58), which results in the Liouville integral

$$e^{-u/2} = C_1|\phi_1|^2 + C_2|\phi_2|^2 \tag{2.61}$$

for C_1 and C_2 defined by (2.59).

Put

$$\psi_1 = C_1^{1/2} 8^{-1/4}\phi_1, \quad \psi_2 = C_1^{-1/2} 8^{1/4}\phi_2, \quad F = \psi_2/\psi_1.$$

Since

$$W(\psi_1, \psi_2) = W(\phi_1, \phi_2) = 1$$

it holds that

$$\left(\frac{1}{8}\right)^{1/2} e^{u/2} = \frac{1}{|\psi_1|^2 + |\psi_2|^2} = \frac{|F'|}{1+|F|^2} \equiv \rho(F). \tag{2.62}$$

Here, F is a meromorphic function of $z \in \Omega \subset \mathbf{C}$ made by a quotient of two linearly independent solutions to (2.56), so that it satisfies

$$\{F; z\} = -\frac{1}{2}s,$$

where

$$\{F; z\} = \frac{3}{4}\left(\frac{F''}{F'}\right)^2 - \frac{1}{2}\frac{F'''}{F'}$$

is the Schwarzian derivative. This $\rho(F)$ stands for the spherical derivative of the conformal mapping $F : \Omega \to S^2$, where $S^2 \subset \mathbf{R}^3$ denotes the Gauss sphere, that is, a sphere with diameter 1. This property means

$$\rho(F) = \frac{d\sigma}{ds},$$

where the left-hand side stands for the ratio of standard line element in Ω and that in S^2 mapped by F.

If $S^2 \subset \mathbf{R}^3$ is thus the sphere with south and north poles at (0, 0, 0) and (0, 0, 1), respectively, and if $\tau : \mathbf{C}\bigcup\{\infty\} \to S^2$ denotes the inverse stereographic projection, then under the conformal mapping $\overline{F} = \tau \circ F : \Omega \to S^2$, the standard metrics on $\Omega \subset \mathbf{C} \cong \mathbf{R}^2$ and S^2 denoted by $ds^2 = dx_1^2 + dx_2^2$ and $d\sigma^2$, respectively, are related as

$$\frac{d\sigma}{ds} = \rho(F). \tag{2.63}$$

If $\omega \subset\subset \Omega$ is a subdomain, the immersed length of the image of $\partial\omega$ and the immersed area of the image of ω under $\overline{F}$, are given by

$$\ell(\partial\omega) = \int_{\partial\omega} \rho(F)\, d\sigma, \quad m(\omega) = \int_{\omega} \rho(F)^2 dx, \tag{2.64}$$

respectively.

The Gel'fand equation,

$$-\Delta v = \lambda e^v \text{ in } \Omega \subset \mathbf{R}^2, \quad v|_{\partial\Omega} = 0, \tag{2.65}$$

similarly, has the integral

$$(\frac{\lambda}{8})^{1/2} e^{v/2} = \rho(F).$$

Therefore, this boundary value problem is reduced to finding a conformal mapping $\overline{F} : \Omega \to S^2$ such that

$$\left.\frac{d\sigma}{ds}\right|_{\partial\Omega} = (\frac{\lambda}{8})^{1/2}.$$

Since $|S^2| = \pi$, this expression induces a quantized blowup mechanism, which exhibits

$$\int_\Omega \rho(F)^2 \, dx = \frac{1}{8} \int_\Omega \lambda e^v \, dx \equiv \frac{\Sigma}{8} \quad \to \quad \pi m, \ m \in \mathbf{N}$$

as the singular limit of the solution to (2.65) approaches. More precisely, ℓ-covering of S^2 by $\overline{F}(\Omega)$ arises as $L \downarrow 0$, where L stands for the immersed length of $\partial\Omega$:

$$L = (\frac{\lambda}{8})^{1/2} |\partial\Omega|.$$

2.3.5 *Radial Solutions*

If $f = f(v)$ is locally Lipschitz continuous and $\Omega = B(0, R) \subset \mathbf{R}^n$ is a ball, any positive classical solution $v = v(x)$ to

$$-\Delta v = f(v), \ v > 0 \quad \text{in } \Omega, \qquad v = 0 \quad \text{on } \partial\Omega, \tag{2.66}$$

is radially symmetric [9]. Since $\lambda e^v > 0$ in the Gel'fand problem (2.65) we obtain $v > 0$ in Ω by the maximum principle. In particular, any solution is radially symmetric if Ω is a ball.

Here we continue to take the case $n = 2$, and classify the solution to (2.65) for $\Omega = B(0, 1)$:

$$-\Delta v = \lambda e^v \quad \text{in } B = B(0, 1), \qquad v = 0 \quad \text{on } \partial B.$$

Since $v = v(r)$ for $r = |x|$, this equation is reduced to

$$v'' + \frac{1}{r} v' + \lambda e^v = 0, \quad r > 0, \qquad v'(0) = 0. \tag{2.67}$$

Following (2.65), equation (2.67) is provided with the scaling invariance. Thus if $v_0 = v_0(r)$, then

$$v_\alpha(r) = v_0(e^{\alpha/2}r) + \alpha \tag{2.68}$$

defined for $\alpha \in \mathbf{R}$ is also a solution. We shall find a solution to (2.67) denoted by $v_0(r)$, and adjust $\alpha \in \mathbf{R}$ such that $v = v_\alpha(r)$ in (2.68) satisfies the boundary condition in (2.65), $v_\alpha(1) = 0$, that is,

$$v_0(e^{\alpha/2}) + \alpha = 0. \tag{2.69}$$

For this purpose we use $s = \log r$ to write (2.67) as

$$\frac{d^2}{ds^2}(v + 2s) + \lambda e^{v+2s} = 0. \tag{2.70}$$

Thus, $u = v + 2s + \log\lambda$ in (2.70) satisfies (2.54) in one space dimension,

$$u'' + e^u = 0, \quad -\infty < s < \infty. \tag{2.71}$$

Since (2.71) implies

$$\left\{u'' - \frac{1}{2}(u')^2\right\}' = 0,$$

we seek $u = u(s)$ satisfying

$$u'' - \frac{1}{2}(u')^2 = -2. \tag{2.72}$$

Then $\ell = \dfrac{2 - u'(s/2)}{4}$ satisfies the logistic equation

$$\ell' = (1 - \ell)\ell. \tag{2.73}$$

Take a solution to (2.73) in the form of

$$\ell(s) = \frac{1}{2}(1 + \tanh\frac{s}{2}).$$

Then a solution to (2.67) is obtained as

$$v_0 = -2\log\cosh s + \log\frac{2}{\lambda} - 2s = \log\{\frac{8/\lambda}{(r^2+1)^2}\}.$$

Equality (2.69) means

$$\frac{8}{\lambda} = \frac{(e^{\alpha}+1)^2}{e^{\alpha}},$$

and therefore, the solution to (2.65) on

$$\Omega = \Omega^* \equiv \left\{ x \in \mathbf{R}^2 \mid |x| < 1 \right\}$$

is given by

$$v = v^*_{\lambda\pm}(x) = \log\{\frac{8\beta_\pm/\lambda}{(1+\beta_\pm |x|^2)^2}\}$$

$$\beta_\pm = \frac{4}{\lambda}\{1 - \frac{\lambda}{4} \pm (1-\frac{\lambda}{2})^{1/2}\}. \tag{2.74}$$

Regarding (2.74), we see that the solution is unique for $\lambda = 2$ and is given by

$$v^*_{\lambda+}(x) = v^*_{\lambda-}(x) = 2\log\frac{2}{1+|x|^2}.$$

Also, there are two solutions for $0 < \lambda < 2$, and no solution for $\lambda > 2$. The total set of solutions, denoted by $\mathscr{C}^* = \{(\lambda, v)\}$, is homeomorphic to $\mathbf{R}$ in the function space $\mathbf{R}_+ \times C(\overline{\Omega}^*)$, and there arises

$$\lim_{\lambda\downarrow 0} v^*_{\lambda-}(x) = 0 \qquad \text{uniformly in } x \in \overline{\Omega}^*$$

$$\lim_{\lambda\downarrow 0} v^*_{\lambda+}(x) = 4\log\frac{1}{|x|} \qquad \text{locally uniformly in } x \in \overline{\Omega}^* \setminus \{0\}.$$

In other words, the endpoints of $\mathscr{C}^*$ are $(0, 0)$ and $(0, v_*)$ with the latter outside the above function space $\mathbf{R}_+ \times C(\overline{\Omega}^*)$, where

$$v_*(x) = 4\log\frac{1}{|x|}. \tag{2.75}$$

This $v_* = v_*(x)$ in (2.75) is called the singular limit of the solution, which is not a solution to (2.65).

Radial solutions in (2.74) take the form

$$(\frac{\lambda}{8})^{1/2}e^{v/2} = (\frac{e^u}{8})^{1/2} = \frac{\mu^{1/2}}{|x|^2+\mu}$$

$$\mu = \mu_\pm = \frac{8}{\lambda}\{1 - \frac{\lambda}{4} \mp \sqrt{1-\frac{\lambda}{2}}\} = \beta_\pm^{-1}, \tag{2.76}$$

and therefore, their Liouville integrals are given by

$$F(z) = C_\pm z, \quad C_\pm = \mu_\pm^{-1/2} = \{\frac{1}{\lambda}\{4 - \lambda \pm 2\sqrt{4 - 2\lambda}\}\}^{1/2}. \tag{2.77}$$

The immersed length of $\overline{F}(\partial\Omega^*)$ and the immersed area of $\overline{F}(\Omega^*)$ given by (2.64) are equal to

$$\ell(\partial\Omega^*) = \int_{\partial\Omega^*} (\frac{\lambda e^v}{8})^{1/2}\, ds = 2\pi(\frac{\lambda}{8})^{1/2}$$

$$m(\Omega^*) = \int_{\Omega^*} \frac{\lambda e^v}{8}\, dx = \frac{\pi}{1 + \mu_\pm} \equiv \frac{\Sigma}{8\pi}.$$

Therefore, as the solution varies from $(0, 0)$ to $(0, v_*)$ along $\mathscr{C}^*$, the value σ increases monotonically from 0 to 8π. The value $\lambda = 2$ on $\mathscr{C}^*$ corresponds to $\Sigma = 4\pi$, while λ increases from 0 to 2 and then decreases from 2 to 0, monotonically.

The Liouville integral for radial solution on the disc is associated with

$$f(z) = cz,$$

while multiplicity of the Liouville integral on the punctured disc is controlled by z^α [7]. This form is used in the study of non-radial bifurcation of the solution to

$$-\Delta u = \lambda e^u \text{ in } A, \quad u|_{\partial A} = 0,$$

where $A = \{a < |x| < 1\} \subset \mathbf{R}^2$ [13, 15].

2.3.6 Differential Geometry of a Sphere

Recall that S^2 is the Riemann sphere with $(0, 0, 0)$ and $(0, 0, 1)$ representing the south and the north poles, respectively, with the canonical metric $d\sigma$. The meromorphic function F in the previous subsection induces the conformal mapping $\overline{F} = \tau \circ F$ from Ω into S^2 through the inverse stereographic mapping $\tau : \mathbf{C} \cup \{\infty\} \to S^2$.

Under $\overline{F}$, the relation (2.63) is valid, where $ds^2 = dx_1^2 + dx_2^2$ denotes the Euclidean metric on Ω. From this fact, the quantity $\rho(F)$ has been called the spherical derivative of F in the complex function theory. In particular, $\rho(F)$ is invariant under the $O(3)$-action on S^2. For each sub-domain $\omega \subset \Omega$ the quantities $\ell(\partial w)$ and $m(\omega)$ in (2.64) indicate the length of $\overline{F}(\partial\omega)$ and the area of $\overline{F}(\omega)$ in S^2, respectively, as immersions so that they enjoy the isoperimetric inequality on the sphere,

$$\ell(\partial w)^2 \geq 4m(\omega)(\pi - m(\omega)).$$

This inequality is nothing but a special form of *Bol's inequality* for surfaces with Gaussian curvatures dominated above by a constant.

A modification of the Schwarz symmetrization is possible by this inequality, to create an isoperimetric inequality on the first eigenvalue $\mu_1(p, \Omega)$ of the linearized operator $A_p \equiv -\Delta_D - p$, where

$$\Delta_D = \frac{\partial^2}{\partial x_1^2} + \frac{\partial^2}{\partial x_2^2}, \quad \cdot|_{\partial\Omega} = 0,$$

$p = \lambda e^u$, and (λ, u) is a solution to (2.65):

$$-\Delta u = \lambda e^u \text{ in } \Omega \subset \mathbf{R}^2, \quad u|_{\partial\Omega} = 0.$$

Putting

$$\nu_1 = \inf\{\int_\Omega |\nabla v|^2 \, dx \mid v \in H_0^1(\Omega), \int_\Omega v^2 p \, dx = 1\},$$

we obtain

$$\nu_1(p, \Omega) \geq \nu_1(p^*, \Omega^*), \tag{2.78}$$

if p satisfies

$$-\Delta \log p \leq p \text{ in } \Omega, \quad \Sigma \equiv \int_\Omega p dx < 8\pi. \tag{2.79}$$

In (2.78), $\Omega^* = \{|x| < 1\} \subset \mathbf{R}^2$ denotes the unit disc, and

$$p^* = \lambda^* e^{u^*}$$

with (λ^*, u^*) satisfying

$$-\Delta u^* = \lambda^* e^{u^*} \text{ in } \Omega^*, \quad u^*|_{\partial\Omega^*} = 0, \quad \int_{\Omega^*} \lambda^* e^{u^*} \, dx = \Sigma. \tag{2.80}$$

Such a pair (λ^*, u^*) actually exists uniquely for each $\Sigma \in (0, 8\pi)$, see Sect. 2.3.5.

Inequality (2.78) is derived by the method of rearrangement. Given a non-negative function $v = v(x)$ in $x \in \Omega$ and $t > 0$, we put

$$D_t = \{v > t\}$$

to obtain an open concentric disc D_t^* of Ω^* satisfying

$$\int_{D_t^*} p^*\, dx = \int_{D_t} p\, dx \in (0, 8\pi).$$

Then the spherically decreasing rearrangement of v, denoted by $v^* = v^*(x)$, $x \in \Omega^*$, is defined by

$$v^*(x) = \sup\{t \mid x \in D_t^*\}.$$

It is an equi-measurable rearrangement and it holds that

$$\int_\Omega v^2 p\, dx = \int_{\Omega^*} v^{*2} p^*\, dx. \tag{2.81}$$

The decrease of the Dirichlet integral,

$$\int_\Omega |\nabla v|^2\, dx \geq \int_{\Omega^*} |\nabla v^*|^2\, dx, \tag{2.82}$$

on the other hand, is valid from the co-area formula and Bol's inequality, if, for example, v is a non-negative C^1 function with the value zero on the boundary $\partial\Omega$.

From what has been described, the condition

$$\Sigma = \int_B \lambda^* e^{u^*} dx < 4\pi$$

implies the strict minimum of the radial solution u^*. By the elliptic theory [8], this property means the positivity of the first eigenvalue $\mu_1(p^*, \Omega^*)$ of the linearized operator

$$A_p^* \equiv -\Delta_D - p^* \text{ on } \Omega^*, \quad p^* = \lambda^* e^{u^*}.$$

Since $\mu_1(p^*, \Omega^*) > 0$ is equivalent to

$$\nu_1(p^*, \Omega^*) \equiv \{\int_{\Omega^*} |\nabla v|^2\, dx \mid v \in H_0^1(\Omega^*),\ \int_{\Omega^*} v^2 p^*\, dx = 1\} > 1. \tag{2.83}$$

we obtain the following fact [1].

Theorem 2.4 *If a C^2 function $p = p(x) > 0$ satisfies (2.79) and*

$$\Sigma = \int_\Omega p\, dx < 4\pi,$$

it holds that $\mu_1(p, \Omega) > 0$.

The analytic proof of (2.83) under $\Sigma = \int_{\Omega^*} p^* dx < 4\pi$ is reduced to the study of the eigenvalue problem

$$\Delta\phi + \frac{1}{\Lambda}\zeta_0(x)\phi = 0 \text{ in } B, \quad \phi|_{\partial B} = 0$$

$$\zeta_0(x) = p^* = \lambda^* e^{u^*} = \frac{8\rho}{(|x|^2+\rho)^2}, \quad \rho > 0. \tag{2.84}$$

After the separation of variables $\phi = \Phi(r)e^{\iota m\theta}$, we use the transformation

$$\xi = (\rho - r^2)/(\rho + r^2) \tag{2.85}$$

to induce a two point boundary value problem for the associated Legendre equation

$$\left[\left(1-\xi^2\right)\Phi_\xi\right]_\xi + \left[2/\Lambda - m^2/\left(1-\xi^2\right)\right]\Phi = 0, \quad \xi_\rho < \xi < 1$$

$$\Phi(1) = 1, \quad \Phi(\xi_\rho) = 0, \ \xi_\rho = (\rho-1)/(\rho+1). \tag{2.86}$$

Hence (2.83) is reduced to $\Phi(0) > 0$ for the solution Φ to (2.86) with $\Lambda = 1$ and $m = 0$, that is, $\Phi(\xi) = \xi$. Actually,

$$\int_B p^* \, dx < 4\pi$$

is equivalent to $\rho > 1$, or $\xi_\rho > 0$, and therefore, $\Sigma < 4\pi$ implies (2.83).

If (λ^*, u^*) satisfies (2.80), it is associated with the conformal mapping

$$\overline{F} = \tau \circ F : \Omega^* \to S^2$$

through $F(z)$ in (2.77), where $\tau : \mathbf{C} \cup \{\infty\} \to S^2$ denotes the inverse stereographic mapping. Under this transformation, the value $\nu_1(p^*, \Omega^*)$ is equal to the first eigenvalue $\hat{\mu}_1(\omega)$ of the Laplace–Beltrami operator under the Dirichlet condition in $\omega \subset S^2$, where ω is a disc of the area $\Sigma/8$. This observation explains the reason why the associated Legendre equation has arisen in the study of (2.84). In fact, this equation appears with the separation of variables of the three-dimensional Laplacian, or, the Laplace–Beltrami operator on the two-dimensional sphere. Then the transformation (2.85) indicates the inversion of the stereographic projection $j : S^2 \to \mathbf{C} \cup \{\infty\}$. Then, inequality (2.83) means that if the area of ω is less than that of the chemi-sphere of S^2, then $\hat{\mu}_1(\omega) > 1$ holds.

The spherically decreasing rearrangement of Bandle is a Schwarz symmetrization on a round sphere in this context. Let $8S^2$ be the two-dimensional sphere of the area 8π, and $d\sigma$ be its canonical metric. Given $p = p(x) > 0$ in a domain $\Omega \subset \mathbf{R}^2$ satisfying (2.79) and

$$\Sigma = \int_\Omega p \, dx < 8\pi,$$

we take a disc ω in $8S^2$ such that $\int_\omega dV = \Sigma$. Then, for each non-negative function $v(x)$ in Ω we define $v^*(x)$ in $x \in \omega$ by

$$v^*(x) = \sup \{t \mid x \in \omega_t\},$$

where ω_t denotes the open concentric disc of ω satisfying

$$\int_{\omega_t} dV = \int_{\{v>t\}} p\, dx.$$

Then the relations (2.81) and (2.82) mean

$$\int_\Omega v^2 p\, dx = \int_\omega v^{*2}\, dV, \quad \int_\Omega |\nabla v|^2\, dx \geq \int_\omega |dv^*|^2 dV,$$

respectively, where d denotes the differentiation with respect to the canonical metric on $8S^2$.

Theorem 2.4, thus, means the following: Given a domain $\omega \subset 8S^2$, the first eigenvalue $\mu_1(\omega)$ of the Laplace-Beltrami operator in ω under the Dirichlet condition is greater than one, provided that

$$\int_\omega dV < 4\pi.$$

2.3.7 Sinh–Poisson Equations

The Wente surface indicates a closed surface with genus 1 and the mean curvature $H = \frac{1}{2}$ [24]. To construct such an object, we use (2.53) for the rectanglar domain $\Omega = (0, a) \times (0, b)$:

$$-\Delta z = \lambda \sinh z, \ z > 0 \ \text{in } \Omega, \quad z|_{\partial\Omega} = 0. \tag{2.87}$$

Taking a doubly periodic function $\tilde{z}$ by the odd extension of $z = z(u, v)$, put

$$\omega(u, v) = \frac{1}{2}\tilde{z}\left(\frac{u}{\sqrt{\lambda}}, \frac{v}{\sqrt{\lambda}}\right),$$

which satisfies

$$\Delta\omega + \sinh\omega\ \cosh\omega = 0 \quad \text{in } \mathbf{R}^2. \tag{2.88}$$

Given such an ω, we construct a conformal mapping $x : \mathbf{R}^2 \to \mathbf{R}^3$ satisfying (2.47) for $E = e^{2\omega}$. In fact, this condition is represented by an over-determined system of (x_u, x_v, n),

$$\begin{aligned}
x_{uu} &= \omega_u x_u - \omega_v x_v + e^{\omega} \sinh \omega\, n \\
x_{uv} &= \omega_v x_u + \omega_u x_v \\
x_{vv} &= -\omega_u x_u + \omega_v x_v + e^{\omega} \cosh \omega\, n \\
n_u &= -e^{-\omega} \sinh \omega\, x_u \\
n_v &= -e^{-\omega} \cosh \omega\, x_v,
\end{aligned}$$

of which solvability follows from (2.88). The problem left is to make the surface $\mathscr{M} = x(\mathbf{R}^2)$ closed-up, which is possible by taking the parameters $(\lambda, b/a)$ appropriately.

As the Gel'fand equation takes the integral in complex variables, the sinh–Gordon equation is associated with quaternions. Here we study the general case,

$$-\Delta u = 2\lambda_1 e^u - 2\lambda_2 e^{-u} \quad \text{in } \Omega \subset \mathbf{R}^2,$$

where $\lambda_i > 0$, $i = 1, 2$, are constants. Then we write

$$z = x_1 + \sqrt{-1}x_2, \quad \overline{z} = x_1 - \sqrt{-1}x_2$$

for $x = (x_1, x_2) \in \Omega$ as before. Confirm

$$\frac{\partial}{\partial z} = \frac{1}{2}\left(\frac{\partial}{\partial x_1} - \sqrt{-1}\frac{\partial}{\partial x_2}\right), \quad \frac{\partial}{\partial \overline{z}} = \frac{1}{2}\left(\frac{\partial}{\partial x_1} + \sqrt{-1}\frac{\partial}{\partial x_2}\right)$$

to deduce

$$u_{z\overline{z}} = -\frac{\lambda_1}{2}e^u + \frac{\lambda_2}{2}e^{-u}.$$

Letting

$$U = \frac{1}{2}\begin{pmatrix} u_z & -\sqrt{\lambda_2}e^{-u/2} \\ \sqrt{\lambda_1}e^{u/2} & 0 \end{pmatrix}, \quad V = \frac{1}{2}\begin{pmatrix} 0 & -\sqrt{\lambda_1}e^{u/2} \\ \sqrt{\lambda_2}e^{-u/2} & u_{\overline{z}} \end{pmatrix},$$

we get

$$U_{\overline{z}} = \frac{1}{2}\begin{pmatrix} u_{z\overline{z}} & \frac{\sqrt{\lambda_2}}{2}e^{-\frac{u}{2}} \\ \frac{\sqrt{\lambda_1}}{2}e^{\frac{u}{2}}u_{\overline{z}} & 0 \end{pmatrix} = \begin{pmatrix} \frac{-\lambda_1 e^u + \lambda_2 e^{-u}}{4} & \frac{\sqrt{\lambda_2}e^{-\frac{u}{2}}u_{\overline{z}}}{4} \\ \frac{\sqrt{\lambda_1}e^{\frac{u}{2}}u_{\overline{z}}}{4} & 0 \end{pmatrix}$$

$$V_z = \frac{1}{2}\begin{pmatrix} 0 & -\frac{\sqrt{\lambda_1}}{2}e^{\frac{u}{2}}u_z \\ -\frac{\lambda_2}{2}e^{-\frac{u}{2}}u_z & u_{z\overline{z}} \end{pmatrix} = -\begin{pmatrix} 0 & \frac{\sqrt{\lambda_1}e^{\frac{u}{2}}u_z}{4} \\ \frac{\sqrt{\lambda_2}e^{-\frac{u}{2}}u_z}{4} & \frac{\lambda_1 e^u - \lambda_2 e^{-u}}{4} \end{pmatrix},$$

which results in

$$U_{\overline{z}} - V_z = \frac{1}{4}\begin{pmatrix} -\lambda_1 e^u + \lambda_2 e^{-u} & \sqrt{\lambda_2} e^{-u/2} u_z + \sqrt{\lambda_1} e^{u/2} u_z \\ \sqrt{\lambda_1} e^{u/2} u_{\overline{z}} + \sqrt{\lambda_2} e^{-u/2} u_z & \lambda_1 e^u - \lambda_2 e^{-u} \end{pmatrix}.$$

It holds also that

$$UV = \frac{1}{4}\begin{pmatrix} -\lambda_2 e^{-u} & -\sqrt{\lambda_1} e^{u/2} u_z - \sqrt{\lambda_2} e^{-u/2} u_{\overline{z}} \\ 0 & -\lambda_1 e^u \end{pmatrix}$$

$$VU = \frac{1}{4}\begin{pmatrix} -\lambda_1 e^u & 0 \\ \sqrt{\lambda_2} e^{-u/2} u_z + \sqrt{\lambda_1} e^{u/2} u_{\overline{z}} & -\lambda_2 e^{-u} \end{pmatrix},$$

and hence

$$U_{\overline{z}} - V_z + [U, V] = 0, \quad [U, V] = UV - VU. \tag{2.89}$$

We recall that the set of quaternions constitutes an algebra $\mathscr{H}$, generated by $\{\mathbf{1}, \mathbf{i}, \mathbf{j}, \mathbf{k}\}$ satisfying

$$\mathbf{ij} = \mathbf{k}, \quad \mathbf{jk} = \mathbf{i}, \quad \mathbf{ki} = \mathbf{j}.$$

This algebra has a representation,

$$\sigma_1 = \sqrt{-1}\mathbf{i}, \quad \sigma_2 = \sqrt{-1}\mathbf{j}, \quad \sigma_3 = \sqrt{-1}\mathbf{k},$$

where

$$\mathbf{1} = \begin{pmatrix} 1 & 0 \\ 0 & 1 \end{pmatrix}, \quad \sigma_1 = \begin{pmatrix} 0 & 1 \\ 1 & 0 \end{pmatrix}, \quad \sigma_2 = \begin{pmatrix} 0 & -\imath \\ \imath & 0 \end{pmatrix}, \quad \sigma_3 = \begin{pmatrix} 1 & 0 \\ 0 & -1 \end{pmatrix}.$$

Thus we regard $\mathscr{H} = \langle \mathbf{1}, \sigma_1, \sigma_2, \sigma_3 \rangle \subset M_2(\mathbf{C})$.

Then the vector $A = (a_1, a_2, a_3) \in \mathbf{R}^3$ is identified with

$$A = -\sqrt{-1} \sum_{i=1}^{3} a_i \sigma_i \in \operatorname{Im} \mathscr{H},$$

while any $\phi \in \mathscr{H}_* = \mathscr{H} \setminus \{0\}$ is represented by

$$\phi = \begin{pmatrix} a & b \\ -\overline{b} & \overline{a} \end{pmatrix}, \quad |a|^2 + |b|^2 \neq 0, \quad a, b \in \mathbf{C}. \tag{2.90}$$

Hence

$$\alpha \equiv U dz + V d\overline{z}$$

$$= \frac{1}{2}\begin{pmatrix} u_z ds & -\sqrt{\lambda_2}e^{-u/2}dz - \sqrt{\lambda_1}e^{u/2}d\overline{z} \\ \sqrt{\lambda_1}e^{u/2}dz + \sqrt{\lambda_2}e^{-u/2}d\overline{z} & u_{\overline{z}}d\overline{z} \end{pmatrix}$$

is a one form in Ω with the value in $\mathscr{H}$.

If Ω is simply-connected then there is $\phi : \Omega \to \mathscr{H}_*$ such that

$$d\phi \cdot \phi^{-1} = \alpha. \tag{2.91}$$

In fact, this equation is reduced to $d\phi = \alpha\phi$, which is equivalent to $\phi_z = U\phi$, $\phi_{\overline{z}} = V\phi$. Then, the Frobenius-Poincaré theorem assures its solvability under the compatibility condition

$$(U\phi)_{\overline{z}} = (V\phi)_z,$$

which is actually the case by (2.89) and

$$(U\phi)_{\overline{z}} = U_{\overline{z}} + U\phi_{\overline{z}} = (U_{\overline{z}} + UV)\phi$$
$$(V\phi)_z = V_z\phi + V\phi_z = (V_z + VU)\phi.$$

It is unique under the constraint $\phi(x_*) = e^{u(x_*)/2}\mathbf{1}$ for fixed $x_* \in \Omega$, and then, $\phi = \phi(z,\overline{z}) \neq 0$ follows from the component-wise analyticity. Hence we obtain $\phi \in \mathscr{H}_*$, which induces (2.91).

By (2.90) we obtain $\det\phi \neq 0$, which may be positive without loss of generality. Then $h = \log(\det\phi)$ satisfies

$$h_z = \mathrm{tr}\,(\phi^{-1}\phi_z) = \mathrm{tr}\,(\phi_z\phi^{-1}) = \mathrm{tr}\,U = \frac{1}{2}u_z, \quad h_{\overline{z}} = \mathrm{tr}\,V = \frac{1}{2}u_{\overline{z}}$$

by (2.91), and hence $h - \frac{u}{2} =$ constant in Ω. Recalling

$$h(x_*) = \det\left(e^{u(x_*)/4}\mathbf{1}\right) = e^{u(x_*)/2},$$

we obtain $h = \frac{u}{2}$, or $\det\phi = e^{u/2}$.

Now we take $\Omega = B \equiv B_1(0)$ and show

$$N \equiv \phi^{-1}\mathbf{k}\phi : B \to S^2.$$

In fact, using (2.90) for $\phi = \phi(x) \in \mathscr{H}_*$, $x \in B$, we obtain

$$\phi^{-1} = \frac{1}{\det\phi}\begin{pmatrix} \overline{a} & -b \\ \overline{b} & a \end{pmatrix}.$$

Then it follows that

$$N = \phi^{-1} \cdot (-\sqrt{-1})\sigma_3 \cdot \phi = \frac{-\sqrt{-1}}{\det \phi} \begin{pmatrix} \overline{a} & -b \\ \overline{b} & a \end{pmatrix} \begin{pmatrix} a & b \\ -\overline{b} & \overline{a} \end{pmatrix}$$

$$= \frac{-\sqrt{-1}}{\det \phi} \begin{pmatrix} |a|^2 - |b|^2 & 2\overline{a}b \\ 2a\overline{b} & |b|^2 - |a|^2 \end{pmatrix} = -\imath \sum_{i=1}^{3} \alpha_i \sigma_i$$

for

$$\alpha_1 = \frac{2(a_1 b_1 + a_2 b_2)}{|a|^2 + |b|^2}, \quad \alpha_2 = \frac{2(a_2 b_1 - a_1 b_2)}{|a|^2 + |b|^2}, \quad \alpha_3 = \frac{|a|^2 - |b|^2}{|a|^2 + |b|^2}$$

from $\det \phi = |a|^2 + |b|^2$, where $a = a_1 + \imath a_2$ and $b = b_1 + \imath b_2$. We thus have

$$N : \Omega \to \mathbf{R}^3 = \mathrm{Im}\, \mathscr{H}.$$

Writing

$$N(x) = \begin{pmatrix} \alpha_1 \\ \alpha_2 \\ \alpha_3 \end{pmatrix} = \frac{1}{|a|^2 + |b|^2} \begin{pmatrix} 2(a_1 b_1 + a_2 b_2) \\ 2(a_2 b_1 - a_1 b_2) \\ |a|^2 - |b|^2 \end{pmatrix},$$

we see $|N(x)| = 1$ for any $x \in B$. Hence $N : B \to S^2$ is a Gauss map on B, a surface with constant mean curvature.

2.4 Boltzmann–Poisson Equations

2.4.1 *Quantized Blowup Mechanism*

An equivalent form of (2.65) is the Boltzmann–Poisson equation

$$-\Delta v = \frac{\lambda e^v}{\int_\Omega e^v dx} \text{ in } \Omega \subset \mathbf{R}^2, \quad v|_{\partial\Omega} = 0. \tag{2.92}$$

There is actually a correspondence between (2.6) and (2.92) by

$$\lambda \leftrightarrow \frac{\lambda}{\int_\Omega e^v dx}.$$

This form of (2.92) arises as a mean field limit of equilibrium point vortices in the theory of L. Onsager, and also a stationary state of the fundamental equation of material transport in the context of canonical statistical mechanics [20]. Then, a recursive control of a Hamiltonian is observed across the material hierarchy [21]. This section is devoted to the quantized blowup mechanism and a uniqueness theorem to (2.92), where the geometric background of the Liouville integral in the previous section provides the driving force.

Let $G = G(x, x')$ be Green's function for the Poisson equation

$$-\Delta G(\cdot, x') = \delta_{x'} \text{ in } \Omega, \quad G(\cdot, x')\big|_{\partial\Omega} = 0, \tag{2.93}$$

where $\Omega \subset \mathbf{R}^2$ is a bounded domain with smooth boundary $\partial\Omega$ and $\delta_{x'} = \delta_{x'}(dx)$ is the delta function with the support at $x = x' \in \Omega$. The Robin function is defined by

$$R(x) = \left[G(x, x') + \frac{1}{2\pi} \log |x - x'| \right]_{x'=x}, \tag{2.94}$$

where

$$H = H_\ell(x_1, \cdots, x_\ell) = \frac{1}{2} \sum_{j=1}^{\ell} R(x_j) + \sum_{1 \leq i < j \leq \ell} G(x_i, x_j) \tag{2.95}$$

denotes the ℓ-th Hamiltonian of point vortices for $\ell \in \mathbf{N}$. The fundamental solution to $-\Delta$ in two space dimensions, $-\Delta\Gamma = \delta$, is given by

$$\Gamma(x) = \frac{1}{2\pi} \log \frac{1}{|x|}, \tag{2.96}$$

and we obtain

$$-\Delta w = 0 \text{ in } \Omega, \quad w = -\Gamma(\cdot - x') \text{ on } \partial\Omega$$

for $x' \in \Omega$ and $w = G(\cdot, x') - \Gamma(\cdot - x')$. Hence this $w = w(x)$ is smooth on $\overline{\Omega}$ and so is $R = R(x)$ in Ω. It holds also that $R|_{\partial\Omega} = -\infty$.

Here we provide the proof of a theorem, using the complex function theory. This theorem corresponds to the case $0 < m < +\infty$ in (2.6), (2.7), and (2.8).

Theorem 2.5 ([14]) *Let $\Omega \subset \mathbf{R}^2$ be a bounded domain with smooth boundary $\partial\Omega$, and (λ_k, v_k), $k = 1, 2, \cdots$, be a sequence of the classical solutions to (2.92):*

$$-\Delta v_k = \frac{\lambda_k e^{v_k}}{\int_\Omega e^{v_k} dx} \text{ in } \Omega, \quad v_k|_{\partial\Omega} = 0, \tag{2.97}$$

satisfying

$$\lim_{k\to\infty} \lambda_k = \lambda_0 \in (0, +\infty), \quad \lim_{k\to\infty} \|v_k\|_\infty = +\infty. \tag{2.98}$$

Then, it holds that $\lambda_0 = 8\pi\ell$, $\ell \in \mathbf{N} \equiv \{1, 2, \cdots, \}$. *Passing to a subsequence, there is a set of* ℓ*-interior points* $x_1^*, \cdots, x_\ell^*$ *satisfying*

$$\nabla_{x_j} H_\ell(x_1^*, \cdots, x_\ell^*) = 0, \quad 1 \leq j \leq \ell. \tag{2.99}$$

Here, $\mathscr{S} = \{x_1^*, \cdots, x_\ell^*\}$ *is the blowup set of* $\{v_k\}$*:*

$$\mathscr{S} = \{x_0 \in \overline{\Omega} \mid \exists x_k \to x_0,\ v_k(x_k) \to +\infty\}.$$

It holds that

$$v_k \to v_0 \text{ locally uniformly in } \overline{\Omega} \setminus \mathscr{S}$$

$$v_0 = v_0(x) = 8\pi \sum_{j=1}^{\ell} G(x, x_j^*). \tag{2.100}$$

Lemma 2.1 *Let* $B \equiv \{|x| < 1\} \subset \mathbf{R}^n$, $n \geq 2$, *and* $v \in W^{1,p}(B)$ *for* $1 < p < \infty$. *Assume, furthermore,* $\Delta v = 0$ *in* $B \setminus \{0\}$. *Then there is* $a \in \mathbf{R}$ *such that* $v - aE$ *is harmonic in* B, *where*

$$E(x) = \begin{cases} |x|^{2-n}, & n > 2 \\ -\log|x|, & n = 2 \end{cases}$$

denotes the fundamental solution: $-\Delta E = \delta$.

Proof Since the support of the distribution $\Delta v \in \mathscr{D}'(B)$ is contained in $\{0\}$, it holds that

$$\Delta v = \sum_{|\alpha| \leq k} C_\alpha D^\alpha \delta, \quad \exists k,\ \exists C_\alpha \in \mathbf{R}.$$

We take $\zeta \in C_0^\infty(B)$ satisfying

$$(-1)^{|\alpha|} D^\alpha \zeta(0) = C_\alpha, \quad |\alpha| \leq k,$$

to put $\zeta_\epsilon(x) = \zeta(x/\epsilon)$ for $0 < \epsilon \ll 1$. Then it holds that

$$-\int_B \nabla v \cdot \nabla \zeta_\epsilon \, dx = \langle \zeta_\varepsilon, \Delta v \rangle_{\mathscr{D},\mathscr{D}'} = \sum_{|\alpha| \leq k} C_\alpha^2 / \epsilon^{|\alpha|}.$$

Since $\|\nabla \zeta_\epsilon\|_\infty = O(\varepsilon^{-1})$ implies

$$\|\nabla \zeta_\varepsilon\|_{p'} = o(\epsilon^{-1}), \quad \frac{1}{p'} + \frac{1}{p} = 1,$$

we have

$$\int_B \nabla v \cdot \nabla \zeta_\epsilon \, dx = o(\epsilon^{-1}), \quad \epsilon \downarrow 0$$

by $\nabla v \in L^p(B)$, which implies $C_\alpha = 0$ for $|\alpha| \geq 1$. Hence the result follows. □

Proof *(Theorem 2.5)* Since L^1 norm of the right-hand side of (2.97) is bounded, the elliptic estimate [8] guarantees

$$\|v_k\|_{W^{1,q}} = O(1), \quad 1 \leq q < 2 = \frac{n}{n-1} \tag{2.101}$$

by $n = 2$. Now we apply the method of moving planes in Sect. 3.3.2 to (2.97) [9].

First, we take the case that Ω is convex. Since $v_k > 0$ in Ω, there is a family $A = \{T\}$ of simplices with uniform shape such that $v_k(x)$ takes the maximum in $T \in A$ at its vertex which is closest to $\partial\Omega$, and furthermore,

$$\omega \equiv \Omega \cap \hat{\omega} \subset \bigcup_{T \in A} T,$$

where $\hat{\omega}$ is an open set containing $\partial\Omega$. This family $A = \{T\}$ is determined by Ω, independent of the nonlinearity. We call this property the monotone decreasing of v_k near $\partial\Omega$. Since $\|v_k\|_1 \leq C_1$ follows from (2.101) this property implies

$$\|v_k\|_{L^\infty(\omega)} = O(1). \tag{2.102}$$

For general Ω, we take a disc outscribing $\partial\Omega$. Then, regarding its center the origin, we apply the Kelvin transformation $T : x \mapsto y = x/|x|^2$ and $v(y) = |x|^{n-2}u(x)$, which results in $\Delta v_y = |x|^{n+2}\Delta_x u$. Due to $n = 2$, we have the monotonicity of v_k near $\partial\Omega$.

Using (2.102) and local elliptic regularity near the boundary to (2.97), we obtain the uniform boundedness of any partial derivatives of $\{v_k\}$ near $\partial\Omega$. In particular, the holomorphic functions

$$s_k = v_{kzz} - \frac{1}{2}v_{kz}^2$$

of $z = x_1 + \sqrt{-1}x_2$ are uniformly bounded near $\partial\Omega$, and then the classical maximum principle and Montel's theorem guarantee that the family $\{s_k(z)\}$ takes a subsequence, denoted by the same symbol, such that $s_k(z) \to s_0(z)$ locally uniformly in Ω as $k \to \infty$. Using

$$u_k = v_k + \log \lambda_k - \log \int_\Omega e^{v_k} dx, \tag{2.103}$$

we have

$$-\Delta u_k = e^{u_k} \quad \text{in } \Omega \tag{2.104}$$

and

$$s_k = u_{kzz} - \frac{1}{2}u_{kz}^2, \tag{2.105}$$

where the argument in Sect. 2.3.4 is applicable.

Let $x_k = (x_{1k}, x_{2k}) \in \Omega$ be the maximum point of $v_k(x)$, and let $\{\varphi_{1k}(z), \varphi_{2k}(z)\}$ be the fundamental system of solutions to

$$\varphi_{zz} + \frac{1}{2}s_k(z)\varphi = 0$$

satisfying

$$(\varphi_{1k}, \frac{\partial \varphi_{1k}}{\partial z})\bigg|_{z=z_k^*} = (1, 0), \quad \left(\varphi_{2k}, \frac{\partial \varphi_{2k}}{\partial z}\right)\bigg|_{z=z_k^*} = (0, 1)$$

for $z_k^* = x_{1k} + \sqrt{-1}x_{2k}$. Then, we obtain

$$e^{-u_k/2} = \tilde{c}_k|\varphi_{1k}|^2 + \frac{\tilde{c}_k^{-1}}{8}|\varphi_{2k}|^2 \tag{2.106}$$

for $\tilde{c}_k = e^{-u_k(x_k)/2}$. From (2.103), equality (2.106) means

$$e^{-v_k/2} = c_k|\varphi_{1k}|^2 + \frac{\sigma_k c_k^{-1}}{8}|\varphi_{2k}|^2 \tag{2.107}$$

for $c_k = e^{-v_k(x_k)/2}$ and $\sigma_k = \dfrac{\lambda_k}{\int_\Omega e^{v_k}}$.

Passing to a subsequence, we have

$$\lim_{k\to\infty} x_k = x_0 \equiv (x_{10}, x_{20}) \in \Omega$$

by (2.102). Then we define the fundamental system of solutions $\{\varphi_{10}(z), \varphi_{20}(z)\}$ to

$$\varphi_{zz} + \frac{1}{2}s_0(z)\varphi = 0$$

by

$$(\varphi_{10}, \frac{\partial \varphi_{10}}{\partial z})\Big|_{z=z_0^*} = (1,0), \quad (\varphi_{20}, \frac{\partial \varphi_{20}}{\partial z})\Big|_{z=z_0^*} = (0,1),$$

where $z_0^* = x_{10}^* + \sqrt{-1}x_{20}^*$. It thus holds that $\varphi_{1k} \to \varphi_{10}$ and $\varphi_{2k} \to \varphi_{20}$ locally uniformly in Ω.

In (2.107) we have $v_k(x_k) = \|v_k\|_\infty \to +\infty$ by (2.98), and therefore,

$$\lim_{k\to\infty} c_k = 0. \tag{2.108}$$

Passing to a subsequence, we have

$$\lim_{k\to\infty} \sigma_k c_k^{-1} = \gamma > 0 \tag{2.109}$$

by (2.102). The blowup set $\mathscr{S}$ of $\{v_k\}$ thus coincides with the zero set of φ_{20} in Ω. Since zeros of the analytic function $\varphi_{20}(z)$ do not take accumulation points in Ω, the set $\mathscr{S}$ is finite. There arises also (2.100) for $v_0 = v_0(x)$ defined by

$$e^{-v_0/2} = \gamma|\varphi_{20}|^2.$$

By the elliptic regularity, this convergence is valid up to their derivatives of any order. We have also $\sigma_k \to 0$ by (2.108) and (2.109), which implies

$$\lim_{k\to\infty} \int_\Omega e^{v_k} = +\infty. \tag{2.110}$$

Taking the limit in (2.97), we obtain

$$-\Delta v_0 = 0 \text{ in } \Omega \setminus \mathscr{S}, \quad v_0 = 0 \text{ on } \partial\Omega.$$

This $v_0(x)$ takes each element in $\mathscr{S} = \{x_1^*, \cdots, x_\ell^*\}$ as an isolated singular point. By Lemma 2.1 we obtain $a_j \in \mathbf{R}$, $1 \le j \le \ell$, for which

$$u_0(x) = v_0(x) + \sum_{j=1}^{\ell} a_j \log|x - x_j^*| \tag{2.111}$$

is harmonic in Ω.

Now we make $k \to \infty$ in (2.105), to obtain

$$s_0 = v_{0zz} - \frac{1}{2}v_{0z}^2.$$

This $s_0(z)$ is thus a holomorphic function in z. The singularities on the right-hand side at $z = x_{1j}^* + \sqrt{-1}x_{2j}^*$ for $x_j^* = (x_{1j}^*, x_{2j}^*)$ are, therefore, removable. By (2.111), we take an expansion of the right-hand side.

Vanishing of the pole $z = z_j^*$ of the second and the first orders implies $a_j = 4$, which is equivalent to $m_j = 8\pi$, and (2.99), respectively. In fact, it holds that

$$\begin{aligned} s_0 \equiv u_{0zz} - u_{0z} &= (w_{jzz} - \frac{1}{2}w_{jz}^2) + (H_{jzz} - \frac{1}{2}H_{jz}^2) - w_{jz}H_{jz} \\ &= \frac{a_j/2 - a_j^2/8}{(z-\kappa_j)^2} + (H_{jzz} - \frac{1}{2}H_{jz}^2) + \frac{a_j}{2(z-\kappa_j)}H_{jz} \end{aligned}$$

by $\frac{\partial}{\partial z}\log|z| = \frac{z}{2}$. □

Remark 2.1 By (2.108)–(2.109) it holds that

$$\|v_k\|_\infty = -2\log\sigma_k + 2\log\gamma + o(1), \quad \sigma_k = \frac{\lambda_k}{\int_\Omega e^{v_k}} \tag{2.112}$$

in (2.97)–(2.98). We can show, on the other hand, the existence of a local maximizer $x = x_k^j$ of $v_k(x)$ such that $\lim_{k\to\infty} x_k^j = x_j^*$ [12]. By the above argument using this $x = x_k^j$ in (2.98), we obtain a result analogous to (2.112), that is,

$$v_k(x_k^j) = -2\log\sigma_k + 2\log\gamma_j + o(1), \quad 1 \le j \le \ell, \tag{2.113}$$

for some $\gamma_j > 0$. Equality (2.113) means the concentration at each blowup point of $\{v_k\}$ with the same magnitude.

Theorem 2.5 is associated with the following theorem concerning the equation without non-local term,

$$-\Delta u = \lambda f(u), \ u > 0 \quad \text{in } \Omega, \quad u|_{\partial\Omega} = 0 \tag{2.114}$$

for $f(u) = e^u$, where λ is a positive constant.

Theorem 2.6 ([14]) *Let $\Omega \subset \mathbf{R}^2$ be a bounded domain with smooth boundary $\partial\Omega$, and let (λ, u) be a solution to (2.114) for $f(u) = e^u$. Put*

$$\Sigma = \lambda \int_\Omega f(u)\,dx.$$

Then, as $\lambda \downarrow 0$, the family $\{\Sigma\}$ accumulates to $8\pi m$, where $m = 0, 1, \dots, +\infty$. According to the sequence converging to this value, the solutions $\{u\}$ behave as follows:

1. *$m = 0$. Uniform convergence to 0: $\|u\|_\infty \to 0$.*
2. *$0 < m < +\infty$. Finite point blowup: there exists a set composed of m points denoted by*

$$\mathscr{S} = \{\kappa_1, \kappa_2, \cdots, \kappa_m\} \subset \Omega$$

such that $\{u(x)\}$ has a limit $u_0(x)$ locally uniformly in $x \in \overline{\Omega} \setminus \mathscr{S}$, while $\mathscr{S}$ is the blowup set of $\{u(x)\}$.
3. *$m = +\infty$. Entire blowup: $u(x) \to +\infty$ for every $x \in \Omega$. This blowup is locally uniform in Ω.*

Theorem 2.7 ([14]) *In the second case of Theorem 2.6, it holds that*

$$u_0(x) = 8\pi \sum_{j=1}^{m} G(x, \kappa_j), \tag{2.115}$$

where $G(x, y)$ denotes the Green function of $-\Delta$ under the Dirichlet condition as in (2.93). The blowup points $\kappa_j \in \Omega$, $1 \leq j \leq m$, satisfy the relation

$$\frac{1}{2}\nabla_x R(\kappa_j) + \sum_{\ell \neq j} \nabla_x G(\kappa_j, \kappa_\ell) = 0, \quad 1 \leq j \leq m, \tag{2.116}$$

where $R(x)$ denotes the Robin function in (2.94).

2.4.2 Bol's Inequality in Analytic Form

We continue to suppose that $\Omega \subset \mathbf{R}^2$ is a simply-connected domain. Let $v = v(x)$ be a solution to (2.92). Then the function $0 < p = p(x) \in C^2(\Omega) \cap C^0(\overline{\Omega})$ defined by

$$p = \frac{\lambda e^v}{\int_\Omega e^v dx}$$

induces a parametric surface with Gaussian curvature bounded above by $1/2$, where Bol's inequality is valid. Here we show an analytic proof [2] based on the following lemma [16].

Lemma 2.2 (Nehari) *Each harmonic function h in Ω admits the inequality*

$$\left(\int_{\partial\omega} e^{h/2}\, ds\right)^2 \geq 4\pi \int_\omega e^h\, dx, \tag{2.117}$$

where $\omega \subset\subset \Omega$ is a subdomain.

Proof Since h is harmonic, there exists a holomorphic function $g = g(z)$ in Ω satisfying $|g'(z)|^2 = e^h$ in Ω. Then we obtain

$$\int_{\partial\omega} e^{h/2}\, ds = \int_{\partial\omega} |g'|\, ds = H^1(\partial\{g(\omega)\})$$
$$\int_{\omega} e^h\, dx = \int_{\omega} |g'|^2\, dx = H^2(g(\omega))$$

as an immersion, and then (2.117) follows from the isoperimetric inequality for the flat Riemannian surface $g(\omega)$. □

Theorem 2.8 (Bandle) *If* $0 < p = p(x) \in C^2(\Omega)$ *satisfies*

$$-\Delta \log p \leq p \text{ in } \Omega, \quad \lambda \equiv \int_{\Omega} p dx < 8\pi, \tag{2.118}$$

it holds that

$$\ell(\partial\omega)^2 \geq \frac{1}{2}(8\pi - m(\omega))m(\omega), \tag{2.119}$$

where $\omega \subset\subset \Omega$ *is a sub-domain and*

$$\ell(\partial\omega) = \int_{\partial\omega} p^{1/2}\, ds, \quad m(\omega) = \int_{\omega} p\, dx.$$

Proof Let h be the harmonic lifting of $\log p$ in ω,

$$-\Delta h = 0 \text{ in } \omega, \quad h = \log p \text{ on } \partial\omega, \tag{2.120}$$

and put $q = pe^{-h}$. Then it holds that

$$-\Delta \log q \leq qe^h \text{ in } \omega, \quad q = 1 \text{ on } \partial\omega. \tag{2.121}$$

We put

$$K(t) = \int_{\{q>t\}} qe^h\, dx, \quad \mu(t) = \int_{\{q>t\}} e^h\, dx,$$

and apply the co-area formula as

$$-K'(t) = \int_{\{q=t\}} \frac{qe^h}{|\nabla q|}\, ds = t\int_{\{q=t\}} \frac{e^h}{|\nabla q|}\, ds = -t\mu'(t) \quad \text{a.e. } t, \tag{2.122}$$

Then we use Green's formula as

$$\int_{\{q>t\}} (-\Delta \log q)\, dx = \int_{\{q=t\}} \frac{|\nabla q|}{q}\, ds = \frac{1}{t}\int_{\{q=t\}} |\nabla q|\, ds \quad \text{a.e. } t > 1$$

regarding Sard's lemma, and then it follows that

$$\frac{1}{t}\int_{\{q=t\}} |\nabla q|\, ds \leq \int_{\{q>t\}} qe^h\, dx = K(t) \quad \text{a.e. } t > 1$$

from (2.121). Schwarz's inequality now implies

$$\begin{aligned} -K'(t)K(t) &\geq \frac{1}{t}\int_{\{q=t\}} |\nabla q|\, ds \cdot t \int_{\{q=t\}} \frac{e^h}{|\nabla q|}\, ds \geq (\int_{\{q=t\}} e^{h/2}\, ds)^2 \\ &\geq 4\pi \int_{\{q>t\}} e^h\, dx = 4\pi\mu(t) \quad \text{a.e. } t > 1 \end{aligned}$$

by (2.117), and hence it follows that

$$\begin{aligned} \frac{d}{dt}(\mu(t)t - K(t) + \frac{1}{8\pi}K(t)^2) &= \mu(t) + \frac{1}{4\pi}K(t)K'(t) \\ &\leq 0 \quad \text{a.e. } t > 1. \end{aligned} \tag{2.123}$$

We obtain, on the other hand, $K(t+0) = K(t) \leq K(t-0)$, while

$$j(t) \equiv K(t) - \mu(t)t = \int_{\{q>t\}} (q-t)e^h\, dx$$

is continuous: $j(t+0) = j(t) = j(t-0)$. Thus (2.123) implies

$$\left[j(t) + \frac{1}{8\pi}K(t)^2 \right]_{t=1}^{t=\infty} = j(1) - \frac{K(1)^2}{8\pi} \leq 0 \tag{2.124}$$

Using

$$j(1) = \int_{\{q>1\}} (q-1)e^h\, dx \geq \int_\omega (q-1)e^h\, dx = m(\omega) - \int_\omega e^h\, dx$$

and $K(1)^2 \leq m(\omega)^2$, we deduce

$$m(\omega) - \frac{1}{8\pi}m(\omega)^2 \leq \int_\omega e^h\, dx \leq \frac{1}{4\pi}(\int_{\partial\omega} p^{1/2}\, ds)^2 = \frac{\ell(\partial\omega)^2}{4\pi}.$$

This inequality means that (2.119) holds. □

2.4.3 Spherically Decreasing Rearrangement

Here we introduce a decreasing rearrangement on a sphere based on Bol's inequality [3], to deduce isoperimetric inequalities concerning eigenvalues on the associated Laplace–Beltrami operator, that is, Theorem 2.4.

First, we provide a metric on $\Omega^* \equiv \{|x| < 1\} \subset \mathbf{R}^2$, using the solution $u^* \in C^2(\Omega^*) \cap C(\overline{\Omega^*})$ to

$$-\Delta u^* = \frac{\lambda e^{u^*}}{\int_{\Omega^*} e^{u^*} dx} \text{ in } \Omega^*, \quad u^*\big|_{\partial\Omega^*} = 0 \tag{2.125}$$

for each $\lambda \in (0, 8\pi)$. Classification of radial solutions in Sect. 2.3.5 ensures the following fact.

Lemma 2.3 *Each $\lambda \in (0, 8\pi)$ admits a unique solution $u^* = u^*(|x|)$ to (2.125).*

Using the explicit form of this u^*, we obtain

$$p^*(x) \equiv \frac{\lambda e^{u^*}(x)}{\int_{\Omega^*} e^{u^*}\, dx} = \frac{8\mu}{(|x|^2 + \mu)^2}, \quad \mu = \mu(\lambda) \in (0, +\infty). \tag{2.126}$$

This $p^* = p^*(|x|)$ is strictly decreasing in $r = |x|$, and it holds that

$$\lim_{\lambda \downarrow 0} \mu(\lambda) = +\infty, \quad \lim_{\lambda \uparrow 8\pi} \mu(\lambda) = 0,$$

and

$$-\Delta \log p^* = p^* \quad \text{in } \Omega^*. \tag{2.127}$$

Then we obtain the equality in (2.119) for $p = p^*$ and $\Omega = \Omega^*$ from the proof of Theorem 2.8.

Let $\Omega \subset \mathbf{R}^2$ be a simply-connected domain, $p = p(x) > 0$ be a C^2 function satisfying

$$-\Delta \log p \leq p \text{ in } \Omega, \quad \lambda \equiv \int_{\Omega} p\, dx < 8\pi. \tag{2.128}$$

We take the solution $u^* = u^*(x)$ to (2.125) on the unit disc Ω^* to define $p^* = p^*(x)$ by (2.126). Given a measurable function $v = v(x)$ of $x \in \Omega$ and

$$0 < t < t_0 = \|v\|_\infty,$$

let $\Omega_t = \{v > t\}$ and define an open concentric disc Ω_t^* of Ω^* by

$$\int_{\Omega_t^*} p^* \, dx = \int_{\Omega_t} p \, dx = a(t) \in (0, 8\pi). \tag{2.129}$$

Then, spherically decreasing rearrangement v^* of v is a non-negative function in Ω^* satisfying $v^*(x) = \sup\{t \mid x \in \Omega_t^*\}$. It is an equi-measurable rearrangement, and in particular, it holds that

$$\int_\Omega v^2 p \, dx = \int t^2 \, d(-a(t)) = \int_{\Omega^*} {v^*}^2 p^* \, dx. \tag{2.130}$$

It is obvious that v^* is radially symmetric.

Lemma 2.4 *If $v(x)$ is C^2 on $\overline{\Omega}$, $v \geq 0$ in Ω, and $v|_{\partial\Omega} = 0$, then it holds that*

$$\int_\Omega |\nabla v|^2 \, dx \geq \int_{\Omega^*} |\nabla v^*|^2 \, dx. \tag{2.131}$$

Proof The function $a(t)$ in (2.129) is right-continuous and strictly decreasing in $t > 0$. Since $v \geq 0$ in Ω, the co-area formula assures

$$-a'(t) = \int_{\{v=t\}} \frac{p}{|\nabla v|} \, ds$$

$$-\frac{d}{dt} \int_{\Omega_t} |\nabla v|^2 \, dx = \int_{\{v=t\}} |\nabla v| \, ds \quad \text{a.e. } t \in (0, t_0).$$

By Sard's lemma, Schwarz's inequality, and Bol's inequality, we now obtain

$$\begin{aligned} -\frac{d}{dt} \int_{\Omega_t} |\nabla v|^2 \, dx &\geq \Big(\int_{\{v=t\}} p^{1/2} \, ds\Big)^2 \Big/ \Big(\int_{\{v=t\}} \frac{p}{|\nabla v|} \, ds\Big) \\ &= \frac{\ell(\{v=t\})^2}{-a'(t)} \\ &\geq \frac{(8\pi - a(t))a(t)}{-a'(t)} \quad \text{a.e. } t \in (0, t_0). \end{aligned} \tag{2.132}$$

The function $j(t) = -\int_{\Omega_t} |\nabla v|^2 \, dx$ is continuous and strictly decreasing in t by

$$j(t) - j(t-0) = \int_{\{v=t\}} |\nabla v|^2 \, dx = 0.$$

Thus this $j(t)$ is absolutely continuous and hence it holds that

$$\int_\Omega |\nabla v|^2 \, dx = \int_0^{t_0} dt \, \Big(-\frac{d}{dt} \int_{\Omega_t} |\nabla v|^2 \, dx\Big)$$

$$\geq \int_0^{t_0} \frac{(8\pi - a(t))a(t)}{-a'(t)}\, dt. \qquad (2.133)$$

The function $v = v^*(x)$, on the other hand, is strictly decreasing in $r = |x|$, and equalities hold at each step in (2.132) for $\Omega = \Omega^*$ and $v = v^*$. We thus obtain the equality in (2.133) for this case, and then inequality (2.131) follows from

$$\begin{aligned}\int_{\Omega^*} |\nabla v^*|^2\, dx &= \int_0^{t_0} dt\, (-\frac{d}{dt}\int_{\Omega^*} |\nabla v^*|^2\, dx) \\ &= \int_0^{t_0} \frac{(8\pi - a(t))a(t)}{-a'(t)}\, dt,\end{aligned}$$

and the proof is complete. □

Proof *(Theorem 2.4)* Similarly to the isoperimetric inequality on the first eigenvalue of $-\Delta$ under the Dirichlet boundary condition, equality (2.130) and inequality (2.131) imply

$$\nu_1(p, \Omega) \geq \nu_1(p^*, \Omega^*) \qquad (2.134)$$

for

$$\begin{aligned}\nu_1(p, \Omega) &= \inf\{\int_\Omega |\nabla v|^2\, dx \mid v \in H_0^1(\Omega),\ \int_\Omega v^2 p\, dx = 1\} \\ \nu_1(p^*, \Omega^*) &= \inf\{\int_{\Omega^*} |\nabla v|^2\, dx \mid v \in H_0^1(\Omega^*),\ \int_{\Omega^*} {v^*}^2 p^*\, dx = 1\},\end{aligned}$$

provided that

$$\lambda \equiv \int_\Omega p\, dx = \int_{\Omega^*} p^* dx\ \in (0, 8\pi).$$

Hence Theorem 2.4 is reduced to the radially symmetric case, and the following lemma concludes the proof. □

Lemma 2.5 *It holds that $\nu_1(p^*, \Omega^*) > 1$ in (2.134) if*

$$\lambda = \int_{\Omega^*} p^*\, dx < 4\pi.$$

Proof By the transformation described in Sect. 2.3.6, this fact is proven by the study of

$$-\Delta\phi = \Lambda p^*(r)\phi \text{ in } \Omega^*, \quad \phi|_{\partial\Omega^*} = 0 \qquad (2.135)$$

for

$$p^*(r) = \frac{8\mu}{(r^2+\mu)^2}, \qquad r = |x|.$$

In fact, from the separation of variables $\phi = \Phi(r)e^{\iota m\theta}$ and the stereographic transformation of variables

$$\xi = (\mu - r^2)/(\mu + r^2),$$

problem (2.135) is reduced to the two-point boundary value problem on the associated Legendre equation

$$[(1-\xi^2)\Psi_\xi]_\xi + [2/\Lambda - m^2/(1-\xi^2)]\Phi = 0, \quad \xi_\mu < \xi < 1 \tag{2.136}$$

with $\Phi(1) = 1$ and $\Phi(\xi_\mu) = 0$ for

$$\xi_\mu = (\mu - 1)/(\mu + 1).$$

Let Φ be the solution to (2.136) for $\Lambda = 1$, $m = 0$, and $\Phi(1) = 1$. Then the relation $\nu_1(p^*, \Omega^*) > 1$ is equivalent to

$$\Phi(\xi) > 0, \quad \xi_\mu < \xi < 1.$$

Since such Φ is given by $P_0(\xi) = \xi$, it means $\xi_\mu > 0$, or $\mu > 1$. Finally, by

$$p^*(r) = 8\mu/(r^2+\mu)^2$$

we have $\mu > 1$ if and only if $\lambda = \int_{\Omega^*} p^*\, dx < 4\pi$. Hence the proof is complete.

□

Lemma 2.5 is proven alternatively by the bending of the set of radially symmetric solutions to the Gel'fand equation (2.65) in the λu plane. Inequality (2.134), furthermore, is an isoperimetric inequality of the Laplace–Beltrami operator on a surface with Gaussian curvature bounded above by $1/2$. More precisely, let S be the two-dimensional sphere with area 8π, and its canonical metric and area element be $d\sigma$ and dV, respectively. Let, furthermore, $\iota : S \setminus \{n\} \to \mathbf{R}^2$ be the stereographic projection from the north pole $n \in S$ onto the whole space $\mathbf{R}^2$, which is tangent to the south pole $s \in S$. Let $\omega^* \subset S$ be a ball (or boul, more precisely) with center $s \in S$ and $-\Delta_S(\omega^*)$ be the Laplace–Beltrami operator in ω^* under the Dirichlet boundary condition on $\partial\omega^*$. Then the injection $\iota : \omega^* \to \mathbf{R}^2$ transforms $-\Delta_S(\omega^*)$ to $-\Delta/\tilde{p}$ in $\tilde{\Omega} = \iota(\omega^*)$ for some $\tilde{p} = \tilde{p}(|x|) > 0$, under the Dirichlet boundary condition on $\partial\tilde{\Omega}$. Since the Gaussian curvature of $\mathscr{S}$ is $1/2$, the radially symmetric function $\tilde{p}(|x|)$ satisfies $-\Delta \log \tilde{p} = \tilde{p}$ in $\tilde{\Omega}$, besides

$$\Sigma = \int_{\tilde{\Omega}} \tilde{p}\, dx = \int_{\omega^*} dV. \tag{2.137}$$

If ι^* denotes the pull-back of ι, the relations

$$\int_{\omega^*} |d(\iota^* v)|^2\, dV = \int_{\tilde{\Omega}} |\nabla v|^2\, dx$$
$$\int_{\omega^*} (\iota^* v)^2\, dV = \int_{\tilde{\Omega}} v^{*2} \tilde{p}\, dx \tag{2.138}$$

hold, and hence $\nu(\tilde{p}, \tilde{\Omega})$ coincides with the first eigenvalue of $-\Delta_S(\omega^*)$ under the Dirichlet boundary condition on $\partial\omega^*$. Here, the operation d in (2.138) denotes the differentiation with respect to the canonical metric on S as in Sect. 2.3.6. This observation reveals the reason why the associated Legendre equation arises in the study of (2.135).

Let R be the radius of $\tilde{\Omega}$, and take $y = x/R$. Then $p^*(y) = R^2\tilde{p}(x) > 0$ is radially symmetric and satisfies

$$-\Delta \log p^* = p^* \quad \text{in } \Omega^* \equiv \{|y| < 1\} \subset \mathbf{R}^2.$$

Furthermore, we have

$$\frac{\int_{\tilde{\Omega}} |\nabla v|^2\, dx}{\int_{\tilde{\Omega}} v^2 \tilde{p}\, dx} = \frac{\int_{\Omega^*} |\nabla v_1|^2\, dy}{\int_{\Omega^*} v_1^2 p^*\, dy}, \qquad v \in H_0^1(\tilde{\Omega}),\ v_1(y) = v(x).$$

Hence the eigenvalue problem for $-\Delta_S(\omega^*)$ is reduced to that of $-\Delta/p^*$ on the unit disc,

$$-\Delta\phi = \Lambda p^* \phi \text{ in } \Omega^*, \qquad \phi|_{\partial\Omega^*} = 0, \tag{2.139}$$

and it holds also that

$$\int_{\Omega^*} p^*\, dx = \int_{\tilde{\Omega}} \tilde{p}\, dx = \Sigma \in (0, 8\pi). \tag{2.140}$$

For the constant $\lambda^* = p^*|_{\partial\Omega^*} > 0$, the function p^* is realized as $p^* = \lambda^* e^{u^*}$, where $u^* = u^*(|y|)$ is a solution to

$$-\Delta u^* = \lambda^* e^{u^*} \text{ in } \Omega^*, \qquad u^*\big|_{\partial\Omega^*} = 0.$$

Hence we obtain

$$p^*(y) = \frac{8\mu}{(|y|^2 + \mu)^2}$$

with $\mu > 0$ determined by (2.140), and (2.139) coincides with (2.135).

We have thus confirmed that the eigenvalue problem for $-\Delta_S(\omega^*)$ is equivalent to (2.135). From this property, standard separation of variables for $\Delta_S(\omega^*)$, or three-dimensional Laplacian,

$$\Delta = \frac{\partial^2}{\partial x_1^2} + \frac{\partial^2}{\partial x_2^2} + \frac{\partial^2}{\partial x_3^2},$$

induces the associated Legendre equation, which is also true in (2.135) if the inverse stereographic transformation $\xi = (\mu - r^2)/(\mu + r^2)$ is applied.

As noticed in Sect. 2.3.6, the above spherically decreasing rearrangement is nothing but the Schwarz symmetrization on the sphere. Given a domain $\Omega \subset \mathbf{R}^2$ with sufficiently smooth boundary, and given $0 < p = p(x) \in C^2(\Omega) \cap C(\overline{\Omega})$ in we take all ball $\omega^* \subset S$ satisfying

$$\int_{\omega^*} dV = \Sigma.$$

Then, for a measurable function $0 \leq v = v(x)$ in $x \in \Omega$ we define $v^* : \omega^* \to (-\infty, +\infty]$ by

$$v^*(x) = \sup\{t \mid x \in \omega_t\},$$

where $\omega_t \subset S$ denotes the open concentric ball of ω^* such that

$$\int_{\omega_t} dV = \int_{\{v>t\}} p\, dx.$$

We thus obtain the following properties.

Lemma 2.6 *If $\psi : \mathbf{R} \to \mathbf{R}$ is a continuous monotone function, it holds that*

$$\int_\Omega (\psi \circ v) p\, dx = \int_{\omega^*} (\psi \circ v^*)\, dV.$$

If $v = v(x)$ is C^2, $v \geq 0$ in Ω, and $v|_{\partial\Omega} = 0$, there arises

$$\int_\Omega |\nabla v|^2\, dx \geq \int_{\omega^*} |dv^*|^2\, dV.$$

2.4.4 A Uniqueness Theorem

Here we show the following theorem.

Theorem 2.9 ([19]) *If $\Omega \subset \mathbf{R}^2$ is simply-connected, then the solution to (2.92) is unique for $0 < \lambda < 8\pi$.*

Since the a priori estimate is valid for this range of λ by Theorem 2.5 and the solution to $\lambda = 0$ is unique, that is, $v = 0$, this theorem is reduced to the non-degeneracy of the linearized operator by a standard bifurcation analysis. Equation (2.92) has a variational structure. Hence it is the Euler–Lagrange equation to

$$J_\lambda(v) = \frac{1}{2}\|\nabla v\|_2^2 - \lambda \log \int_\Omega e^v dx, \quad v \in H_0^1(\Omega) \tag{2.141}$$

because the elliptic estimate associated with the Trudinger–Moser inequality guarantees the regularity of the weak solution to $v \in H_0^1(\Omega)$ satisfying

$$\left.\frac{d}{ds} J_\lambda(v + s\varphi)\right|_{s=0} = 0, \quad \forall \varphi \in C_0^\infty(\Omega),$$

which ensures (2.92) for this $v = v(x)$. From this variational structure, the linearized operator to (2.92) is realized as a self-adjoint operator in $L^2(\Omega)$.

Given the solution $v = v(x)$ to (2.92), we thus put

$$p = \frac{\lambda e^v}{\int_\Omega e^v dx} \tag{2.142}$$

to introduce the bilinear form

$$\begin{aligned} Q(\varphi, \varphi) &= \left.\frac{1}{2}\frac{d^2}{ds^2} J_\lambda(v + s\varphi)\right|_{s=0} \\ &= (\nabla\varphi, \nabla\varphi) - \lambda \frac{\int_\Omega e^v \varphi^2 dx}{\int_\Omega e^v dx} + \lambda \frac{(\int_\Omega e^v \varphi\, dx)^2}{(\int_\Omega e^v dx)^2} \\ &= (\nabla\varphi, \nabla\varphi) - \int_\Omega p\varphi^2 dx + \frac{1}{\lambda}(\int_\Omega p\varphi\, dx)^2 \end{aligned}$$

defined for $\varphi \in C_0^\infty(\Omega)$, where $(\ ,\ )$ denotes the L^2 inner product. This $Q = Q(\varphi, \varphi)$ is extended to a bounded bilinear form on $H_0^1(\Omega)$, and then the self-adjoint operator on $L^2(\Omega)$ denoted by $\mathscr{L}$ is defined by

$$Q(\varphi, \psi) = (\mathscr{L}\varphi, \psi), \quad \varphi \in D(\mathscr{L}) \subset H_0^1(\Omega), \ \psi \in H_0^1(\Omega)$$

with the domain $D(\mathscr{L}) = H_0^1(\Omega) \cap H^2(\Omega)$. It is realized by

$$\mathscr{L}\psi = -\Delta\psi - p\psi + \frac{1}{\lambda}(\int_\Omega p\psi\, dx)p, \quad \psi \in D(\mathscr{L}).$$

This $\mathscr{L}$ is the linearized operator of (2.92) around its solution $v = v(x)$.

It is degenerate if and only if the eigenvalue problem

$$-\Delta\varphi = \nu p\varphi \text{ in } \Omega, \quad \varphi = \text{constant on } \partial\Omega, \quad \int_{\partial\Omega} \frac{\partial\varphi}{\partial\nu} = 0 \tag{2.143}$$

takes the eigenvalue $\nu = 1$. In fact, if

$$\mathscr{L}\psi = 0, \quad \exists\psi \in H_0^1(\Omega) \cap H^2(\Omega) \setminus \{0\},$$

there arises

$$-\int_{\partial\Omega} \frac{\partial\psi}{\partial\nu} dS = -\int_\Omega \Delta\psi \, dx = 0$$

and therefore,

$$\varphi = \psi - \frac{1}{\lambda}\int_\Omega p\psi \, dx \tag{2.144}$$

satisfies (2.143) for $\nu = 1$. In the case of $\varphi = 0$ in (2.144), we obtain $\psi =$ constant, which contradicts $\psi \in H_0^1(\Omega) \setminus \{0\}$. Hence (2.143) takes the eigenvalue $\nu = 1$.

Assume, conversely, that (2.143) takes the eigenvalue $\nu = 1$, and let the corresponding eigenfunction and its boundary value be $\varphi \neq 0$ and φ_Γ, respectively. Then

$$\psi = \varphi - \varphi_\Gamma \in H_0^1(\Omega) \cap H^2(\Omega) \tag{2.145}$$

satisfies

$$\int_\Omega p\psi \, dx = \int_\Omega p\varphi \, dx - \varphi_\Gamma \int_\Omega p \, dx = -\varphi_\Gamma \lambda$$

by (2.143) with $\nu = 1$, and hence

$$\mathscr{L}\psi = -\Delta\varphi - p\psi - \frac{1}{\lambda}(\int_\Omega p\psi \, dx)p = 0$$

by (2.143) with $\nu = 1$ and (2.145). If $\psi = 0$, furthermore, there arises $\varphi =$ constant and then (2.143) with $\nu = 1$ implies $\varphi = 0$, a contradiction.

The eigenvalue problem (2.143) is equivalent to finding $(\phi, \nu) \in (V \setminus \{0\}) \times \mathbf{R}$ such that

$$a(\phi, \psi) = \nu b(\phi, \psi) \quad \forall\psi \in V,$$

where $X = L^2(\Omega)$, $V = H_c^1(\Omega) \equiv \left\{v \in H^1(\Omega) \mid v = \text{constant on } \partial\Omega\right\}$, and

$$a(\varphi, \psi) = (\nabla\varphi, \nabla\psi), \quad b(\varphi, \psi) = \int_\Omega \varphi\psi \; pdx.$$

The eigenvalues of this problem are denoted by

$$0 \leq \nu_1 \leq \nu_2 \leq \cdots \to +\infty$$

with the associated eigenfunctions φ_j, $j = 1, 2, \cdots$, normatized by $b(\varphi_i, \varphi_j) = \delta_{ij}$.

The first eigenvalue of (2.143) is obviously $\nu_1 = 0$ with the eigenfunction $\varphi_1 = \lambda^{-1/2}$, and hence the second eigenvalue is given by

$$\nu_2(p, \Omega) = \inf\left\{\int_\Omega |\nabla v|^2 \mid v \in H_c^1(\Omega), \int_\Omega vp = 0, \int_\Omega v^2 p = 1\right\} \tag{2.146}$$

attained by the second eigenfunction $\varphi = \varphi_2$ from the mini-max theorem. Since $v = v(x)$ is a solution to (2.92), $p = p(x) > 0$ in (2.142) satisfies (2.118):

$$-\Delta \log p \leq p \text{ in } \Omega, \quad \lambda = \int_\Omega pdx < 8\pi.$$

Theorem 2.9 is thus reduced to the following theorem.

Theorem 2.10 ([19]) *If $\Omega \subset \mathbf{R}^2$ is a simply-connected domain and $0 < p = p(x) \in C^2(\Omega) \cap C(\overline{\Omega})$ satisfies (2.118), the second eigenvalue $\nu_2(p, \Omega)$ of (2.143) given by (2.146) satisfies $\nu_2(p, \Omega) > 1$.*

Proof Let $\phi \in H_c^1(\Omega) \setminus \{0\}$ be the second eigenfunction and $K = \nu_2(p, \Omega)$. It holds that

$$-\Delta\phi = Kp\phi \text{ in } \Omega, \quad \phi|_{\partial\Omega} = \text{constant}, \quad \int_{\partial\Omega} \frac{\partial\phi}{\partial\nu} ds = 0 \tag{2.147}$$

and

$$\int_\Omega \phi^2 p \, dx = 1, \quad \int_\Omega \phi p \, dx = 0 \tag{2.148}$$

where ν denotes the outer unit normal vector on $\partial\Omega$.

Let $\{\Omega_i\}_{i\in I}$ be the nodal domains of ϕ, the set of connected components of $\{\phi \neq 0\}$. Each $\partial\Omega_i$ consists of a number of piecewise C^2 Jordan curves [10, 11]. It holds that

$$\int_{\partial\Omega_i} \frac{\partial\phi}{\partial\nu}\phi \, ds = 0, \quad i \in I \tag{2.149}$$

by (2.147). In fact, each $\partial\Omega_i$ is a portion of nodal lines $\{\phi = 0\}$ or the boundary $\partial\Omega$, which implies $\partial\Omega_i = \overline{\gamma_0} \cup \overline{\gamma_1}$ with $\gamma_0 \subset \partial\Omega$ and $\gamma_1 \subset \{\phi = 0\}$ satisfying

$\gamma_0 \cap \gamma_1 = \emptyset$. Since

$$\int_{\partial\Omega_i} \frac{\partial\phi}{\partial\nu}\phi \, ds = \int_{\gamma_0} \frac{\partial\phi}{\partial\nu}\phi \, ds,$$

we have only to take the case $\gamma_0 \neq \emptyset$ to show (2.149). If $\overline{\gamma_0} \cap \overline{\gamma_1} \neq \emptyset$, then $\gamma_0 \subset \{\phi = 0\}$ because ϕ is constant on $\partial\Omega$, which implies

$$\int_{\gamma_0} \frac{\partial\phi}{\partial\nu}\phi \, ds = 0$$

and hence (2.149).

The other case of $\overline{\gamma_0} \cap \overline{\gamma_1} = \emptyset$ induces $\gamma_0 = \partial\Omega$ by $\gamma_0 \neq \emptyset$ because Ω is simply-connected and hence $\partial\Omega$ consists of one component, which is to be used in accordance with the process of rearrangement. Then it follows that

$$\int_{\gamma_0} \frac{\partial\phi}{\partial\nu}\phi \, ds = \int_{\partial\Omega} \frac{\partial\phi}{\partial\nu} \, ds \cdot \phi|_{\partial\Omega} = 0,$$

and hence (2.149).

By the proof of the nodal domain theorem [18], property (2.149) implies $\sharp\, I \leq 2$ for the second eigenfunction ϕ by the mini-max principle, while it cannot be of definite sign from the second equality of (2.148). Therefore, each $\Omega_\pm = \{\pm\phi > 0\}$ is a nodal domain of ϕ, and hence is an open connected set with the boundary composed of a number of piecewise smooth Jordan curves.

Let $\kappa = \phi|_{\partial\Omega} \in \mathbf{R}$. If $\kappa = 0$, this $\phi \in H_c^1(\Omega)$ satisfies

$$-\Delta\phi = Kp\phi, \ \pm\phi > 0 \ \text{in} \ \Omega_\pm, \quad \phi|_{\partial\Omega_\pm} = 0.$$

Since

$$\int_\Omega p \, dx < 8\pi,$$

there arises either $\int_{\Omega_+} p \, dx < 4\pi$ or $\int_{\Omega_-} p \, dx < 4\pi$, which implies $K > 1$ by Theorem 2.4. If $\kappa \neq 0$, on the other hand, any nodal line of ϕ does not touch $\partial\Omega$, and hence either Ω_+ or Ω_- is simply-connected. Without loss of generality, suppose that Ω_- is simply-connected, that is, $\partial\Omega \subset \partial\Omega_+$, to put

$$\Sigma_\pm = \int_{\Omega_\pm} p \, dx.$$

Since $\Sigma_+ + \Sigma_- = \Sigma < 8\pi$, we have either $\Sigma_- < 4\pi$ or $\Sigma_+ < 4\pi \leq \Sigma_-$. In the first case of $\Sigma_- < 4\pi$, we obtain $K > 1$ again by Theorem 2.4, because

$$\phi_- \equiv \phi|_{\Omega_-} \in H_0^1(\Omega_-)$$

holds by the above topological assumption on $\Omega_\pm$. The proof is thus reduced to the case $\Sigma_+ < 4\pi \leq \Sigma_-$ with simply-connected Ω_-.

Putting

$$\Gamma = \partial\Omega,\ \gamma = \partial\Omega_-,\ \phi_+ = \phi|_{\Omega_+},$$

we obtain

$$-\Delta\phi_+ = Kp\phi_+,\ \phi_+ > 0 \text{ in } \Omega,\ \ \phi_+|_\gamma = 0,\ \ \phi_+|_\Gamma = \text{constant},\ \int_\Gamma \frac{\partial\phi_+}{\partial\nu}\, ds = 0.$$

This property means that K is its first eigenvalue. Hence we obtain

$$K = \inf\{\int_{\Omega_+} |\nabla v|^2\, dx \mid v \in \tilde{H}_c^1(\Omega_+),\ \int_{\Omega_+} v^2 p\, dx = 1\} \tag{2.150}$$

for

$$\tilde{H}_c^1(\Omega_+) = \{v \in H^1(\Omega) \mid v|_\gamma = 0,\ \ v|_\Gamma = \text{constant}\},$$

and the minimum K in (2.150) is attained by a constant times ϕ_+. We take a zero extension of this minimizer ψ to Ω_-, and show the desired conclusion $K > 1$ in (2.150) by reducing it to the radially symmetric case.

Let

$$\tau = \psi|_\Gamma,\ \Omega_1 = \{\psi \leq \tau\},\ \Omega_2 = \{\psi > \tau\}.$$

It holds that $\Omega = \Omega_1 \cup \Omega_2$. The open set Ω_2 may be empty, but otherwise we take the spherically decreasing rearrangement ψ_2^* of $\psi_2 = \psi|_{\Omega_2}$ in Sect. 2.4.3. Henceforth, S denotes a sphere with area 8π, and $d\sigma$ and dV stand for its canonical metric and volume element, respectively.

Then we take a disc $B \subset S$ satisfying

$$\int_B dV = \int_{\Omega_2} p\, dx. \tag{2.151}$$

The function ψ_2^* on B is defined by

$$\psi_2^*(x) = \sup\{t \mid x \in \omega_t\},$$

with ω_t standing for the open concentric disc of B such that

$$\int_{\omega_t} dV = \int_{\{\psi_2 > t\}} p\, dx.$$

It holds that

$$\int_{\Omega_2} \psi^2 p\, dx = \int_B (\psi_2^*)^2\, dV, \quad \int_{\Omega_2} |\nabla \psi_2|^2\, dx \geq \int_B |d\psi_2^*|^2\, dV, \quad \psi_2^*|_{\partial B} = \tau.$$

We use the following procedure for $\psi_1 = \psi|_{\Omega_1}$, which may be called an annular increasing rearrangement. First, we take open concentric discs $B_0 \subset B_1 \subset S$ such that

$$\int_{B_0} dV = \int_{\Omega_-} p\, dx = \Sigma_-, \quad \int_{B_1} dV = \int_{\Omega_1} p dx, \tag{2.152}$$

noting $\Omega_- \subset \Omega_1$. Second, the function ψ_{1*} on the annulus $A \equiv B_1 \setminus \overline{B_0}$ is defined by

$$\psi_{1*} = \inf\{t \mid x \in A_t\},$$

where A_t is the closed concentric annulus of A satisfying that $A_t \cup \overline{B_0} \subset S$ is a closed ball and

$$\int_{A_t \cup \overline{B_0}} dV = \int_{\{\psi \leq t\}} p\, dx.$$

Since this process is an equi-measurable rearrangement it holds that

$$\int_{\Omega_+ \cap \Omega_1} \psi^2 p\, dx = \int_A (\psi_{1*})^2\, dV.$$

Decrease of the Dirichlet integral, on the other hand, is achieved as in the proof of Theorem 2.4,

$$\int_{\Omega_+ \cap \Omega_1} |\nabla \psi|^2\, dx \geq \int_A |d\psi_{1*}|^2\, dV,$$

because $0 \leq \psi \leq \tau$ in Ω_1, $\psi|_\gamma = 0$, and $\psi|_{\partial\Omega_1 \setminus \gamma} = \tau$ imply

$$\partial\{x \in \Omega_1 \cup \overline{\Omega_-} \mid \psi(x) \leq t\} = \{x \in \Omega_1 \cup \overline{\Omega_-} \mid \psi(x) = t\} \quad \text{a.e. } t \in (0, \tau).$$

The relations $\psi_{1*}\big|_{\gamma^*} = 0$ and $\psi_{1*}|_{\Gamma^*} = \tau$, finally, are obvious, where $\gamma^* = \partial B_0$ and $\Gamma^* = \partial B_1$. Regarding

$$\Sigma_+ < 4\pi \leq \Sigma_-, \quad \Sigma_+ + \Sigma_- < 8\pi,$$

we arrange the disc B and the annulus A so as to be concentric, with the center on the south pole s on S. Thus B_0 and B_1 are concentric discs with centers on the north and south poles, respectively, and $B = \{\psi_{1*} > \tau\}$ and $A = \{0 < \psi_{1*} < \tau\}$ are concentric to B_1. Furthermore, $\omega_+ = A \cup B$ is a disjoint sum contained in the south hemisphere of S, and γ^* and Γ^* are the outscribing and inscribing boundaries of A, respectively. We thus obtain

$$K \geq K^* \equiv \inf\{\int_{\omega_+} |dv|^2\, dV \mid v \in \tilde{H}_c^1(\omega_+)\}, \tag{2.153}$$

where

$$\tilde{H}_c^1(\omega_+) = \{v \in H^1(\omega_+) \mid v|_{\gamma^*} = 0,\ v = \tau \in \mathbf{R} \text{ on } \Gamma^* \cup \partial B\}.$$

This theorem is thus reduced to $K^* > 1$.

Through the stereographic projection $\iota : S \to \mathbf{R}^2 \cup \{\infty\}$ used in Sect. 2.4.3, this K^* is realized as the first eigenvalue of

$$\begin{aligned} -\Delta\phi = Kp^*\phi \text{ in } \Omega^* = A^* \cup B^*, \quad & \phi|_{\Gamma_1} = 0, \\ \phi = \tau \in \mathbf{R} \text{ on } \Gamma_2 \cup \Gamma_3, \quad & \int_{\Gamma_2 \cup \Gamma_3} \frac{\partial \phi}{\partial \nu}\, ds = 0, \end{aligned} \tag{2.154}$$

where $A^* = \iota(A)$, $B^* = \iota(B)$, $\Gamma_1 = \iota(\gamma^*)$, $\Gamma_2 = \iota(\Gamma^*)$, and $\Gamma_3 = \iota(\partial B)$. The above function p^* is derived from the projection ι of the Laplace-Beltrami operator $-\Delta_S$ on S. The annulus A^* and the disc B^* are in the flat plane, concentric, and disjoint. Through the scalling transformation introduced in the previous section, the outer radius of A^* may be equal to 1. Then, it holds that

$$p^*(x) = \frac{8\mu}{(|x|^2 + \mu)^2},\ \mu > 0, \quad \int_{\Omega^*} p^* dx = \int_{\omega_+} dV < 4\pi. \tag{2.155}$$

The first eigenfunction $\phi(x)$ of (2.154) is radially symmetric and positive. Therefore, in terms of

$$\xi = (\mu - r^2)/(\mu + r^2)$$

this $\Psi(\xi) = \phi(x)$ satisfies, with some a and b in $a < b < 1$,

$$\begin{aligned} &[\left(1 - \xi^2\right) \Phi_\xi]_\xi + (2/K^*)\Phi = 0,\ \Phi(\xi) > 0, \quad \xi_\mu < \xi < a,\ b < \xi < 1 \\ &\Phi(1) = 1,\ \Phi(b) = \Phi(a),\ \Phi'(b) = \Phi'(a),\ \Phi(\xi_\mu) = 0 \end{aligned}$$

if $\Omega_2 \neq \emptyset$, and

$$[\left(1-\xi^2\right)\Phi_\xi]_\xi + (2/K^*)\Phi = 0,\ \Phi(\xi) > 0, \quad \xi_\mu < \xi < a$$
$$\Phi(1) = 1,\ \Phi'(a) = 0,\ \Phi(\xi_\mu) = 0$$

if $\Omega_2 = \emptyset$. As is shown in Sect. 2.4.3, the last inequality in (2.155) is equivalent to

$$\xi_\mu \equiv \frac{\mu - 1}{\mu + 1} > 0. \tag{2.156}$$

Inequality $K^* > 1$, therefore, follows if

$$\hat{\Phi}(\xi) > 0, \quad \xi_\mu < \xi < a,\ b < \xi < 1 \tag{2.157}$$

for $\hat{\Phi} = \hat{\Phi}(\xi)$ satisfying

$$[\left(1-\xi^2\right)\hat{\Phi}_\xi]_\xi + 2\hat{\Phi} = 0,\ -1 < \xi < a$$
$$b < \xi < 1, \quad \hat{\Phi}(1) = 1,\ \hat{\Phi}(b) = \hat{\Phi}(a),\ \hat{\Phi}'(b) = \hat{\Phi}'(a), \tag{2.158}$$

and if

$$\hat{\Phi}(\xi) > 0, \quad \xi_\mu < \xi < a \tag{2.159}$$

for $\hat{\Phi} = \hat{\Phi}(\xi)$ satisfying

$$[\left(1-\xi^2\right)\hat{\Phi}_\xi]_\xi + 2\hat{\Phi} = 0,\ -1 < \xi < a, \quad \hat{\Phi}(a) = 1,\ \hat{\Phi}'(a) = 0. \tag{2.160}$$

Since the fundamental solutions to (2.158) or (2.160) are

$$P_1(\xi) = \xi, \quad Q_1(\xi) = -1 + \frac{\xi}{2}\log\frac{1+\xi}{1-\xi},$$

the solutions to (2.158) is $\hat{\Phi}(\xi) = \xi$ for $b < \xi < 1$ and

$$\hat{\Phi}(\xi) = (1 + (b-a)(\frac{1-a^2}{2}\log\frac{1+a}{1-a} + a))\xi$$
$$- (1-a^2)(b-a)(-1 + \frac{\xi}{2}\log\frac{1+\xi}{1-\xi})$$

for $-1 < \xi < a$, and that to (2.160) is

$$\hat{\Phi}(\xi) = (1-a^2)(-\frac{\xi}{2}\log\frac{(1-a)(1+\xi)}{(1+a)(1-\xi)} + 1) + a\xi, \quad -1 < \xi < a.$$

Then we confirm (2.157) and (2.159) by (2.156), using

$$\hat{\Phi}(\xi) \geq \xi > 0, \quad 0 < \xi < a, \ b < \xi < 1 \tag{2.161}$$

and

$$\hat{\Phi}(\xi) \geq \hat{\Phi}(0) = 1 - a^2 > 0, \quad 0 < \xi < a \tag{2.162}$$

for the former and the latter, respectively. In fact, inequality (2.161) is easy to see in the first case. For (2.162) we note

$$\hat{\Phi}(\xi) = \eta(\xi)a + \left(1 - a^2\right)$$

with $\eta'(\xi) < 0$ for $0 < \xi < a$ and

$$\eta(0) = -\frac{1-a^2}{2} \log \frac{1-a}{1+a} + a > 0.$$

□

Remark 2.2 The reverse theory to Theorem 2.5 is known. If $x_* = (x_1^*, \cdots, x_\ell^*)$ is a non-degenerate critical point of the ℓ-th Hamiltonian, $H_\ell = H_\ell(x_1, \cdots, x_\ell)$, there is a family of solutions (λ_k, v_k), $k = 1, 2, \cdots$, to (2.97) satisfying (2.98) and (2.100) [4].

Remark 2.3 The quantized blowup mechanism is not restricted to the Boltzmann–Poisson equation in two space dimension. The exact correspondence is the equation arising in plasma confinement [20],

$$-\Delta u = u_+^{\frac{n}{n-2}} \text{ in } \Omega, \quad u|_{\partial\Omega} = \text{constant}, \quad -\int_{\partial\Omega} \frac{\partial u}{\partial \nu}\, ds = \lambda, \tag{2.163}$$

where $\Omega \subset \mathbf{R}^n$, $n \geq 3$, is a bounded domain with smooth boundary $\partial\Omega$. Hence the sequence (λ_k, u_k), $k = 1, 2, \cdots$, to (2.163) with $\lambda_k = O(1)$, If (2.98) is the case, passing to a subsequence, we obtain $\lambda_0 = \lambda_* \ell$, $\ell \in \mathbf{N}$, with $\lambda_* > 0$ uniquely defined by

$$\lambda_* = \int_{\mathbf{R}^n} U_+^{\frac{n-2}{2}},$$

using the entire solution $U = U(x)$ satisfying

$$-\Delta U = U_+^{\frac{n}{n-2}}, \ U \leq U(0) = 1 \ \text{in } \mathbf{R}^n, \quad \int_{\mathbf{R}^n} U_+^{\frac{n}{n-2}} < +\infty.$$

The blowup set $\mathscr{S}$ is composed of ℓ-points denoted by $x_1^*, \cdots, x_\ell^*$, and then, $x_* = (x_1^*, \cdots, x_\ell^*)$ is a critical point of the ℓ-th Hamiltonian $H_\ell = H_\ell(x_1, \cdots, x_\ell)$ defined by (2.95). It holds also that (2.100) with 8π replaced by λ_* [21]. This profile is localized in the context of blowup analysis developed in the next chapter [22, 23].

The quantized blowup mechanism is widely observed in nonlinear elliptic and parabolic equations other than Remark 2.3. To approach these problems, evidently, completely new ideas other than the complex variables are necessary. We turn to the elliptic theory in the following chapter, where the other Liouville property takes a role.

References

1. C. Bandle, Existence theorems, qualitive results and a priori bounds for a class of nonlinear Dirichlet problems. Arch. Rational Mech. Anal. **58**, 219–238 (1975)
2. C. Bandle, On a differential inequality and its applications to geometry. Math. Z. **147**, 253–261 (1976)
3. C. Bandle, *Isoperimetric Inequalities and Applications* (Pitman, Boston, 1980)
4. S. Baraket, F. Pacard, Construction of singular limits for a semilinear elliptic equation in dimension 2. Calc. Var. **6**, 1–38 (1998)
5. D. Bartolucchi, C.-S. Lin, Existence and uniqueness for mean field equation on multiply connected domain at the critical parameter. Math. Ann. **359**, 1–44 (2014)
6. E. Cagilioti, P.L. Lions, C. Marchioro, M. Pulvirenti, A special class of stationary flows for two-dimensional Euler equations: a statistical mechanics description. Commun. Math. Phys. **143**, 501–525 (1992); II, ibid, **174**, 229–260 (1995)
7. K.S. Chow, T.Y.H. Wan, Asymptotic radial symmetry for solutions of$\Delta u + e^u = 0$ on a punctured disc. Pac. J. Math. **163**, 269–276 (1994)
8. M.G. Crandall, P.H. Rabinowitz, Some continuation and variational methods for positive solutions of non-linear elliptic eigenvalue problems. Arch. Rational Mech. Anal. **58**, 207–218 (1975)
9. B. Gidas, W.-M. Ni, L. Nirenberg, *Symmetry and related properties via the maximum principle.* Commun. Math. Phys. **68**, 209–243 (1979)
10. P. Hartman, A. Wintner, On the local behavior of solutions of non-parabolic partial differential equations. Amer. J. Math. **75**, 449–476 (1953)
11. T. Kuo, On C^∞ sufficiency of sets of potential functions. Topology **8**, 167–171 (1969)
12. Y.Y. Li, I. Shafrir, Blow-up analysis for solutions of $-\Delta u = Ve^u$ in dimension two. Indiana Univ. Math. J. **43**, 1255–1270 (1994)
13. S.S. Lin, On non-radially symmetric bifurcation in the annulus. J. Differ. Equations **80**, 251–279 (1989)
14. K. Nagasaki, T. Suzuki, Asymptotic analysis for two-dimensional elliptic eigenvalue problems with exponentially-dominated nonlinearities. Asymptot. Anal. **3**, 173–188 (1990)
15. K. Nagasaki, T. Suzuki, Radial and nonradial solutions for the nonlinear eigenvalue problem$\Delta u + \lambda e^u = 0$ on annului in $\mathbf{R}^2$. J. Differ. Equations **87**, 144–168 (1990)
16. Z. Nehari, On the principal frequency of a membrane. Pacific J. Math. **8**, 285–293 (1958)
17. L. Onsager, Statistical hydrodynamics. Suppl. Nuovo Cimento **6**, 279–287 (1949)
18. A. Pleijel, Remarks on Courant's nodal line theorem. Commun. Pure Appl. Math. **9**, 543–550 (1956)

19. T. Suzuki, Global analysis for a two-dimensional elliptic eigenvalue problem with the exponential nonlinearity. Ann. Inst. Henri Poincaré, Analyse non linéaire **9**, 367–398 (1992)
20. T. Suzuki, *Mean Field Theories and Dual Variation, Mathematical Structures of the Mesoscopic Model*, 2nd edn. (Atlantis Press, Paris, 2015)
21. T. Suzuki, *Semilinear Elliptic Equations, Modern and Classical Theories*, 2nd edn. (De Gruyter, Berlin, 2020)
22. T. Suzuki, R. Takahashi, Critical blowup exponent to a class of semilinear elliptic equations with constraints in higher dimensions – local properties. Ann. Mat. Pura Appl. **195**, 1123–1151 (2016)
23. G. Wang, D. Ye, On a nonlinear elliptic equation arising in a free boundary problem. Mat. Z. **244**, 531–548 (2003)
24. H.C. Wente, Counter example to a conjecture of H. Hopf. Pac. J. Math. **121**, 193–243 (1986)

Chapter 3
Elliptic Theory: A Classical and Modern Study

Abstract The Liouville integral for surfaces of constant Gaussian curvature induces a quantized blowup mechanism of solutions to the Gel'fand or Boltzmann–Poisson equation, as is clarified in the previous chapter. The standard argument to approach this phenomenon, however, is the blowup analysis based on the self-similarity of the equation, which is commonly provided in the fundamental equations of mathematical physics. This property is the invariance under transformation involving independent and dependent variables, which causes a lack of compactness of the family of (approximate) solutions. Breaking down of compactness becomes clear by self-similar transformation, to reach a profile of the structure of total set of solutions from the microscopic viewpoint. Meanwhile, a crucial role is taken by the classification of entire solutions, which traces back to Liouville's theory on harmonic functions. We thus encounter the third of Liouville's theorems. The first section is a summary. We begin with several fundamental properties of harmonic functions in the second section, and turn to recent results on semilinear elliptic equations with power nonlinearity in the third section, in accordance with the Liouville property concerning entire solutions. We then generalize the result on exponential nonlinearity obtained in the previous chapter in the final section.

3.1 Introduction

3.1.1 Linear Elliptic Equations

The elliptic equation is concerned with the stationary state of things, which can be linear or nonlinear. An equation of second order provided with boundary conditions is a base; typical examples of linear equations are the *Poisson equation*

$$-\Delta u = f \text{ in } \Omega, \quad u = 0 \text{ on } \partial\Omega, \tag{3.1}$$

and the *Laplace equation*

T. Suzuki, *Liouville's Theory in Linear and Nonlinear Partial Differential Equations*, Springer Asia Pacific Mathematics Series 8,
https://doi.org/10.1007/978-981-95-1975-0_3

$$\Delta u = 0 \text{ in } \Omega, \quad u = g \text{ on } \partial\Omega, \tag{3.2}$$

where $\Omega \subset \mathbf{R}^n$ is a bounded domain with smooth boundary $\partial\Omega$ and

$$\Delta = \frac{\partial^2}{\partial x_1^2} + \cdots + \frac{\partial^2}{\partial x_n^2}$$

is the Laplacian.

The fundamental question of the unique existence of the solution has been studied in detail, accompanied with the profile of regularity. There may be at least three approaches, function theory, integral equation, and variational method. Comparison theorems may be applicable in all of them, relying on the qualitative feature of the solution. In the first approach one uses a fundamental function or parametrix. Perron's method for (3.2) is a combination of a fundamental function and comparison theorem, directed to the mean value theorem and Harnack principle.

The third of Liouville's theorems is then the classification of solutions on the whole space, that is, entire solutions. The original form is on the harmonic function,

$$\Delta u = 0 \quad \text{in } \mathbf{R}^n. \tag{3.3}$$

This $u = u(x)$ is thus a constant if it is bounded above or below. Especially, any subharmonic, bounded above function in $\mathbf{R}^2$ is a constant, as is shown in Theorem 3.17 below.

Dimension control of the blowup set D arises in the spirit of Liouville's third theorem. If, for example, $\Omega \subset \mathbf{R}^n$ is a bounded open set, $u : \overline{\Omega} \to (-\infty, +\infty]$ is continuous and regular near $\partial\Omega$,

$$D \equiv \{x \in \overline{\Omega} \mid u(x) = +\infty\} \subset \Omega,$$

and $\Delta u \leq 0$ in $\Omega \setminus D$ in the sense of distributions, 2-capacity of D, denoted by $\mathrm{Cap}_2(D)$ vanishes, and hence the Hausdorff dimension of D is less than or equal to $(n-2)$ [32]. In its parabolic version [38], if $\Omega \subset \mathbf{R}^n$ is a bounded open set, $u : \Omega \times [0, T] \to (-\infty, +\infty]$ is continuous and sufficiently regular near $\partial\Omega \times [0, T]$, $D \subset \Omega \times [0, T]$, and

$$u_t - \Delta u \geq 0 \quad \text{in } \Omega \times (0, T) \setminus D \tag{3.4}$$

in the sense of distributions for

$$D = \bigcup_{t \in [0,T]} D(t) \times \{t\}, \quad D(t) = \overline{\{x \in \Omega \mid u(x, t) = +\infty\}},$$

it holds that

$$\int_0^T \mathrm{Cap}_2(D(t))\, dt \le \frac{|\Omega|}{2}. \tag{3.5}$$

See [37] for related results.

The method employed for the proof of (3.5) is valid to induce a class of L^1 ill-posed quasilinear parabolic equations [30]. The simplest example is the equation

$$x_t = (x^{-2} x_u)_u \tag{3.6}$$

in $n = 1$ [29]. We note that the Bäcklund transformation $x = x(u, t)$ implies (3.6) if $u = u(x, t)$ is a solution to

$$u_t = u_{xx}.$$

Interface vanishing of the Maxwell equation is the other issue describing the feature of solutions to piecewisely defined equations [36].

3.1.2 Nonlinear Elliptic Equations

Study of (2.6) in the previous chapter has been the motivation for several qualitative studies on nonlinear partial differential equations, particularly semilinear elliptic equations. Since this nonlinearity comes from that between a flat plane and a sphere, one reaches the mean value theorem and bubbled Harnack principle for spherical subharmonic functions [37].

The other aspect of (2.6) is the control of a blowup family of solutions. Using scaling invariance of the model one induces the method of blowup analysis, where a hierarchical argument brushes the profile of the solution up, to reach, for example, the quantized blowup mechanism observed in the previous chapter. This argument of blowup analysis works well if the scaling limit of the solution is classified, that is, under the above described Liouville's third theorem in nonlinear problems.

Here we take

$$-\Delta u = u^p, \ u > 0 \ \text{ in } \Omega, \quad u|_{\partial\Omega} = 0, \tag{3.7}$$

as an example. This equation is invariant under the self-similar transformation

$$u_\mu(x) = \mu^{\frac{2}{p-1}} u(\mu x), \quad \mu > 0.$$

To show the existence of an a priori bound of the solution to (3.7), for example, we assume the contrary, the existence of a sequence of solutions $u_k = u_k(x)$, $k = 1, 2, \cdots$, such that

$$\|u_k\|_\infty \to +\infty. \tag{3.8}$$

Define $x_k \in \Omega$ by $u_k(x_k) = M_k \equiv \|u_k\|_\infty$, and put

$$\tilde{u}_k(x) = \mu_k^{\frac{2}{p-1}} u(\mu_k x + x_k), \quad \mu_k^{\frac{2}{p-1}} M_k = 1.$$

We obtain $\mu_k \to 0$ by (3.8).

Under this transformation it holds that

$$-\Delta \tilde{u}_k = \tilde{u}_k^p, \ 0 < \tilde{u}_k \leq \tilde{u}_k(0) = 1 \ \text{in } \tilde{\Omega}_k, \quad \tilde{u}_k|_{\partial \tilde{\Omega}_k} = 0 \tag{3.9}$$

for $\tilde{\Omega}_k = \mu_k^{-1}(\Omega - \{x_k\})$, and then, the elliptic regularity assures the existence of a subsequence, denoted by the same symbol, such that either $\tilde{u}_k \to \tilde{u}$ locally uniformly in $\mathbf{R}^n$, with $\tilde{u} = \tilde{u}(x)$ satisfying

$$-\Delta \tilde{u} = \tilde{u}^p, \ 0 \leq \tilde{u} \leq \tilde{u}(0) = 1 \quad \text{in } \mathbf{R}^n, \tag{3.10}$$

or $\tilde{u}_k \to \tilde{u}$ locally uniformly in $\{x = (x', x_n) \mid x_n > -c\}$ under suitable rotation of variables, with $c > 0$ and $\tilde{u} = \tilde{u}(x)$ satisfying

$$-\Delta \tilde{u} = \tilde{u}^p, \ 0 \leq \tilde{u} \leq \tilde{u}(0) = 1 \ \text{in } x_n > -c, \quad \tilde{u}|_{x_n=-c} = 0. \tag{3.11}$$

Liouville's property for

$$-\Delta u = u^p, \ u \geq 0 \quad \text{in } \mathbf{R}^n, \tag{3.12}$$

however, excludes both cases of (3.10)–(3.11) for $1 < p < \frac{n+2}{n-2}$ and hence (3.8) does not occur in this case (Theorem 3.23).

In such arguments of blowup analysis, control of the rescaled solution at infinity, called the *tail*, sometimes takes an essential role.

3.1.3 Organization of Chap. 3

Section 3.2 is a classical theory of the Laplace equation (3.2). We describe the *method of Perron*, first, in two space dimensions, and then in higher space dimensions. The Liouville property of entire harmonic functions is derived from the *Poisson integral*. This property is converted to the origin by the *Kelvin transformation* to induce the theory of removable singularities.

Section 3.3 is devoted to the Liouville property and its consequence on (3.12) accompanied with the blowup analysis. Here we show two approaches to establishing Liouville's property for this equation; integral identity and moving planes. Then blowup analysis is introduced and several related results are stated.

An extension of Theorem 2.5 in the previous chapter is executed in Sect. 3. We set the problem to the study of the sequence of solutions to

$$-\Delta u_k = V_k(x)e^{u_k}, \ 0 \leq V_k(x) \leq b \ \text{ in } \Omega, \quad \int_\Omega e^{u_k} \leq C. \tag{3.13}$$

This condition is free from boundary conditions, and u_k can take negative values.

The first result, Theorem 3.29 below, is a rough estimate. It is stated in the form of alternatives for compactness, (negatively) entire blowup, and concentration. In the second result, the last case of the above alternatives, concentration, is quantized under the assumption that $\{V_k(x)\}$ is compact in $C(\Omega)$ (Theorem 3.29). Here, self-similar transformation on

$$-\Delta u = e^u,$$

formulated as

$$u_\mu(x) = u(\mu x) + 2\log\mu, \quad \mu > 0$$

is used to set up blowup analysis.

More precisely, successive bubbles of u_k, $k = 1, 2, \cdots$, are detected according to their concentration rates, under the scaling by the Liouville property of the limit equation. Here the classification of the solution to the Liouville equation

$$-\Delta\tilde{u} = e^{\tilde{u}}, \ \tilde{u} \leq \tilde{u}(0) = 0 \ \text{ in } \mathbf{R}^2, \quad \int_{\mathbf{R}^2} e^{\tilde{u}} < +\infty, \tag{3.14}$$

is required. It is Theorem 3.32, which ensures

$$\tilde{u}(x) = 2\log\frac{1}{1+\frac{|x|^2}{8}}$$

for the solution $\tilde{u} = \tilde{u}(x)$ to (3.14). Hence we apply Liouville's property to the Liouville equation.

This process of detecting bubbles is exhausted by the $\sup + \inf$ inequality. It is Theorem 3.36 of this chapter and controls the tail of the rescaled solution, to induce residual vanishing. Blowup quantization thus arises even without boundary conditions, provided by possible collision of bubbles (Theorem 3.30).

Here, we observe that each bubble mass and the total mass of the entire solution to (3.14) coincide as 8π. Under the presence of boundary conditions, we can control the tail, particularly, the transient region in the rescaled variable by this fact, to reach a locally uniform estimate near the blowup point (Theorem 3.30).

The application of this estimate is a generalization of Theorem 2.5 with variable coefficients. This result is Theorem 3.38 below. Here we mention an argument other than blowup analysis applicable to the proof of this theorem, that is, the *method of duality* described in [37].

3.2 Harmonic Functions

3.2.1 Laplace Equation

The function $u = u(x)$ satisfying $\Delta u = 0$ is said to be *harmonic*. It is a fundamental problem in mathematical physics to solve

$$\Delta u = 0 \quad \text{in } \Omega, \qquad u = f \quad \text{on } \partial\Omega, \tag{3.15}$$

where Ω is a bounded domain and $f = f(\xi)$ is a continuous function on its boundary $\partial\Omega$.

Theorem 3.1 *Problem (3.15) admits a solution $u \in C^2(\Omega) \cap C(\overline{\Omega})$ to (3.15) for any arbitrary given $f \in C(\partial\Omega)$ if and only if any point on $\partial\Omega$ is regular.*

Here the boundary point $\xi \in \partial\Omega$ is said to be *regular* if it has a *barrier* $w(x)$, which means that this $w(x)$ is continuous on $\overline{\Omega}$, superharmonic there, positive in $\overline{\Omega} \setminus \{\xi\}$, and zero at $x = \xi$. Superharmonicity of the continuous function $w(x)$, furthermore, is defined through the *mean value property*

$$\frac{1}{|B(x_0, r)|} \int_{\partial B(x_0,r)} w \, dS \leq w(x_0), \quad x_0 \in \Omega, \ 0 < r \ll 1, \tag{3.16}$$

which is actually equivalent to $\Delta w \leq 0$ in Ω if $w(x)$ is twice differentiable, recalling that $B(x_0, r)$ denotes an open ball with center x_0 and radius r. If $u \in C^2(\Omega) \cap C(\overline{\Omega})$ satisfies (3.15) it is called the classical solution. A sufficient condition for the regularity of a boundary point is the outer circumscribing ball condition.

Theorem 3.1 is proven by Perron's method. First, the subharmonicity of continuous function $w(x)$ is defined by the reverse inequality of (3.16). Second, we take

$$\begin{aligned} &\underline{\mathscr{S}}(f) = \left\{ v \in C(\overline{\Omega}) \mid v \text{ is subharmonic in } \Omega \text{ and } v \leq f \text{ on } \partial\Omega \right\} \\ &u(x) = \sup \left\{ v(x) \mid v \in \underline{S}(f) \right\}. \end{aligned}$$

It is proven that the above $u = u(x)$ is harmonic in Ω, and, furthermore, if $\xi \in \partial\Omega$ is a regular point it holds that

$$\lim_{x \in \Omega \to \xi} u(x) = f(\xi).$$

A rough description of this theory is as follows. First, using the potential and the Kelvin transformation, we represent the solution to (3.15) by the Poisson integral when Ω is a ball. By this formula we obtain the mean value theorem and Harnack's inequality for harmonic function, which imply weak and the strong maximum principles, and the Harnack principle, respectively. Then the method of harmonic lifting and the barrier provide the solution to (3.15).

Henceforth the *harmonic function* indicates a C^2 function u satisfying $\Delta u = 0$ in a domain Ω in $\mathbf{R}^n$. If Δ is taken in the sense of distribution, this notion is extended to $\mathscr{D}'(\Omega)$. *Weyl's lemma*, however, guarantees that any harmonic distribution is regarded as a harmonic function, and, furthermore, any harmonic function is real analytic, which means that it is represented by a Taylor series around any point in Ω.

3.2.2 *Poisson Integral*

For the moment we concentrate on the two-dimensional case, using the complex function theory where $x = (x_1, x_2) \in \mathbf{R}^2$ is identified with $z = x_1 + \sqrt{-1}x_2 \in \mathbf{C}$.

Theorem 3.2 *If $\Omega \subset \mathbf{R}^2$ is simply-connected and a real-valued function $u = u(x)$ is harmonic in Ω then there is a holomorphic function $f = f(z)$, $z \in \Omega$, such that Re $f = u$.*

Here $f(z)$ is unique up to an additive pure imaginary constant. Even if Ω is not simply-connected, such an $f(z)$ is taken locally and hence can be an analytic function in Ω. Harmonic functions are thus associated with the complex function theory in two space dimensions. Below, any harmonic function is taken to be real-valued.

Let $u = u(z)$ be harmonic and continuous in $|z| < R$ and on $|z| \leq R$, respectively, and take a holomorphic function $f = f(z)$ in $|z| < R$ satisfying Re $f = u$. Then *Cauchy's integral formula* guarantees

$$f(z) = \frac{1}{2\pi\imath}\int_{|\zeta|=\rho} \frac{f(\zeta)}{\zeta - z}d\zeta, \qquad |z| < \rho,$$

where $\rho \in (0, R)$. Taking the mirror image of $z = re^{\imath\theta}$ for $r \in (0, \rho)$ with respect to the circle $|\zeta| = \rho$, $z^* = \rho^2 e^{\imath\theta}/r$, we have

$$0 = \frac{1}{2\pi\imath}\int_{|\zeta|=\rho} \frac{f(\zeta)}{\zeta - z^*}d\zeta$$

by $|z^*| > \rho$, which implies

$$f(re^{\iota\theta}) = \frac{1}{2\pi\iota}\int_{|\zeta|=\rho} f(\zeta)\left\{\frac{1}{\zeta - z} - \frac{1}{\zeta - z^*}\right\} d\zeta$$

$$= \frac{1}{2\pi\iota}\int_{|\zeta|=\rho} f(\zeta)\left\{\frac{1}{\zeta - re^{\iota\theta}} - \frac{1}{\zeta - \rho^2 e^{\iota\theta}/r}\right\} d\zeta.$$

Then putting $\zeta = \rho e^{\iota\varphi}$, we obtain

$$f(re^{\iota\theta}) = \frac{1}{2\pi}\int_0^{2\pi} f(\rho e^{\iota\varphi})\left\{\frac{1}{\zeta - re^{\iota\theta}} - \frac{1}{\zeta - \rho^2 e^{\iota\theta}/r}\right\}\zeta\, d\varphi$$

by $d\zeta = \iota\rho e^{\iota\varphi} d\varphi$. It holds that

$$\left\{\frac{1}{\zeta - re^{\iota\theta}} - \frac{1}{\zeta - \rho^2 e^{\iota\theta}/r}\right\}\zeta$$

$$= \left\{1 + \frac{re^{\iota\theta}}{\zeta - re^{\iota\theta}}\right\} - \left\{1 + \frac{\rho^2 e^{\iota\theta}/r}{\zeta - \rho^2 e^{\iota\theta}/r}\right\}$$

$$= \frac{r}{\rho e^{\iota(\varphi-\theta)} - r} - \frac{\rho^2/r}{\rho e^{\iota(\varphi-\theta)} - \rho^2/r}$$

$$= \frac{r}{\rho e^{\iota(\varphi-\theta)} - r} - \frac{\rho}{re^{\iota(\varphi-\theta)} - \rho}$$

$$= \frac{r\left(\rho e^{-\iota(\varphi-\theta)} - r\right)}{\rho^2 - 2r\rho\cos(\varphi-\theta) + r^2} - \frac{\rho\left(re^{-\iota(\varphi-\theta)} - \rho\right)}{r^2 - 2r\rho\cos(\varphi-\theta) + \rho^2}$$

$$= \frac{\rho^2 - r^2}{\rho^2 - 2\rho r\cos(\varphi-\theta) + r^2} = \frac{\rho^2 - r^2}{\rho^2 - 2\rho r\cos(\theta-\varphi) + r^2}$$

and hence

$$f(re^{\iota\theta}) = \frac{1}{2\pi}\int_0^{2\pi} f(\rho e^{\iota\varphi})\frac{\rho^2 - r^2}{\rho^2 - 2\rho r\cos(\theta-\varphi) + r^2} d\varphi.$$

The real part of the above equality reads as

$$u(re^{\iota\theta}) = \frac{1}{2\pi}\int_0^{2\pi} u(\rho e^{\iota\varphi})\frac{\rho^2 - r^2}{\rho^2 - 2\rho r\cos(\theta-\varphi) + r^2} d\varphi, \quad 0 < r < \rho < R,$$

which guarantees *Poisson's integral formula* with $\rho \uparrow R$.

Theorem 3.3 *If* $n = 2$ *and* $u(z)$ *is harmonic and continuous in* $|z| < R$ *and on* $|z| \leq R$*, respectively, it holds that*

$$u(re^{\imath\theta}) = \frac{1}{2\pi}\int_0^{2\pi} u(Re^{\imath\varphi})\frac{R^2 - r^2}{R^2 - 2Rr\cos(\theta-\varphi) + r^2}d\varphi, \tag{3.17}$$

$0 \le r < R$.

A consequence of equality (3.17) is the *mean value theorem*,

$$u(0) = \frac{1}{2\pi}\int_0^{2\pi} u(Re^{\imath\varphi})d\varphi. \tag{3.18}$$

In the case of $u(Re^{\imath\varphi}) \ge 0$ this equality implies

$$\frac{R-|z|}{R+|z|}u(0) \le u(z) \le \frac{R+|z|}{R-|z|}u(0)$$

by

$$\frac{R-r}{R+r} \le \frac{R^2-r^2}{R^2-2Rr\cos(\theta-\varphi)+r^2} \le \frac{R+r}{R-r},$$

and, hence, *Harnack's inequality*

$$\frac{1}{3}u(0) \le u(z) \le 3u(0), \qquad |z| < R/2. \tag{3.19}$$

.

Theorem 3.4 (strong maximum principle) *Any non-constant harmonic function defined in a domain $\Omega \subset \mathbf{R}^2$ cannot attain the maximum or the minimum.*

Proof It suffices to show that any non-constant harmonic function $u = u(z)$ in Ω cannot attain the maximum. Suppose the contrary that there is $z_0 \in \Omega$ such that

$$u(z) \le u(z_0), \quad z \in \Omega.$$

Then the mean value theorem

$$m \equiv u(z_0) = \frac{1}{|\partial B(z_0,R)|}\int_{\partial B(z_0,R)} u\, dS \le \max_{\partial B(z_0,R)} u \le m$$

implies $u = m$ on $\partial B(z_0, R)$. This equality is the case near z_0 because $0 < R \ll 1$ is arbitrary, which means that the non-empty set

$$\{z \in \Omega \mid u(z) = m\}$$

is open. It is obviously closed and hence coincides with Ω by its connectivity. Thus this $u = u(z)$ is identical to the constant m. □

Theorem 3.5 (weak maximum principle) *If $u = u(x)$ is harmonic and continuous in Ω and on $\overline{\Omega}$, respectively, it holds that*

$$\sup_{\Omega} u = \max_{\partial\Omega} u$$

where $\Omega \subset \mathbf{R}^2$ is a bounded domain.

Proof It is a direct consequence of Theorem 3.4. □

Theorem 3.6 (general form of Harnack's inequality) *If $\Omega \subset \mathbf{R}^2$ is a domain and $E \subset \Omega$ is a bounded closed set, there is $K = K(\Omega, E) > 0$ such that any non-negative harmonic function $u = u(z)$, $z \in \Omega$, admits*

$$\sup_{E} u \leq K \inf_{E} u. \tag{3.20}$$

Proof The compact set E is covered by a finite number of discs with the radius $R/2 > 0$, where

$$0 < R < \text{dist}(E, \partial\Omega) = \inf_{x \in E,\ y \in \partial\Omega} \text{dist}(x, y),$$

which means that there is an integer m and $z_1, \cdots, z_m \in E$ such that

$$E \subset \bigcup_{i=1}^{m} B(z_i, R/2) \subset \Omega.$$

Inequality (3.19) guarantees

$$\frac{1}{3}u(z) \leq u(z_i) \leq 3u(z), \quad z \in B(z_i, R/2),\ i = 1, \cdots, m,$$

and hence

$$3^{-m}u(w_1) \leq 3^m u(w_2), \quad \forall w_1,\ w_2 \in E.$$

We obtain

$$3^{-m} \sup_{E} u \leq 3^m \inf_{E} u$$

or (3.20) with $K = 3^{2m}$. □

Theorem 3.7 (Harnack principle) *Let $\{u_k(z)\}_{k=1}^{\infty}$ be a sequence of harmonic functions in the domain $\Omega \subset \mathbf{R}^2$, non-decreasing in k at each point $z \in \Omega$, and bounded at some $z_0 \in \Omega$. Then there is a harmonic function $u = u(z)$, $z \in \Omega$, such* *that $u_k(z) \to u(z)$ locally uniformly in $z \in \Omega$, that is, this convergence is uniform on any compact set in Ω.*

Proof The function $u_{k+1}(z) - u_k(z)$ is non-negative and harmonic in Ω and hence any compact set $E \subset \Omega$ containing z_0 admits $K > 0$ such that

$$0 \le u_{k+1}(z) - u_k(z) \le K\left\{u_{k+1}(z_0) - u_k(z_0)\right\}, \quad z \in E,\ k = 1, 2, \cdots.$$

From the assumption we have

$$\sum_{k=1}^{\infty} \{u_{k+1}(z_0) - u_k(z_0)\} < +\infty,$$

which implies

$$\sum_{k=1}^{\infty} \{u_{k+1}(z) - u_k(z)\} < +\infty$$

uniformly in $z \in E$. We obtain the limit function

$$u(z) = \lim_{k\to\infty} u_k(z) = u_1(z) + \sum_{k=1}^{\infty} \{u_{k+1}(z) - u_k(z)\}$$

with the convergence uniform in E, and in particular, $\{u_k(z)\}$, $z \in E$, is bounded. Applying this argument finite times we see that the above convergence is locally uniform in Ω. Then we obtain $u \in C(\Omega)$.

Since $u_k(z)$, $z \in \Omega$, is harmonic we have

$$u_k(z_1 + re^{\imath\theta}) = \frac{1}{2\pi}\int_0^{2\pi} u_k(z_1 + Re^{\imath\varphi})\frac{R^2 - r^2}{R^2 - 2Rr\cos(\theta - \varphi) + r^2}d\varphi$$

for $z_1 \in \Omega$, $0 \le r < R \ll 1$. Sending $k \to \infty$, we have

$$u(z_1 + re^{\imath\theta}) = \frac{1}{2\pi}\int_0^{2\pi} u(z_1 + Re^{\imath\varphi})\frac{R^2 - r^2}{R^2 - 2Rr\cos(\theta - \varphi) + r^2}d\varphi,$$

which implies that $u = u(z)$ is continuously differentiable twice and is harmonic in Ω. See Lemma 3.1 below. □

Bol's inequality in the previous chapter induces a spherical mean value theorem which ensures bubbled Harnack principle for spherical sub-harmonic functions [37].

3.2.3 Perron Solution

A real-valued continuous function $u = u(z)$ defined on the domain $\Omega \subset \mathbf{R}^2$ is said to be superharmonic if

$$u(z) \geq \frac{1}{|\partial B(z, R)|} \int_{\partial B(z,R)} u \, dS, \quad \forall z \in \Omega, \ 0 < \forall R \ll 1.$$

Similarly to the case of harmonic functions, any non-constant superharmonic function cannot attain the infimum in the interior. If $-u(z)$ is superharmonic we say that $u(z)$ is subharmonic. Any non-constant subharmonic function cannot attain the interior maximum, and therefore, a continuous subharmonic function attains a maximum on the boundary. A continuous function is harmonic if and only if it is sub- and superharmonic.

Given a bounded domain $\Omega \subset \mathbf{R}^2$ and $f \in C(\partial\Omega)$, we suppose that the Dirichlet problem (3.15) has a solution $u \in C^2(\Omega) \cap C(\overline{\Omega})$. If $v \in C(\overline{\Omega})$ is subharmonic in Ω satisfying

$$v \leq f \quad \text{on } \partial\Omega,$$

then the function $v - u \in C(\overline{\Omega})$ is subharmonic in Ω, satisfying

$$v - u \leq 0 \quad \text{on } \partial\Omega.$$

Then the weak maximum principle guarantees

$$v \leq u \quad \text{on } \overline{\Omega},$$

which means that $u(z)$ attains the maximum of $\{v(z) \mid v \in \underline{\mathscr{L}}(f)\}$for any $z \in \overline{\Omega}$, where

$$\underline{\mathscr{L}}(f) = \left\{ v \in C(\overline{\Omega}) \mid \text{subharmonic in } \Omega, \ v \leq f \text{ on } \partial\Omega \right\}.$$

This observation leads us to define the Perron solution to (3.15):

$$u(z) = \sup \left\{ v(z) \mid v \in \underline{\mathscr{L}}(f) \right\}, \quad z \in \Omega. \tag{3.21}$$

The Poisson integral $\mathscr{P}(g)$ for $g \in C(\partial B)$ is defined by

$$(\mathscr{P}g)(z) = \frac{1}{2\pi} \int_0^{2\pi} g(z_0 + Re^{\imath\varphi}) \frac{R^2 - r^2}{R^2 - 2Rr\cos(\theta - \varphi) + r^2} d\varphi,$$

where

$$B = B(z_0, R),\ z = z_0 + re^{\imath\theta},\ 0 \le r < R.$$

Here, we use a counterpart of Theorem 3.3, the proof of which is left to the reader.

Lemma 3.1 *Given $g \in C(\partial B)$, we have*

$$\mathscr{P}g \in C(\overline{B}) \cap C^2(B), \quad \Delta(\mathscr{P}g) = 0 \text{ in } B, \quad \mathscr{P}g = g \text{ on } \partial B.$$

The harmonicity of $u = u(z)$, $z \in \Omega$, defined by (3.21) follows from the fact that each element in $\underline{\mathscr{L}}(f)$ can lift as a harmonic function on any disc contained in Ω. Since the set $\underline{\mathscr{L}}(f)$ is flexible, such a process is closed.

Lemma 3.2 *Given $B = B(z_0, R)$, $\overline{B} \subset \Omega$, and $v \in \underline{\mathscr{L}}(f)$, we take*

$$v_B(z) = \begin{cases} (\mathscr{P}g)(z), & z = z_0 + re^{\imath\theta} \in B \\ v(z), & z \in \Omega \setminus B \end{cases} \tag{3.22}$$

for $g = v|_{\partial B}$. Then it holds that $v_B \in \underline{\mathscr{L}}(f)$.

Proof By Lemma 3.1 the function v_B defined by (3.22) is continuous on $\overline{\Omega}$. The relation $v_B \le f$ on $\partial\Omega$ is obvious. To show that $w = v_B$ is subharmonic in Ω it suffices to prove

$$w_D(z) \ge w(z), \quad z \in \Omega, \tag{3.23}$$

for any disc D, $\overline{D} \subset \Omega$.

In fact $v(z)$ is subharmonic and hence the weak maximum principle guarantees $w \ge v$ in B. Consequently, there arises

$$w \ge v \quad \text{on } \overline{\Omega}$$

because the other case is obvious, which implies $w_D \ge v_D$ in Ω. We have $v_D \ge v$ in Ω similarly, and hence

$$w_D \ge v_D \ge v \quad \text{in } \Omega. \tag{3.24}$$

If $z \notin D$, first, then $w_D(z) = w(z)$. If $z \notin B$, second, then $v(z) = v_B(z) = w(z)$ and hence

$$w_D(z) \ge w(z)$$

by (3.24). Hence we have only to examine the case $z \in D \cap B$.

In fact, since

$$v = v_B = w \text{ on } \partial B \cap D, \quad w_D = w \text{ on } \partial D \cap B,$$

we have

$$w_D \geq w \quad \text{on } \partial(D \cap B)$$

by (3.24). Here, both w_D and $w = v_B$ are harmonic in $D \cap B$, so it holds that

$$w_D \geq w \quad \text{in } D \cap B$$

by the weak maximum principle and hence (3.23) in Ω. □

The function v_B defined by (3.22) is called the *harmonic lifting* of v on B.

Lemma 3.3 *If the continuous functions v and w are subharmonic in Ω, so is* $\max\{v, w\}$.

Proof It is a direct consequence of the definition. □

Theorem 3.8 *The function $u(z)$ defined by (3.21) is harmonic in Ω.*

Proof Since $-m \in \underline{\mathscr{L}}(f)$ for a constant $m \gg 1$, it holds that $\underline{\mathscr{L}}(f) \neq \emptyset$. The weak maximum principle, on the other hand, guarantees

$$v(z) \leq \max_{\partial\Omega} v \leq \max_{\partial\Omega} f < +\infty, \quad v \in \underline{\mathscr{L}}(f),$$

and hence

$$u(z) = \sup\left\{v(z) \mid v \in \underline{\mathscr{L}}(f)\right\} < +\infty, \quad z \in \Omega.$$

We take $B = B(y, R)$, $\overline{B} \subset \Omega$, and $\{v_k\} \subset \underline{\mathscr{L}}(f)$ such that

$$v_k(y) \uparrow u(y), \; k \to \infty.$$

Define $\tilde{v}_k$, $k = 1, 2, \cdots$, by $\tilde{v}_1 = v_1$, $\tilde{v}_2 = \max\{v_1, v_2\}$, and so forth. Since $\{\tilde{v}_k\} \subset \underline{\mathscr{L}}(f)$ and

$$\tilde{v}_k(y) = v_k(y) \uparrow u(y)$$

by Lemma 3.3, we may suppose that $\{v_k\}$ is non-decreasing at any point in Ω.

If $V_k = (v_k)_B$ is the harmonic lifting of $v_k(x)$ on $B = B(y, R)$ then

$$\{V_k\} \subset \underline{\mathscr{L}}(f)$$

is non-decreasing at any point in Ω. Furthermore V_k is harmonic in B and hence $v_k \leq V_k$ in Ω by the weak maximum principle. This inequality implies

$$v_k(y) \leq V_k(y) \quad \to \quad u(y) < +\infty \tag{3.25}$$

by $v_k(y) \to u(y)$ and (3.21) and hence the Harnack principle, Theorem 3.7, is applicable. Then we obtain a harmonic function $v(z)$, $z \in B$, such that $V_k \to v$ locally uniformly in B. If $v = u$ in B is the case then $u(z)$ is harmonic in B, and, therefore, it is harmonic in Ω because B is arbitrary.

To show this equality, first we note that $V_k \leq u$ holds in Ω by (3.21), which implies

$$v \leq u \quad \text{in } B.$$

We have, on the other hand,

$$v(y) = u(y)$$

by (3.25). If there is $z \in B$, $u(z) > v(z)$, we have $\tilde{u} \in \underline{\mathscr{L}}(f)$ such that

$$u(z) > \tilde{u}(z) > v(z).$$

Let

$$w_k = \max\{v_k, \tilde{u}\}.$$

It holds that $\{w_k\} \subset \underline{\mathscr{L}}(f)$ and is non-decreasing in Ω. Let, furthermore,

$$W_k = (w_k)_B.$$

Similarly to $\{V_k\}$ we have a harmonic function w in B such that $W_k \to w$ locally uniformly in B and $W_k(y) \to u(y) = w(y)$. Since $w_k \geq v_k$ we have $W_k \geq V_k$ in Ω, which implies

$$w \geq v \quad \text{in } B.$$

We have, on the other hand, $w(y) = u(y) = v(y)$ and hence $w = v$ in B by the strong maximum principle. It holds that $w(z) = v(z)$ in particular.

We have, however, $w_k \geq \tilde{u}$ in Ω and hence $w_k(z) \geq \tilde{u}(z)$. Then $w(z) \geq \tilde{u}(z) > v(z)$, a contradiction. □

3.2.4 Boundary Regularity

We call $\xi \in \partial\Omega$ a regular point if there is a barrier $w = w(z) \in C(\overline{\Omega})$ that is superharmonic in Ω and satisfies

$$w(\xi) = 0, \quad w(x) > 0, \ \forall x \in \overline{\Omega} \setminus \{\xi\}. \tag{3.26}$$

It is obvious that if the Dirichlet problem (3.15) has a solution $u \in C^2(\Omega) \cap C(\overline{\Omega})$ for any $f \in C(\partial\Omega)$, any point on $\partial\Omega$ is regular.

Theorem 3.9 *If f is continuous on $\partial\Omega$ and any point on $\partial\Omega$ is regular, the Perron solution $u(z)$ defined by (3.21) is continuous on $\overline{\Omega}$ and satisfies $u|_{\partial\Omega} = f$.*

Proof We show

$$\xi \in \partial\Omega, \ \{x\} \subset \Omega, \ x \to \xi \quad \Rightarrow \quad u(x) \to f(\xi).$$

First, since f is continuous on $\partial\Omega$ it holds that

$$M = \max_{\partial\Omega} |f| < +\infty$$

and any $\varepsilon > 0$ admits $\delta > 0$ such that

$$|x - \xi| < \delta, \ x \in \partial\Omega \quad \Rightarrow \quad |f(x) - f(\xi)| < \varepsilon.$$

There is a barrier $w(x)$ at ξ from the assumption. Since $w \in C(\overline{\Omega})$ satisfies (3.26), there is $k > 0$ such that

$$kw(x) \geq 2M, \qquad x \in \partial\Omega, \ |x - \xi| \geq \delta.$$

This $w = w(x)$ is superharmonic and hence

$$v_1(x) = f(\xi) - \varepsilon - kw(x)$$

is subharmonic and continuous in Ω and on $\overline{\Omega}$, respectively. For $x \in \partial\Omega$ in $|x - \xi| < \delta$ it holds that

$$v_1(x) = f(x) + (f(\xi) - f(x) - \varepsilon) - kw(x) \leq f(x).$$

For $x \in \partial\Omega$ in $|x - \xi| \geq \delta$, on the other hand, we have

$$v_1(x) \leq f(\xi) - kw(x) \leq f(\xi) - 2M \leq -M \leq f(x).$$

There arises $v_1 \in \underline{\mathscr{L}}(f)$ and hence $u \geq v_1$ in Ω.

Now we take

$$v_2(x) = f(\xi) + \varepsilon + kw(x),$$

which is superharmonic and continuous in Ω and on $\overline{\Omega}$, respectively. Similarly we obtain $v_2 \geq f$ on $\partial\Omega$. Since any $v \in \underline{\mathscr{L}}(f)$ is subharmonic in Ω and satisfies

$v \leq f$ on $\partial\Omega$ by the definition, it follows that $v \leq v_2$ in Ω from the weak maximum principle. We thus obtain $u \leq v_2$ in Ω because $v \in \underline{\mathscr{S}}(f)$ is arbitrary.

These relations are summarized as

$$|u(x) - f(\xi)| \leq \varepsilon + kw(x), \qquad x \in \Omega.$$

Since $w \in C(\overline{\Omega})$ and $w(\xi) = 0$ it holds that

$$\limsup_{x \in \Omega \to \xi} |u(x) - f(\xi)| \leq \varepsilon$$

and hence $\lim_{x \in \Omega \to \xi} u(x) = f(\xi)$ with $\varepsilon \downarrow 0$. □

3.2.5 *Green's Function*

Given a bounded domain $\Omega \subset \mathbf{R}^2$, the function $G(x, x_0)$, $(x, x_0) \in \overline{\Omega} \times \Omega$, is called Green's function, if, as a function of x, it is harmonic in $\Omega \setminus \{x_0\}$,

$$K(x, x_0) = G(x, x_0) + \frac{1}{2\pi} \log |x - x_0|$$

is harmonic in a neighborhood of x_0, and

$$\lim_{x \in \Omega \to \xi} G(x, x_0) = 0, \quad \xi \in \partial\Omega.$$

Green's function exists if any boundary point of $\partial\Omega$ is regular and is unique if it exists. If $r > 0$ is small and $|x - x_0| = r$ we have $G(x, x_0) > 0$ by the second requirement to $G(x, x_0)$, and, therefore, $G(x, x_0) > 0$, $x \in \Omega \setminus B(x_0, r)$, from the third requirement and the maximum principle. Since $r > 0$ is arbitrary, it holds that

$$G(x, x_0) > 0, \quad x \in \Omega \setminus \{x_0\},$$

and, therefore,

$$\frac{\partial G(\xi, x_0)}{\partial \nu_\xi} \leq 0, \quad \xi \in \partial\Omega,$$

if the left-hand side exists, where $\nu = \nu_\xi$ denotes the outer unit normal vector. The following theorem is a generalization of Poisson's formula (3.17).

Theorem 3.10 *If $\partial\Omega$ is C^2 and $u(x)$ is harmonic and continuous in Ω and on $\overline{\Omega}$, respectively, it holds that*

$$u(x) = -\int_{\partial\Omega} u(\xi)\frac{\partial G(\xi, x)}{\partial \nu_\xi} dS_\xi, \quad x \in \Omega, \tag{3.27}$$

where dS_ξ denotes the line element.

Proof By Green's formula in $\Omega \setminus \overline{B(x_0, \rho)}$ for $x_0 \in \Omega$, $0 < \rho \ll 1$, it holds that

$$\int_{\partial B(x_0,\rho)} u(\xi)\frac{\partial G(\xi, x_0)}{\partial \nu_\xi} - G(\xi, x_0)\frac{\partial u(x_0)}{\partial \nu_\xi}\, dS_\xi + \int_{\partial\Omega} u(\xi)\frac{\partial G(\xi, x_0)}{\partial \nu_\xi} - G(\xi, x_0)\frac{\partial u(\xi)}{\partial \nu_\xi}\, dS_\xi = 0. \tag{3.28}$$

Given $\xi \in \partial B(x_0, \rho)$, we have

$$G(\xi, x_0) = \frac{1}{2\pi}\log\frac{1}{r} + K(\xi, x_0), \quad r = \rho,$$
$$\frac{\partial G(\xi, x_0)}{\partial \nu_\xi} = -\frac{\partial G(\xi, x_0)}{\partial r} = \frac{1}{2\pi r} - \frac{\partial K(\xi, x_0)}{\partial r}$$

and hence the first term on the left-hand side of (3.28) converges,

$$\int_{\partial B(x_0,\rho)} u(\xi)\frac{\partial G(\xi, x_0)}{\partial \nu_\xi} - G(\xi, x_0)\frac{\partial u(\xi)}{\partial \nu_\xi} dS_\xi$$
$$= \int_0^{2\pi} \Big[u(x_0 + re^{\iota\theta})(\frac{1}{2\pi r} - \frac{\partial K(x_0 + re^{\iota\theta}, x_0)}{\partial r})$$
$$+ (\frac{1}{2\pi}\log\frac{1}{r} + K(x_0 + re^{\iota\theta}, x_0))\frac{\partial u(x_0 + re^{\iota\theta})}{\partial r}\Big]_{r=\rho} \rho d\theta \to u(x_0)$$

as $\rho \downarrow 0$. Then (3.27) follows from $G(\xi, x_0) = 0$. □

Complex function theory provides with a relation between the Riemann mapping and Green's function.

Theorem 3.11 *Let $D, \Omega \subset \mathbf{R}^2$ be domains and let $w = f(z) : D \to \Omega$ be a conformal homeomorphism. If $G = G(w, w_0)$ is Green's function on Ω then*

$$G_D(z, z_0) = G(f(z), f(z_0)) \tag{3.29}$$

is that on D.

Proof Since $f : D \to \Omega$ is conformal, if $u = u(w)$ is harmonic in Ω then

$$v = v(z) \equiv u(f(z))$$

is so in D. We show that $G_D = G_D(z, z_0)$ defined by (3.29) satisfies the second requirement of Green's function. In fact

$$G(w, w_0) + \frac{1}{2\pi} \log |w - w_0|$$

is harmonic at $w = w_0 \in \Omega$, which implies

$$\begin{aligned} & G_D(z, z_0) + \frac{1}{2\pi} \log |z - z_0| \\ &= G(f(z), f(z_0)) + \frac{1}{2\pi} \log |f(z) - f(z_0)| - \frac{1}{2\pi} \log \left| \frac{f(z) - f(z_0)}{z - z_0} \right|. \end{aligned}$$

The last term on the right-hand side is harmonic in a neighborhood of $z = z_0$ by $f'(z_0) \neq 0$. Thus we obtain the conclusion. □

Remark 3.1 The domain surrounded by a regular Jordan curve is called a Jordan region. Show that Green's function on the unit disc D, $|x| < 1$, is given by

$$G(x, x_0) = \frac{1}{2\pi} \log \left| \frac{1 - \overline{z}_0 z}{z - z_0} \right|, \qquad z = x_1 + \imath x_2, \ z_0 = x_{01} + \imath x_{02},$$

where $x = (x_1, x_2)$, $x_0 = (x_{01}, x_{02})$. Then,

$$G(x, x_0) = \frac{1}{2\pi} \log \left| \frac{1 - \overline{f(z_0)} f(z)}{f(z) - f(z_0)} \right|$$

is Green's function on the Jordan region Ω, where $f : \Omega \to D$ denotes the Riemann mapping, that is, a conformal homeomorphism.

3.2.6 *Newton Potential*

The theory of two-dimensional harmonic functions is based on the representation formula of the solution to the Dirichlet problem on the disc, that is, the Poisson integral. The function $\log |x|$ is a two-dimensional harmonic function in $x \neq 0$ because it is the real part of the analytic function $\log z$. This harmonic function depends only on $r = |x|$. We can find such functions even in higher dimensions.

Let n be the space dimension. We take the polar coordinates

$$x = r\omega \in \mathbf{R}^n, \ r = |x|, \ \omega \in S^{n-1} = \left\{ x \in \mathbf{R}^n \mid |x| = 1 \right\},$$

recalling Sect. 1.4.1. It holds that

$$\begin{aligned}\Delta &= \frac{\partial^2}{\partial x_1^2} + \cdots + \frac{\partial^2}{\partial x_n^2} = \frac{\partial^2}{\partial r^2} + \frac{n-1}{r}\frac{\partial}{\partial r} + \frac{1}{r^2}\Lambda \\ &= \frac{1}{r^{n-1}}\frac{\partial}{\partial r}\left(r^{n-1}\frac{\partial}{\partial r}\right) + \frac{1}{r^2}\Lambda,\end{aligned}$$

where Λ is the Laplace–Beltrami operator on S^{n-1}. If $n = 2$ we take

$$x_1 = r\cos\varphi,\ x_2 = r\sin\varphi, \quad \varphi \in [0, 2\pi),$$

to obtain

$$\Lambda = \frac{\partial^2}{\partial\varphi^2}.$$

If $n = 3$ we take

$$x_1 = r\sin\theta\cos\varphi,\ x_2 = r\sin\theta\sin\varphi,\ x_3 = r\cos\theta, \quad \theta \in [0, \pi],\ \varphi \in [0, 2\pi).$$

It holds that

$$\Lambda = \frac{1}{\sin\theta}\frac{\partial}{\partial\theta}\left(\sin\theta\frac{\partial}{\partial\theta}\right) + \frac{1}{\sin^2\theta}\frac{\partial^2}{\partial\varphi^2}.$$

If $\Gamma = \Gamma(|x|)$ is harmonic then

$$\frac{1}{r^{n-1}}\frac{\partial}{\partial r}\left(r^{n-1}\frac{\partial\Gamma}{\partial r}\right) = 0,$$

which implies

$$\Gamma(r) = c_1 r^{-n+2} + c_2$$

for $n \geq 3$, where c_1, c_2 are constants. Such a Γ satisfies

$$\Delta\Gamma(x) = 0, \quad x \neq 0,$$

where the singularity at $x = 0$ of Γ is important in later arguments. In this context we take

$$\Gamma(x) = \frac{1}{\omega_n(2-n)}\,|x|^{2-n},$$

where ω_n denotes the area of S^{n-1} given by $2\pi^{n/2}/\gamma(n/2)$ using the Gamma function

$$\gamma(z) = \int_0^\infty e^{-s} s^{z-1} ds.$$

In the case of $n = 3$ we have the Newton potential

$$\Gamma(x) = -\frac{1}{4\pi\,|x|}.$$

Let $u \in C^2(\overline{B})$ be harmonic in B, $B = \{x \in \mathbf{R}^n \mid |x| < 1\}$, and take

$$x \in B,\ 0 < \varepsilon \ll 1,\ \Omega = B \setminus \overline{B(x, \varepsilon)}.$$

If ν and dS denote the outer normal vector and the area element, respectively, we obtain

$$\begin{aligned} &\int_\Omega \Delta u(y) \cdot \Gamma(x-y) - u(y)\Delta_y \Gamma(x-y)\, dy \\ &= \int_{\partial\Omega} \frac{\partial u}{\partial \nu_y}(y)\Gamma(x-y) - u(y)\frac{\partial \Gamma}{\partial \nu_y}(x-y)\, dS_y \end{aligned} \tag{3.30}$$

by Green's formula. Here the left-hand side vanishes by

$$\Delta u(y) = \Delta_y \Gamma(x-y) = 0, \quad y \in \Omega = B \setminus \overline{B(x, \varepsilon)},$$

while the right-hand side is divided into

$$\int_{\partial B} + \int_{\partial B(x,\varepsilon)}.$$

It holds that

$$\begin{aligned} &\left| \int_{\partial B(x,\varepsilon)} \frac{\partial u}{\partial \nu_y}(y)\Gamma(x-y) dS_y \right| \\ &\leq \|\nabla u\|_{L^\infty(B)} \cdot \int_{|y-x|=\varepsilon} |\Gamma(y-x)|\, dS_y \\ &= \|\nabla u\|_{L^\infty(B)} \cdot \frac{1}{\omega_n(n-2)} \cdot \varepsilon^{2-n} \cdot \omega_n \varepsilon^{n-1} \\ &= \|\nabla u\|_{L^\infty(B)} \frac{\varepsilon}{n-2} \to 0. \end{aligned}$$

Since

$$\frac{\partial}{\partial \nu_y} = -\frac{\partial}{\partial \rho}, \quad \rho = |y-x| \text{ on } \partial B(x, \varepsilon),$$

we obtain

$$\int_{\partial B(x,\varepsilon)} u(y)\frac{\partial \Gamma}{\partial \nu_y}(x-y)\, dS_y$$
$$= \int_{\partial B(x,\varepsilon)} u(y)\cdot \frac{1}{\omega_n(2-n)}\cdot (2-n)\,|x-y|^{1-n}\,(-1)\, dS_y$$
$$= -\frac{1}{\omega_n \varepsilon^{n-1}}\int_{\partial B(x,\varepsilon)} u(y)\, dS_y$$
$$= -\frac{1}{|\partial B(x,\varepsilon)|}\int_{\partial B(x,\varepsilon)} u(y) dS_y \to -u(x), \quad \varepsilon \downarrow 0.$$

These relations are summarized as

$$\int_{\partial B} \frac{\partial u}{\partial r}(y)\Gamma(x-y) - u(y)\frac{\partial \Gamma}{\partial r_y}(x-y)\, dS_y + u(x) = 0, \quad x \in B. \tag{3.31}$$

Lemma 3.4 *The mirror image with respect to ∂B is associated with the Kelvin transformation as follows:*

$$y = x/\,|x|^2\,,\ U(y) = |x|^{n-2}\,u(x) \quad \Rightarrow \quad \Delta_y U = |x|^{n+2}\,\Delta_x u.$$

Proof In the polar coordinate $x = r\omega$ it holds that $y = r^{-1}\omega$. Putting $\rho = r^{-1}$, we obtain

$$\Delta_y U = \frac{1}{\rho^{n-1}}\frac{\partial}{\partial \rho}(\rho^{n-1}\frac{\partial U}{\partial \rho}) + \frac{1}{\rho^2}\Lambda U = -r^{n-1}(\rho^{n-1}\frac{\partial U}{\partial \rho})_r\cdot r^2 + r^2\cdot \Lambda U$$
$$= r^{n-1}(r^{-n+1}(r^{n-2}u)_r\cdot r^2)_r\cdot r^2 + r^2\cdot \Lambda\left(r^{n-2}u\right)$$
$$= r^{n+1}(r^{-n+3}(r^{n-2}u)_r)_r + r^n\Lambda u = r^{n+1}(ru_r + (n-2)u)_r + r^n\Lambda u$$
$$= r^{n+1}(ru_{rr} + (n-1)u_r) + r^n\Lambda u = r^{n+2}(u_{rr} + \frac{n-1}{r}u_r) + r^n\Lambda u$$
$$= r^{n+2}\{\frac{1}{r^{n-1}}(r^{n-1}u_r)_r + \frac{1}{r^2}\Lambda u\}$$

and hence the conclusion. □

Now we take the Kelvin transformation of $\Gamma(x-\cdot)$ denoted by

$$\Gamma^x(y) = \Gamma(|y|(x-y/|y|^2)) = \Gamma(\frac{1}{|z|}(x-z))$$

$$= \frac{1}{\omega_n(2-n)}|z|^{n-2}|x-z|^{-(n-2)} = |z|^{n-2}\Gamma(x-z)$$

where $z = y/|y|^2$. Then Lemma 3.4 implies

$$\Delta_y \Gamma^x = |z|^{n+2}\Delta_z \Gamma(x-z) = 0, \qquad z \neq x.$$

If $x, y \in B$ then $z \notin \overline{B}$ and hence $x \neq z$. Thus Γ^x is harmonic on $\overline{B}$. Replacing Γ by Γ^x in (3.30), we obtain

$$\int_{\partial B} \frac{\partial u}{\partial r_y}(y)\Gamma^x(y) - u(y)\frac{\partial \Gamma^x}{\partial r_y}(y)\, dS_y = 0.$$

If $y \in \partial B$, on the other hand, then $z = y$ and hence $\Gamma^x(y) = \Gamma(x-y)$. It holds that

$$\int_{\partial B} \frac{\partial u}{\partial r_y}(y)\Gamma(x-y) - u(y)\frac{\partial \Gamma^x}{\partial r_y}(y)\, dS_y = 0. \tag{3.32}$$

Equalities (3.31) and (3.32) imply

$$u(x) = \int_{\partial B} \{\frac{\partial \Gamma}{\partial r_y}(x-y) - \frac{\partial \Gamma^x}{\partial r_y}(y)\}u(y)\, dS_y. \tag{3.33}$$

This equality is extended to $u \in C^2(B) \cap C(\overline{B})$ satisfying $\Delta u = 0$, by replacing $B(0,r)$ to $B = B(0,1)$ and making $r \uparrow 1$ in analogous equalities. We thus obtain the *Poisson integral* for $n \geq 3$.

Theorem 3.12 *If $u \in C^2(B)$ is harmonic and continuous in B and on $\overline{B}$, respectively, then it holds that*

$$u(x) = \int_{\partial B} P(x,y)u(y)dS_y, \qquad x \in B, \tag{3.34}$$

where

$$P(x,y) = \frac{1-|x|^2}{\omega_n |y-x|^n}$$

is the Poisson kernel.

Proof We have

$$\frac{\partial \Gamma}{\partial x_j}(x) = \frac{1}{\omega_n}\cdot\frac{1}{|x|^{n-1}}\cdot\frac{x_j}{|x|}, \qquad \frac{\partial^2 \Gamma}{\partial x_j \partial x_k}(x) = \frac{1}{\omega_n}\left\{\delta_{jk} - n\frac{x_j x_k}{|x|^2}\right\}|x|^{-n}. \tag{3.35}$$

Since $y \cdot \nabla_y = r\frac{\partial}{\partial r}$, $r = |y|$, it holds that

$$\begin{aligned}\frac{\partial \Gamma}{\partial r_y}(x-y) &= \sum_{j=1}^{n} \frac{y_j}{|y|}\frac{\partial}{\partial y_j}\Gamma(x-y) \\ &= \sum_{j=1}^{n} \frac{y_j}{|y|} \cdot \frac{1}{\omega_n} \cdot \frac{1}{|x-y|^{n-1}} \cdot \frac{-(x_j-y_j)}{|x-y|} \\ &= \frac{1}{\omega_n} \cdot \frac{1}{|y|} \cdot \frac{(y, y-x)}{|y-x|^n}. \end{aligned} \tag{3.36}$$

For $\rho = |z| = 1/r$, $z = y/|y|^2$, we have

$$\frac{\partial \Gamma^x}{\partial r_y}(y) = -\left(\rho^{n-2}\Gamma(x-z)\right)_\rho \cdot r^{-2} = -(n-2)\Gamma(x-z) - \frac{\partial}{\partial \rho}\Gamma(x-z)$$

for $\rho = 1$. Here, the right-hand side is equal to

$$-(n-2)\Gamma(x-z) - \frac{1}{\omega_n} \cdot \frac{(z, z-x)}{|z-x|^n} = \frac{1}{\omega_n}|x-y|^{2-n} - \frac{1}{\omega_n} \cdot \frac{(y, y-x)}{|y-x|^n}$$

for $\rho = 1$ by (3.36). It holds that (3.34) by

$$\begin{aligned}\frac{\partial \Gamma}{\partial r_y}(x-y) - \frac{\partial \Gamma^x}{\partial r_y}(y) &= \frac{2(y, y-x) - |y-x|^2}{\omega_n |y-x|^n} \\ &= \frac{(y+x, y-x)}{\omega_n |y-x|^n} = \frac{|y|^2 - |x|^2}{\omega_n |y-x|^n}.\end{aligned}$$

The proof is complete. □

A formula similar to (3.34) holds on $B = B(x_0, R)$ for the case where $P(x, y)$ is replaced by

$$P(x, y) = \frac{R^2 - |x-x_0|^2}{R\omega_n |y-x|^n},$$

as in the next subsection.

We can reproduce the arguments in Sects. 3.2.2, 3.2.3, and 3.2.4, that is, the mean value theorem, the maximum principle, Harnack's inequality, the Harnack principle, sub- and superharmonic functions, Perron's solution, harmonic lifting, and barrier, and consequently, Theorem 3.1.

3.2.7 *Entire Harmonic Functions*

Liouville's theorem on entire bounded holomorphic functions is a consequence of Cauchy's inequality in the following form obtained by Cauchy's integral formula,

$$f^{(k)}(0) = \frac{k!}{2\pi\sqrt{-1}} \int_{|z|=r} \frac{f(z)}{z^{k+1}} dz, \quad 0 < r < R.$$

Theorem 3.13 (Cauchy) *If $f(z)$ is a holomorphic function of $z = x_1 + \sqrt{-1}x_2 \in B(0, R)$ satisfying $|f(z)| \leq M$, it follows that*

$$|f^{(k)}(0)| \leq \frac{k!}{R^k} M, \quad \forall k = 1, 2, \cdots.$$

Theorem 3.14 (Liouville) *Let $f(z)$ is a holomorphic function of $z \in \mathbf{C}$, satisfying $|f(z)| \leq M$. Then this $f(z)$ is a constant.*

Proof By the previous theorem it holds that

$$|f^{(k)}(a)| \leq \frac{k!}{R^k} M$$

for any $a \in \mathbf{C}$, $k = 1, 2, \cdots$, and $R > 0$. Sending $R \uparrow +\infty$, we obtain $f' \equiv 0$, and hence f is constant in $\mathbf{C}$. □

Recall Theorem 3.12 and assume that $u(x)$ is harmonic in $B_R = B(0, R)$ and continuous on $\overline{B_R}$. Then it holds that

$$\tilde{u}(x') = \int_{\partial B} \frac{1}{\omega_n} \frac{1 - |x'|^2}{|y' - x'|^n} \tilde{u}(y) dS_{y'}$$

for $\tilde{u}(x') = u(Rx')$, $x = Rx'$, and $y = Ry'$, which implies

$$u(x) = \int_{\partial B_R} \frac{1}{\omega_n} \frac{1 - |\frac{x}{R}|^2}{|\frac{y}{R} - \frac{x}{R}|^n} u(y) \frac{dS_y}{R^{n-1}} = \frac{1}{R} \int_{\partial B_R} \frac{1}{\omega_n} \frac{R^2 - |x|^2}{|y - x|^n} u(y) dS_y. \tag{3.37}$$

Let $u(x)$ be harmonic in $\mathbf{R}^n$ satisfying $|u(x)| \leq M$. Since

$$\left.\frac{\partial}{\partial x_i} \frac{R^2 - |x|^2}{|y - x|^n}\right|_{x=0} = \left. R^2 \frac{\partial}{\partial x_i} |x - y|^{-n} \right|_{x=0} = \left. - R^2 n |x - y|^{-n-1} \frac{x_i - y_i}{|x - y|} \right|_{x=0}$$
$$= nR^2 |y|^{-n-2} y_i$$

it holds that

$$\left|\frac{\partial u}{\partial x_i}(0)\right| \leq \frac{1}{R}\frac{1}{\omega_n} MnR^2 \cdot \frac{1}{R^{n+1}}\omega_n R^{n-1} = \frac{Mn}{R} \longrightarrow 0, \quad R \uparrow +\infty,$$

for $i = 1, 2, \cdots, n$. Hence $u(x)$ is constant because $\frac{\partial u}{\partial x_i} \equiv 0$ by this calculation.

To refine the result, we use

$$\frac{R-r}{(R+r)^{n-1}} \leq \frac{R^2 - |x|^2}{|y-x|^n} \leq \frac{R+r}{(R-r)^{n-1}}, \quad |x| = r < R = |y|,$$

and the mean value theorem derived from (3.37),

$$u(0) = \frac{1}{R}\frac{1}{\omega_n}\frac{R^2}{R^n}\int_{\partial B_R} u \, dS_y = \frac{1}{\omega_n R^{n-1}}\int_{\partial B_R} u \, dS_y.$$

If $\Delta u = 0$, $u \geq 0$ in B_R, and $u(x)$ is continuous on $\overline{B_R}$, therefore, there arises

$$\begin{aligned} u(x) &= \frac{1}{\omega_n R}\int_{\partial B_R} \frac{R^2 - |x|^2}{|y-x|^n} u(y) \, dS_y \\ &\leq \frac{1}{\omega_n R}\frac{R+r}{(R-r)^{n-1}}\int_{\partial B_R} u \, dS \\ &= \frac{R+r}{(R-r)^{n-1}} R^{n-2} u(0), \quad |x| = r < R, \end{aligned}$$

and similarly,

$$u(x) \geq \frac{1}{\omega_n R}\frac{R+r}{(R-r)^{n-1}}\int_{\partial B_R} u \, dS = \frac{R-r}{(R+r)^{n-1}} R^{n-2} u(0).$$

These inequalities induce the Harnack inequality, which results in the following fact.

Theorem 3.15 (Liouville) *If $u(x)$ is harmonic function and bounded below or above in $\mathbf{R}^n$, it is a constant.*

Proof We may assume $u(x) \geq 0$ in $\mathbf{R}^n$. Then it holds that

$$\frac{R-r}{(R+r)^{n-1}} R^{n-2} u(0) \leq u(x) \leq \frac{R+r}{(R-r)^{n-1}} R^{n-2} u(0), \quad r = |x|$$

for any $R >$ and hence $u(x) = u(0)$ with $R \uparrow +\infty$. □

Theorem 3.16 (Hadamard) *If u is sub-harmonic in $A \equiv \{x \in \mathbf{R}^2 \mid r_1 < |x| < r_2\}$ and is continuous on $\overline{A}$, it holds that*

$$M(r) \equiv \max_{|x|=r} u(x) \leq \frac{M(r_1)\log(r_2/r) + M(r_2)\log(r/r_1)}{\log(r_2/r_1)} \tag{3.38}$$

for $r \in (r_1, r_2)$.

Proof From the assumption there arises

$$\Delta u \geq 0 \quad \text{in } A.$$

We use

$$\varphi(r) = a + b \log r, \ r = |x|,$$

satisfying

$$\Delta \varphi = 0, \quad r > 0.$$

Define a and b by

$$\varphi(r_1) = M(r_1), \quad \varphi(r_2) = M(r_2),$$

that is,

$$\varphi(r) = \frac{M(r_1)\log(r_2/r) + M(r_2)\log(r/r_1)}{\log(r_2/r_1)}.$$

Then $v(x) \equiv u(x) - \varphi(|x|)$ satisfies

$$\Delta v \geq 0 \text{ in } A, \quad v|_{r=r_1,r_2} \leq 0.$$

There arises, therefore, $v \leq 0$ in A from the maximum principle, which implies

$$u(x) \leq \varphi(r), \quad |x| = r \in (r_1, r_2),$$

and hence $M(r) \leq \varphi(r)$. □

Theorem 3.17 (Liouville) *If* $\Delta u \geq 0$, $u \leq M$ *in* $\mathbf{R}^2 \setminus \{0\}$, *then* u *is a constant.*

Proof From the assumption, inequality (3.38) implies

$$M(r) \leq M(r_1)\frac{\log(r_2/r)}{\log(r_2/r_1)} + M\frac{\log(r/r_1)}{\log(r_2/r_1)}, \quad r_1 < r < r_2$$

for any $0 < r_1 < r_2$. Letting $r_2 \uparrow +\infty$, we obtain

$$M(r) \leq M(r_1), \quad 0 < r_1 < r.$$

We have, on the other hand,

$$M(r) \leq M\frac{\log(r_2/r)}{\log(r_2/r_1)} + M(r_2)\frac{\log(r/r_1)}{\log(r_2/r_1)}, \quad r_1 < r < r_2$$

again by (3.38). Making $r_1 \downarrow 0$ now implies

$$M(r) \leq M(r_2) \quad 0 < r < r_2$$

and therefore,

$$M(r_1) \leq M(r_2) \leq M(r_1).$$

Hence $M(r)$ is a constant, which implies that u takes a maximum in A. Then the strong maximum principle guarantees that u is a constant. □

Remark 3.2 From the proof, the requirement $u \leq M$ in $\mathbf{R}^2 \setminus \{0\}$ is replaced by a weaker assumption

$$M(r) \leq C(1 + |\log r|) \quad r > 0. \tag{3.39}$$

Remark 3.3 If

$$\Delta u \geq 0 \quad \text{in } \mathbf{R}^n \setminus \{0\}$$

for $n \geq 3$, it holds that

$$M(r) = \max_{|x|=r} u(x) \leq \frac{M(r_1)(r^{2-n} - r_2^{2-n}) + M(r_2)(r_1^{2-n} - r^{2-n})}{r_1^{2-n} - r_2^{2-n}}$$

for $0 < r_1 < r < r_2$. Consequently, we obtain the following properties:

1. If

$$\liminf_{r_1 \downarrow 0} r_1^{n-2} M(r_1) \leq 0,$$

then $M(r)$ is non-decreasing. In particular, u is bounded above near 0.

2. If

$$\liminf_{r_2 \uparrow +\infty} r_2^{n-2} M(r_2) \leq 0,$$

then $r^{n-2} M(r)$ is non-increasing. In particular, if

$$\lim_{r \to \infty} M(r) \leq 0,$$

then $r^{n-2} M(r)$ is bounded above.

The above described Liouville property, however, is not valid for subharmonic functions bounded above if $n \geq 3$.

3.2.8 Removable Singularities

Here, we take the case of $n \geq 3$.

Theorem 3.18 *Let $B = B(0,1) \subset \mathbf{R}^n$ for $n \geq 3$, and $0 \leq u \in C^2(B \setminus \{0\})$ satisfy*

$$-\Delta u \geq 0 \quad \text{in } B \setminus \{0\}.$$

Then it holds that $u \in L^1_{loc}(B)$ and

$$-\Delta u \geq 0 \quad \text{as } \mathscr{D}'(B),$$

i.e., in the sense of distributions.

Proof Each $k = 1, 2, \cdots$ admits $G_k \in C^2[0, \infty)$ such that

$$\begin{aligned} & G_k(s) = s, && 0 \leq s \leq k \\ & G_k(s) \leq s, && s \geq 0 \\ & G_k(s) = k+1, && s \gg 1 \\ & G_k' \geq 0 \geq G_k'', && G_k \leq G_{k+1}. \end{aligned}$$

Then $u_k = G_k(u)$, $k = 1, 2, \cdots$ is non-decreasing and $u_k \to u$ pointwise as $k \to \infty$. It holds that

$$-\Delta u_k = -G_k'(u)\Delta u - G_k''(u)|\nabla u|^2 \geq 0 \quad \text{in } B \setminus \{0\}. \tag{3.40}$$

Given $\varphi, \xi \in C_0^\infty(B)$, let

$$\xi_j(x) = \xi(jx), \ \varphi_j = (1 - \xi_j)\varphi, \quad j = 1, 2, \cdots.$$

We multiply (3.40) by $(1 + u_k)^{-\alpha}$ for $\alpha > 0$, which results in

$$\begin{aligned} 0 &\leq \int_B \nabla u_k \cdot \nabla(\varphi_j^2 (1 + u_k)^{-\alpha}) \\ &= \int_B -\alpha \varphi_j^2 (1 + u_k)^{-1-\alpha} |\nabla u_k|^2 + 2\varphi_j (1 + u_k)^{-\alpha} \nabla u_k \cdot \nabla \varphi_j \, dx, \end{aligned}$$

and therefore,

$$\begin{aligned}
&\frac{\alpha}{2}\int_B \varphi_j^2(1+u_k)^{-1-\alpha}|\nabla u_k|^2 \\
&\leq \int_B \varphi_j(1+u_k)^{-\alpha}\nabla u_k\cdot\nabla\varphi_j \\
&\leq \int_B \frac{\alpha}{4}\varphi_j^2(1+u_k)^{-1-\alpha}|\nabla u_k|^2+\frac{1}{\alpha}|\nabla\varphi_j|^2(1+u_k)^{1-\alpha}\,dx.
\end{aligned}$$

Hence it holds that

$$\int_B \varphi_j^2(1+u_k)^{-1-\alpha}|\nabla u_k|^2 \leq C_\alpha\int_B |\nabla\varphi_j|^2(1+u_k)^{1-\alpha}.$$

Since $|\nabla\varphi_j|^2 \to |\nabla\varphi|^2$ in $L^1(B)$ and $(1+u_k)^{1-\alpha} \in L^\infty(B)$, Fatou's lemma ensures

$$\int_B \varphi^2(1+u_k)^{-1-\alpha}|\nabla u_k|^2 \leq C_\alpha\int_B |\nabla\varphi|^2(1+u_k)^{1-\alpha}.$$

Taking $\alpha = 1$, we have $u_k \in H^1_{loc}(B)$ because $1 \leq 1+u_k \leq k+2$ and $\varphi \in C_0^\infty(B)$ is arbitrary. With $\alpha = 2/n$, next, it follows that

$$\int_B \varphi^2(1+u_k)^{-1-2/n}|\nabla u_k|^2 \leq C\int_B |\nabla\varphi|^2(1+u_k)^{1-\frac{2}{n}}. \tag{3.41}$$

Assume $\varphi = 1$ for $|x| \leq 1/4$ and $\omega \equiv \operatorname{supp}\varphi \subset B/2$. Then, Sobolev's inequality implies

$$\begin{aligned}
\left(\int_{|x|<\frac{1}{4}} 1+u_k\,dx\right)^{\frac{n-2}{n}} &= \|(1+u_k)^{\frac{n-2}{2n}}\|^2_{L^{\frac{2n}{n-2}}(|x|<1/4)} \\
&\leq C\|(1+u_k)^{\frac{n-2}{2n}}\|^2_{H^1(|x|<1/4)} \\
&= C\int_{|x|<1/4} |\nabla(1+u_k)^{\frac{n-2}{2n}}|^2+(1+u_k)^{\frac{n-2}{n}}\,dx \\
&\leq C\int_\omega (1+u_k)^{\frac{n-2}{n}}
\end{aligned} \tag{3.42}$$

by (3.41), $1-2/n = (n-2)/n$, and

$$|\nabla(1+u_k)^{\frac{n-2}{2n}}|^2 = \left(\frac{n-2}{2n}\right)^2(1+u_k)^{-1-\frac{2}{n}}|\nabla u_k|^2.$$

Apply now Hölder's inequality to the right-hand side of (3.42), taking $0<\rho<1/4$. Since

$$\frac{1}{\frac{n}{n-2}} + \frac{1}{\frac{n}{2}} = 1, \quad |B_\rho| = \omega_n \rho^n, \quad n \cdot \frac{2}{n} = 2,$$

it holds that

$$\left(\int_{|x|<\frac{1}{4}} 1 + u_k\, dx\right)^{\frac{n-2}{n}} \le C \int_{\rho<|x|<1/2} (1+u_k)^{\frac{n-2}{n}} + C\rho^2 \left(\int_{|x|<\rho} 1 + u_k dx\right)^{\frac{n-2}{n}}.$$

Make $0 < \rho \ll 1$, note that u is bounded in $\rho < |x| < 1/2$ and $u_k \le u$, and obtain

$$\int_{|x|<1/4} u_k\, dx \le C,$$

independent of k. Then we have $u \in L^1_{loc}(B)$ by Fatou's lemma. It holds also that $u_k \to u$ in $L^1_{loc}(B)$ by the dominated convergence theorem.

Take $\varphi \ge 0$, finally, and multiply (3.40) by φ_j to deduce

$$\int_B \nabla u_k \cdot \nabla \varphi_j \ge 0.$$

We note $u_k \in H^1_{loc}(B)$ and make $j \to \infty$, which results in

$$\int_B \nabla u_k \cdot \nabla \varphi = -\int_B u_k \Delta \varphi \ge 0.$$

Then $-\Delta u \ge 0$ in $\mathscr{D}'(B)$ follows with $k \to \infty$. □

Remark 3.4 Lemma 2.1 of the previous chapter says that if $n \ge 2$,

$$v \in W^{1,p}(B),\ 1 < p < \infty, \quad \Delta v = 0 \text{ in } B \setminus \{0\},$$

there is $a \in \mathbf{R}$ such that $v - aE$ is harmonic in B, where

$$E = \begin{cases} |x|^{2-n},\ n > 2 \\ \log |x|,\ n = 2 \end{cases}$$

A related result is in [1] on

$$-\Delta u = |x|^\sigma u^{\frac{n+\sigma}{n-2}},\ u \ge 0 \quad \text{in } B \setminus \{0\}$$

for $-2 < \sigma < 2$.

3.3 Entire Solutions to Semilinear Equations

3.3.1 *Sub-critical Case*

This section is devoted to several results on existence, uniqueness, and non-existence of the solution to

$$-\Delta u = u^p,\ u \geq 0 \quad \text{in } \mathbf{R}^n \tag{3.43}$$

in accordance with the exponent

$$p_s = \begin{cases} +\infty, & n = 2 \\ \frac{n+2}{n-2}, & n \geq 3 \end{cases}$$

associated with the Sobolev embedding $H^1(\mathbf{R}^n) \hookrightarrow L^{p_s+1}(\mathbf{R}^n)$. This subsection is devoted to the following theorem [17].

Theorem 3.19 (Gidas–Spruck) *If* $1 < p < p_s$, *any solution* u *to (3.43) is identically zero,* $u \equiv 0$.

Lemma 3.5 *Let* $\Omega \subset \mathbf{R}^n$ *be a bounded domain,* $0 \leq \varphi \in C_0^\infty(\Omega)$, *and* $0 < u \in C^2(\Omega)$. *Given* $q \in \mathbf{R}$, *let*

$$I \equiv \int_\Omega \varphi u^{q-2}|\nabla u|^4, \quad J \equiv \int_\Omega \varphi u^{q-1}|\nabla u|^2 \Delta u, \quad K \equiv \int_\Omega \varphi u^q (\Delta u)^2,$$

and

$$\alpha = -\frac{n-1}{n}k^2 + (q-1)k - \frac{q(q-1)}{2}, \quad \beta = \frac{n+2}{n}k - \frac{3q}{2}, \quad \gamma = \frac{n-1}{n}.$$

Then, it holds that

$$\alpha I + \beta J + \gamma K \leq \frac{1}{2}\int_\Omega u^q|\nabla u|^2\Delta\varphi + u^q(\Delta u + (q-k)u^{-1}|\nabla u|^2)\nabla u \cdot \nabla\varphi \, dx. \tag{3.44}$$

Proof We begin with the inequality

$$\begin{aligned} &\int_\Omega \frac{m}{2}(1-m)\varphi v^{m-2}|\nabla v|^4 - \frac{3m}{2}\varphi v^{m-1}|\nabla v|^2\Delta v + \frac{n-1}{n}\varphi v^m(\Delta v)^2 \, dx \\ &\leq \int_\Omega \frac{1}{2}v^m|\nabla v|^2 + (v^m\Delta v + mv^{m-1}|\nabla v|^2)\nabla v \cdot \nabla\varphi \, dx, \end{aligned} \tag{3.45}$$

valid for $0 < v \in C^2(\Omega)$ and $m \in \mathbf{R}$. In fact, since

$$\frac{1}{2}\Delta|\nabla v|^2 = \nabla(\Delta v)\cdot\nabla v + |\nabla^2 v|^2, \quad |\nabla^2 v|^2 = \sum_{i,j}\left(\frac{\partial^2 v}{\partial x_i \partial x_j}\right)^2$$

we have

$$\tilde{I} \equiv \int_\Omega \frac{1}{2}\Delta|\nabla v|^2 \cdot \varphi v^m = II + III,$$

with

$$\begin{aligned} II &= \int_\Omega \varphi v^m \nabla(\Delta v)\cdot v = -\int_\Omega (\Delta v)\nabla\cdot(\varphi v^m)\nabla v \\ &= -\int_\Omega v^m(\Delta v)\nabla v\cdot\nabla\varphi + m\varphi v^{m-1}|\nabla v|^2\Delta v + \varphi v^m(\Delta v)^2\, dx. \end{aligned}$$

Hence it holds that

$$\begin{aligned} \tilde{I} &= \int_\Omega |\nabla v|^2\frac{1}{2}\Delta(\varphi v^m) = \int_\Omega |\nabla v|^2\frac{1}{2}\{v^m\Delta\varphi + 2\nabla\varphi\cdot\nabla v^m + \varphi\Delta v^m\} \\ &= \int_\Omega |\nabla v|^2\{\frac{1}{2}v^m\Delta\varphi + m v^{m-1}\nabla v\cdot\nabla\varphi + \frac{m}{2}\varphi v^{m-1}\Delta v + \frac{m}{2}(m-1)\varphi v^{m-2}|\nabla v|^2\} \\ &= \int_\Omega \frac{1}{2}v^m|\nabla v|^2\Delta\varphi + m v^{m-1}|\nabla v|^2\nabla v\cdot\nabla\varphi + \frac{m}{2}\varphi v^{m-1}|\nabla v|^2\Delta v \\ &\quad + \frac{m(m-1)}{2}\varphi v^{m-2}|\nabla v|^4\, dx \end{aligned}$$

and these inequalities imply

$$\begin{aligned} &\int_\Omega \frac{m}{2}(1-m)\varphi v^{m-2}|\nabla v|^4 - \frac{3m}{2}\varphi v^{m-1}|\nabla v|^2\Delta v + \varphi v^m|\nabla^2 v|^2\, dx \\ &= \int_\Omega \varphi v^m(\Delta v)^2 + \frac{1}{2}v^m|\nabla v|^2\Delta\varphi + (v^m\Delta v + m v^{m-1}|\nabla v|^2)\nabla v\cdot\nabla\varphi\, dx. \end{aligned}$$

Now we use

$$(\Delta v)^2 = (\mathrm{tr}\,\nabla^2 v)^2 \le \mathrm{tr}[(\nabla^2 v)\cdot{}^T(\nabla^2 v)]\cdot\mathrm{tr} I_n = n|\nabla^2 v|^2$$

for the unit matrix I_n, which implies

$$|\nabla^2 v|^2 - (\Delta v)^2 \ge (\frac{1}{n} - 1)(\Delta v)^2.$$

Then it follows that (3.45).

To complete the proof, we apply (3.45) for

$$v = u^{k+1}, \quad m = (k+1)^{-1}(q-2k), \quad q = (k+1)m + 2k.$$

First, it holds that

$$\begin{aligned}\int_\Omega \varphi v^{m-2}|\nabla v|^4 &= (k+1)^4 \int_\Omega \varphi u^{(k+1)(m-2)+4k}|\nabla u|^4 \\ &= (k+1)^4 \int_\Omega \varphi u^{q-2}|\nabla u|^4 = (k+1)^4 I\end{aligned}$$

by $\nabla v = (k+1)u^k$ and $(k+1)(m-2)+4k = q-2$. Second, we have

$$\Delta v = (k+1)(ku^{k-1}|\nabla u|^2 + u^k \Delta u) = (k+1)u^k(ku^{-1}|\nabla u|^2 + \Delta u)$$

and hence

$$\begin{aligned}\int_\Omega \varphi v^m (\Delta v)^2 &= (k+1)^2 \int_\Omega \varphi u^{(k+1)m+2k}\{(\Delta u)^2 + 2k(\Delta u)u^{-1}|\nabla u|^2 + k^2 u^{-2}|\nabla u|^4\} \\ &= (k+1)^2(k^2 I + 2kJ + K).\end{aligned}$$

Finally, we use

$$\begin{aligned}\int_\Omega v^m(\Delta v)\nabla v\cdot\nabla\varphi &= (k+1)^2 \int_\Omega u^{(k+1)m+2k}(\Delta u + ku^{-1}|\nabla u|^2)\nabla u\cdot\nabla\varphi \\ \int_\Omega v^{m-1}|\nabla v|^2\nabla v\cdot\nabla\varphi &= (k+1)^3 \int_\Omega u^{(k+1)m+2k-1}|\nabla u|^2\nabla u\cdot\nabla\varphi\end{aligned}$$

to end up with

$$\begin{aligned}&\left[\frac{1-m}{m}(k+1)^2 - \frac{3m}{2}k(k+1) - \frac{n-1}{n}k^2\right] I \\ &\quad + \left[-\frac{3m}{2}(k+1) - 2k\frac{n-1}{n}\right] J - \frac{n-1}{n} K \\ &\le \int_\Omega \frac{1}{2}u^q|\nabla u|^2\Delta\varphi + u^q(\Delta u + (k+m(k+1))u^{-1}|\nabla u|^2)\nabla u\cdot\nabla\varphi\, dx,\end{aligned}$$

which means that (3.44) holds. □

Lemma 3.6 *Let $\Omega \subset \mathbf{R}^n$ be a bounded domain and assume that $0 < u \in C^2(\Omega)$ satisfies*

$$-\Delta u = u^p \quad \text{in } \Omega.$$

Given $0 \le \varphi \in C_0^\infty(\Omega)$ *and* $q, k \in \mathbf{R}$ *in* $q \ne -p$ *and* $k \ne -1$, *let*

$$I = \int_\Omega \varphi u^{q-2}|\nabla u|^4, \quad K = \int_\Omega \varphi u^{2p+q}.$$

Then it holds that

$$\alpha I + \delta K \le \int_\Omega \frac{1}{2} u^q |\nabla u|^2 \Delta\varphi + C(u^{p+q} + u^{q-1}|\nabla u|^2)|\nabla u \cdot \nabla\varphi| \, dx, \tag{3.46}$$

where $C = C(n, p, q, k) > 0$ *and*

$$\alpha = -\frac{n-1}{n}k^2 + (q-1)k - \frac{q(q-1)}{2}, \quad \delta = \frac{1}{p+q}(\frac{3q}{2} - \frac{n+2}{n}k) - \frac{n-1}{n}.$$

Proof By $-\Delta u = u^p$ we obtain

$$\begin{aligned}
(p+q)J &= -\int_\Omega \varphi(p+q)u^{p+q-1}|\nabla u|^2 = -\int_\Omega \varphi\nabla u \cdot \varphi u^{p+q} \\
&= \int_\Omega \varphi(\Delta u)u^{p+q} + (\nabla\varphi \cdot \nabla u)u^{p+q} \, dx \\
&= \int_\Omega -\varphi u^{2p+q} + (\nabla\varphi \cdot \nabla u)u^{p+q} \, dx = -K + \int_\Omega (\nabla\varphi \cdot \nabla u)u^{p+q},
\end{aligned}$$

and hence the result. □

Lemma 3.7 *If* $1 < p < p_s$ *there arise* $q \ne -p$ *and* $k \ne -1$ *such that*

$$\alpha > 0, \quad \delta > 0, \quad 2p + q > \frac{n}{2}(p-1) \tag{3.47}$$

in the previous lemma.

Proof We have $\delta > 0$ if and only if

$$k < k_0(q) \equiv \frac{q}{2} - \frac{n-1}{n+2}p, \quad 2p + q > \frac{n}{2}(p-1)$$

if and only if $q > q_0(p) \equiv \frac{(n-4)p-n}{2}$. For $k = k_0(q)$ we have

$$\alpha = \alpha(k_0(q)) = \frac{n-1}{n}\left(-\frac{q^2}{4} - \frac{pq}{n+2} + \frac{n(n+2)p - (n-1)^2p^2}{(n+2)^2}\right)$$

and $\alpha(k_0(q) > 0$ if and only if

$$-\frac{2p}{n+2}-2\sqrt{D}<q<-\frac{2p}{n+2}+2\sqrt{D},\ D=\frac{p^2+n(n+2)p-(n-1)^2p^2}{(n+2)^2}>0.$$

Since $p<p_s$, we have $-\frac{2p}{n+2}>q_0(p)$. Therefore, it holds that (3.47) for

$$q=-\frac{2p}{n+2},\quad k=k_0(q)_-(\neq -1).$$

□

Lemma 3.8 *If* $1<p<p_s$ *and* $B=B(0,1)\subset\mathbf{R}^n$, *there is* $r=r(n,p)>\max\{n(p-1)/2,p\}$ *such that*

$$\int_{B/2}u^r\leq C(n,p)$$

for any solution $u>0$ *to* $-\Delta u=u^p$ *in* B.

Proof By Lemmas 3.6 and 3.7 for $\Omega=B$, there exist $q\neq -p$ and $k\neq -1$ such that

$$\alpha>0,\quad \delta>0,\quad 2p+q>\frac{n}{2}(p-1).$$

Let $0\leq\xi\in C_0^\infty(\Omega)$ be such that $\xi(x)=1$ for $|x|<\frac{1}{2}$ and $0\leq\xi\leq 1$. Put

$$\theta=\frac{3p+1+2q}{2(2p+q)}\in(0,1),\quad \varphi=\xi^m,\quad m=\frac{2}{1-\theta}.$$

It holds that

$$|\nabla\varphi|\leq C\xi^{m-1}\leq C\varphi^\theta,\quad |\Delta\varphi|\leq C\xi^{m-2}\leq C\varphi^\theta.$$

Young's inequality implies

$$\int_\Omega u^q|\nabla u|^2\Delta\varphi\leq\varepsilon\int_\Omega \varphi u^{q-2}|\nabla u|^4+\varphi u^{2p+q}dx+C_\varepsilon,$$

$$C\int_\Omega u^{q-1}|\nabla u|^2|\nabla\cdot\nabla\varphi|\leq\varepsilon\int_\Omega\varphi u^{q-2}|\nabla u|^4+\varphi u^{q+2q}\,dx+C_\varepsilon$$

for $\varepsilon>0$, and hence

$$\alpha I+\delta K\leq C(n,p,q,k)\cdot\varepsilon(I+K)+C(\varepsilon).$$

We thus obtain $I,K\leq C$ and hence the result holds for

$$r = 2p + q > \max\{\frac{n}{2}(p-1), p\}.$$

□

Proof *(Theorem 3.19)* Assume the contrary, we have

$$-\Delta u = u^p, \quad u > 0 \text{ in } \mathbf{R}^n, \quad 1 < p < p_s$$

by the strong maximum principle. Given $\mu > 0$, define u_μ by

$$u_\mu(x) = \mu^{\frac{2}{p-1}} u(\mu x),$$

which satisfies

$$-\Delta u_\mu = u_\mu^p, \; u_\mu > 0 \quad \text{in } \mathbf{R}^n.$$

Then Lemma 3.8 implies

$$\int_{|x|<\mu/2} u^r = \mu^n \cdot \mu^{-\frac{2r}{p-1}} \int_{B/2} u_\mu^r \leq C(n,p)\mu^{n-\frac{2r}{p-1}}$$

of which the right-hand side converges to zero as $\mu \uparrow +\infty$. Hence we obtain $u \equiv 0$, a contradiction. □

Theorem 3.20 *Let* $1 < p < p_{sg}$ *and assume*

$$-\Delta u \geq u^p, \; u \geq 0 \quad \text{in } \mathbf{R}^n.$$

Then it holds that $u \equiv 0$, *where*

$$p_{sg} = \begin{cases} +\infty, & n = 2 \\ \frac{n}{n-2}, & n \geq 3 \end{cases}$$

This property holds even to $p = p_{sg}$ *for* $n \geq 3$.

Proof Take $\xi \in C_0^\infty(B)$, $B = B(0,1) \subset \mathbf{R}^n$, satisfying $0 \leq \xi = \xi(x) \leq 1$, and $\xi = 1$ for $|x| \leq \frac{1}{2}$. Let

$$m = \frac{2p}{p-1}, \quad R > 0, \quad \varphi_R(x) = \xi(\frac{x}{R})^m.$$

It holds that

$$\Delta \varphi_R = mR^{-2}(\xi^{m-1}\Delta\xi + (m-1)\xi^{m-2}|\nabla\xi|^2)(\frac{x}{R}),$$

and hence

$$|\Delta\varphi_R| \le CR^{-2}\xi^{m-2}\chi_{\frac{R}{2}<|x|<R} = CR^{-2}\varphi_R^{\frac{1}{p}}\chi_{\frac{R}{2}<|x|<R}$$

by $m\frac{1}{p} = \frac{2}{p-1}$ and $m-2 = \frac{2p}{p-1} - 1 = \frac{2}{p-1}$. We then obtain

$$\begin{aligned}
\int_{\mathbf{R}^n} u^p\varphi_R &\le -\int_{\mathbf{R}^n} \Delta u\cdot\varphi_R^p = \int_{\mathbf{R}^n} u\cdot(-\Delta\varphi_R^p) \\
&\le CR^{-2}\int_{\frac{R}{2}<|x|<R} u\varphi_R^{\frac{1}{p}} CR^{-2}(R^n)^{\frac{1}{p'}}\left(\int_{\frac{R}{2}<|x|<R} u^p\varphi_R\right)^{\frac{1}{p}} \\
&\le CR^{-2}R^{\frac{n}{p'}}\left(\int_{\mathbf{R}^n} u^p\varphi_R\right)^{\frac{1}{p}} \qquad (3.48)
\end{aligned}$$

with $\frac{1}{p} + \frac{1}{p'} = 1$, which implies

$$\left(\int_{\mathbf{R}^n} u^p\varphi_R\right)^{\frac{1}{p'}} \le CR^{-2}R^{\frac{n}{p'}},$$

or,

$$\int_{\mathbf{R}^n} u^p\varphi_R \le CR^{n-\frac{2}{p'}} = CR^{n-\frac{2p}{p-1}}.$$

If $p < p_{sg}$ we have $n - \frac{2p}{p-1} < 0$, which implies

$$\int_{\mathbf{R}^n} u^p = 0$$

and hence $u \equiv 0$ with $R \uparrow +\infty$. If $p = p_{sg}$ for $n \ge 3$, we have

$$\int_{\mathbf{R}^n} u^p < +\infty.$$

Then (3.48) implies

$$\int_{\mathbf{R}^n} u^p\varphi_R \le C\left(\int_{\frac{R}{2}<|x|<R} u^p\varphi_R\right)^{\frac{1}{p}},$$

of which the right-hand side tends to zero as $R \uparrow +\infty$, and the result follows. □

3.3.2 *Moving Planes*

This subsection is devoted to the method of moving planes. We show the following theorem [16], an alternative proof of Theorem 3.19, and a related result [11].

Theorem 3.21 (Gidas–Spruck) *Let $1 < p \leq p_s$ for $n \geq 3$ and $1 < p < p_s = +\infty$ for $n = 2$. Then any solution u to*

$$-\Delta u = u^p, \ u \geq 0 \ \text{in } \mathbf{R}^n_+ \equiv \{(x', x_n) \mid x_n > 0\}, \quad u|_{\partial \mathbf{R}^n_+} = 0 \tag{3.49}$$

is identically zero, $u \equiv 0$.

Proof Assuming the contrary, we have $u > 0$ in $\mathbf{R}^n_+$ from the strong maximum principle. We use the Kelvin transformation with center $e_n = (0, \cdots, 0, -1)$, so that

$$v(y) = |x + e_n|^{n-2} u(x), \quad y = \frac{x + e_n}{|x + e_n|^2}.$$

It holds that

$$-\Delta v = |y|^\sigma v^p \quad \text{in } \overline{D} \setminus \{0\}$$

for $D = B(\frac{e_n}{2}, \frac{1}{2})$ and $\gamma = (n-2)p - (n+2) \leq 0$.

Given $0 \leq \lambda < 1/2$, let

$$\Sigma_\lambda = \{y \in D \mid y_1 > \lambda\}, \quad y^\lambda = (2\lambda - y_1, y_2, \cdots, y_n),$$

and $w = w_\lambda(y) \equiv v(y^\lambda) - v(y)$ for $y \in \overline{\Sigma_\lambda}$. It holds that

$$\begin{aligned} \Delta w &= \Delta v(y^\lambda) - \Delta v(y) = -|y^\lambda|^\gamma v(y^\lambda)^p + |y|^\gamma v(y)^p \\ &= (|y|^\gamma - |y^\lambda|^\gamma) v(y^\lambda)^p - |y|^\gamma (v(y^\lambda)^p - v(y)^p) \end{aligned}$$

with

$$v(y^\lambda)^p - v(y)^p = p\xi^{p-1} w$$

for either $\xi \in [v(y^\lambda), v(y)]$ or $\xi \in [v(y), v(y^\lambda)]$. We have, therefore,

$$\Delta w + |y|^\gamma p \xi^{p-1} w = (|y|^\gamma - |y^\lambda|^\gamma) v(y^\lambda)^p \geq 0 \quad \text{in } \Sigma_\lambda$$

because the maximum principle implies $v > 0$ in D. It holds also that $\frac{\partial v}{\partial \nu} < 0$ on ∂D, and hence $w > 0$ for $0 < 1/2 - \lambda \ll 1$.

Assume

$$\overline{\mu} \equiv \inf\{\mu > 0 \mid w \geq 0 \text{ in } \Sigma_\lambda,\ \forall \lambda \geq \mu\} > 0.$$

It holds that $w \geq 0$ in $\Sigma_{\overline{\mu}}$ and there exists $0 < \lambda_i \uparrow \overline{\mu}$ such that $\inf_{\Sigma_{\lambda_i}} w_{\lambda_i} < 0$. We have

$$w_{\lambda_i}\big|_{D\cap\partial\Sigma_{\lambda_i}} = 0, \quad w_{\lambda_i}\big|_{\partial D\cap\overline{\Sigma_{\lambda_i}}} = v(y^\lambda) > 0,$$

and hence $w_{\lambda_i}\big|_{\partial\Sigma_{\lambda_i}} \geq 0$. There is thus $q_i \in \Sigma_{\lambda_i}$ such that

$$\nabla w_{\lambda_i}(q_i) = 0, \quad w_{\lambda_i}(q_i) < 0.$$

It holds also that $w_{\lambda_i} \geq 0$ in a neighborhood of $\partial D \cap \overline{\Sigma_{\lambda_i}}$ independent of i.

Assuming $q_i \to \overline{q} \in \overline{\Sigma_{\overline{\mu}}} \setminus \partial D$, we obtain

$$w_{\overline{\mu}}(\overline{q}) = 0, \quad \nabla w_{\overline{\mu}}(\overline{q}) = 0.$$

Then the strong maximum principle and the Hopf lemma imply $w_{\overline{\mu}} \equiv 0$ in $\Sigma_{\overline{\mu}}$, while we have

$$w_{\overline{\mu}}\big|_{\partial D} = v(y^{\overline{\mu}}) > 0,$$

a contradiction. We thus obtain $\overline{\mu} = 0$ and $w_0 \geq 0$ on Σ_0. Similarly, there arise $w_0 \leq 0$ on Σ_0, and therefore, v is symmetric with respect to the y_1 axis,

$$v(-y_1, y_2, \cdots, y_n) = v(y_1, y_2, \cdots, y_n).$$

This property holds also for any y_i axis with $2 \leq i \leq n-1$, which implies $v = v(|y'|, y_n)$ and hence

$$u = u(|x'|, x_n).$$

By sliding u in the x_n direction, we reach $u = u(x_n) = u(t)$, and hence

$$u'' + u^p = 0, \ u > 0 \ \text{in } t > 0, \quad u(0) = 0.$$

This $u = u(t)$ satisfies, therefore, $u' \geq 0$ in $t \geq 0$, and hence

$$u''(t) = -u(t)^p \leq -u(t_1)^p, \quad t \geq t_1.$$

We obtain

$$
\begin{aligned}
0 < u(t) &= u(t_1) + (t - t_1)u'(t) + \int_{t_1}^{t} (t-s)u''(s)ds \\
&\le u(t_1) + (t - t_1)u'(t) - \frac{1}{2}(t - t_1)^2 u(t_1)^p \to -\infty, \quad t \uparrow +\infty,
\end{aligned}
$$

a contradiction. □

The following lemma implies the conclusion of Theorem 3.21 for $n \ge 3$ and also a classification of the solution for $p = p_s$ by an ODE approach.

Lemma 3.9 (Gidas–Spruck) *Assume $n \ge 3$, $1 < p \le p_s$, and*

$$
-\Delta u = u^p, \; u \ge 0 \quad \text{in } \mathbf{R}^n.
$$

Then there is $x_0 \in \mathbf{R}^n$ such that $u = u(|x - x_0|)$.

For the proof, assume $u \not\equiv 0$, and obtain

$$
-\Delta u = u^p, \; u > 0 \quad \text{in } \mathbf{R}^n. \tag{3.50}
$$

Letting

$$
v(y) = |x|^{n-2}u(x), \quad y = \frac{x}{|x|^2},
$$

we have

$$
-\Delta v = |y|^\gamma v^p, \; v > 0 \text{ in } \mathbf{R}^n \setminus \{0\}, \quad 0 < v(y) \le C|y|^{2-n} \text{ in } |y| \ge 1 \tag{3.51}
$$

for $\gamma = (n-2)p - (n+2) \le 0$. In particular,

$$
v \ge 0, \; -\Delta v \ge 0 \quad \text{in } \mathbf{R}^n \setminus \{0\},
$$

and hence Theorem 3.18 implies $-\Delta v \ge 0$ as $\mathscr{D}'(\mathbf{R}^n)$. The maximum principle now implies

$$
v \ge \eta(R) \equiv \min_{|y|=R} v \quad \text{in } B_R \setminus \{0\} \tag{3.52}
$$

for $B_R = B(0, R)$.

Given $\lambda \le 0$, let

$$
\begin{aligned}
& y^\lambda = (2\lambda - y_1, y_2, \cdots, y_n) \\
& \Sigma_\lambda = \{z = (z_1, z') \in \mathbf{R}^n \mid z_1 < \lambda\}, \quad \Sigma'_\lambda = \Sigma_\lambda \setminus \{0^\lambda\} \\
& w_\lambda(y) = v(y^\lambda) - v(y), \quad y \in \Sigma_\lambda \setminus \{0^\lambda\}.
\end{aligned}
$$

There arises, similarly,

$$\Delta w + |y|^{\gamma} p \xi^{p-1} w \leq 0 \quad \text{in } \Sigma_\lambda' \tag{3.53}$$

for $\gamma = (n-2)p - (n+2) \leq 0$ with

$$\xi = \xi_\lambda(y)$$

either in $\xi \in [v(y^\lambda), v(y)]$ or $\xi \in [v(y), v(y^\lambda)]$.

Then

$$\tilde{w}_\lambda(y) = |y|^{\alpha} w_\lambda(y), \quad \alpha = \frac{n-2}{2}$$

satisfies

$$\Delta \tilde{w}_\lambda(y) - \frac{n-2}{|y|^2} y \cdot \nabla \tilde{w}_\lambda(y) + c_\lambda(y) \tilde{w}_\lambda(y) \leq 0 \quad \text{in } \Sigma_\lambda' \tag{3.54}$$

for

$$c_\lambda(y) = -\frac{(n-2)^2}{4|y|^2} + |y|^{\gamma} p \xi^{p-1}.$$

Lemma 3.10 *It holds that $\tilde{w}_\lambda(y) \geq 0$ in Σ_λ' for $\lambda \ll -1$.*

Proof Assume the contrary,

$$\exists \lambda_i \downarrow -\infty, \quad \inf_{\Sigma_{\lambda_i}'} \tilde{w}_{\lambda_i} < 0. \tag{3.55}$$

By (3.51) and (3.52) with $R = 1$, we have

$$w_{\lambda_i}(y) \geq 0, \quad |\forall y - 0^{\lambda_i}| < 1, \ i \gg 1,$$

and hence $\tilde{w}_{\lambda_i}(y) \geq 0$, there. It holds also that

$$\tilde{w}_{\lambda_i} = 0 \text{ on } \partial \Sigma_{\lambda_i}, \quad \lim_{|y| \to \infty} \tilde{w}_{\lambda_i}(y) = 0 \tag{3.56}$$

by $\alpha + 2 - n = -\frac{\alpha}{2} < 0$. Hence

$$\inf_{\Sigma_{\lambda_i}'} \tilde{w}_{\lambda_i}$$

is attained at some $q_i \in \Sigma_{\lambda_i}'$ in $|q_i - 0^{\lambda_i}| \geq 1$, which satisfies $\lim_{i \to \infty} |q_i| = +\infty$. We thus obtain

$$\lim_{i\to\infty} v(q_i) = 0$$

by (3.51).

If $\{q_i^{\lambda_i}\}$ were bounded, there exists $c_1 > 0$ such that $v(q_i^{\lambda_i}) \geq c_1$ and hence

$$w_{\lambda_i}(q_i) > 0, \quad i \gg 1,$$

which is equivalent to $\tilde{w}_{\lambda_i}(q_i) > 0$. This property contradicts (3.55), and therefore, a subsequence satisfies $\lim_{i\to\infty} |q_i^{\lambda_i}| = +\infty$.

Since

$$\lim_{|y|\to\infty} v(y)|y|^{n-2} = u(0),$$

we have

$$\begin{aligned} w_\lambda(y) &= \frac{1}{|y_\lambda|^{n-2}} \cdot |y_\lambda|^{n-2} v(y^\lambda) - \frac{1}{|y|^{n-2}} |y|^{n-2} v(y) \\ &= \left(\frac{1}{|y_\lambda|^{n-2}} - \frac{1}{|y|^{n-2}} \right) (u(0) + o(1)), \quad |y| \to \infty. \end{aligned}$$

If $\lim_{i\to\infty} \frac{|q_i^{\lambda_i}|}{|q_i|} = 0$ were the case, there arises $w_{\lambda_i}(q_i) > 0$ for $i \gg 1$, a contradiction again. Hence a subsequence admits

$$\frac{|q_i^{\lambda_i}|}{|q_i|} \geq \exists c_2 > 0.$$

We thus obtain

$$0 < v(q_i^{\lambda_i}) \leq C q_i^{2-n},$$

besides $0 < v(q_i) \leq C q_i^{2-n}$, which guarantees

$$\xi_{\lambda_i}(q_i) \leq c q_i^{2-n}.$$

Then it follows that

$$\begin{aligned} c_{\lambda_i}(q_i) &= -\frac{(n-2)^2}{4|q_i|^2} + |q_i|^\gamma p \xi^{\gamma-1} \\ &\leq -\frac{(n-2)^2}{4|q_i|^2} + \frac{C^p}{|q_i|^4} < 0 \quad i \gg 1. \end{aligned} \tag{3.57}$$

Since $\tilde{w}_{\lambda_i}$ attains a minimum in Σ'_{λ_i} at $q_i \in \Sigma'_{\lambda_i}$, we have

$$\Delta \tilde{w}_{\lambda_i}(q_i) \geq 0, \quad \nabla \tilde{w}_{\lambda_i}(q_i) = 0$$

in (3.54). Then it follows that $\tilde{w}_{\lambda_i}(q_i) > 0$, contradiction. □

By Lemma 3.10 the value

$$\overline{\mu} \equiv \sup\{\mu \leq 0 \mid \overline{w}_\lambda \geq 0 \text{ in } \Sigma'_\lambda, \ \forall \lambda \leq \mu\}$$

is well-defined. It holds that $\overline{w}_{\overline{\mu}} \geq 0$ in $\Sigma'_{\overline{\mu}}$. Assume $\overline{\mu} < 0$, and obtain

$$\exists \lambda_i \downarrow \overline{\mu}, \quad \inf_{\Sigma'_{\lambda_i}} \overline{w}_{\lambda_i} < 0.$$

Lemma 3.11 *It holds that $\tilde{w}_{\overline{\mu}} \equiv 0$ in $\Sigma'_{\overline{\mu}}$.*

Proof Assume the contrary, and obtain

$$w_{\overline{\mu}} \geq 0, \ w_{\overline{\mu}} \not\equiv 0 \quad \text{in } \Sigma'_{\overline{\mu}}.$$

Inequality (3.53) then implies

$$\Delta w_{\overline{\mu}} \leq -|y|^\gamma p \xi^{p-1} w_{\overline{\mu}} \leq 0 \quad \text{in } \Sigma'_{\overline{\mu}},$$

by $0 \leq \xi \in [v(y^\lambda), v(y)]$ or $0 \leq \xi \in [v(y), v(y^\lambda)]$, and hence

$$w_{\overline{\mu}} > 0 \quad \text{in } \Sigma'_{\overline{\mu}}$$

from the strong maximum principle.

Then it follows that

$$w_{\overline{\mu}} \geq \exists c_3 > 0 \quad \text{in } U = B(0^{\overline{\mu}}, \frac{\overline{\mu}}{2}) \setminus \{0^{\overline{\mu}}\},$$

which implies

$$w_{\lambda_i}(y) = w_{\overline{\mu}}(y - 2(\lambda_i - \overline{\mu})e_1) + v(y - 2(\lambda_i - \overline{\mu})e_1) - v(y) \geq 0$$

in $y \in B(0^{\lambda_i}, \overline{\mu}/4) \setminus \{0^{\lambda_i}\}$ for $i \gg 1$, because the contribution of the last two terms on the right-hand side becomes zero as $i \to \infty$ uniformly in y. We thus obtain

$$\tilde{w}_{\lambda_i} \geq 0 \quad \text{in } B(0^{\lambda_i}, \overline{\mu}/4) \setminus \{0^{\lambda_i}\}, \ i \gg 1,$$

and hence $\inf_{\Sigma'_\lambda} \tilde{w}_{\lambda_i} < 0$ is attained at some $q_i \in \Sigma'_\lambda$ satisfying $|q_i - 0^{\lambda_i}| \geq \frac{\overline{\mu}}{4}$ by (3.56).

If $\lim_{i\to\infty} |q_i| = +\infty$ were the case, we have

$$\lim_{i\to\infty} \frac{|q_i|^{\lambda_i}}{|q_i|} = 1$$

this time of $\lambda_i \downarrow \overline{\mu}$, which, however, implies a contradiction similarly, by (3.57). A subsequence thus satisfies

$$q_i \to \overline{q} \in \overline{\Sigma'_{\overline{\mu}}} \setminus \{0^{\overline{\mu}}\}.$$

It holds that $\tilde{w}_{\overline{\mu}} \geq 0$ in $\Sigma'_{\overline{\mu}}$, $\tilde{w}_{\overline{\mu}}(\overline{q}) = 0$, $\nabla \tilde{w}_{\overline{\mu}}(\overline{q}) = 0$, and hence

$$w_{\overline{\mu}} \geq 0 \text{ in } \Sigma'_{\overline{\mu}}, \quad w_{\overline{\mu}}(\overline{q}) = 0, \quad \nabla w_{\overline{\mu}}(\overline{q}) = 0.$$

Inequality (3.53) and the strong maximum principle, however, imply $w_{\overline{\mu}} \equiv 0$, i.e.

$$\tilde{w}_{\overline{\mu}} \equiv 0 \quad \text{in } \Sigma'_{\overline{\mu}},$$

a contradiction. □

Proof *(Lemma 3.9 by [23])* We assume $u \not\equiv 0$. By Lemma 3.11 we have

$$w_{\overline{\mu}} \equiv 0 \quad \text{in } \Sigma'_{\overline{\mu}},$$

and hence v is symmetric with respect to $y_1 = \overline{\mu} \leq 0$.

If $1 < p < p_s$, we have

$$-\Delta v / v^p = |y|^\gamma, \quad \gamma < 0,$$

which implies $\overline{\mu} = 0$. Hence $v = v(y)$ is symmetric with respect to the hyperplane $y_1 = 0$. Changing directions, we obtain $v = v(|x|)$ and hence $u = u(|x|)$. Since this equation is invariant under the translation, it follows that $u = u(|x - x_0|)$, $\forall x_0$, and hence $u =$ constant > 0. Using $-\Delta u = u^p$, we obtain a contradiction, and hence $u \equiv 0$.

Even if $p = p_s$, the function $v = v(y)$ is symmetric with respect to the hyperplane $y_1 = \overline{\mu}$. Changing the direction e_1 to $e_2, e_3, \cdots, e_n$, we obtain

$$\overline{y} = (\overline{y_1}, \cdots, \overline{y}_n) \in \mathbf{R}^n$$

such that $v = v(y)$ is symmetric with respect to $y_k = \overline{y_k}$, $1 \leq k \leq n$. Then it holds that $v = v(|y - \overline{y}|)$. If $\overline{y} \neq 0$, the singular point $y = 0$ of $v = v(y)$ is removable, which implies

$$0 < u(x) \leq C|x|^{2-n}, \quad |x| \geq 1.$$

Then the above proof ensures

$$u = u(|x - \overline{x}|), \quad \exists \overline{x} \in \mathbf{R}^n.$$

In the other case of $\overline{y} = 0$, we have $v = v(|y|)$ and hence $u = u(|x|)$. □

Theorem 3.22 (Dancer) *Any bounded solution* u *to (3.49) for* $1 < p < p_s'$ *is identically zero,* $u \equiv 0$*, where*

$$p_s' = \begin{cases} \infty, & n = 2, 3 \\ \frac{n+1}{n-3}, & n \geq 4. \end{cases}$$

Proof Continue to write $x = (x', x_n) \in \mathbf{R}^n_+$ for $x' = (x_1, \cdots, x_{n-1})$, $x_n > 0$, and assume

$$-\Delta u = u^p, \; u > 0 \quad \text{in } \mathbf{R}^n_+, \quad u|_{\partial \mathbf{R}^n_+} = 0$$

for $1 < p < p_s$.

The moving planes method in the x_n direction then guarantees the monotonicity,

$$\frac{\partial u}{\partial x_n} > 0 \quad \text{in } \mathbf{R}^n_+.$$

Hence there arises

$$0 < U(x') = \lim_{x_n \uparrow +\infty} u(x', x_n)$$

because u is bounded.

Take $0 \leq \psi = \psi(x_n) \in C_0^\infty(0, 1)$ in $\int_0^1 \psi \, dx_n = 1$, $\varphi \in C_0^\infty(\mathbf{R}^{n-1})$, and $k > 0$. It holds that

$$-\int\int_{\mathbf{R}^{n-1} \times \mathbf{R}} u^p \varphi(x') \psi(x_n - k) \, dx' dx_n = \int\int_{\mathbf{R}^{n-1} \times \mathbf{R}} u \Delta(\varphi(x') \psi(x_n - k)) \, dx' dx_n,$$

or,

$$\begin{aligned} &-\int\int_{\mathbf{R}^{n-1} \times \mathbf{R}} u^p(x', s+k) \varphi(x') \psi(s) \, dx' ds \\ &= \int\int_{\mathbf{R}^{n-1} \times \mathbf{R}} u(x', s+k) \Delta(\varphi(x') \psi(s)) \, dx' ds. \end{aligned}$$

Sending $k \to \infty$, we have

$$\begin{aligned} -\int_{\mathbf{R}^{n-1}} U(x')^p \varphi(x') dx' &= \int\int_{\mathbf{R}^{n-1} \times \mathbf{R}} U(x') \Delta(\varphi(x') \psi(s)) dx' ds \\ &= \int_{\mathbf{R}^{n-1}} U(x') \Delta \varphi(x') dx' \end{aligned}$$

by the dominated convergence theorem and

$$\int_{\mathbf{R}} \Delta\psi(s)\, ds = 0.$$

It holds that

$$-\Delta U = U^p, \ U \geq 0 \quad \text{in } \mathbf{R}^{n-1}$$

as $\mathscr{D}'(\mathbf{R}^{n-1})$, and we obtain the result by Theorem 3.19. □

3.3.3 Existence and Non-existence of the Solution

Scaling invariance of the equation sometimes ensures the existence of the solution on the bounded domain from the non-existence of the solution on the whole space [17]. Here we describe the details of the argument in Sect. 3.1.2.

Theorem 3.23 (Gidas–Spruck) *Let $\Omega \subset \mathbf{R}^n$ be a bounded domain with smooth boundary $\partial\Omega$ and $1 < p < p_s$. Then there is $C = C(\Omega, p) > 0$ such that any solution $u = u(x)$ to*

$$-\Delta u = u^p, \ u > 0 \ \text{in } \Omega, \quad u|_{\partial\Omega} = 0 \tag{3.58}$$

admits $\|u\|_\infty \leq C$.

Proof Assume the contrary, and get a sequence $u = u_k(x)$, $k = 1, 2, \cdots$, of the solution to (3.58) such that

$$M_k \equiv \|u_k\|_\infty \uparrow +\infty.$$

Letting $M_k = u_k(x_k)$ with $x_k \in \Omega$, we have a subsequence satisfying

$$x_k \to \exists x_0 \in \overline{\Omega}.$$

If $x_0 \in \Omega$, we put $d = \text{dist}\,(x_0, \partial\Omega)/2 > 0$ and take $\lambda_k > 0$ such that

$$\lambda_k^{\frac{2}{p-1}} M_k = 1.$$

It holds that $\lambda_k \downarrow 0$. Then we put

$$v_k(x) = \lambda_k^{\frac{2}{p-1}} u_k(x_k + \lambda_k x), \quad x \in B(0, \frac{d}{\lambda_k})$$

and obtain

$$-\Delta v_k = v_k^p, \; v_k > 0 \text{ in } B(0, \frac{d}{\lambda_k}), \quad \|v_k\|_{L^\infty(B(0,\frac{d}{\lambda_k}))} = v_k(0) = 1.$$

From the elliptic regularity, there is a subsequence such that $v_k \to \exists v$ locally uniformly in $\mathbf{R}^n$. It holds that

$$-\Delta v = v^p, \; 0 \leq v = v(x) \leq v(0) = 1 \quad \text{in } \mathbf{R}^n,$$

contradicting Theorem 3.19.

If $x_0 \in \partial\Omega$, we have $d_k = \text{dist}\,(x_k, \partial\Omega) \to 0$ besides $\lambda_k \downarrow 0$. Without loss of generality, we write $\partial\Omega$ near x_0 as $x_n = \varphi(x')$ with

$$B(x, R) \cap \Omega \subset \{x_n > \varphi(x')\}, \quad 0 < R \ll 1,$$

where $x = (x', x_n)$. Then

$$v_k(x) = \lambda_k^{\frac{2}{p-1}} u_k(x_k + \lambda_k x), \quad x \in B(0, \frac{\delta}{\lambda_k}) \cap \{x_n > \varphi_k(x')\} \equiv D_k$$

is defined for

$$0 < \delta \ll 1, \quad \varphi_k(x') = \frac{1}{\lambda_k}(-x_{kn} + \varphi(x_k' + \lambda_k x'))$$

because

$$x_{kn} + \lambda_k x_n > \varphi(x_k' + \lambda_k x')$$

is equivalent to $x_n > \varphi_k(x')$. There arises

$$-\Delta v_k = v_k^p, \; 0 < v_k \leq v_k(0) = 1 \text{ in } D_k, \quad v_k|_{\Gamma_k} = 0$$

for $\Gamma_k = B(0, \frac{\delta}{\lambda_k}) \cap \{x_n = \varphi_k(x')\}$.

Since

$$\frac{1}{\lambda_k}(-x_{kn} + \varphi(x_k')) < 0,$$

a subsequence satisfies

$$\frac{1}{\lambda_k}(-x_{kn} + \varphi(x_k')) \to c \in [-\infty, 0].$$

If $c = -\infty$, the elliptic regularity again ensures a subsequence such that $v_k \to v$ locally uniformly in $\mathbf{R}^n$ with $v = v(x)$ satisfying

$$-\Delta v = v^p, \ 0 \leq v \leq v(0) = 1 \quad \text{in } \mathbf{R}^n,$$

a contradiction again.

If $-\infty < c \leq 0$, we have a subsequence again, such that $v_k \to v$ locally uniformly in $\mathbf{R}^n$ with $v = v(x)$ satisfying

$$-\Delta v = v^p, \ 0 \leq v \leq v(0) = 1 \ \text{ in } x_n > c, \quad v|_{x_n = c} = 0.$$

Hence we have $c < 0$ and $v \equiv 0$ by Theorem 3.21, a contradiction. □

A standard proof of the following theorem relies on the variational method [37]. Here we show it in a different way, according to [12]. Hence a subcritical nonlinearity ensures the non-existence of entire solutions called the Liouville property, which implies an apriori estimate of the solution on the bounded domain with suitable boundary conditions by the blowup analysis, and then the existence of its solution is assured.

Theorem 3.24 *If $\Omega \subset \mathbf{R}^n$ is a bounded domain with smooth boundary $\partial\Omega$ and $1 < p < p_s$ there exists a solution $u = u(x)$ to (3.58).*

Lemma 3.12 *Let $\Omega \subset \mathbf{R}^n$ be a bounded domain with smooth boundary $\partial\Omega$ and $1 < p < p_s$. Then there is $s_0 > 0$ such that*

$$-\Delta u = u^p + s, \ u > 0 \ \text{ in } \Omega, \quad u|_{\partial\Omega} = 0 \tag{3.59}$$

admits no solution, provided that $s \geq s_0$.

Proof If not, there is a sequence $s_k \uparrow +\infty$ which admits $u_k = u_k(x)$ satisfying

$$-\Delta u_k = u_k^p + s_k \geq s_k \text{ in } \Omega, \quad u_k|_{\partial\Omega} = 0.$$

Then a comparison theorem guarantees $\|u_k\|_\infty \to +\infty$. Let

$$M_k \equiv \|u_k\|_\infty = u_k(x_k), \ x_k \in \Omega, \quad \mu_k^{\frac{2}{p-1}} M_k = 1, \quad \tilde{u}_k(x) = \mu_k^{\frac{2}{p-1}} u_k(x_k + \mu_k x).$$

It holds that $\mu_k \to 0$ and

$$-\Delta\tilde{u}_k = \tilde{u}_k^p + \mu_k^{\frac{2}{p-1}} s_k, \ 0 < \tilde{u}_k \leq \tilde{u}_k(0) = 1 \ \text{ in } \Omega_k, \quad \tilde{u}_k|_{\partial\Omega_k} = 0$$

for $\Omega_k = \mu_k^{-1}(\Omega - \{x_k\})$, and a subsequence admits

$$\mu_k^{\frac{2}{p-1}} s_k \to c_0 \in [0, \infty].$$

We obtain a contradiction if $c_0 = 0$ by the proof of Theorem 3.23. The case $c_0 = +\infty$ is also excluded because a comparison theorem for

$$-\Delta \tilde{u}_k \geq \mu_k^{\frac{2}{p-1}} s_k, \quad \tilde{u}_k|_{\partial \Omega_k} = 0$$

implies $\|\tilde{u}_k\|_\infty \to +\infty$ by $\mu_k^{\frac{2}{p-1}} s_k \to +\infty$.

Hence we obtain a subsequence such that either $\tilde{u}_k \to \exists u$ locally uniformly in $\mathbf{R}^n$ and

$$-\Delta u \geq c_0, \; 0 \leq u \leq u(0) = 1 \quad \text{in } \mathbf{R}^n, \tag{3.60}$$

or $\tilde{u}_k \to \exists u$ locally uniformly in $\{x_n > c\}$ and

$$-\Delta u \geq c_0, \; 0 \leq u \leq u(0) = 1 \text{ in } x_n > c, \quad u|_{x_n = c} = 0 \tag{3.61}$$

for some $c < 0$.

In the first case of (3.60), we take $0 \leq \varphi = \varphi(x) \in C_0^\infty(\mathbf{R}^n)$ satisfying $\int_{\mathbf{R}^n} \varphi = 1$, to put

$$\varphi_R(x) = R^{-n} \varphi(x/R).$$

It holds that

$$c_0 \leq \int_{\mathbf{R}^n} (-\Delta u)\varphi_R = \int_{\mathbf{R}^n} u(-\Delta \varphi_R) \leq R^{-2} \|\Delta \varphi\|_\infty |B_1| \to 0$$

as $R \uparrow +\infty$, a contradiction.

For the second case of (3.61) to treat we use

$$\tilde{x} = (x, x_n + c), \quad \tilde{u}(\tilde{x}) = u(x).$$

Given $\varphi \in C_0^\infty(\mathbf{R}^n)$ with $\int_{x_n > 0} \varphi = 1$ and $\varphi|_{x_n = 0} = 0$, let

$$\varphi_R(x) = R^{-n} \varphi(\frac{x}{R}).$$

We obtain

$$c_0 \leq \int_{x_n > 0} (-\Delta u)\varphi_R = \int_{x_n > 0} u(-\Delta \varphi_R) \leq C R^{-2} \to 0$$

similarly, a contradiction. □

Remark 3.5 Similarly to Theorem 3.23, if $1 < p < p_s$ there is $C = C(\Omega, p) > 0$ such that any solution $u = u(x)$ to

$$-\Delta u = u^p + s, \ u \geq 0 \quad \text{in } \Omega, \qquad u|_{\partial\Omega} = 0$$

for $s \geq 0$ admits $\|u\|_\infty \leq C$.

Proof *(Theorem 3.24)* Equation (3.48) is equivalent to

$$u = \Phi u, \quad u \in C \equiv \{C(\overline{\Omega}) \mid u \geq 0 \text{ in } \Omega\},$$

where

$$\Phi u \equiv (-\Delta_D)^{-1} u^p$$

for Δ_D, standing for the Laplacian provided with the Dirichlet boundary condition.

Let $\lambda_1 > 0$ be the first eigenvalue of $-\Delta_D$ and take $0 < \alpha < \lambda_1$ and $r > 0$ satisfying

$$s^p \leq \alpha s, \quad 0 < \forall s < r.$$

It holds that

$$\Phi u \leq \alpha(-\Delta_D)^{-1} u, \quad \|u\| \leq r$$

and

$$\sigma(\alpha(-\Delta_D)^{-1}) = \alpha\sigma(-\Delta_D)^{-1}) < 1,$$

where $\|\cdot\| = \|\cdot\|_\infty$ and $\sigma(T)$ denotes the spectral radius of the bounded linear operator $T : C(\overline{\Omega}) \to C(\overline{\Omega})$. Then it holds that

$$u \neq t\Phi(u), \quad 0 \leq \forall t \leq 1, \ \forall u \in C, \ \|u\| = r.$$

Let, next,

$$F(u, s) \equiv (-\Delta_D)^{-1}(u^p + s).$$

By Lemma 3.12, there is $s_0 \geq 0$ such that

$$u \neq F(u, s), \quad \forall s \geq s_0, \ \forall u \in C.$$

There exists also $R > r$ such that

$$u \neq F(u, s), \quad \forall s \geq 0, \ \forall u \in C, \ \|u\| \geq R.$$

by Remark 3.5. Then the result follows from the following lemma [12]. □

To state the lemma, let X be a Banach space over $\mathbf{R}$. A closed set $C \subset X$ which admits the following property is called a *cone*:

1. $x, y \in C, \alpha, \beta \geq 0 \Rightarrow \alpha x + \beta y \in C$.
2. $x \in C \setminus \{0\} \Rightarrow -x \notin C$.

Let $B_\rho = \{x \in C \mid \|x\| < \rho\}$, $\mathscr{O} \subset X$ be an open set, and $W = \mathscr{O} \cap C$. If $f : \overline{W} \to C$ is a compact mapping satisfying

$$f(x) \neq x, \quad \forall x \in \partial W,$$

there is a compact mapping $g : \overline{O} \to C$ such that

$$g|_{\overline{W}} = f.$$

Then we can define the *fixed point index* of f by the *Leray–Schauder degree*:

$$i_c(f, W) = \deg\,(I - g, \mathscr{O}, 0).$$

There is a fixed point of f on W if $i_c(f, W) \neq 0$ [13].

Lemma 3.13 (Krasnoselskii) *Let C be a closed cone in a Banach space X over $\mathbf{R}$, and $\Phi : C \to C$ be a compact mapping with $\Phi(0) = 0$ satisfying the following property for some $0 < r < R$ and $s_0 > 0$:*

1. $0 \leq t \leq 1$, $\|x\| = r \Rightarrow x \neq t\Phi(x)$.
2. *There is a compact mapping $F : \overline{B_R} \times [0, \infty) \to C$ such that*

$$\begin{aligned} F(x, 0) &= \Phi(x), && \|x\| = R \\ F(x, s) &\neq x, && \|x\| = R,\ s \geq 0 \\ F(x, s) &\neq x, && x \in \overline{B_R},\ t \geq t_0. \end{aligned}$$

Then, it holds that $i_c(\Phi, B_R) = 0$, $i_c(\Phi, B_r) = 1$, and $i_c(\Phi, U) = -1$.

3.3.4 Structure of Radial Solutions

We summarize the result for

$$-\Delta u = u^p,\ u > 0 \quad \text{in } \mathbf{R}^n. \tag{3.62}$$

If $1 < p < p_s$, first, there is no solution to (3.62), where

$$p_s = \begin{cases} +\infty, \ n = 2 \\ \frac{n+2}{n-2}, \ n \geq 3. \end{cases}$$

If $n \geq 3$ and $p = p_s$, second, it holds that $u(x) = U(|x - x_0|)$ for some $x_0 \in \mathbf{R}^n$ and $U = U(r)$. Third, from the scaling invariance of (3.62),

$$U_\mu(r) = \mu U(\mu^{\frac{p-1}{2}} r), \quad \mu > 0, \tag{3.63}$$

is also a solution. Then an ODE approach ensures the following theorem [6].

Theorem 3.25 (Caffarelli–Gidas–Spruck) *If $n \geq 3$ and $p = p_s$ each solution u to (3.43) admits $x_0 \in \mathbf{R}^n$ and $\mu > 0$, satisfying*

$$u(x) = U_\mu(|x - x_0|)$$

for $U_\mu = U_\mu(r)$ defined by (3.63) and

$$U(x) = \{n(n-2)\}^{\frac{4}{n-2}} / (1 + |x|^2)^{\frac{n-2}{2}}.$$

Hence it holds that

$$u(x) = \left(\frac{\mu\sqrt{n(n-2)}}{\mu^2 + |x - x_0|^2} \right)^{(n-2)/2}, \quad \mu > 0, \ x_0 \in \mathbf{R}^n.$$

Non-radial solutions arise in (3.62) for p large. If $p = \frac{n+3}{n-1}$, for example, there is a solution as in Theorem 3.25 with n and x replaced by $n-1$ and $x' = (x_2, \cdots, x_n)$, respectively, denoted by $v = v(x')$, which is not radially symmetric with respect to any point.

Here we show several results on the solution to (3.62) for $p \geq p_s$, in the form of

$$u(x) = U(r), \quad r = |x|, \quad U(0) = \alpha > 0.$$

It holds that $U = U(r) \in C^2[0, \infty)$ and

$$U'' + \frac{n-1}{r} U' + U^p = 0 \text{ in } r > 0, \quad U(0) = \alpha, \quad U'(0) = 0. \tag{3.64}$$

By

$$(r^{n-1} U')' = -r^{n-1} U^p$$

we have

$$U(r) = \alpha - \int_0^r ds \int_0^s \left(\frac{t}{s} \right)^{n-1} U^p(s) \, dt \tag{3.65}$$

which is uniquely solvable for $0 < r \ll 1$ by Banach's fixed point theorem. Henceforth, this solution is denoted by $U_\alpha(r)$. From the scaling invariance and uniqueness of the solution, it holds that

$$U_\alpha(r) = \alpha U_1(\alpha^{\frac{p-1}{2}} r).$$

A singular solution to (3.64) in the form of

$$U_*(r) = c_p r^{-\frac{2}{p-1}}, \quad c_p^{p-1} = \frac{2}{(p-1)^2}((n-2)p - n)$$

is well-defined for $p > p_{sg}$, $n \geq 3$, where

$$p_{sg} = \begin{cases} +\infty, \; n = 2 \\ \frac{n}{n-2}, \; n \geq 3. \end{cases}$$

Let

$$p_{jl} = \begin{cases} +\infty, & n \leq 10 \\ 1 + 4\frac{n-4+2\sqrt{n-1}}{(n-2)(n-10)}, & n \geq 11. \end{cases}$$

Theorem 3.26 *If $p \geq p_s$ each $\alpha > 0$ admits a solution $0 < U = U(r) \in C^2[0, \infty)$ to 3.64). Let $\alpha_1 > \alpha_2 > 0$.*

1. *If $p = p_s$ it holds that*

$$U_1 = \left(\frac{n(n-2)}{n(n-2) + r^2}\right)^{(n-2)/2}.$$

   The solutions U_{α_1} and U_{α_2} intersect once, while U_{α_1} and U_* intersect twice.
2. *If $p > p_s$ it holds that*

$$\lim_{r \uparrow +\infty} r^{\frac{2}{p-1}} U_\alpha(r) = c_p.$$

3. *If $p_s < p < p_{jl}$, the solutions U_{α_1} and U_{α_2} intersect infinitely many times. It holds also that U_{α_1} and U_* intersect infinitely many times.*
4. *If $p \geq p_{jl}$, there arises*

$$U_*(r) > U_{\alpha_1}(r) > U_{\alpha_2}(r), \quad \forall r > 0.$$

Remark 3.6 By the Emden transformation

$$w(s) = r^{\frac{2}{p-1}} U(r), \quad s = \log r,$$

equation (3.64) is transformed into the autonomous system

$$w'' + \beta w' + w^p - \gamma w = 0 \text{ in } s \in \mathbf{R}, \quad \lim_{s \downarrow -\infty} w(s) = \lim_{s \downarrow -\infty} w'(s) = 0 \tag{3.66}$$

for

$$\beta = \frac{1}{p-1}((n-2)p - (n+2)) \geq 0$$
$$\gamma = c_p^{p-1} = \frac{2}{(p-1)^2}((n-2)p - n) > 0.$$

Then it holds that

$$\frac{dE}{ds} = -\beta(w')^2 \leq 0, \tag{3.67}$$

where

$$E \equiv \frac{1}{2}(w')^2 - \frac{\gamma}{2}w^2 + \frac{1}{p+1}w^{p+1}.$$

To develop the phase plane analysis let $x = w$ and $y = w'$. We have

$$\frac{d}{ds}\begin{pmatrix} x \\ y \end{pmatrix} = \begin{pmatrix} y \\ -\beta y - x^p + \gamma x \end{pmatrix} \equiv F(x, y). \tag{3.68}$$

It holds that $x > 0$,

$$\lim_{s \downarrow -\infty} (x, y) = (0, 0),$$

and $F(x, y) = 0$ if and only if $(x, y) = (0, 0)$ or $(x, y) = (c_p, 0)$. The linearized matrix around these stationary points are given by

$$A_1 \equiv \nabla F(0, 0) = \begin{pmatrix} 0 & 1 \\ \gamma & -\beta \end{pmatrix}, \quad A_2 \equiv \nabla F(c_p, 0) = \begin{pmatrix} 0 & 1 \\ \gamma(p-1) & -\beta \end{pmatrix}.$$

By (3.67) and $\lim_{s \downarrow -\infty} E = 0$, the trajectory does not hit $x = 0$. We obtain also that w and w' remain bounded in s. Finally, the stationary state $(c_p, 0)$ corresponds to the singular solution U_*.

If $p > p_s$ we have $\beta > 0$ and hence A_1 has two real eigenvalues denoted by

$$\nu_1 < 0 < \nu_2 = \frac{2}{\beta - 1}.$$

Two eigenvalues of A_2 are real if and only if $p \geq p_{jl}$. If $p_s < p < p_{j\ell}$ two eigenvalues of A_2 are complex conjugate each other, with a negative real part. Hence $(c_p, 0)$ is a stable focal.

3.3.5 Exterior Domains

Theorem 3.19 is equivalent to the following theorem. This fact is applied to the Hénon equation

$$-\Delta u = |x|^\alpha u^p, \ u > 0 \quad \text{in } \mathbf{R}^n$$

by [27].

Theorem 3.27 *Given $n \geq 2$ and $1 < p < \frac{n+2}{n-2}$, there is $C = C(p, n) > 0$ such that any solution to*

$$-\Delta u = u^p, \ u > 0 \ \textit{in} \ \Omega_\rho = \mathbf{R}^n \setminus \overline{B(0, \rho)}$$

admits

$$u(x) \leq C|x|^{-\frac{2}{p-1}}, \ |\nabla u(x)| \leq C|x|^{-\frac{p+1}{p-1}} \quad \textit{in } |x| > 2\rho.$$

Proof *(of Theorem 3.19 by Theorem 3.27)* By the Rellich–Pohozaev identity [37] it holds that

$$\left(\frac{n}{p+1} - \frac{n-2}{2}\right) \int_{B_r} u^{p+1} = \int_{\partial B_r} r\frac{u^{p+1}}{p+1} + \frac{1}{r}|(x \cdot \nabla)u|^2 - \frac{r}{2}|\nabla u|^2 + \frac{n-2}{2} u u_r \, d\sigma$$

in (3.43). Since $1 < p < \frac{n+2}{n-2}$ and

$$\begin{aligned} \frac{1}{r}|(x \cdot \nabla)u|^2 - \frac{r}{2}|\nabla u|^2 + \frac{n-2}{2} u u_r &\leq \frac{r}{2}|\nabla u|^2 + \frac{n-1}{2} u u_r \\ &\leq C(r|\nabla u|^2 + r^{-1}u^2) \end{aligned}$$

we have

$$\begin{aligned} f(r) \equiv \int_{B_r} u^{p+1} &\leq C \int_{\partial B_r} r u^{p+1} + r|\nabla u|^2 + r^{-1}u^2 \, d\sigma \\ &= Cr^n \int_{S^{n-1}} u^{p+1} + |\nabla u|^2 + r^{-2}u^2 \, ds. \end{aligned}$$

Theorem 3.27 implies

$$u(x) \leq C|x|^{-\frac{2}{p-1}}, \ |\nabla u(x)| \leq C|x|^{-\frac{p+1}{p-1}}, \quad x \neq 0,$$

and therefore,

$$f(r) \leq Cr^n\{r^{-\frac{2(p+1)}{p-1}} + r^{-2} \cdot r^{-\frac{4}{p-1}}\}$$

$$= Cr^{n-\frac{2(p+1)}{p-1}} \to 0, \quad r \uparrow +\infty,$$

which implies $u \equiv 0$. □

Below we show that, conversely, Theorem3.19 implies Theorem3.27.

Lemma 3.14 (doubling lemma) *Let (X, d), $D \subset X$, and $\Sigma \subset X$ be a complete metric space, a non-empty set, and a closed set containing D, respectively, and put $\Gamma \equiv \Sigma \setminus D$. Let, furthermore, $M : D \to (0, \infty)$ be a mapping bounded on any compact set of D and $k > 0$, and assume the existence of $y \in D$ such that*

$$M(y) dist\ (y, \Gamma) > 2k.$$

Then there is $x \in D$ such that

$$M(x) dist\ (x, \Gamma) > 2k, \ M(x) \geq M(y)$$

$$M(z) \leq 2M(x), \quad \forall z \in D \cap \overline{B_X(x, kM^{-1}(x))}.$$

Proof Assuming the contrary, we show the existence of $x_j \in D$, $j = 1, 2, \cdots$, such that

$$M(x_j)\text{dist}\ (x_j, \Gamma) > 2k, \quad M(x_{j+1}) > 2M(x_j), \quad d(x_j, x_{j+1}) \leq kM^{-1}(x_j). \tag{3.69}$$

To confirm this property, first, let $x_0 = y \in D$. Since the conclusion does not arise, there is $x_1 \in D$ such that

$$M(x_1) > 2M(x_0), \quad d(x_0, x_1) \leq kM^{-1}(x_0).$$

Having found $x_0, \cdots, x_i$ such that (3.69) holds for $j = 0, \cdots, i-1$, we obtain

$$\text{dist}(x_i, \Gamma) \geq \text{dist}\ (x_{i-1}, \Gamma) - d(x_{i-1}, x_i) > (2k - k)M^{-1}(x_{i-1}) > 2kM^{-1}(x_i)$$

and hence

$$M(x_i)\text{dist}\ (x_i, \Gamma) > 2k.$$

It is obvious that

$$M(x_i) \geq M(y)$$

by $M(x_i) > 2M(x_0)$, $x_0 = y$. We thus find $x_{i+1} \in D$ such that

$$M(x_{i+1}) > 2M(x_i), \quad d(x_i, x_{i+1}) \leq kM^{-1}(x_i),$$

which means that (3.69) holds for $j = i$, and we obtain the conclusion.

We thus have a sequence $x_j \in D$, $j = 1, 2, \cdots$, satisfying (3.69), which ensures

$$M(x_i) \geq 2^i M(x_0), \; d(x_i, x_{i+1}) \leq k2^{-i} M^{-1}(x_0), \quad i = 1, 2, \cdots. \tag{3.70}$$

Hence this $\{x_i\}$ is a Cauchy sequence, and there is $a \in \overline{D} \subset \Sigma$ such that $x_i \to a$. It holds also that

$$d(x_0, x_i) \leq \sum_{j=0}^{i-1} d(x_j, x_{j+1}) \leq kM^{-1}(x_0) \sum_{j=0}^{i-1} 2^{-i} \leq 2kM^{-1}(x_0),$$

and therefore,

$$\text{dist}\,(x_i, \Gamma) \geq \text{dist}\,(x_0, \Gamma) - 2kM^{-1}(x_0) \equiv \delta > 0, \quad i = 1, 2, \cdots. \tag{3.71}$$

It thus holds that

$$K \equiv \{x_i \mid i = 1, 2, \cdots\} \cup \{a\}$$

is a compact set contained in $\Sigma \setminus \Gamma = D$. We have, on the other hand, $M(x_i) \to +\infty$ by (3.70) and hence M is unbounded on K. This property violates the assumption, and we obtain the result. □

Assuming Theorem 3.19, we show the following lemma [28].

Lemma 3.15 *If $n \geq 2$ and $1 < p < p_s$, there exists $C = C(p, n) > 0$ such that any solution $u = u(x)$ to*

$$-\Delta u = u^p, \; u > 0 \quad in \; B = B(0, 1) \tag{3.72}$$

admits

$$u(x)^{\frac{p-1}{2}} + |\nabla u(x)|^{\frac{p-1}{p+1}} \leq C(1 + dist^{-1}(x, \partial B)).$$

Proof Assume the contrary, the existence of a sequence $u_k = u_k(x)$, $k = 1, 2, \cdots$, of solutions to (3.69) and $y_k \in B$ satisfying

$$M_k(y_k) > 2k(1 + \mathrm{dist}^{-1}(x, \partial B)), \quad k = 1, 2, \cdots,$$

where

$$M_k(x) \equiv u_k(x)^{\frac{p-1}{2}} + |\nabla u_k(x)|^{\frac{p-1}{p+1}}.$$

By the doubling lemma, there is $x_k \in B$ such that

$$M_k(x_k) \geq M_k(y_k), \ M_k(x_k) > 2k \ \mathrm{dist}^{-1}(x_k, \partial B)$$

$$M_k(z) \leq 2M_k(x_k), \ \forall z \text{ in } |z - x_k| \leq kM_k^{-1}(x_k). \tag{3.73}$$

Then it holds that $\lambda_k \equiv M_k(x_k)^{-1} \to 0$ as $k \to \infty$.

Letting

$$\tilde{u}_k(x) \equiv \lambda_k^{\frac{2}{p-1}} u_k(\lambda_k x + x_k),$$

we obtain

$$\tilde{u}_k(0)^{\frac{p-1}{2}} + |\nabla \tilde{u}_k(0)|^{\frac{p-1}{p+1}} = 1$$

and

$$-\Delta \tilde{u}_k = \tilde{u}_k^p, \ \tilde{u}_k > 0, \ \tilde{u}_k^{\frac{p-1}{2}} + |\nabla \tilde{u}_k|^{\frac{p-1}{p+1}} \leq 2 \quad \text{in } |x| < k.$$

In fact we have $B(x_k, kM_k^{-1}(x_k)) \subset B$ because

$$z = \lambda_k x + x_k, \quad |z - x_k| < kM_k^{-1}(x_k)$$

means $|x| < k$, which implies $|z - x_k| < \frac{1}{2}\mathrm{dist}\,(x_k, \partial B)$.

From the elliptic regularity, a subsequence satisfies $\tilde{u}_k \to \tilde{u}$ locally uniformly in $\mathbf{R}^n$, with $\tilde{u} = \tilde{u}(x)$ satisfying

$$-\Delta \tilde{u} = \tilde{u}^p, \ \tilde{u} > 0 \ \text{in } \mathbf{R}^n, \quad \tilde{u}(0)^{\frac{p-1}{2}} + |\nabla \tilde{u}(0)|^{\frac{p-1}{p+1}} = 1,$$

contradicting Theorem 3.19. □

Proof *(of Theorem 3.27 by Theorem 3.19)* Take $|x_0| > 2\rho$ and put $R = \frac{1}{2}|x_0|$. Any $y \in B = B(0, 1)$ admits

$$\frac{|x_0|}{2} < |x_0 + Ry| < \frac{3}{2}|x_0|$$

and hence $x_0 + Ry \in \Omega_\rho$.

Letting

$$U(y) \equiv R^{\frac{2}{p-1}} u(x_0 + Ry),$$

we obtain

$$-\Delta U = U^p, \; U > 0 \quad \text{in } B.$$

Then it follows that $U(0) + |\nabla U(0)| \leq C$ by Lemma 3.15, which implies

$$u(x_0) \leq C R^{-\frac{2}{p-1}}, \quad |\nabla u(x_0)| \leq C R^{-\frac{p+1}{p-1}}$$

by $R = \frac{1}{2}|x_0|$, and the proof is complete. □

3.4 Blowup Analysis

3.4.1 Scaling

Scaling invariance is observed commonly in fundamental equations in mathematical physics [35]. Turning to the object treated in the previous chapter, we take a model arising in the theories of point vortex turbulence and Abelian Higgs theory [20],

$$-\Delta u = e^u.$$

If $u(x)$ is a solution and $\mu > 0$ is a constant, then

$$u^\mu(x) = u(\mu x) + 2\log\mu$$

satisfies the same equation, although the domains where they are defined may be different. This scaling invariance causes a lack of compactness of the family of (approximate) solutions, and this mechanism is clarified by blowup analysis, the ingredients of which are summarized as follows:

1. scaling invariance of the problem.
2. classification of the entire solution.
3. control at infinity of the rescaled solution.
4. hierarchical argument.

The following theorem [5], free from boundary conditions, is useful in such a study because the effect of boundary conditions is lost in the scaling argument. It deals with the nonhomogeneous coefficient case with the lack of the complex structure in contrast with Theorem 2.5, which is associated with the following theorem for

$$v_k - \log \sigma_k, \quad \sigma_k = \frac{\lambda_k}{\int_\Omega e^{v_k} dx}.$$

Actually, we obtain $v > 0$ in (2.92).

Theorem 3.28 (Brezis–Merle) *Let $\Omega \subset \mathbf{R}^2$ be a bounded domain and $v_k = v_k(x)$, $k = 1, 2, \ldots$, be a sequence of solutions to*

$$-\Delta v_k = V_k(x) e^{v_k}, \ 0 \leq V_k(x) \leq C_1 \ \text{in } \Omega, \quad \int_\Omega e^{v_k} \leq C_2, \tag{3.74}$$

where C_1, C_2 are constants. Then, passing to a subsequence, there arises the following alternatives:

1. *$\{v_k\}$ is locally uniformly bounded in Ω.*
2. *$v_k \to -\infty$ locally uniformly in Ω.*
3. *There is a finite set $\mathscr{S} = \{x_j^*\} \subset \Omega$ and $m_j \geq 4\pi$ such that $v_k \to -\infty$ locally uniformly in $\Omega \setminus \mathscr{S}$ and*

$$V_k(x) e^{v_k} dx \rightharpoonup \sum_j m_j \delta_{x_j^*}(dx) \quad \text{in } \mathscr{M}(\Omega). \tag{3.75}$$

Furthermore, $\mathscr{S}$ is the blowup set of $\{v_k\}$ in Ω.

The following theorem[22] describes a mass quantization without boundary conditions, with possible collision of delta functions. It is applied to create non-trivial solutions to the Boltzmann – Poisson equation for disquantized total mass [14, 34].

Theorem 3.29 (Y.Y. Li–Shafrir) *In the third case of the above theorem there arises $m_j = 8\pi n_j$ for some $n_j \in \mathbf{N}$, provided that $V_k \to V$ in $C_{loc}(\Omega)$.*

A rough estimate derived from the prescaled analysis is sufficient for the proof of Theorem 3.28. It is based on an inequality concerning linear equations.

Lemma 3.16 *If $\Omega \subset \mathbf{R}^2$ is a bounded domain, $f \in L^1(\Omega)$, and*

$$-\Delta v = f(x) \ \text{in } \Omega, \quad v|_{\partial\Omega} = 0,$$

it holds that

$$\int_\Omega \exp\left(\frac{4\pi - \delta}{\|f\|_1} |v(x)| \right) dx \leq \frac{4\pi^2}{\delta} (\text{diam } \Omega)^2, \quad 0 < \delta < 4\pi.$$

This lemma implies ε-regularity in the form of the following lemma, and then a standard covering argument guarantees Theorem 3.28.

Lemma 3.17 *Let $\Omega \subset \mathbf{R}^2$ be a bounded domain, $K \subset \Omega$ a compact set, $c_1, c_2 > 0$, and $\varepsilon_0 \in (0, 4\pi)$. Then, there is $C > 0$ such that*

$$-\Delta v = V(x)e^v, \ 0 \leq V(x) \leq c_1 \ \text{in } \Omega, \quad \|v_+\|_1 \leq c_2, \quad \int_\Omega V(x)e^v dx \leq \varepsilon_0$$

implies $\|v_+\|_{L^\infty(K)} \leq C$.

Once Theorem 3.28 is proven, Theorem 3.29 is localized as follows [22]. Let $B = B(0, R) \subset \mathbf{R}^2$ and $B_r = B(0, r)$.

Theorem 3.30 (Y.Y. Li-Shafrir) *If*

$$-\Delta v_k = V_k(x)e^{v_k}, \ V_k(x) \geq 0 \ \text{in } B, \quad V_k \to V \quad \text{in } C(\overline{B})$$

$$\max_{\overline{B}} v_k \to +\infty, \ \max_{\overline{B}\setminus B_r} v_k \to -\infty, \quad 0 < \forall r < R$$

$$\int_B V_k(x)e^{v_k} dx \to \alpha, \quad \int_B e^{v_k} dx \leq C,$$

it holds that $\alpha \in 8\pi\mathcal{N}$.

In spite of the case $\alpha = 8\pi\ell$, $\ell \geq 2$ in [9], there is the following fact [21] under the boundary condition, with $C > 0$ standing for a generic constant. The original proof is done by the moving planes method, observing that the profile of $v_k = v_k(x)$ in the outer region $x \in B \setminus B(0, \delta_k)$, $\delta_k = e^{-u_k(x_k)/2}$ for $u_k(x_k) = \|u_k\|_\infty$, is approximated by that of the Kelvin transformation of $v_k = v_k(x)$ on $B(0, \delta_k)$, under the assumption. An alternative proof is based on the mass identity, where the collapse mass 8π is equal to the total mass of the entire solution, to control the outer region [24].

Theorem 3.31 (Y.Y. Li) *It holds that $\alpha = 8\pi$ in the previous theorem, provided that*

$$\max_{\partial B} v_k - \min_{\partial B} v_k \leq C, \quad \|\nabla V_k\|_\infty \leq C, \tag{3.76}$$

and furthermore,

$$\left| v_k(x) - \log \frac{e^{v_k(0)}}{\left(1 + \frac{V_k(0)}{8} e^{v_k(0)} |x|^2\right)^2} \right| \leq C, \quad \forall k, \ x \in B. \tag{3.77}$$

Theorem 3.30 is proven by blowup anlaysis. We take $x_k \in B$ satisfying

$$v_k(x_k) = \|v_k\|_\infty, \quad x_k \to 0,$$

to put

$$\tilde{v}_k(x) = v_k(\delta_k x + x_k) + 2\log\delta_k, \quad \delta_k = e^{-v_k(x_k)/2} \to 0.$$

It holds that

$$-\Delta\tilde{v}_k = V_k(\delta_k x + x_k)e^{\tilde{v}_k}, \ \tilde{v}_k \leq \tilde{v}_k(0) = 0 \text{ in } B(0, \frac{R}{2\delta_k})$$

$$\int_{B(0, \frac{R}{2\delta_k})} e^{\tilde{v}_k} dx \leq C_0,$$

and Theorem 3.28 is applicable to this $\{\tilde{v}_k\}$. Thus, $\{\tilde{v}_k\}$ is locally uniformly bounded in $\mathbf{R}^2$, and passing to a subsequence, we have $\tilde{v}_k \to \tilde{v}$ locally uniformly in $\mathbf{R}^2$ with

$$-\Delta\tilde{v} = V(0)e^{\tilde{v}}, \ \tilde{v} \leq \tilde{v}(0) = 0 \text{ in } \mathbf{R}^2, \ \int_{\mathbf{R}^2} e^{\tilde{v}} dx \leq C_0$$

by the elliptic regularity. Then we get $V(0) > 0$, and assume

$$0 < a \leq V_k(x) \leq b < +\infty, \quad x \in B$$

without loss of generality, with a, b independent of k.

We obtain, furthermore,

$$\tilde{v} = \tilde{v}(|x|)$$

by the method of moving planes, which results in

$$\tilde{v}(x) = \log\frac{1}{\left(1 + \frac{V(0)}{8}|x|^2\right)^2}, \quad \int_{\mathbf{R}^2} V(0)e^{\tilde{v}} = 8\pi$$

by the following theorem [8], shown in Sect. 3.4.4 below.

Theorem 3.32 (W. Chen–C. Li) *If*

$$-\Delta v = e^v \ in \ \mathbf{R}^2, \quad \int_{\mathbf{R}^2} e^v \, dx < +\infty, \tag{3.78}$$

it holds that

$$v(x) = \log\{\frac{8\mu^2}{(1 + \mu^2|x - x_0|^2)^2}\}, \quad \exists x_0 \in \mathbf{R}^2, \ \exists\mu > 0, \tag{3.79}$$

and hence

$$\int_{\mathbf{R}^2} e^v dx = 8\pi.$$

3.4.2 Brezis–Merle Inequality

This subsection is devoted to the proof of Theorem 3.28 and an extension of Theorems 2.6–2.7 as its application.

Proof *(Lemma 3.16)* We take $B = B(x_0, R)$ containing Ω for $R = \frac{1}{2}\mathrm{diam}\,\Omega$, zero extension of $f(x)$ outside Ω, and

$$\overline{v}(x) = \frac{1}{2\pi}\int_B \log\frac{2R}{|x-x'|}\cdot\left|f(x')\right|dx'.$$

It holds that

$$-\Delta\overline{v} = |f| \text{ in } \mathbf{R}^2, \quad \overline{v} \geq 0 \text{ in } B$$

by

$$2R \geq |x-x'|, \quad x, x' \in B,$$

and, therefore,

$$|v| \leq \overline{v} \quad \text{in } \Omega$$

from the maximum principle. We thus obtain

$$\int_\Omega \exp(\frac{4\pi-\delta}{\|f\|_1}|v(x)|)dx \leq \int_\Omega \exp(\frac{4\pi-\delta}{\|f\|_1}\overline{v}(x))dx. \tag{3.80}$$

Since Jensen's inequality implies

$$\begin{aligned}\exp(\frac{4\pi-\delta}{\|f\|_1}\overline{v}(x)) &= \exp(\int_B \frac{4\pi-\delta}{2\pi}\cdot\log\frac{2R}{|x-x'|}\cdot\frac{|f(x')|}{\|f\|_1}dx') \\ &\leq \int_B (\frac{2R}{|x-x'|})^{2-\frac{\delta}{2\pi}}\frac{|f(x')|}{\|f\|_1}dx',\end{aligned}$$

the right-hand side of (3.80) is estimated above by

$$\int_B \frac{|f(x')|}{\|f\|_1} dx' \cdot \int_B \left(\frac{2R}{|x-x'|}\right)^{2-\frac{\delta}{2\pi}} dx.$$

Here, $B = B(x_0, R) \subset B(x', 2R)$ for any $x' \in B$, and it holds that

$$\begin{aligned}\int_\Omega \exp\left(\frac{4\pi - \delta}{\|f\|_1}|v(x)|\right)dx &\leq \int_B \frac{|f(x')|}{\|f\|_1}dx' \cdot \int_{B(x',2R)} \left(\frac{2R}{|x-x'|}\right)^{2-\frac{\delta}{2\pi}} dx \\ &= \frac{4\pi^2}{\delta}(2R)^2 = \frac{4\pi^2}{\delta}(\text{diam } \Omega)^2.\end{aligned}$$

The proof is complete. □

Corollary 3.1 *Let $\Omega \subset \mathbf{R}^2$ be a bounded domain with smooth boundary $\partial\Omega$, and $u = u(x)$ is a classical solution to*

$$-\Delta u = V(x)e^u \text{ in } \Omega, \quad u|_{\partial\Omega} = 0.$$

Take $1 < p \leq \infty$ and

$$0 < \gamma < 4\pi/p', \quad \frac{1}{p'} + \frac{1}{p} = 1.$$

Then each $M > 0$ admits $C > 0$ such that

$$\|e^u\|_1 \leq \gamma, \ \|V\|_p \leq M \quad \Rightarrow \quad \|u\|_\infty \leq C.$$

Proof We take $\delta > 0$ satisfying

$$4\pi - \delta > \gamma(p' + \delta).$$

For $f = V(x)e^u$, there arises

$$\frac{4\pi - \delta}{\|f\|_1} \geq \frac{4\pi - \delta}{\gamma} > p' + \delta$$

and hence

$$\int_\Omega e^{(p'+\delta)|u|}dx \leq \int_\Omega \exp\left(\frac{4\pi - \delta}{\|f\|_1}|u(x)|\right) dx \leq \frac{4\pi^2}{\delta}(\text{diam } \Omega)^2,$$

which results in

$$\|e^u\|_{p'+\delta} \leq C, \ \|V\|_p \leq M,$$

and hence

$$\|Ve^u\|_q = \| - \Delta u\|_q \le C, \quad \exists q > 1.$$

Then the standard elliptic regularity ensures

$$\|u\|_{W^{2,q}(\Omega)} \le C'$$

and hence $\|u\|_\infty \le C$ by Sobolev's embedding theorem. □

Remark 3.7 In spite of Theorem 3.29, the upper bound $4\pi/p'$ of γ is the best possible. To see this property, we take

$$f_k(x) = \begin{cases} \frac{4}{p'}k^2, & |x| < \frac{1}{k} \\ 0, & \text{otherwise}, \end{cases} \qquad k = 1, 2, \cdots,$$

and define $u_k = u_k(x)$ by

$$- \Delta u_k = f_k \text{ in } B = B(0, 1) \subset \mathbf{R}^2, \qquad u_k|_{\partial B} = 0.$$

There arises

$$u_k(x) = \begin{cases} \frac{1}{p'}(-k^2r^2 + 1 + 2\log k), & 0 < r = |x| < \frac{1}{k} \\ -\frac{2}{p'}\log r, & \frac{1}{k} < r < 1 \end{cases}$$

and hence

$$V_k(x) = \begin{cases} \frac{4}{p'}k^{2/p}e^{(-1+k^2r^2)/p'}, & 0 < r = |x| < \frac{1}{k} \\ 0, & \frac{1}{k} < r < 1 \end{cases}$$

for $V_k = e^{-u_k} f_k$. Although the values

$$\|V_k\|_p^p = \left(\frac{4}{p'}\right)^p k^2 \int_{|x|<\frac{1}{k}} e^{(-1+k^2r^2)/p'}\, dx = \left(\frac{4}{p'}\right)^p \pi p'(1 - e^{-1/p'})$$

$$\|V_k e^{u_k}\|_1 = \|f_k\|_1 = \frac{4\pi}{p'}$$

are independent of k, it holds that

$$\|u_k\|_\infty = u_k(0) = \frac{1}{p'}(1 + 2\log k) \uparrow +\infty, \qquad k \uparrow +\infty.$$

The following theorem [39] is a generalization of Theorems 2.6–2.7.

Theorem 3.33 (Ye) *Let $\Omega \subset \mathbf{R}^2$ be a bounded domain with smooth boundary $\partial\Omega$, and*

$$0 \leq f(u) = e^u + o(e^u), \quad u \uparrow +\infty.$$

Let, furthermore, $\{\lambda_k, u_k)\}$, *is a family of solutions to*

$$-\Delta u = \lambda f(u) \ in \ \Omega, \quad u|_{\partial\Omega} = 0 \tag{3.81}$$

for $(\lambda, u) = (\lambda_k, u_k)$, *such that* $\lambda_k > 0$, $\|u_k\| \to +\infty$, *and*

$$\Sigma_k = \int_\Omega \lambda_k e^{u_k} \, dx = O(1).$$

Then it holds that $\lambda_k \to 0$, *and the conclusion of Theorem 2.7 holds.*

Proof By the proof of Theorem 2.6 in Chap. 2, there is a neighborhood of $\partial\Omega$ denoted by ω, such that

$$\|u_k\|_{L^\infty(\Omega\cap\omega)} \leq C.$$

Writing $V_k = e^{-u_k} f(u_k)$ and $v_k = u_k + \log \lambda_k$, we obtain

$$-\Delta v_k = V_k(x) e^{v_k} \quad \text{in } \Omega$$

by (3.81). Then there arises

$$0 \leq V_k(x) \leq C, \quad \int_\Omega e^{v_k} \, dx = \Sigma_k = O(1),$$

and hence Theorem 3.28 is applicable, so that $\lambda_k \to 0$ follows from the boundary estimate of $\{u_k\}$ above. It holds also that

$$V_k(x) e^{v_k} = \lambda_k f(u_k) \rightharpoonup \sum_{i=1}^{\ell} a_i \delta_{\kappa_i} \quad \text{in } \mathcal{M}(\overline{\Omega}) \tag{3.82}$$

with $a_i > 0$, $\kappa_i \in \Omega$, and $1 \leq i \leq \ell$.

Let

$$f(u) = e^u + p(u).$$

Since $\lambda_k \downarrow 0$, we obtain

$$\|\lambda_k e^{u_k} p(u_k)\|_1 \to 0$$

by $p(u) = o(e^u)$ as $u \uparrow +\infty$, and therefore, the right-hand side of (3.82) may be replaced by $\lambda_k e^{u_k}$:

$$\lambda_k e^{u_k} \rightharpoonup \sum_{i=1}^{\ell} a_i \delta_{\kappa_i} \quad \text{in } \mathcal{M}(\overline{\Omega}).$$

Let, furthermore, $H = u_z^2$ in (3.81), using the complex variable $z = x_1 + \sqrt{-1}x_2$ for $x = (x_1, x_2)$. It holds that

$$H_{\overline{z}} = -\frac{1}{2}\lambda u_z f(u) = -\frac{1}{2}\lambda(e^u + P(u))_z \text{ in } \Omega, \quad P(u) = \int_0^u p(u)\, du.$$

Here we put

$$T = \frac{1}{\pi z^2} = (\frac{2}{\pi}\log|z|)_{zz}, \quad W = (\lambda e^u + P(u))\chi_\Omega \in L^1(\mathbf{R}^2),$$

and $\alpha = W * T$ in the sense of distributions.

Since

$$(\frac{2}{\pi}\log|z|)_{z\overline{z}} = \delta,$$

there arises

$$\alpha_z = W * T_{\overline{z}} = W_z * \delta = W_z \quad \text{in } \mathbf{R}^2.$$

We thus end up with

$$\tilde{H}_{\overline{z}} = 0 \text{ in } \Omega, \quad \tilde{H} = H + \frac{\alpha}{2}.$$

Then we can argue similarly as in the proof of Theorem 2.7, to reach (2.115)–(2.116). □

Theorem 3.34 *Let* $1 < p \leq \infty$ *and* $\frac{1}{p} + \frac{1}{p'} = 1$. *If*

$$v_- \in L^1_{loc}(\mathbf{R}^2), \quad V \in L^p(\mathbf{R}^2), \quad e^v \in L^{p'}(\mathbf{R}^2),$$

and

$$-\Delta v = V(x)e^v \quad \text{in } \mathbf{R}^2,$$

it holds that $v_+ \in L^\infty(\mathbf{R}^2)$, *where* $v_\pm = \max(\pm v, 0)$.

Proof Given $V(x)e^v \in L^1(\mathbf{R}^2)$ and $\varepsilon > 0$, we have

$$V(x)e^v = f_1 + f_2, \quad \|f_1\|_{L^1(\mathbf{R}^2)} < \varepsilon, \quad f_2 \in L^\infty(\mathbf{R}^2). \tag{3.83}$$

Assume $\varepsilon \in (0, 1/p')$.

Take $x_0 \in \mathbf{R}^2$, put $B_r = B(x_0, r)$, and define $v_i = v_i(x)$, $i = 1, 2$, by

$$-\Delta v_i = f_i \text{ in } B_1, \quad v_i = 0 \text{ on } \partial B_1.$$

Then, Lemma 3.16 for $\delta = 4\pi - 1$ implies

$$\int_{B_1} \exp\left(|v_1| / \varepsilon\right) \leq C, \tag{3.84}$$

and hence

$$\|v_1\|_{L^1(B_1)} \leq C. \tag{3.85}$$

We have, on the other hand,

$$\|v_2\|_{L^\infty(B_1)} \leq C \tag{3.86}$$

by the elliptic regularity, while

$$v_3 = v - v_1 - v_2$$

is harmonic in B_1. By the mean value theorem for harmonic functions it holds that

$$\|v_{3+}\|_{L^\infty(B_{1/2})} \leq C \|v_{3+}\|_{L^1(B_1)}. \tag{3.87}$$

Since

$$\int_{\mathbf{R}^2} v_+ \leq \int_{\mathbf{R}^2} (e^{v_+} - 1) \leq \int_{\{v>0\}} e^v \leq \int_{\mathbf{R}^2} e^v \leq C, \tag{3.88}$$

therefore, we obtain

$$\|v_{3+}\|_{L^1(B_1)} \leq C$$

by

$$v_{3+} \leq v_+ + |v_1| + |v_2|$$

and (3.85)–(3.86). Then, inequality (3.87) implies

$$\|v_{3+}\|_{L^\infty(B_{1/2})} \leq C. \tag{3.89}$$

Apply (3.84), (3.86), and (3.89) to

$$-\Delta v = V(x)e^{v} = V(x)e^{v_1} \cdot e^{v_2+v_3} \equiv g. \tag{3.90}$$

We have

$$e^{v_2+v_3} \in L^\infty(B_{1/2}), \quad V = V(x) \in L^p(B_1), \quad e^{v_1} \in L^{1/\varepsilon}(B_1).$$

Since $1/\varepsilon > p'$, furthermore, there is $\delta > 0$ such that

$$\|g\|_{L^{1+\delta}(B_{1/2})} \leq C. \tag{3.91}$$

Now we define the harmonic function h by

$$\Delta h = 0 \text{ in } B_{1/2}, \quad h = v \text{ on } \partial B_{1/2},$$

to apply the elliptic regularity to (3.90). Since the space dimension is two: $n = 2$, there arises

$$v = w + h, \quad \|w\|_{L^\infty(B_{1/2})} \leq C\,\|g\|_{L^{1+\delta}(B_{1/2})}. \tag{3.92}$$

By the mean value theorem for harmonic functions it follows that

$$\begin{aligned}\|h_+\|_{L^\infty(B_{1/4})} &\leq C\,\|h_+\|_{L^1(B_{1/2})} \\ &\leq C\left(\|v_+\|_{L^1(B_{1/2})} + \|w\|_{L^1(B_{1/2})}\right) \leq C',\end{aligned}$$

and then inequalities (3.91) and (3.92) imply

$$\|v_+\|_{L^\infty(B_{1/4})} \leq C,$$

independent of x_0. □

Remark 3.8 The value $\|f_2\|_\infty$ is not estimated by the known quantities in (3.83). Hence a uniform estimate of $\|v_+\|_\infty$ does not arise in Theorem 3.34, while Lemma 3.17 describes an ε regularity.

Proof *(Lemma 3.17)* By the covering argument, we assume

$$\Omega = B_R, \quad K = \overline{B_{R/2}},$$

where $B_r = B(0, r)$ for $r = R, R/2$. We divide v into

$$v = v_1 + v_2,$$

using the harmonic function v_2 in Ω:

$$-\Delta v_1 = V(x)e^v \quad \text{in } \Omega, \qquad v_1 = 0 \quad \text{on } \partial\Omega.$$

As for v_2, the mean value theorem for harmonic functions and the inequality

$$v_{2+} \le v_+ + |v_1|$$

are applicable. Hence we obtain

$$\|v_{2+}\|_{L^\infty(B_{3R/4})} \le C_R \|v_{2+}\|_{L^1(B_R)} \le C_R \|v_1\|_{L^1(B_R)} + C. \tag{3.93}$$

From the assumption, on the other hand, there is $\delta > 0$ in

$$4\pi - \delta > \varepsilon_0(p' + \delta),$$

and then Lemma 3.16 implies

$$\left\| e^{|v_1|} \right\|_{L^{p'+\delta}(B_R)} \le C. \tag{3.94}$$

Similarly to (3.88), inequality (3.94) implies

$$\|v_1\|_{L^1(B_R)} \le C, \tag{3.95}$$

and then

$$\|e^v\|_{L^{p'+\delta}(B_{3R/4})} \le C$$

follows from (3.93)–(3.95). Hence there is $q > 1$ such that

$$\|Ve^v\|_{L^q(B_{3R/4})} \le C.$$

Now we divide v to

$$v = w_1 + w_2,$$

using

$$-\Delta w_1 = V(x)e^v, \ \Delta w_2 = 0 \ \text{ in } B_{3R/4}, \quad w_1 = 0 \text{ on } \partial B_{3R/4}$$

to apply elliptic regularity to w_1 and mean value theorem and the inequality

$$w_{2+} \le v_+ + |w_1|$$

to w_2. There arises

$$\|v_+\|_{L^\infty(B_{R/2})} \leq C$$

and the proof is complete. □

Here is a generalization of Theorem 3.28 [5].

Theorem 3.35 *Let $\Omega \subset \mathbf{R}^2$ be a bounded domain, $v_k = v_k(x)$, $k = 1, 2, \cdots$ be a sequence of solutions to*

$$-\Delta v_k = V_k(x)e^{v_k}, \ V_k = V_k(x) \geq 0 \quad \text{in } \Omega,$$

satisfying

$$\|V_k\|_p \leq C, \ \left\|e^{v_k}\right\|_{p'} \leq C, \quad k = 1, 2, \cdots$$

for $1 < p \leq \infty$. Then, passing to a subsequence, we have the following alternatives:

1. *$\{v_k\}$ is locally uniformly bounded in Ω.*
2. *$v_k \to -\infty$ locally uniformly in Ω.*
3. *There arise a finite set $\mathscr{S} = \{a_i\} \subset \Omega$ and $\alpha_i \geq 4\pi/p'$ such that $v_k \to -\infty$ locally uniformly in $\Omega \setminus \mathscr{S}$ and*

$$V_k(x)e^{v_k}dx \rightharpoonup \sum_i \alpha_i \delta_{a_i}(dx) \quad \text{in } \mathscr{M}(\Omega).$$

Furthermore, $\mathscr{S}$ is the blowup set of $\{v_{k+}\}$ in Ω:

$$\mathscr{S} = \{x_0 \in \Omega \mid \exists x_k \to x_0 \text{ such that } v_k(x_k) \to +\infty\}.$$

Proof Since $\{V_k(x)e^{v_k}\} \subset L^1(\Omega)$ is bounded, there is a measure $\mu(dx)$ in Ω such that

$$V_k e^{v_k} dx \rightharpoonup \mu(dx) \quad \text{in } \mathscr{M}(\Omega)$$

for a subsequence. We have $\mu(\Omega) \leq C$ and hence

$$\Sigma = \left\{x_0 \in \Omega \mid \mu(\{x_0\}) \geq 4\pi/p'\right\}$$

is a finite set.

Given $x_0 \notin \Sigma$, take $\psi \in C_0^\infty(\Omega)$ such that $0 \leq \psi \leq 1$ and $\psi = 1$ near x_0. Since

$$\int_\Omega \psi d\mu < 4\pi/p'$$

there is $R_0 > 0$ such that

$$\|v_{k+}\|_{L^\infty(B(x_0,R_0))} = O(1), \quad k \to \infty$$

by Lemma 3.17, which implies $\mathscr{S} \subset \Sigma$. If $x_0 \notin \mathscr{S}$, conversely, $\{(v_k)_+\}$ is uniformly bounded in $B(x_0, R_0)$ for some $R_0 > 0$, which implies $\mu(\{x_0\}) = 0$ and hence $x_0 \notin \Sigma$. We thus obtain $\mathscr{S} = \Sigma$, and furthermore,

$$\mu(\{x_0\}) < 4\pi/p' \quad \Rightarrow \quad x_0 \notin \mathscr{S}.$$

Here we confirm that $\Sigma = \emptyset$ implies the first and the second alternatives. To this end we assume $\Omega = B_R$, using the covering and diagonal argument. Since

$$\Sigma = \mathscr{S} = \emptyset,$$

the sequence $\{v_{k+}\}$ is uniformly bounded in $\omega = B_{R/2}$. Then we define w_k by

$$-\Delta w_k = f_k \equiv V_k e^{v_k} \text{ in } \omega, \quad w_k = 0 \text{ on } \partial\omega.$$

Since $\{f_k\} \subset L^p(\omega)$ is bounded, the sequence $\{w_k\}$ is uniformly bounded in ω, and hence $\{\tilde{v}_{k+}\}$ is uniformly bounded in ω, where

$$\tilde{v}_k = v_k - w_k$$

is a harmonic function in ω. By the Harnack principle for harmonic functions, this $\{\tilde{v}_k\}$ is either uniformly bounded in ω or $\tilde{v}_k \to -\infty$ uniformly in ω, which induces the first and the second alternatives.

If $\Sigma \neq \emptyset$, on the contrary, there arises the third alternatives. In fact, since Σ is finite, each $x_0 \in \Sigma$ admits $R > 0$ such that

$$\omega = B(x_0, R) \subset\subset \Omega, \quad B(x_0, R) \cap \Sigma = \{x_0\}.$$

By the above argument, the sequence $\{v_k\}$ in $B(x_0, R) \setminus \{x_0\}$ is either locally uniformly bounded or diverses to $-\infty$ locally uniformly.

In the first case, we have

$$v_k \geq -C \quad \text{on } \partial\omega.$$

Hence z_k defined by

$$-\Delta z_k = f_k \quad \text{in } \omega, \qquad z_k = -C_{27} \quad \text{on } \partial\omega \tag{3.96}$$

satisfies $v_k \geq z_k$ in ω from the maximum principle. Up to a subsequence, on the other hand, we have $\alpha \geq 4\pi/p'$ and $0 \leq f \in L^1(\omega)$ such that

$$f_k(x)dx \rightharpoonup \alpha\delta_{x_0}(dx) + f(x)dx \quad \text{in } \mathscr{M}(\overline{\omega}). \tag{3.97}$$

By (3.96) and (3.97) we obtain $z_k \to z$ locally uniformly in $\overline{\omega} \setminus \{x_0\}$ and also

$$z(x) \geq \frac{4\pi}{p'} \cdot \frac{1}{2\pi} \log \frac{1}{|x - x_0|} - C, \quad x \in \overline{\omega} \setminus \{x_0\}.$$

Then Fatou's lemma guarantees

$$+\infty = \int_{B(x_0,R)} e^{p'z} \leq \liminf_k \int_{B(x_0,R)} e^{p'v_k} \leq \liminf_k \|e^{v_k}\|_{p'}^{p'} < +\infty,$$

a contradiction.

Therefore, a subsequence satisfies $v_k \to -\infty$ locally uniformly in $B(x_0, R) \setminus \{x_0\}$, and then

$$V_k(x)e^{v_k} \to 0 \quad \text{in } L^p_{loc}(\Omega \setminus \Sigma)$$

by a covering argument. Hence it holds that

$$\mu(dx) = \sum_i \alpha_i \delta_{a_i}(dx),$$

and the proof is complete. □

3.4.3 sup + inf *Inequality*

In this subsection we show a key inequality for the proof of Theorem 3.30, using an isoperimetric inequality. We recall that sharp forms of geometric isoperimetric inequalities depend on the topology of the domain. Here we use the simplest form of Alexandroff's inequality, a generalization of Bol's inequality. See [2] for the proof.

Lemma 3.18 (Alexsandroff–Bol) *Let $B = B(0, 1) \subset \mathbf{R}^2$ be the unit disc and $V(x) > 0$ be a continuous function in Ω. Assume that $v = v(x)$ satisfies*

$$-\Delta v = V(x)e^v \quad \text{in } B$$

and put

$$\ell = \int_{\partial B} e^{v/2}\, ds, \quad m = \int_B e^v\, dx.$$

Then each $K_0 \in \mathbf{R}$ admits

$$\ell^2 \geq (2\alpha - K_0 m)m, \tag{3.98}$$

provided that $\alpha = 2\pi - \omega_{K_0}^+(B) > 0$, *where*

$$\omega_{K_0}^+(B) = \int_{K>K_0} (K(x) - K_0)e^v dx, \quad K(x) = V(x)/2.$$

Let $p = e^v$, $0 < p = p(x) \in C^2(B) \cap C(\overline{B})$, in Lemma 3.18. We introduce the metric

$$d\sigma^2 = p(x)ds^2$$

to $B = B(0, 1) \subset \mathbf{R}^2$. Then the Gauss curvature K, the total area m, and the length ℓ of the boundary are given by

$$K = -\frac{\Delta \log p}{2p}, \quad m = \int_B p\, dx, \quad \ell = \int_{\partial B} p^{1/2}\, ds. \tag{3.99}$$

Then this lemma guarantees for

$$m_\mu^+(B) = \int_{\{K>\mu\}} (K(x) - \mu)\, p\, dx, \quad \mu \in \mathbf{R}$$

that

$$\alpha = 2\pi - m_\mu^+(B) > 0 \quad \Rightarrow \quad \ell^2 \geq (2\alpha - \mu m)m. \tag{3.100}$$

If $0 < p = p(x) \in C^2(B) \cap C(\overline{B})$ satisfies (2.118) in $\Omega = B$,

$$-\Delta \log p \leq p \quad \text{in } B, \tag{3.101}$$

it holds that $K = K(x) \leq 1/2$, and therefore, $m_\mu^+(B) = 0$ for $\mu = 1/2$. Thus, (3.100) implies Bol's inequality,

$$\ell^2 \geq \frac{1}{2}m(8\pi - m). \tag{3.102}$$

Using (3.100), we can make a sharper form of inequality (3.102) under (3.101). For this purpose, we take μ_* in

$$\int_{\{K>\mu_*\}} p\, dx \leq m/2, \quad \int_{\{K<\mu_*\}} p\, dx \leq m/2.$$

Since $\{K > \mu_*\} \neq B$ and $\{K < \mu_*\} \neq B$ we have

$$\{K \leq \mu_*\} \neq \emptyset, \quad \{K \geq \mu_*\} \neq \emptyset.$$

It holds also that

$$\frac{a}{2} \leq K(x) \leq \frac{b}{2} \quad \Rightarrow \quad \frac{a}{2} \leq \mu_* \leq \frac{b}{2} \tag{3.103}$$

where $a, b > 0$ are constants.

We define $C, D \geq 0$ by

$$\int_{\{K>\mu_*\}} p\, dx = \frac{m}{2} - C, \quad \int_{\{K<\mu_*\}} p\, dx = \frac{m}{2} - D.$$

Then

$$\alpha \equiv 2\pi - m^+_{\mu_*}(B) = 2\pi - \int_{\{K>\mu_*\}} (K - \mu_*) p\, dx$$

satisfies

$$\begin{aligned} 2\alpha - \mu_* m &= 4\pi - 2\int_{\{K>\mu_*\}} (K - \mu_*) p\, dx - \mu_* m \\ &= 4\pi - 2\int_{\{K\geq\mu_*\}} (K - \mu_*) p\, dx - \mu_* m \\ &= 4\pi - 2\int_{\{K\geq\mu_*\}} Kp\, dx + 2\mu_* D. \end{aligned} \tag{3.104}$$

Since

$$\begin{aligned} \int_{\{K<\mu_*\}} Kp\, dx &\geq \frac{a}{2}\int_{\{K<\mu_*\}} p\, dx = \frac{a}{2}\left(\frac{m}{2} - D\right) \\ &\geq \frac{a}{2b}\int_B Kp\, dx - \frac{a}{2}D, \end{aligned}$$

on the other hand, it holds that

$$\int_{\{K\geq\mu_*\}} Kp\, dx \leq \left(1 - \frac{a}{2b}\right)\int_B Kp\, dx + \frac{a}{2}D. \tag{3.105}$$

By (3.104) and (3.105) we obtain

$$2\alpha - \mu_* m \geq 4\pi - 2\left(1 - \frac{a}{2b}\right)\int_B Kp\, dx + (2\mu_* - a)D$$

$$\geq 4\pi - 2\left(1 - \frac{a}{2b}\right) \int_B Kp\, dx. \tag{3.106}$$

Here we put

$$4\pi < \alpha_0 < \frac{4\pi}{1 - \frac{a}{2b}}, \quad \gamma_0 = 4\pi - \left(1 - \frac{a}{2b}\right)\alpha_0 > 0.$$

Then inequality (3.106) guarantees

$$\int_B Kp\, dx \leq \alpha_0/2 \quad \Rightarrow \quad 2\alpha - \mu_* m \geq \gamma_0. \tag{3.107}$$

We obtain the following lemma by this inequality (3.107).

Lemma 3.19 *Let $B = B(0,1) \subset \mathbf{R}^2$ be the unit disc, and $a, b > 0$ be constants. Then there are $\alpha_0 > 4\pi$ and $C_1 > 0$ such that*

$$-\Delta v = V(x)e^v,\ a \leq V(x) \leq b \ \text{ in } B, \qquad \int_B V(x)e^v \leq \alpha_0 \tag{3.108}$$

implies $v(0) \leq C_1$.

Proof We assume (3.108), and apply (3.107) for $p = e^v$ and $K = V/2$. In fact, the function

$$f(r) = 4\pi - 2\int_{\{K>\mu_*\}\cap B_r} (K - \mu_*)p\, dx - \mu_* \int_{B_r} p\, dx$$

is non-increasing in $r \in [0, 1)$, and it holds that

$$\begin{aligned} f(r) \geq f(1) &= 4\pi - 2\int_{\{K>\mu_*\}} (K - \mu_*)p\, dx - \mu_* \int_B p\, dx \\ &= 2\alpha - \mu_* m \geq \gamma_0. \end{aligned} \tag{3.109}$$

Let

$$A(r) = \int_0^r dr' \int_{\partial B(0,r')} p\, ds = \int_{B(0,r)} p\, dx.$$

From (3.100) for each $0 \leq r < 1$ it follows that

$$A'(r) = \int_{\partial B(0,r)} p\, ds \geq \frac{1}{2\pi r}\{\int_{\partial B(0,r)} p^{1/2} ds\}^2 \geq \frac{1}{2\pi r} f(r)A(r). \tag{3.110}$$

Then (3.110) implies, for each $0 < r_0 < 1$,

$$
\begin{aligned}
-\frac{1}{2\pi}\log r_0 &\le \int_{r_0}^{1} \frac{A'(r)}{f(r)A(r)}\,dr = \int_{r_0}^{1} \frac{1}{f(r)}\{\log A(r)\}'\,dr \\
&= \frac{\log A(1)}{f(1)} - \frac{\log A(r_0)}{f(r_0)} + \int_{r_0}^{1} \frac{\log A(r)}{f(r)^2} f'(r)\,dr. \qquad (3.111)
\end{aligned}
$$

Here we have

$$
A(r_0) = \pi r_0^2 p(0)(1+o(1)), \quad f(r_0) = 4\pi - O\left(r_0^2\right), \quad r_0 \downarrow 0,
$$

and hence

$$
\begin{aligned}
\lim_{r_0\downarrow 0}\left\{\frac{\log A(r_0)}{f(r_0)} - \frac{1}{2\pi}\log r_0\right\} &= \lim_{r_0\downarrow 0} \frac{1}{4\pi}\log\frac{A(r_0)}{r_0^2} \\
&= \frac{1}{4\pi}\left(\log p(0) + \log \pi\right). \qquad (3.112)
\end{aligned}
$$

By (3.111)–(3.112) it holds that

$$
\begin{aligned}
v(0) &= \log p(0) \\
&\le \frac{4\pi \log A(1)}{f(1)} - \log\pi + 4\pi \int_0^1 \frac{\log A(r)}{f(r)^2} f'(r)\,dr. \qquad (3.113)
\end{aligned}
$$

For the third term of the right-hand side of (3.113) we use

$$
A(r) \le A(1) = \int_B p\,dx \le \frac{1}{a}\int_B Vp\,dx \le \frac{\alpha_0}{a}
$$

derived from (3.108). Then (3.109) implies

$$
\begin{aligned}
\frac{\log A(r)}{f(r)^2} f'(r) &= \frac{\log\frac{\alpha_0}{A(r)} + \log\frac{1}{\alpha_0}}{f(r)^2}\left(-f'(r)\right) \\
&\le \gamma_0^{-2}\log\frac{\alpha_0}{A(r)} \cdot \left(-f'(r)\right). \qquad (3.114)
\end{aligned}
$$

Here we have

$$
\begin{aligned}
0 \le -f'(r) &= 2\int_{\{K>\mu_*\}\cap\partial B_r} (K-\mu_*)p\,ds + \mu_*\int_{\partial B_r} p\,ds \\
&\le \left(\frac{3}{2}b - a\right)\int_{\partial B_r} p\,ds = \left(\frac{3}{2}b - a\right)A'(r) \qquad (3.115)
\end{aligned}
$$

by (3.103). Writing $t = A(r)/\alpha_0$ in (3.114)–(3.115), we obtain

$$\int_0^1 \frac{\log A(r)}{f(r)^2} \cdot f'(r)\, dr \leq \gamma_0^{-2} \left(\frac{3}{2} b - a \right) \cdot \alpha_0 \cdot \int_0^{1/a} \log \frac{1}{t}\, dt < +\infty$$

and the result follows. □

We emphasize that $\alpha_0 > 4\pi$ in (3.107) is assured by the estimate of $V(x) > 0$ below. Without this condition, we have a similar conclusion but with $\alpha_0 < 4\pi$, which is not sufficient to guarantee residual vanishing for the blowup family of solutions. This result is concerned with the case that $V(x)$ is restricted to a compact set in $C(\overline{\Omega})$, where α_0 can be arbitrary close to 8π. For the proof of this property of quantization we apply the following theorem [33] for the rescaled solution, which actually follows from Lemma 3.19.

Theorem 3.36 (Shafrir) *Let $\Omega \subset \mathbf{R}^2$ be a bounded domain, $V = V(x) \in C(\overline{\Omega})$, and $K \subset \Omega$ be a compact set. Then, given $a, b > 0$, we have*

$$c_1 = c_1(a, b) \geq 1, \quad c_2 = c_2(a, b, dist\,(K, \partial\Omega)) > 0$$

such that

$$-\Delta v = V(x)e^v, \ a \leq V(x) \leq b \quad in\ \Omega \tag{3.116}$$

implies

$$\sup_K v + c_1 \inf_\Omega v \leq c_2. \tag{3.117}$$

Proof By the covering argument, this theorem is reduced to the case of $\Omega = B(x_0, r)$, $K = \{x_0\}$, and $c_2 = c_2(a, b, r)$. We assign $x_0 = 0$ and take

$$\tilde{v}(x) = v(rx) + 2 \log r,$$

which satisfies (3.116) for $\Omega = B(0, 1)$ and $\tilde{V}(x) = V(rx)$. Thus it suffices to take the case that $\Omega = B = B(0, 1)$ and $K = \{0\}$. We shall show

$$v(0) + \frac{c_1}{2\pi} \int_{\partial B} v\, ds \leq c_2, \tag{3.118}$$

which implies (3.117) in this case. Namely, below we show (3.118) if (3.116) holds for $\Omega = B$.

Noting $\alpha_0 > 4\pi$ in Lemma 3.19, we take $c_1 \geq 1$ in $4\pi(c_1 + 1)/c_1 = \alpha_0$. In the case of

$$\int_B V e^v\, dx > 4\pi(c_1 + 1)/c_1 \tag{3.119}$$

we define $r_0 \in (0, 1)$ by

$$\int_{B(0,r_0)} Ve^v dx = 4\pi(c_1+1)/c_1,$$

and $r_0 = 1$ if (3.119) does not hold. Thus we have $0 < r_0 \leq 1$ satisfying

$$\int_{B(0,r_0)} Ve^v dx \leq 4\pi(c_1+1)/c_1. \tag{3.120}$$

Then, we have (3.116) for

$$\tilde{u}(x) = u(r_0 x) + 2\log r_0, \quad \tilde{V}(x) = V(r_0 x), \quad \Omega = B = B(0,1),$$

together with

$$\int_B \tilde{V} e^{\tilde{v}} dx \leq 4\pi(c_1+1)/c_1 = \alpha_0.$$

Hence Lemma 3.19 implies

$$\tilde{v}(0) = v(0) + 2\log r_0 \leq c_0. \tag{3.121}$$

Let

$$\begin{aligned} G(r) &= v(0) + \frac{c_1}{2\pi r}\int_{\partial B(0,r)} v\, ds + 2(c_1+1)\log r \\ &= v(0) + c_1 \int_0^{2\pi} v(re^{\imath\theta})\, d\theta + 2(c_1+1)\log r. \end{aligned}$$

Since v is superharmonic, we have

$$\frac{1}{2\pi r}\int_{\partial B(0,r)} v\, ds \leq v(0),$$

which implies

$$G(r_0) \leq (c_1+1)v(0) + 2(c_1+1)\log r_0 \leq (c_1+1)c_0 \tag{3.122}$$

by (3.121). If $r_0 = 1$ in (3.122) then it follows that (3.118) from

$$G(1) = v(0) + \frac{c_1}{2\pi}\int_{\partial B} v\, ds.$$

If $r_0 < 1$ we have (3.120), and hence

$$\int_{B(0,r)} Ve^v dx \geq 4\pi(c_1+1)/c_1, \quad r_0 \leq r \leq 1. \tag{3.123}$$

Here we obtain

$$\begin{aligned} G'(r) &= c_1 \int_0^{2\pi} v_r(r,\theta)\, d\theta + \frac{2(c_1+1)}{r} \\ &= \frac{1}{r}\{\frac{c_1}{2\pi}\int_{\partial B(0,r)} v_r\, ds + 2(c_1+1)\} \end{aligned}$$

and

$$\int_{\partial B(0,r)} v_r\, ds = \int_{B(0,r)} \Delta v\, dx = -\int_{B(0,r)} Ve^v\, dx,$$

and therefore,

$$G'(r) \leq 0, \quad r_0 \leq r \leq 1$$

by (3.123). Now, (3.122) implies

$$v(0) + \frac{1}{2\pi}\int_{\partial B} v\, ds = G(1) \leq G(r_0) \leq (c_1+1)c_0$$

and hence (3.118). □

Remark 3.9 If $\|\nabla V\|_\infty \leq C$ in (3.116), there arises $c_1 = 1$ in (3.117). The proof is based on the blowup analysis, without relying on the Alexandroff–Bol inequality [4].

3.4.4 Chen–Li Theorem

Provided with the scaling invariance of the model, classification of the scaling limit is the second fundamental element of the blowup analysis. The limit takes the form of the original model defined in the whole space. This solution is called the *entire solution*, and its classification is called the *Liouville property*, succeeding the case of harmonic functions.

In many cases this entire solution is radially symmetric with respect to a point, and then it is subject to an ordinally differential equation with respect to the radial vector. Since this ordinally differential equation is also provided with the scaling invariance, its solution takes a normal form of which scalings and translations give all the entire solutions.

This subsection is devoted to the proof of Theorem 3.32 in Sect. 3.4.1. We take several preliminaries to apply the method of moving planes.

Lemma 3.20 *It holds that*

$$\int_{\mathbf{R}^2} e^v dx \geq 8\pi \tag{3.124}$$

if $v = v(x)$ is a solution to (3.78).

Proof If $t \in \mathbf{R}$ is not a critical value of $v(x)$, there arises

$$\int_{\Omega_t} e^v dx = -\int_{\Omega_t} \Delta v \, dx = \int_{\partial\Omega_t} |\nabla v| \, dx, \quad -\frac{d}{dt}|\Omega_t| = \int_{\partial\Omega_t} \frac{dS}{|\nabla v|}$$

by Green's and co-area formulas, where $\Omega_t = \left\{x \in \mathbf{R}^2 \mid v(x) > t\right\}$ denotes the level set. Then we obtain

$$(-\frac{d}{dt}|\Omega_t|) \cdot \int_{\Omega_t} e^v dx \geq |\partial\Omega_t|^2 \geq 4\pi \, |\Omega_t|$$

by Schwarz' and the isoperimetric inequalities, and hence

$$\begin{aligned} \frac{d}{dt}(\int_{\Omega_t} e^v dx)^2 &= 2\int_{\Omega_t} e^v (-\int_{\partial\Omega_t} \frac{e^v}{|\nabla v|} dS) \, dx \\ &= 2e^t \cdot \frac{d}{dt}|\Omega_t| \cdot \int_{\Omega_t} e^v dx \leq -8\pi e^t \cdot |\Omega_t| . \end{aligned} \tag{3.125}$$

Since $v = v(x)$ is real-analytic, its critical points are isolated and the measure of the set of critical values is zero. Operating $\int_{-\infty}^{\infty} \cdot dt$ to (3.125) then induces

$$\begin{aligned} -(\int_{\mathbf{R}^2} e^v dx)^2 &\leq -8\pi \int_{-\infty}^{\infty} e^t \, |\Omega_t| \, dt = 8\pi \int_{-\infty}^{\infty} e^t \cdot \frac{d}{dt}|\Omega_t| \, dt \\ &= -8\pi \int_{-\infty}^{\infty} e^t dt \cdot \int_{\partial\Omega_t} \frac{dS}{|\nabla v|} = 8\pi \int_{-\infty}^{\infty} \frac{d}{dt} \int_{\Omega_t} e^v dx = -8\pi \int_{\mathbf{R}^2} e^v, \end{aligned}$$

which implies (3.124). □

Lemma 3.21 *It holds that*

$$\frac{v(x)}{\log|x|} \to -\frac{1}{2\pi} \int_{\mathbf{R}^2} e^v dx \leq -4 \tag{3.126}$$

as $|x| \to \infty$*, uniformly.*

Proof By the second inequality of (3.78), we get the absolute convergence of

$$w(x) = \frac{1}{2\pi} \int_{\mathbf{R}^2} (\log|x-y| - \log|y|) e^{v(y)}\, dy. \tag{3.127}$$

In fact, we have

$$|y| > 2|x| \quad \Rightarrow \quad |\log|x-y| - \log|y|| \le \log 2 \tag{3.128}$$

by

$$\log|x-y| - \log|y| = \log\left|\frac{x}{|y|} - \frac{y}{|y|}\right|,$$

while $v = v(y)$ is a classical solution and hence

$$\int_{|y| \le 2|x|} (|\log|x-y|| + |\log|y||)\, e^{v(y)} dy < +\infty$$

for each $x \in \mathbf{R}^2$. Then there arises the absolute convergence of (3.127). The same result is true for the derivatives in x of the integrand, which induces

$$\Delta w = e^v \quad \text{in } \mathbf{R}^2$$

by the dominated convergence theorem.

Now, we show

$$\lim_{|x|\to\infty} \frac{w(x)}{\log|x|} = \frac{1}{2\pi} \int_{\mathbf{R}^2} e^v\, dx, \tag{3.129}$$

using

$$\lim_{|x|\to\infty} \frac{1}{\log|x|} (\log|x-y| - \log|y|) = 1, \quad \forall y \in \mathbf{R}^2 \setminus \{0\}. \tag{3.130}$$

In fact, we have

$$\frac{\log|x-y|}{\log|x|} = 1 + \frac{\log\left|\frac{x}{|x|} - \frac{y}{|x|}\right|}{\log|x|} \tag{3.131}$$

and hence (3.130). To apply the dominated convergence theorem for the proof of (3.129), we estimate the integrand of (3.127) uniformly in $|x| \ge 2$.

In fact, first, we have

$$|x| \geq 2, \ \frac{|x|}{|y|} < \frac{1}{2} \quad \Rightarrow \quad \frac{1}{\log|x|} \, |\log|x-y| - \log|y|| \leq 1 \tag{3.132}$$

by (3.128). Second, equality (3.131) implies

$$|x| \geq 2, \ \frac{|y|}{|x|} < \frac{1}{2} \quad \Rightarrow \quad \left| \frac{\log|x-y|}{\log|x|} \right| \leq 2,$$

and therefore,

$$\begin{aligned} &|x| \geq 2, \ |y| \geq \delta, \ \frac{|y|}{|x|} < \frac{1}{2} \quad \Rightarrow \\ &\frac{1}{\log|x|} |\log|x-y| - \log|y|| \\ &\leq 2 + \frac{1}{\log|x|} \max\{\log|x| - \log 2, -\log\delta\} \leq C \end{aligned} \tag{3.133}$$

for each $\delta > 0$. By (3.132)–(3.133) we obtain

$$\begin{aligned} &|x| \geq 2, \ |y| \geq \delta, \ |y| < \frac{1}{2}|x| \quad \text{or} \quad |y| > 2|x| \\ &\Rightarrow \quad \frac{1}{\log|x|} \, |\log|x-y| - \log|y|| \leq C. \end{aligned} \tag{3.134}$$

If

$$\frac{1}{2}|x| \leq |y| \leq 2|x|,$$

finally, there arises

$$|\log|x| - \log|y|| \leq \log 2,$$

and hence

$$|x| \geq 2, \ \frac{1}{2}|x| \leq |y| \leq 2|x| \quad \Rightarrow \quad \left| \frac{\log|y|}{\log|x|} \right| \leq 2.$$

In this case, furthermore, equality (3.131) implies

$$\left| \frac{\log|x-y|}{\log|x|} \right| \leq 1 + \frac{\log(1 + \frac{|y|}{|x|})}{\log|x|} \leq 1 + \frac{\log 3}{\log 2}$$

and we end up with

$$|x| \geq 2, \ \frac{1}{2}|x| \leq |y| \leq 2|x| \ \Rightarrow \ \frac{1}{\log|x|}\left|\log|x-y| - \log|y|\right| \leq C. \tag{3.135}$$

By (3.127) and $e^v \in L^1(\mathbf{R}^2)$, inequalities (3.134)–(3.135) imply

$$\lim_{|x|\to\infty} \frac{1}{\log|x|} \int_{|y|\geq\delta} \left|\log|x-y| - \log|y| - 1\right| e^{v(y)} dy = 0. \tag{3.136}$$

Since

$$\sup_{|x|\geq 2, \ |y|<1} \frac{\left|\log|x-y|\right|}{\log|x|} \leq \sup_{|x|\geq 2} \frac{\log(|x|+1)}{\log|x|} < +\infty$$

and $v = v(x)$ is a classical solution, there arises

$$\begin{aligned} &|x| \geq 2, \ 0 < |y| < \delta < 1 \quad \Rightarrow \\ &\frac{1}{\log|x|} \int_{|y|<\delta} \left|\log|x-y| - \log|y| - 1\right| e^{v(y)} dy \\ &\quad \leq C \int_0^\delta r(1 + |\log r|) dr. \end{aligned} \tag{3.137}$$

Relations (3.136)–(3.137) then imply

$$\limsup_{|x|\to\infty} \frac{1}{\log|x|} \int_{\mathbf{R}^2} \left|\log|x-y| - \log|y| - 1\right| e^{v(y)} dy \leq C \int_0^\delta r(1 + |\log r|) dr$$

and hence (3.129) with $\delta \downarrow 0$.

To apply the Liouville property for harmonic functions, we note that $v(x)$ is bounded above by Lemma 3.17, while (3.129) is applicable to $w(x)$. It thus holds that

$$\Delta u = 0, \ u(x) \leq C(1 + \log(|x|+1)) \quad \text{in } \mathbf{R}^2$$

for $u = v + w$, and hence u is a constant by Remark 3.2. Then, (3.129) implies (3.126), and the proof is complete. □

In the proof of Theorem 3.32, we find two directions of symmetry of $v(x)$, the angle of which does not belong to $2\pi\mathbf{Q}$. The crossing point of these two lines is x_0 in (3.79), and then $v(x)$ is radially symmetric with respect to x_0. By the $O(2)$ symmetry and the translation invariance of (3.78), it suffices to show that v is symmetric in the x_1 direction.

By Lemma 3.21, the function v attains a maximum in $\mathbf{R}^2$. By the translation, we suppose that this maximum point is in $x_1 < -3$. Given $\lambda \in \mathbf{R}$, let

$$\Sigma_\lambda = \{(x_1, x_2) \mid x_1 < \lambda\}, \quad T_\lambda = \partial \Sigma_\lambda = \{(x_1, x_2) \mid x_1 = \lambda\}.$$

Let, furthermore, $x^\lambda = (2\lambda - x_1, x_2)$ for $x = (x_1, x_2)$, and

$$v^\lambda(x) = v(x^\lambda), \quad w_\lambda(x) = v^\lambda(x) - v(x).$$

Since $-\Delta v^\lambda = \exp(v^\lambda)$, there is $\psi(x) \in [v(x), v(x^\lambda)]$ such that

$$-\Delta w_\lambda = e^\psi w_\lambda.$$

Regarding (3.126), we put

$$\overline{w}_\lambda(x) = w_\lambda(x)/g(x), \ g(x) = \log(|x| - 1). \tag{3.138}$$

This $\overline{w}_\lambda(x)$ is defined for $x \in \Sigma_\lambda$, $\lambda < -2$, and satisfies

$$\Delta \overline{w}_\lambda + \frac{2}{g} \nabla g \cdot \nabla \overline{w}_\lambda + (\exp \psi + \frac{\Delta g}{g}) \overline{w}_\lambda = 0 \tag{3.139}$$

there.

Lemma 3.22 *There is $R_0 > 2$ independent of $\lambda < -2$ such that if $x_0 \in \Sigma_\lambda$ attains a negative minimum of $\overline{w}_\lambda(x)$ in $x \in \Sigma_\lambda$ it holds that $|x_0| < R_0$.*

Proof Since

$$\Delta g/g = \frac{-1}{|x|\,(|x| - 1)^2 \log(|x| - 1)}$$

Lemma 3.21 ensures $R_0 > 2$ such that

$$\exp v + \frac{\Delta g}{g} < 0 \quad \text{in } \mathbf{R}^2 \setminus B(0, R_0).$$

By the assumption $\overline{w}_\lambda(x_0) < 0$ it holds that

$$\psi(x_0) \leq \max\left\{v(x_0), v^\lambda(x_0)\right\} = v(x_0).$$

In the case of $|x_0| \geq R_0$, therefore, there arises

$$\exp \psi(x_0) + \frac{\Delta g}{g}(x_0) < 0 \tag{3.140}$$

and hence $\overline{w}_\lambda(x)$ in $x \in \Sigma_\lambda$ does not attain a minimum at $x = x_0$ by (3.139). □

Proof *(Theorem 3.32)* We show the existence of $\lambda_0 \in \mathbf{R}$ such that

$$v^{\lambda_0} \equiv v. \tag{3.141}$$

We confirm

$$\lambda < -R_0 \quad \Rightarrow \quad \overline{w}_\lambda \geq 0 \qquad \text{in } \Sigma_\lambda. \tag{3.142}$$

In fact, there arises

$$\lim_{|x|\to\infty} \overline{w}_\lambda(x) = 0$$

by Lemma 3.21. If (3.142) is not the case, therefore, the function $\overline{w}_\lambda$ takes a negative minimum in Σ_λ, contradicting Lemma 3.22.

Since v attains a maximum in $x_1 < -3$, the maximal λ_0 of λ_0' that admits the conclusion of (3.142) for any $\lambda \leq \lambda_0'$ is finite. We show that this λ_0 ensures

$$\lambda < \lambda_0 \quad \Rightarrow \quad w_\lambda > 0 \quad \text{in } \Sigma_\lambda \tag{3.143}$$

and

$$\frac{\partial v}{\partial x_1} > 0, \ w_{\lambda_0} \equiv 0 \quad \text{in } \Sigma_{\lambda_0}. \tag{3.144}$$

To begin with, we note that

$$\frac{\partial v}{\partial x_1} \geq 0, \quad x_1 \leq \lambda_0, \tag{3.145}$$

and that if $\lambda \leq \lambda_0$ it holds that

$$w_\lambda \equiv 0 \quad \Leftrightarrow \quad \frac{\partial v}{\partial x_1} = 0 \text{ somewhere on } T_\lambda. \tag{3.146}$$

In fact, inequality (3.145) is obvious, and (3.146) is a consequence of the strong maximum principle and the Hopf lemma.

Assume the contrary to (3.143), and the strong maximum principle ensures $\delta > 0$ such that

$$w_{\lambda_0-\delta} \equiv 0 \quad \text{in } \Sigma_{\lambda_0-\delta}, \tag{3.147}$$

which means

$$v(\lambda_0 - 2\delta, x_2) = v(\lambda_0, x_2). \tag{3.148}$$

Then, (3.148) and (3.145) imply

$$\frac{\partial v}{\partial x_1} = 0, \quad \lambda_0 - 2\delta \le x_1 \le \lambda_0, \tag{3.149}$$

and hence

$$\frac{\partial v}{\partial x_1} = 0 \quad \text{on } T_{\lambda_0 - \delta}.$$

Therefore, (3.146) ensures

$$w_{\lambda_0 - 2\delta} \equiv 0. \tag{3.150}$$

Repeating the process that (3.147) implies (3.149) and (3.150), we have the independence of $v(x)$ on x_1, which contradicts the second assumption of (3.78).

Having (3.143), we obtain

$$\lambda < \lambda_0 \quad \Rightarrow \quad \frac{\partial w_\lambda}{\partial x_1} < 0 \quad \text{on } T_\lambda \tag{3.151}$$

by the Hopf lemma. Then (3.151) implies $\frac{\partial v}{\partial x_1} > 0$ for $x_1 < \lambda_0$, which means

$$\frac{\partial v}{\partial x_1} > 0 \quad \text{in } \Sigma_{\lambda_0}.$$

The first inequality of (3.144) is thus verified, and hence v is strictly increasing in $x_1 < \lambda_0$, which implies $\lambda_0 < -3$. Then $\overline{w}_\lambda(x)$ is well-defined for $x \in \Sigma_\lambda$ and $\lambda < \lambda_0 + 1$. It is also shown that $\overline{w}_{\lambda_0} \ge 0$.

Assume the contrary to the second equality in (3.144), that is,

$$\overline{w}_{\lambda_0} \not\equiv 0,$$

and apply the strong maximum principle and the Hopf lemma to (3.139). There arises

$$\overline{w}_{\lambda_0} > 0 \text{ in } \Sigma_{\lambda_0}, \quad \frac{\partial \overline{w}_{\lambda_0}}{\partial x_1} < 0 \text{ on } T_{\lambda_0}, \tag{3.152}$$

while we have $\lambda_k \downarrow \lambda_0$ such that

$$\overline{w}_{\lambda_k}(x_k') < 0, \quad \exists x_k' \in \Sigma_{\lambda_k} \tag{3.153}$$

by the definition of λ_0. By (3.153) and Lemma 3.21, there is a minimum point of $\overline{w}_{\lambda_k}(x)$ in $x \in \Sigma_{\lambda_k}$, denoted by $x_k \in \Sigma_{\lambda_k}$. Then we have

$$|x_k| \leq R_0$$

by Lemma 3.22.

A subsequence satisfies $x_k \to x_0$, and then it holds that

$$x_0 \in \Sigma_{\lambda_0} \cup T_{\lambda_0}, \quad \overline{w}_{\lambda_0}(x_0) \leq 0, \quad \nabla \overline{w}_{\lambda_0}(x_0) = 0,$$

contradicting (3.152). Thus we obtain the second equality of (3.144), which means that (3.141) holds. □

3.4.5 Baraket–Pacard Theorem

The solution to (3.78) takes the form (3.79) with the parameters $x_0 \in \mathbf{R}^2$ and $\mu > 0$. Differentiating it with respect to these parameters, we obtain the solution $w = w(x)$ to its linearized equation

$$-\Delta w = e^v w \quad \text{in } \mathbf{R}^2, \tag{3.154}$$

which, however, exhausts all the bounded solution to (3.154). By the scaling invariance of (3.78), this assertion is reduced to the normal form of $u = u(x)$, that is, the following theorem [3].

Theorem 3.37 (Baraket–Pacard) *The linear space of bounded solutions to*

$$-\Delta w = \frac{8}{(1+|x|^2)^2} w \ \textit{in } \mathbf{R}^2, \quad u(x) = \log \frac{8}{(1+|x|^2)^2}$$

is generated by

$$\frac{\partial u}{\partial x_1}, \quad \frac{\partial u}{\partial x_2}, \quad x \cdot \nabla u + 2.$$

Proof We show

$$w_n(r) = 0, \quad |n| \geq 2 \tag{3.155}$$

in the Fourier expansion of $w = w(re^{\imath\theta})$ for $r > 0$, $\theta \in S^1$:

$$w(x) = \sum_{n \in \mathbf{Z}} w_n(r) e^{\imath n\theta}, \quad w_n(r) = \frac{1}{2\pi} \int_0^{2\pi} w(re^{\imath\theta}) e^{-\imath n\theta} \, d\theta \tag{3.156}$$

to ensure the result.

By (3.156), the function $w_n = w_n(r)$ is real-analytic at $r = 0$, and it holds that

$$(w_n)_{rr} + \frac{1}{r}(w_n)_r - \frac{n^2}{r^2}w_n = -\frac{8}{(1+r^2)^2}w_n. \tag{3.157}$$

Let

$$w_n(r) = \sum_{k=0}^{\infty} a_k r^k$$

be its Taylor expansion at $r = 0$. We recall $w(x) = O(1)$ to conclude that $r = 0$ is a *regular singular point* of $w_n = w_n(r)$ in (3.157), which implies

$$a_m = 0, \quad m > |n|,$$

that is,

$$w_n(r) = O(r^{|n|}), \quad (w_n)_r(r) = O(r^{|n|-1}), \qquad r \downarrow 0. \tag{3.158}$$

See [10]. Using the Kelvin transformation $t = 1/r$, we obtain, similarly,

$$w_n(r) = O(r^{-|n|}), \quad (w_n)_r(r) = O(r^{-|n|-1}), \qquad r \uparrow +\infty. \tag{3.159}$$

Then, (3.157), (3.158), and (3.159) imply

$$\int_0^\infty \{(w_n)_r^2 + (\frac{n^2}{r^2} - \frac{8}{(1+r^2)^2})w_n\}rdr = 0 \tag{3.160}$$

We have

$$|n| \geq 2 \quad \Rightarrow \quad \frac{n^2}{r^2} - \frac{8}{(1+r^2)^2} > 0, \ r > 0$$

in (3.160), and therefore, (3.155). □

3.4.6 Li–Shafrir Theorem

Theorem 3.28 localizes the proof of Theorem 3.29 around the blowup point. Hence Theorem 3.29 is reduced to Theorem 3.30. Recall $B = B(0, R) \subset \mathbf{R}^2$, $B_r = B(0, r)$, and

$$-\Delta v_k = V_k(x)e^{v_k}, \ V_k = V_k(x) \geq 0 \text{ in } B, \quad V_k \to V \text{ in } C(\overline{B})$$

$$\max_{\overline{B}} v_k \to +\infty, \quad \max_{\overline{B}\setminus B_r} v_k \to -\infty, \; 0 < r < R$$

$$\lim_{k\to\infty} \int_B V_k e^{v_k} = \alpha, \quad \int_B e^{v_k} \leq C_1, \; k = 1, 2, \cdots. \tag{3.161}$$

Then we show $\alpha \in 8\pi\mathbf{N}$.

Lemma 3.23 *It holds that $V(0) > 0$.*

Proof There is $x_k \in B$ such that

$$v_k(x_k) = \|v_k\|_\infty \to +\infty, \quad x_k \to 0.$$

Letting

$$\tilde{v}_k(x) = v_k(\delta_k x + x_k) + 2\log\delta_k, \quad \delta_k = e^{-v_k(x_k)/2} \to 0, \tag{3.162}$$

we obtain

$$-\Delta\tilde{v}_k = V_k(\delta_k x + x_k)e^{\tilde{v}_k}, \;\; \tilde{v}_k \leq 0 = \tilde{v}_k(0) \quad \text{in } B(0, R/2\delta_k)$$
$$\int_{B(0,R/2\delta_k)} e^{\tilde{v}_k} \leq C.$$

Hence Theorem 3.28 is applicable to $\{\tilde{v}_k\}$.

Since the second and the third alternatives are excluded, the family $\{\tilde{v}_k\}$ is locally uniformly bounded in $\mathbf{R}^2$ by the diagonal argument. Hence the elliptic estimate assures a subsequence satisfying

$$\tilde{v}_k \to \tilde{v} \quad \text{locally uniformly in } \mathbf{R}^2. \tag{3.163}$$

The limit function $\tilde{v} = \tilde{v}(x)$ is a solution to

$$-\Delta\tilde{v} = V(0)e^{\tilde{v}}, \; \tilde{v} \leq 0 = \tilde{v}(0) \;\text{ in } \mathbf{R}^2, \quad \int_{\mathbf{R}^2} e^{\tilde{v}} \leq C. \tag{3.164}$$

Then, Liouville's theorem to harmonic functions in $\mathbf{R}^2$ implies $V(0) > 0$. □

By Lemma 3.23, we may assume $a, b > 0$ independent of $k = 1, 2, \ldots$, such that

$$a \leq V_k(x) \leq b, \quad x \in B.$$

By Theorem 3.32 applied to $v(x) = \tilde{v}(x) + \log V(0)$, furthermore, it holds that

$$\tilde{v}(x) = \log \frac{1}{(1 + \frac{V(0)}{8}|x|^2)^2}, \quad \int_{\mathbf{R}^2} V(0)e^{\tilde{v}} = 8\pi \tag{3.165}$$

in (3.164).

Convergence (3.163) does not control the behavior of $\tilde{v}_k$ far from the origin, but a diagonal argument refines (3.163) and (3.165) as follows.

Lemma 3.24 *Assume (3.161),*

$$\|v_k\|_\infty = v_k(x_k^0) \to +\infty, \quad x_k^0 \to 0, \quad \delta_k^0 = e^{-v_k(x_k^0)/2} \to 0, \tag{3.166}$$

and

$$r_k^0 \downarrow 0, \quad \lim_{k\to\infty} r_k^0/\delta_k^0 = +\infty. \tag{3.167}$$

Then it holds that

$$\lim_{k\to\infty} \int_{B(x_k^0, 2r_k^0)} V_k(x)e^{v_k} = 8\pi.$$

Since $v_k \to -\infty$ locally uniformly in $\overline{B} \setminus \{0\}$, if

$$\lim_{k\to\infty} \int_{B_{R/2}\setminus B(x_k^0, 2r_k^0)} V_k(x)e^{v_k} = 0 \tag{3.168}$$

it follows that

$$\lim_{k\to\infty} \int_B V_k(x)e^{v_k} = 8\pi,$$

and hence $\alpha = 8\pi$. Equality (3.168), called the residual vanishing, is a requirement of the behavior of $\tilde{v}_k$ far away from the origin. A sufficient condition is obtained by the $\sup + \inf$ inequality. The following lemma is a scaling of Theorem 3.36.

Lemma 3.25 *Given $a, b > 0$, there are $c_1 \geq 1$ and $c_2 > 0$ independent of $R > 0$ such that*

$$-\Delta v = V(x)e^v, \ a \leq V(x) \leq b \quad \text{in } B_R$$

implies

$$v(0) + c_1 \inf_{\partial B_r} v + 2(c_1 + 1)\log r \leq c_2, \quad 0 < r \leq R. \tag{3.169}$$

Proof Put

$$\tilde{v}(x) = v(rx) + 2\log r, \quad r \in (0, R],$$

and get

$$-\Delta\tilde{v} = V(rx)e^{\tilde{v}} \quad \text{in } B_1.$$

Then apply Theorem 3.36 to $\Omega = B_1$ and $K = \{0\}$. It holds that

$$\tilde{v}(0) + c_1 \inf_{B_1} \tilde{v} \leq c_2$$

which means

$$v(0) + 2(c_1 + 1)\log r + c_1 \inf_{B_r} v \leq c_2.$$

Since v is superharmonic, we obtain

$$\inf_{B_r} v = \inf_{\partial B_r} v,$$

and hence (3.169). □

The following lemma is obtained by the Harnack inequality for non-negative harmonic functions. Then $\inf_{\partial B_r} v$ in (3.169) is replaced by $\sup_{\partial B_r} v$.

Lemma 3.26 *There are $\beta \in (0, 1)$ and $c_3 > 0$ independent of $R > 0$, $R_0 > 0$, and $0 < R_0 \leq R/4$ such that*

$$-\Delta v = V(x)e^{v}, \; |V(x)| \leq C_2, \; v(x) + 2\log|x| \leq C_3 \quad \textit{in } B_R \setminus \overline{B_{R_0}}$$

implies

$$\sup_{\partial B_r} v \leq c_3 + \beta \inf_{\partial B_r} v + 2(\beta - 1)\log r, \quad 2R_0 \leq r \leq R/2. \tag{3.170}$$

Proof We take

$$\tilde{v}(x) = v(rx) + 2\log r, \quad r \in [2R_0, R/2],$$

to obtain

$$-\Delta\tilde{v} = V(rx)e^{\tilde{v}}$$
$$\tilde{v}(x) = v(rx) + 2\log(r\,|x|) - 2\log|x| \leq C_3 + 2\log 2$$

$$\left|V(rx)e^{\tilde{v}}\right| \leq C_2 \exp\left(C_3 + 2\log 2\right) \quad \text{in } B_2 \setminus \overline{B_{1/2}} \tag{3.171}$$

From the third inequality of (3.171) the solution $w = w(x)$ to

$$-\Delta w = V(rx)e^{\tilde{v}} \text{ in } B_2 \setminus \overline{B_{1/2}}, \quad w = 0 \text{ on } \partial(B_2 \setminus \overline{B_{1/2}})$$

satisfies

$$|w| \leq C_4 \quad \text{in } B_2 \setminus \overline{B_{1/2}}.$$

The second inequality of (3.171), on the other hand, ensures that

$$h = w - \tilde{v} + C_5, \quad C_5 = C_4 + C_3 + 2\log 2$$

is a non-negative harmonic function in $B_2 \setminus \overline{B_{1/2}}$. Then the Harnack inequality admits $\beta \in (0, 1)$ such that

$$\beta \sup_{\partial B_1} h \leq \inf_{\partial B_1} h. \tag{3.172}$$

Both sides of (3.172) are estimated by

$$C_5 + C_4 - \sup_{\partial B_1} \tilde{v} = C_5 + C_4 - 2\log r - \sup_{\partial B_r} v$$

$$\beta\left(C_5 - C_4 - \inf_{\partial B_1} \tilde{v}\right) = \beta\left(C_5 - C_4 - 2\log r - \inf_{\partial B_r} v\right).$$

Then we obtain

$$\sup_{\partial B_r} v \leq (1-\beta)C_5 + (1+\beta)C_4 + 2(\beta - 1)\log r + \beta \inf_{\partial B_r} v$$

and hence (3.170). □

Lemma 3.27 *Given $a, b > 0$, we have $\gamma > 0$ independent of $0 < R_0 \leq R/4$ such that*

$$\begin{aligned} -\Delta v = V(x)e^{v},\ a \leq V(x) \leq b \ &\text{ in } B_R \\ v(x) + 2\log|x| \leq C_6 \qquad &\text{ in } B_R \setminus \overline{B_{R_0}} \end{aligned} \tag{3.173}$$

implies

$$e^{v(x)} \leq C_7 e^{-\gamma v(0)} \cdot |x|^{-2(\gamma+1)}, \quad 2R_0 \leq |x| \leq R/2. \tag{3.174}$$

Proof By Lemmas 3.25 and 3.26, we obtain

$$\inf_{\partial B_r} v \le \frac{c_2}{c_1} - \frac{1}{c_1} v(0) - 2(1 + \frac{1}{c_1}) \log r, \quad 0 < r \le R$$

$$\sup_{\partial B_r} v \le (c_3 + \beta \cdot \frac{c_2}{c_1}) - \frac{\beta}{c_1} v(0) - 2(\frac{\beta}{c_1} + 1) \log r, \quad 2R_0 < r \le R/2$$

by (3.173). Then inequality (3.174) holds for

$$\gamma = \beta / c_1, \ C_7 = \exp(c_3 + \beta \cdot \frac{c_2}{c_1}).$$

□

Lemma 3.27 assures a sufficient condition for the residual vanishing (3.168). In fact, if

$$\sup_{x \in B \setminus B(x_k^0, r_k^0)} \left\{ v_k(x) + 2 \log \left| x - x_k^0 \right| \right\} \le C_8 \tag{3.175}$$

holds besides (3.166), then Lemma 3.27 is applicable. Hence we have

$$e^{v_k(x)} \le C_9 e^{-\gamma v_k(x_k)} \left| x - x_k^0 \right|^{-2(\gamma+1)} \quad \text{in } B_{R/2} \setminus B(x_k, 2r_k^0) \tag{3.176}$$

and then (3.176) implies

$$\begin{aligned} \int_{B_{R/2} \setminus B(x_k^0, 2r_k^0)} V_k(x) e^{v_k} &\le b \cdot C_9 \cdot (\delta_k^0)^{2\gamma} \cdot 2\pi \cdot \int_{2r_k^0}^{+\infty} r^{-2(\gamma+1)} r dr \\ &= \frac{\pi b C_9}{\gamma} (\delta_k^0 / 2r_k^0)^{2\gamma} \to 0, \end{aligned} \tag{3.177}$$

which means that (3.168) holds.

If (3.175) is not the case, there is $\{x_k^1\} \subset \overline{B}$ such that

$$\sup_{x \in B \setminus B(x_k^0, r_k^0)} \left\{ v_k(x) + 2 \log \left| x - x_k^0 \right| \right\} = v_k(x_k^1) + 2 \log \left| x_k^1 - x_k^0 \right| \to +\infty,$$

which implies

$$v_k(x_k^1) \to +\infty, \quad x_k^1 \to 0, \quad \sigma_k^1 = d_k / \delta_k^1 \to +\infty$$

for

$$d_k = |x_k^1 - x_k^0|, \quad \delta_k^1 = e^{-v_k(x_k^1)/2}.$$

Given $|x| \le \sigma_k^1/2$, we have

$$\left|\delta_k^1 x + x_k^1 - x_k^0\right| \ge \left|x_k^1 - x_k^0\right| - \delta_k^1 |x| \ge \frac{1}{2}\left|x_k^1 - x_k^0\right|$$

and therefore,

$$\begin{aligned}
\tilde{v}_k^1(x) &\equiv v_k(\delta_k^1 x + x_k^1) + 2\log \delta_k^1 \\
&\le v_k(x_k^1) + 2\log\left|x_k^1 - x_k^0\right| - 2\log\left|\delta_k^1 x + x_k^1 - x_k^0\right| + 2\log\delta_k^1 \\
&\le v_k(x_k^1) + 2\log\delta_k^1 + 2\log\left|x_k^1 - x_k^0\right| - 2\log\frac{1}{2}\left|x_k^1 - x_k^0\right| \\
&= 2\log 2.
\end{aligned}$$

We thus obtain

$$-\Delta\tilde{v}_k^1 = V_k(\delta_k^1 x + x_k^1)e^{\tilde{v}_k^1}, \ \tilde{v}_k^1 \le 2\log 2 \ \text{ in } B_{\sigma_k^1/2}, \quad \tilde{v}_k^1(0) = 0,$$

and passing to a subsequence, there arises $\tilde{v}_k^1 \to \tilde{v}^1$ in $C^{1,\alpha}_{loc}(\mathbf{R}^2)$ with $\tilde{v}^1 = \tilde{v}^1(x)$ satisfying

$$\tilde{v}^1(x) = \log\frac{a^2}{(1+\mu^2 a^2|x-\overline{x}|^2)^2}, \ \tilde{v}^1(x) \le 2\log 2, \ \ x \in \mathbf{R}^2, \quad \tilde{v}^1(0) = 0$$

for some $\mu, a > 0$ and $\overline{x} \in \mathbf{R}^2$. This convergence allows us to reformulate x_k^1 and δ_k^1, denoted by the same symbol, satisfying

$$v_k(x_k^1) = \|v_k\|_{L^\infty(B(x_k^1, 2r_k^1))} \to +\infty, \quad \delta_k^1 = e^{-v_k(x_k^1)/2} \to 0, \quad r_k^1/\delta_k^1 \to +\infty$$

for $r_k^1 = d_k/4$. Similarly to the above case, it follows that

$$\int_{B(x_k^1, 2r_k^1)} V_k(x)e^{v_k} \to 8\pi, \quad B(x_k^1, 2r_k^1) \cap B(x_k^0, 2r_k^0) = \emptyset.$$

We shall show $\alpha = 16\pi$, if

$$\sup\{v_k(x) + 2\log\min_{j=0,1}|x - x_k^j| \mid x \in B \setminus \bigcup_{j=0}^{1} B(x_k^j, r_k^j)\} < +\infty. \tag{3.178}$$

It suffices to prove

$$\int_{B(x_k^0,2d_k)} V_k(x)e^{v_k} \to 16\pi \tag{3.179}$$

because

$$\int_{B\setminus B(x_k^0,2d_k)} V_k(x)e^{v_k} \to 0$$

follows from (3.178) similarly. For this purpose, we take

$$\tilde{v}_k(x) = v_k(d_k x + x_k^0) + 2\log d_k,$$

to obtain

$$-\Delta\tilde{v}_k = V_k(d_k x + x_k^0)e^{\tilde{v}_k} \text{ in } d_k^{-1}\left(B - \left\{x_k^0\right\}\right).$$

Put, furthermore,

$$\tilde{x}_k^j = \frac{x_k^j - x_k^0}{d_k},\ \tilde{\delta}_k^j = e^{-\tilde{v}_k(\tilde{x}_k^j)/2} = \frac{\delta_k^j}{d_k},\ \tilde{r}_k^j = \frac{r_k^j}{d_k}, \quad j = 0, 1.$$

Then there arises

$$\frac{\tilde{r}_k^j}{\tilde{\delta}_k^j} = \frac{r_k^j}{\delta_k^j} \to +\infty, \quad B(\tilde{x}_k^0, 2\tilde{r}_k^0) \cap B(\tilde{x}_k^1, 2\tilde{r}_k^1) = \emptyset$$

$$\sup\{\tilde{v}_k(x) + 2\log \min_{j=0,1} |x - \tilde{x}_k^j| \mid x \in B_{R/d_k} \setminus \bigcup_{j=0}^{1} B(\tilde{x}_k^j, \tilde{r}_k^j)\} < +\infty$$

$$\int_{B(\tilde{x}_k^j,2\tilde{r}_k^j)} \tilde{V}_k(x)e^{\tilde{v}_k} \to 8\pi, \quad j = 0, 1 \tag{3.180}$$

for $\tilde{V}_k(x) = V_k(d_k x + x_k^0)$.

We obtain $\tilde{x}_k^0 = 0$ and $\left|\tilde{x}_k^1 - \tilde{x}_k^0\right| = 1$, and therefore, $\tilde{x}_k^1 \to \tilde{x}^1$ with $\left|\tilde{x}^1\right| = 1$, passing to a subsequence. The third relation of (3.180) and Theorem 3.28 now imply $\tilde{v}_k \to -\infty$ locally uniformly in $\mathbf{R}^2 \setminus \left\{0, \tilde{x}^1\right\}$. Therefore, if $\tilde{r}_k^1 \to \tilde{r}^1 > 0$, passing to a subsequence, then

$$\int_{B(\tilde{x}_k^1,1/2)} \tilde{V}_k(x)e^{\tilde{v}_k} \to 8\pi \tag{3.181}$$

and hence (3.179). If $\tilde{r}_k^1 \to 0$, we apply the scaling around $\tilde{x}_k^1$. It holds that (3.181) by the third relation of (3.180).

If (3.178) is not the case, we continue the process and obtain $x_k^2 \to 0$ and $r_k^2 \to 0$ satisfying

$$v_k(x_k^2) = \|v_k\|_{L^\infty(B(x_k^2, 2r_k^2))} \to +\infty, \quad r_k^2/\delta_k^2 \to 0, \quad B(x_k^i, 2r_k^i) \cap B(x_k^j, 2r_k^j) = \emptyset$$

for $0 \le i < j \le 2$, and

$$\int_{B(x_k^2, 2r_k^2)} V_k(x) e^{v_k} \to 8\pi, \quad \delta_k^2 = e^{-v_k(x_k^2)/2}.$$

To show $\alpha = 24\pi$ in the case of

$$\sup\{v_k(x) + 2\log \min_{0 \le j \le 2} |x - x_k^j| \mid x \in B \setminus \bigcup_{j=0}^{2} B(x_k^j, r_k^j)\} < +\infty,$$

we classify the rate

$$d_{i,j} = \left| x_k^i - x_k^j \right|$$

of concentration to the origin for $0 \le i < j \le 2$. First, we show the residual vanishing inside the ball containing $B(x_k^j, 2r_k^j)$ with a proportional rate. These balls are contained in a larger ball, where the residual vanishing occurs similarly. We repeat this procedure finitely many times, and obtain the conclusion.

In the following subsection we justify this process, using an induction.

3.4.7 Li–Shafrir Theorem Continued

Given (3.161), we continue the process in the previous subsection. The scaling limit around each center of the bubble is realized in the whole space $\mathbf{R}^2$ under

$$\lim_{k \to \infty} \sigma_k^j = +\infty$$

in (3.182), while such centers are exhausted if the last inequality of (3.183) arises. To complete the proof of Theorem 3.30, we begin with the following lemma.

Lemma 3.28 *Under the assumption of Theorem 3.30, there is a subsequence, denoted by the same symbol, such that any $m \in \mathbf{N}$ in $1 \le m \le V(0) \cdot C_1/(8\pi)$ admits*

$$x_k^j \in B_R, \ \lim_{k\to\infty} x_k^j = 0, \quad \sigma_k^j > 0, \ \lim_{k\to\infty} \sigma_k^j = +\infty, \quad 0 \le j \le m-1, \qquad (3.182)$$

and furthermore,

$$
\begin{aligned}
& v_k(x_k^j) = \max_{B(x_k^j, \sigma_k^j \delta_k^j)} v_k \to +\infty \\
& B(x_k^j, 2\sigma_k^j \delta_k^j) \cap B(x_k^i, 2\sigma_k^i \delta_k^i) = \emptyset, \quad i \neq j \\
& \frac{\partial}{\partial t} v_k(tx + x_k^j)\bigg|_{t=1} < 0, \qquad \delta_k^j \leq |x| \leq 2\sigma_k^j \delta_k^j \\
& \lim_k \int_{B(x_k^j, 2\sigma_k^j \delta_k^j)} V_k e^{v_k} = \lim_k \int_{B(x_k^j, \sigma_k^j \delta_k^j)} V_k e^{v_k} = 8\pi \\
& \max_{x \in \overline{B_R} \setminus \bigcup_{j=1}^{m-1} B(x_k^j, \sigma_k^j \delta_k^j)} \{v_k(x) + 2 \log \min_j |x - x_k^j|\} \leq C_1
\end{aligned} \tag{3.183}
$$

for $\delta_k^j = e^{-v_k(x_k^j)/2}$.

Proof This lemma for the case of $m = 1$ follows from the argument in the previous paragraph. By Lemma 3.24 there arises

$$
x_k^0 \in B_R, \ v_k(x_k^0) = \max_{\overline{B_R}} v_k, \ x_k^0 \to 0, \ v_k(x_k^0) \to +\infty
$$

and

$$
\tilde{v}_k^0 \to \tilde{v} \quad \text{in } C^{1,\alpha}_{loc}(\mathbf{R}^2), \quad 0 < \alpha < 1
$$

for

$$
\tilde{v}_k^0(x) = v_k(\delta_k^0 x + x_k^0) + 2 \log \delta_k^0, \quad \tilde{v}(x) = \log \frac{1}{(1 + \frac{V(0)}{8}|x|^2)^2},
$$

and furthermore,

$$
\sigma_k^0 = r_k^0 / \delta_k^0 \to +\infty, \quad \lim_{k \to \infty} \left\| \tilde{v}_k^0 - \tilde{v} \right\|_{C^{1,\alpha}(B_{2\sigma_k^0})} = 0.
$$

Then it follows that

$$
\begin{aligned}
& \int_{B(x_k^0, 2\sigma_k^0 \delta_k^0)} V_k e^{v_k} = \int_{B_{2\sigma_k^0}} V_k(\delta_k^0 \cdot + x_k^0) e^{\tilde{v}_k^0} \to 8\pi \\
& \int_{B(x_k^0, \sigma_k^0 \delta_k^0)} V_k e^{v_k} = \int_{B_{\sigma_k^0}} V_k(\delta_k^0 \cdot + x_k^0) e^{\tilde{v}_k^0} \to 8\pi
\end{aligned}
$$

$$\frac{\partial}{\partial t} v_k(tx + x_k^0)\Big|_{t=1} < 0, \quad \delta_k^0 \le |x| \le 2\sigma_k^0 \delta_k^0$$

which means the first, the third, and the fourth relations of (3.183) for $m = 1$.

Assume the first, the second, the third, and the fourth relations in (3.183) for $m = \ell$. The proof is done if the fifth relation in (3.183) holds. Assume the contrary,

$$M_k \equiv \max_{x \in \overline{B_R}} \{v_k(x) + 2\log \min_{0 \le j \le \ell-1} |x - x_k^j|\} \to +\infty \tag{3.184}$$

up to a subsequence. Since

$$\max_{\overline{B_R} \setminus B_r} v_k \to -\infty,$$

this M_k is attained at an interior point, denoted by $\overline{x}_k^\ell \in B_R$. It follows that $v_k(\overline{x}_k^\ell) \to +\infty$, which implies

$$\overline{x}_k^\ell \to 0, \quad \overline{\delta}_k^\ell \equiv e^{-v_k(\overline{x}_k^\ell)/2} \to 0.$$

Then (3.184) means

$$m_k \equiv \min_{0 \le j \le \ell-1} \frac{|\overline{x}_k^\ell - x_k^j|}{\overline{\delta}_k^\ell} \to +\infty. \tag{3.185}$$

For $|x| \le m_k/2$ it holds that

$$\min_{0 \le j \le \ell-1} |\overline{x}_k^\ell + \overline{\delta}_k^\ell x - x_k^j| \ge \min_{0 \le j \le \ell-1} |\overline{x}_k^\ell - x_k^j| - \overline{\delta}_k^\ell |x| \ge \frac{1}{2} \min_{0 \le j \le \ell-1} |\overline{x}_k^\ell - x_k^j|,$$

while

$$v_k(x) + 2\log \min_{0 \le j \le \ell-1} |x - x_k^j| \le v_k(\overline{x}_k^\ell) + 2\log \min_{0 \le j \le \ell-1} |\overline{x}_k^\ell - x_k^j|$$

implies

$$\begin{aligned}
\overline{v}_k(x) &\equiv v_k(\overline{\delta}_k^\ell x + \overline{x}_k^\ell) + 2\log \overline{\delta}_k^\ell \le v_k(\overline{x}_k^\ell) + 2\log \min_{0 \le j \le \ell-1} |\overline{x}_k^\ell - x_k^j| \\
&+ 2\log \overline{\delta}_k^\ell - 2\log \min_{0 \le j \le \ell-1} |\overline{\delta}_k^\ell x + \overline{x}_k^\ell - x_k^j| \le v_k(\overline{x}_k^\ell) + 2\log \overline{\delta}_k^\ell \\
&+ 2\log \min_{0 \le j \le \ell-1} |\overline{x}_k^\ell - x_k^j| - 2\log \frac{1}{2} \min_{0 \le j \le \ell-1} |\overline{x}_k^\ell - x_k^j| = 2\log 2.
\end{aligned}$$

Then the blowup argument as in the previous subsection ensures $\overline{\mu} > 0$ and $\overline{x} \in \mathbf{R}^2$ such that

$$\overline{v}_k \to \overline{v} \quad \text{in } C^{1,\alpha}_{loc}(\mathbf{R}^2)$$

up to a subsequence for

$$\overline{v}(x) = \log \frac{\overline{\mu}^2}{(1 + a^2\overline{\mu}^2|x - \overline{x}|^2)^2},$$

satisfying

$$\overline{v} \leq 2\log 2 \ \text{ in } \mathbf{R}^2, \quad \overline{v}(0) = 0, \quad a = (V(0)/8)^{1/2}.$$

Hence it holds that

$$1 \leq \overline{\mu} \leq 2, \quad |\overline{x}| \leq 1/(2a). \tag{3.186}$$

Let

$$L = \max_{|x|\leq 1/a} \overline{v}(x) - \min_{|x|\leq 1/a} \overline{v}(x). \tag{3.187}$$

Given $\sigma_k^\ell \to +\infty$, there is a subsequence satisfying $\|\overline{v}_k - \overline{v}\|_{C^{1,\alpha}(B_{4\sigma_k^\ell})} \to 0$ by a diagonal argument, which implies

$$\frac{\partial}{\partial t}\overline{v}_k(ty + \overline{x})\Big|_{t=1} < 0, \quad \frac{1}{2} \leq |y| \leq 4\sigma_k^\ell,$$

and therefore, there is $y_k^\ell \in B_1$ such that

$$\overline{v}_k(y_k^\ell + \overline{x}) = \max_{y \in \overline{B_{4\sigma_k^\ell}}} \overline{v}_k(y + \overline{x}).$$

In particular, we obtain

$$x_k^\ell \equiv \overline{\delta}_k^\ell(y_k^\ell + \overline{x}) + \overline{x}_k^\ell \to 0, \quad v_k(\overline{x}_k^\ell) \leq v_k(x_k^\ell) \leq v_k(\overline{x}_k^\ell) + L$$

by (3.186) and (3.187), which implies

$$\delta_k^\ell = e^{-v_k(x_k^\ell)/2} \leq \overline{\delta}_k^\ell \leq e^{L/2} \cdot \delta_k^\ell, \quad v_k(x_k^\ell) = \max_{|x - x_k^\ell| \leq 4\sigma_k^\ell \delta_k^\ell} v_k(x) \to +\infty.$$

Hence the scaling

$$\tilde{v}_k^\ell(x) = v_k(\delta_k^\ell x + x_k^\ell) + 2\log\delta_k^\ell, \quad \delta_k^\ell = e^{-v_k(x_k^\ell)/2} \to 0$$

is available, and then Theorems 3.28 and 3.32 imply

$$\int_{B(x_k^\ell, 2\sigma_k^\ell\delta_k^\ell)} V_k e^{v_k} = \int_{B_{2\sigma_k^\ell}} V_k(\delta_k^\ell \cdot + x_k^\ell) e^{\tilde{v}_k^\ell} \to 8\pi$$

$$\int_{B(x_k^\ell, \sigma_k^\ell\delta_k^\ell)} V_k e^{v_k} = \int_{B_{\sigma_k^\ell}} V_k(\delta_k^\ell \cdot + x_k^\ell) e^{\tilde{v}_k^\ell} \to 8\pi$$

$$\frac{\partial}{\partial t} v_k(tx + x_k^\ell)\bigg|_{t=1} < 0, \quad \delta_k^\ell \le |x| \le 2\sigma_k^\ell\delta_k^\ell.$$

By this property and (3.185), we obtain the first, the second, the third, and the fourth relations of (3.183) for $m = \ell + 1$. By the last inequality of (3.161), this process ends in finitely many times with the fifth relation of (3.183), to conclude the proof. □

Theorem 3.30 is now reduced to the *residual vanishing*,

$$\lim_{k\to\infty} \int_{B_R \setminus \cup_{\ell=1}^m B(x_k^\ell, \sigma_k^\ell\delta_k^\ell)} V_k e^{v_k} = 0$$

by Lemma 3.28, where the induction is to be set up according to the order of concentration rates. Hence Theorem 3.30 is a consequence of Lemmas 3.28 and 3.29 below.

Lemma 3.29 *Under the assumption of Theorem 3.30, if $m \ge 1$, $x_k^j \in B_R$, $r_k^j > 0$ appear for $k = 1, 2, \cdots$ and $0 \le j \le m-1$, satisfying*

$$x_k^j \to 0, \quad v_k(x_k^j) \to +\infty, \quad \lim_{k\to\infty} \frac{r_k^j}{\delta_k^j} = +\infty$$

$$B(x_k^j, r_k^j) \cap B(x_k^i, r_k^i) = \emptyset, \quad i \ne j$$

$$\max_{x \in \overline{B_R} \setminus \cup_{j=0}^{m-1} B(x_k^j, r_k^j)} \{v_k(x) + 2\log \min_{0 \le j \le m-1} |x - x_k^j|\} \le C_2$$

$$\lim_{k\to\infty} \int_{B(x_k^j, 2r_k^j)} V_k e^{v_k} = \lim_{k\to\infty} \int_{B(x_k^j, r_k^j)} V_k e^{v_k} = \beta_j \ge 4\pi \tag{3.188}$$

for $\delta_k^j = e^{-v_k(x_k^j)/2}$, it holds that

$$\lim_{k\to\infty}\int_{B_R} V_k e^{v_k} = \sum_{j=0}^{m-1}\beta_j.$$

Proof If $m = 1$ we may assume a subsequence such that

$$x_k^0 = 0, \quad \lim_{k\to\infty} r_k^0 = 0. \tag{3.189}$$

In fact, from the assumption of Theorem 3.30, this lemma is obvious if the second equality of (3.189) does not hold.

Based on (3.189) and the third relation of (3.188), apply Lemma 3.27 to obtain

$$e^{v_k(x)} \le C_3\cdot(\delta_k^0)^{2\gamma}\cdot|x|^{-2(\gamma+1)}, \quad 2r_k^0 \le |x| \le R/2.$$

Hence the first relation of (3.188) ensures

$$\int_{B_{R/2}\setminus B_{2r_k^0}} V_k e^{v_k} \le b\cdot C_2\cdot(\delta_k^0)^{2\gamma}\cdot 2\pi\cdot\int_{2r_k^0}^{\infty} r^{-2(\gamma+1)}\cdot r dr = C_4(\frac{\delta_k^0}{2r_k^0})^{2\gamma} \to 0$$

and then (3.188) implies

$$\lim_{k\to\infty}\int_{B_R} V_k e^{v_k} = \beta_0.$$

Let $m \ge 2$ and assume the assertion for $1, \cdots, m-1$. It may be supposed that

$$x_k^0 = 0, \quad d_k \equiv |x_k^1 - x_k^0| = \min\{|x_k^i - x_k^j| \mid i \ne j\}.$$

First, we take the case

$$|x_k^i - x_k^j| \le Ad_k, \quad 0 \le i, j \le m-1 \tag{3.190}$$

for some $A \ge 1$, to deduce

$$\lim_{k\to\infty}\int_{B_{4Ad_k}} V_k e^{v_k} = \lim_{k\to\infty}\int_{B_{2Ad_k}} V_k e^{v_k} = \sum_{j=0}^{m-1}\beta_j. \tag{3.191}$$

In fact, by (3.190) this lemma is applicable to

$$m = 1, \quad (r_k^0)' = 2Ad_k, \quad (\beta_0)' = \sum_{j=0}^{m-1}\beta_j, \quad (x_k^0)' = 0,$$

and then equality (3.191) implies

$$\lim_{k\to\infty}\int_{B_R} V_k e^{v_k} = \sum_{j=0}^{m-1}\beta_j.$$

To confirm (3.191), we use

$$\tilde{v}_k(x) = v_k(d_k x) + 2\log d_k,\ \tilde{V}_k(x) = V_k(d_k x), \quad |x| \le R/d_k.$$

Put

$$\tilde{x}_k^j = x_k^j/d_k, \quad \tilde{\delta}_k^j = e^{-\tilde{v}_k(\tilde{x}_k^j)/2} = \frac{\delta_k^j}{d_k}, \quad \tilde{r}_k^j = \frac{r_k^j}{d_k}$$

for $0 \le j \le m-1$. From the assumption, we obtain

$$\tilde{x}_k^0 = 0, \quad \lim_{k\to\infty} d_k = 0, \quad \frac{\tilde{r}_k^j}{\tilde{\delta}_k^j} = \frac{r_k^j}{\delta_k^j} \to +\infty$$

$$B(\tilde{x}_k^i, \tilde{r}_k^i) \cap B(\tilde{x}_k^j, \tilde{r}_k^j) = \emptyset, \quad i \neq j$$

$$\max_{x\in \overline{B_{R/d_k}}\setminus \cup_{j=0}^{m-1} B(\tilde{x}_k^j, \tilde{r}_k^j)} \{\tilde{v}_k(x) + 2\log \min_{0\le j\le m-1} |x - \tilde{x}_k^j|\} \le C_5$$

$$\lim_{k\to\infty}\int_{B(\tilde{x}_k^j, 2\tilde{r}_k^j)} \tilde{V}_k e^{\tilde{v}_k} = \lim_{k\to\infty}\int_{B(\tilde{x}_k^j, \tilde{r}_k^j)} \tilde{V}_k e^{\tilde{v}_k} = \beta_j.$$

We assume also $\lim_{k\to\infty} \tilde{x}_k^j = \tilde{x}^j$ for $0 \le j \le m-1$, regarding $|\tilde{x}_k^j| \le A$.

We obtain

$$\tilde{v}_k \to -\infty \quad \text{locally uniformly in } \mathbf{R}^2 \setminus \bigcup_{j=0}^{m-1}\{\tilde{x}^j\}$$

$$1 \le |\tilde{x}^i - \tilde{x}^j| \le A, \quad i \neq j \tag{3.192}$$

by Theorem 3.35. If $\lim_{k\to\infty} \tilde{r}_k^j > 0$ up to a subsequence, (3.192) implies

$$\lim_{k\to\infty}\int_{B_{1/2}(\tilde{x}^j)} \tilde{V}_k e^{\tilde{v}_k} = \beta_j. \tag{3.193}$$

There arises $\lim_{k\to\infty} \tilde{r}_k^j = 0$, otherwise, and hence the lemma is applicable with $m = 1$. We then obtain (3.193), again, and then (3.192)–(3.193) imply

$$\lim_{k\to\infty}\int_{B_{4A}} \tilde{V}_k e^{\tilde{v}_k} = \lim_{k\to\infty}\int_{B_{2A}} \tilde{V}_k e^{\tilde{v}_k} = \sum_{j=0}^{m-1}\beta_j,$$

and hence (3.191).

Second, assume the contrary to (3.190), and obtain $\emptyset \neq J_\ell \subset \{0, 1, \ldots, \ell - 1\}$ for $2 \leq \ell \leq m - 1$ such that

$$|x_k^j - x_k^i| \leq Ad_k, \quad i, j \in J_\ell$$
$$\lim_{k\to\infty} \frac{\left|x_k^i - x_k^j\right|}{d_k} = +\infty, \quad i \in J_\ell, \ j \notin J_\ell.$$

Each $0 \leq j \leq \ell - 1$ admits

$$\lim_{k\to\infty} \int_{B_{2Ad_k}} V_k e^{v_k} = \lim_{k\to\infty} \int_{B_{Ad_k}} V_k e^{v_k} = \sum_{j\in J_\ell} \beta_j,$$

similarly. To reduce this lemma for the case of $m - \ell + 1 \leq m - 1$, let

$$(x_k^0)' = x_k^0 = 0, \quad (r_k^0)' = 2Ad_k, \quad \beta_0' = \sum_{j\in J_\ell} \beta_j.$$

Here we justify

$$B_{Ad_k} \cap B(x_k^j, r_k^j) = \emptyset, \quad \ell \leq j \leq m - 1. \tag{3.194}$$

If (3.194) does not hold, there arises $\ell \leq j \leq m - 1$ such that

$$Ad_k + r_k^j \geq |x_k^j| > d_k + r_k^j,$$

which implies

$$\lim_{k\to\infty} \frac{r_k^j}{d_k} = +\infty, \quad \lim_{k\to\infty} \frac{r_k^j}{|x_k^j|} = 1,$$

and hence $B_{Ad_k} \subset B(x_k^j, 2r_k^j)$ for $k \gg 1$. This property, however, contradicts (3.188) by $\beta_0 > 0$.

We thus end up with

$$\lim_{k\to\infty} \int_{B_R} V_k e^{v_k} = \beta_0' + \sum_{j=\ell}^{m-1} \beta_j = \sum_{j=0}^{m-1} \beta_j$$

and the conclusion of the lemma for any m. □

3.4.8 Y.Y. Li Estimate

Due to the lack of the boundary condition, the location of blowup points is not prescribed by Green's function in Theorem 3.30, differently from Theorem 2.7. The other difference is the possible collision of bubbles [9] indicated by $\alpha = 8\pi m$, $m \geq 2$, which is excluded if v_k is controlled near the blowup point [21].

In Theorem 3.31, the additional inequality (3.76) below takes a role of boundary condition, which prevents the collision of bubbles. Then the behavior of $v_k = v_k(x)$ is locally uniformly controlled by $v_k(0)$. This subsection is thus devoted to the proof of Theorem 3.31 via the method of [24].

Let $R = 1$ in Theorem 3.31: $B = B(0,1) \subset \mathbf{R}^2$. We may assume $V = V(x) > 0$, $x \in \overline{B}$ by Lemma 3.23. By the assumption of this theorem it holds that (3.76) besides the assumption of Theorem 3.30. Then we show $\alpha = 8\pi$ and

$$\left| v_k(x) - \log \frac{e^{v_k(0)}}{(1 + \frac{V_k(0)}{8} e^{v_k(0)} |x|^2)^2} \right| \leq C, \quad x \in B \tag{3.195}$$

for $k = 1, 2, \ldots$.

Letting $m_k = \inf_{\partial B} v_k$, we obtain

$$-\Delta(v_k - m_k) = V_k(x) e^{v_k} \geq 0 \text{ in } B, \quad 0 \leq v_k - m_k \leq C \text{ on } \partial B \tag{3.196}$$

by the first inequality of (3.76), and therefore,

$$v_k \geq m_k \quad \text{in } B \tag{3.197}$$

from the maximum principle. Then the assumption

$$\lim_{k \to \infty} \max_{\overline{B} \setminus B_r} v_k = -\infty, \quad 0 < r < 1 \tag{3.198}$$

implies

$$\lim_{k \to \infty} m_k = -\infty.$$

Let $A_r = B \setminus \overline{B_r}$ and define $w_k = w_k(x)$ by

$$-\Delta w_k = V_k(x) e^{v_k} \text{ in } A_r, \quad w_k = 0 \text{ on } \partial A_r.$$

It holds that

$$\|w_k\|_\infty = O(1).$$

We obtain also

$$-\Delta(v_k - m_k) = o(1) \text{ in } A_r, \quad 0 \le v_k - m_k \le C(r) \text{ on } \partial A_r$$

by (3.197), (3.198), and the Harnack inequality for non-negative harmonic functions. Since

$$\lim_{k\to\infty} \max_{\overline{B}} v_k = +\infty, \quad \lim_{k\to\infty} \int_B V_k(x)e^{v_k}\, dx = \alpha$$

there is a harmonic function $\psi_0 = \psi_0(x)$ on $\overline{B}$ satisfying

$$\begin{aligned} & v_k(x) - m_k \to \frac{\alpha}{2\pi} \log \frac{1}{|x|} + \psi_0(x) \quad \text{locally uniform in } \overline{B} \setminus \{0\} \\ & -\Delta v_k = V_k(x)e^{v_k} \rightharpoonup \alpha\delta_0 \quad \text{in } \mathcal{M}(\overline{B}). \end{aligned} \tag{3.199}$$

Control of the behavior of the rescaled solution far away from the origin is essential in Theorem 3.31. Here we apply the method of [24] based on the fact that the collapse mass $\alpha = 8\pi$ and the total mass 8π of the entire solution in Theorem 3.32 coincide, differently from the original proof [21] using the moving planes method.

Here we use the Pohozaev identity below the proof of which is left to the reader.

Lemma 3.30 *Let $V = V(x)$ be a C^1 function and suppose*

$$-\Delta v = V(x)e^v \text{ in } B = B(0,1) \subset \mathbf{R}^2.$$

Then each $0 < r \le 1$ admits

$$\int_{B_r} (2V(x)+x\cdot\nabla V(x))e^v\, dx = r\int_{\partial B_r} V(x)e^v - \frac{1}{2}|\nabla v|^2 + \left(\frac{\partial v}{\partial \nu}\right)^2\, d\sigma \tag{3.200}$$

for $B_r = B(0,r)$.

Lemma 3.31 *It holds that $\alpha = 8\pi$ under the assumption of Theorem 3.31.*

Proof It holds that

$$\begin{aligned} & \int_B (2V_k(x) + x\cdot\nabla V_k(x))e^{v_k}\, dx \\ & = \int_{\partial B} V_k(x)e^{v_k} - \frac{1}{2}|\nabla v_k|^2 + \left(\frac{\partial v_k}{\partial \nu}\right)^2\, d\sigma \end{aligned} \tag{3.201}$$

by (3.200), where (3.199) is applicable. We use the second relation of (3.199) and

$$\lim_{k\to\infty} V_k = V \quad \text{in } C(\overline{B}), \qquad V = V(x) > 0 \tag{3.202}$$

for the left-hand side of (3.201). Concerning the right-hand side, the first relation of (3.199) and $\Delta\psi_0 = 0$ on $\overline{B}$ are available. It holds that

$$2\alpha = \frac{1}{2}\cdot\left(\frac{\alpha}{2\pi}\right)^2\cdot 2\pi$$

and hence $\alpha = 8\pi$ by $\alpha > 0$. The proof is complete. □

Recall the argument for the proof of Theorem 3.29 in Sect. 3.4.6. From the assumption there is $x_k \in \overline{B}$ such that

$$v_k(x_k) = \max_{\overline{B}} v_k \to +\infty, \qquad \lim_{k\to\infty} x_k = 0.$$

Take $V_k(x)$ for $V_k(x + x_k)$, and assume $x_k = 0$. Putting

$$u_k(x) = v_k(\delta_k x) + 2\log\delta_k, \quad \delta_k = e^{-v_k(0)/2}, \quad \tilde{V}_k(x) = V_k(\delta_k x), \tag{3.203}$$

furthermore, we obtain

$$-\Delta u_k = \tilde{V}_k(x)e^{u_k}, \ u_k \le u_k(0) = 0, \quad |x| < \delta_k^{-1}.$$

Then Theorem 3.28 assures a subsequence satisfying

$$u_k \to u = u(x) \quad \text{locally uniformly in } \mathbf{R}^2. \tag{3.204}$$

Similarly to (3.164), it follows that

$$-\Delta u = ce^u, \ u \le u(0) = 0 \ \text{ in } \mathbf{R}^2, \qquad \int_{\mathbf{R}^2} e^u < +\infty$$

for $c = V(0) > 0$, which implies

$$\int_{\mathbf{R}^2} ce^u = 8\pi, \quad u(x) = \log\frac{1}{(1+\frac{c}{8}|x|^2)^2} \tag{3.205}$$

in accordance with (3.165).

We thus obtain

$$\beta_k \equiv \int_B V_k(x)e^{v_k} = \int_{|x|<\delta_k^{-1}} \tilde{V}_k(x)e^{u_k} \to 8\pi \tag{3.206}$$

by Lemma 3.31. It holds also that

$$\tilde{V}_k(x)e^{u_k} \to ce^u \quad \text{locally uniformly in } \mathbf{R}^2 \tag{3.207}$$

by (3.202) and (3.204). Then, (3.207), (3.205), and (3.206) imply that any $\varepsilon > 0$ admits $R > 0$ and k_0 such that

$$\frac{1}{2\pi}\int_{R<|x|<\delta_k^{-1}} \tilde{V}_k(x)e^{u_k} < \varepsilon/2, \quad k \geq k_0. \tag{3.208}$$

We have, on the other hand,

$$v_k(x) - m_k = \frac{1}{2\pi}\int_B \log\frac{1}{|x-y|} \cdot V_k(y)e^{v_k(y)} + O(1)$$

by (3.196), and hence

$$\begin{aligned} u_k(x) &= m_k + 2\log\delta_k + \frac{1}{2\pi}\int_B \log\frac{1}{|\delta_k x - y|} \cdot V_k(y)e^{v_k(y)} + O(1) \\ &= m_k + 2\log\delta_k + \frac{1}{2\pi}\int_{|y|<\delta_k^{-1}} \{\log\frac{1}{|x-y|} + \log\frac{1}{\delta_k}\} \cdot \tilde{V}_k(y)e^{u_k(y)} + O(1) \\ &= m_k + (2 - \frac{\beta_k}{2\pi})\log\delta_k + O(1) + \frac{1}{2\pi}\int_{|y|<\delta_k^{-1}} \log\frac{1}{|x-y|} \cdot \tilde{V}_k(y)e^{u_k(y)} \end{aligned} \tag{3.209}$$

uniformly in $x \in B_{\delta_k^{-1}}$. In particular, we obtain

$$0 = u_k(0) = m_k + (2 - \frac{\beta_k}{2\pi})\log\delta_k + O(1) + \frac{1}{2\pi}\int_{|y|<\delta_k^{-1}} \log\frac{1}{|y|} \cdot \tilde{V}_k(y)e^{u_k(y)} \tag{3.210}$$

and then (3.209)–(3.210) implies

$$u_k(x) = \frac{1}{2\pi}\int_{|y|<\delta_k^{-1}} \log\frac{|y|}{|x-y|} \cdot \tilde{V}_k(y)e^{u_k(y)} + O(1)$$
$$\text{uniformly in } x \in B_{\delta_k^{-1}}. \tag{3.211}$$

We use (3.211) to control u_k far away from the origin. Note that $u_k(x)$ in (3.203) is defined for $x \in B_{\delta_k^{-1}}$.

Lemma 3.32 *Each $\varepsilon > 0$ admits $R > 0$ and k_0 such that*

$$u_k(x) \leq (\frac{\beta_k}{2\pi} - 3\varepsilon)\log\frac{1}{|x|} + O(1), \quad 2R < |x| < \delta_k^{-1}, \ k \geq k_0. \tag{3.212}$$

Proof First, we have $R \gg 1$ and k_0 such that

$$\frac{1}{2\pi}\int_{|y|<R}\tilde{V}_k(y)e^{u_k(y)} \geq \frac{\beta_k}{2\pi} - \varepsilon, \quad k \geq k_0 \tag{3.213}$$

by (3.206) and (3.208). Second, there arises

$$y \notin B(x, |x|/2) \quad \Rightarrow \quad \frac{|y|}{|x-y|} \leq C \tag{3.214}$$

in (3.211). In fact, we have

$$z_0 = \frac{x}{|x|}, \ z = \frac{y}{|x|}, \ y \notin B(x, |x|/2) \quad \Rightarrow \quad |z_0| = 1, \ |z - z_0| > \frac{1}{2},$$

which implies

$$\frac{|z|}{|z-z_0|} \leq \frac{|z-z_0| + |z_0|}{|z-z_0|} < 1 + 2 = 3$$

and hence (3.214).

Now we divide the integration on the right-hand side of (3.211) into B_R,

$$B(x, |x|/2),$$

and the other. Since (3.206) and (3.214) it follows that

$$\begin{aligned} u_k(x) \leq & \frac{1}{2\pi}\int_{|y|<R} \log\frac{|y|}{|x-y|}\tilde{V}_k(y)e^{u_k(y)} \\ & + \frac{1}{2\pi}\int_{B(x,|x|/2)} \log\frac{|y|}{|x-y|} \cdot \tilde{V}_k(y)e^{u_k(y)} + O(1). \end{aligned} \tag{3.215}$$

The main part of (3.212), converging to $-\infty$ as $|x| \to \infty$, is taken from the first term on the right-hand side of (3.215).

To treat the second term in advance, we note

$$2R > \sqrt{2}, \ |x| > 2R \quad \Rightarrow \quad B(x, |x|^{-1}) \subset B(x, |x|/2)$$

to divide this integral into $B(x, |x|^{-1})$ and $B(x, |x|/2) \setminus B(x, |x|^{-1})$. Inequality (3.208) is available as for the latter because of

$$|x|^{-1} < |x-y| < |x|/2 \quad \Rightarrow \quad \frac{1}{|x-y|} < |x|, \ |y| < \frac{3}{2}|x|.$$

For $|x| > 2R$ and $k \geq k_0$, we thus obtain

$$
\begin{aligned}
&\frac{1}{2\pi}\int_{B(x,|x|/2)\setminus B(x,|x|^{-1})} \log\frac{|y|}{|x-y|}\cdot \tilde{V}_k(y)e^{u_k(y)} \\
&\quad \leq \frac{1}{2\pi}\int_{B(x,|x|/2)} \left\{2\log|x| + \log\frac{3}{2}\right\} \tilde{V}_k(y)e^{u_k(y)} \\
&\quad < \varepsilon \log|x| + O(1).
\end{aligned} \tag{3.216}
$$

Concerning the former, we use

$$
|x-y| < |x|^{-1} \quad \Rightarrow \quad |y| < |x| + |x|^{-1} < |x| + (2R)^{-1}, \ |x| > 2R,
$$

to deduce

$$
\log|y| < \log|x| + C, \quad |x| > 2R. \tag{3.217}
$$

By (3.217), (3.208), and $u_k \leq 0$, it holds that

$$
\begin{aligned}
&\frac{1}{2\pi}\int_{B(x,|x|^{-1})} \log\frac{|y|}{|x-y|}\cdot \tilde{V}_k(y)e^{u_k(y)} \\
&= \frac{1}{2\pi}\int_{B(x,|x|^{-1})} \log\frac{1}{|x-y|}\cdot \tilde{V}_k(y)e^{u_k(y)} + \frac{1}{2\pi}\int_{B(x,|x|^{-1})} \log|y|\cdot \tilde{V}_k(y)e^{u_k(y)} \\
&\leq \frac{C}{2\pi}\int_{B(x,|x|^{-1})} \left|\log\frac{1}{|x-y|}\right| + \frac{\varepsilon}{2}\{\log|x| + C\} \\
&\leq C'\int_0^{|x|^{-1}} r|\log r|\,dr + \frac{\varepsilon}{2}\{\log|x| + C\} \\
&\leq \varepsilon\log|x| + C''
\end{aligned} \tag{3.218}
$$

for $2R < |x| < \delta_k^{-1}$ with $R \gg 1$. Combining (3.216) and (3.218), we obtain

$$
\frac{1}{2\pi}\int_{B(x,|x|/2)} \log\frac{|y|}{|x-y|}\cdot \tilde{V}_k(y)e^{u_k(y)} \leq 2\varepsilon\log|x| + C_{13}. \tag{3.219}
$$

Turning to the main part, the second term on the right-hand side of (3.215), we use

$$
|y| < R, \ |x| > 2R \ \Rightarrow \ \frac{1}{|x|} < \frac{1}{2R}, \ \frac{1}{|y|} > \frac{1}{R} \ \Rightarrow \ \left|\frac{1}{|x|} - \frac{1}{|y|}\right| > \frac{1}{2R}
$$

and hence

$$
|y| < R, \ |x| > 2R \quad \Rightarrow \quad \frac{|x|\cdot|y|}{|x-y|} < 2R, \quad \log\frac{|y|}{|x-y|} < \log\frac{2R}{|x|}.
$$

By (3.213) it holds that

$$\frac{1}{2\pi}\int_{|y|<R}\log\frac{|y|}{|x-y|}\cdot\tilde{V}_k(y)e^{u_k(y)} \leq \frac{1}{2\pi}\log\frac{2R}{|x|}\int_{|y|<R}\tilde{V}_k(y)e^{u_k(y)}$$
$$< (\varepsilon - \frac{\beta_k}{2\pi})\log|x| + O(1) \qquad (3.220)$$

for $2R < |x| < \delta_k^{-1}$ and $k \geq k_0$. Then we obtain (3.212) by (3.219) and (3.220). □

Lemma 3.33 *It holds that*

$$\left|u_k(x) + \frac{\beta_k}{2\pi}\log|x|\right| \leq C, \quad \log\frac{1}{\delta_k} < |x| < \delta_k^{-1} \qquad (3.221)$$

as $k \to \infty$.

Proof By (3.204), (3.205), (3.206), and (3.212), the dominated convergence theorem ensures

$$\int_{B_{\delta_k^{-1}}} (|y| + |\log|y||)\,\tilde{V}_k(y)e^{u_k(y)} \leq C. \qquad (3.222)$$

By Lemmas 3.31 and 3.32, on the other hand, each $0 < r_0 < 1$ admits

$$\int_{r_0\log\frac{1}{\delta_k}<|y|<\delta_k^{-1}} \tilde{V}_k(y)e^{u_k(y)} \leq C\int_{|y|>r_0\log\frac{1}{\delta_k}} |y|^{-\frac{\beta_k}{2\pi}+3\varepsilon}\,dy$$
$$= O((\log\frac{1}{\delta_k})^{-2+4\varepsilon}), \qquad (3.223)$$

which implies

$$|\tilde{\beta}_k(x) - \beta_k| \leq \int_{r_0|x|<|y|<\delta_k^{-1}} \tilde{V}_k(y)e^{u_k(y)} = O((\log\frac{1}{\delta_k})^{-2+4\varepsilon})$$
$$\log\frac{1}{\delta_k} < |x| < \delta_k^{-1}, \quad \tilde{\beta}_k(x) \equiv \int_{|y|<r_0|x|} \tilde{V}_k(y)e^{u_k(y)}. \qquad (3.224)$$

Given $z = x/|y|$, $z_0 = y/|y|$, and $c = r_0^{-1} > 1$, on the other hand, we have

$$|z| < c^{-1} = r_0 \quad \Rightarrow \quad 0 < 1 - r_0 < |z - z_0| < r_0 + 1,$$

and therefore,

$$\left|\int_{r_0|x|<|y|<\delta_k^{-1}} \log\frac{|y|}{|x-y|}\cdot\tilde{V}_k(y)e^{u_k(y)}\right| \leq \int_{r_0|x|<|y|<\delta_k^{-1}} |$$

$$\log \frac{1}{|\frac{x}{|y|}-\frac{y}{|y|}|}| \cdot \tilde{V}_k(y)e^{u_k(y)} \le \int_{c^{-1}<\left|\frac{x}{y}\right|<c,\ |y|<\delta_k^{-1}} |\log \frac{1}{|\frac{x}{|y|}-\frac{y}{|y|}|}|\tilde{V}_k(y)e^{u_k(y)}$$

$$+C\int_{r_0|x|<|y|<\delta_k^{-1}} \tilde{V}_k(y)e^{u_k(y)} \tag{3.225}$$

in (3.211). By (3.223), the second term on the right-hand side of (3.225) is estimated above by

$$\int_{r_0|x|<|y|<\delta_k^{-1}} \tilde{V}_k(y)e^{u_k(y)}\, dy = O((\log\frac{1}{\delta_k})^{-2+4\varepsilon}),\ \log\frac{1}{\delta_k} < |x| < \delta_k^{-1}. \tag{3.226}$$

Lemma 3.32, on the other hand, is applicable for the first term on the right-hand side of (3.225). Hence we have

$$\int_{c^{-1}<\left|\frac{x}{y}\right|<c,\ |y|<\delta_k^{-1}} |\log \frac{1}{|\frac{x}{|y|}-\frac{y}{|y|}|}|\tilde{V}_k(y)e^{u_k(y)}$$

$$\le C\int_{c^{-1}<|\frac{x}{y}|<c} |\log \frac{1}{|\frac{x}{|y|}-\frac{y}{|y|}|}||y|^{-4+4\varepsilon}\, dy. \tag{3.227}$$

On the right-hand side of (3.227), we use

$$\log \frac{1}{|\frac{x}{|y|}-\frac{y}{|y|}|} = \log\frac{|y|}{|x-y|} = \log\left|\frac{\frac{y}{|x|}}{\frac{x}{|x|}-\frac{y}{|x|}}\right|$$

to put $z = y/|x|$ and $z_0 = x/|x|$. Then this quantity is estimated above by

$$|x|^{-2+4\varepsilon}\int_{c^{-1}<|z|<c} \left|\log|\frac{z}{z-z_0}|\right| |z|^{-4+4\varepsilon}\, dz.$$

Since $|z_0| = 1$ we obtain

$$\int_{c^{-1}<|\frac{x}{y}|<c,\ |y|<\delta_k^{-1}} \left|\log \frac{1}{|\frac{x}{|y|}-\frac{y}{|y|}|}\right| \tilde{V}_k(y)e^{u_k(y)} = O((\log\frac{1}{\delta_k})^{-2+4\varepsilon}) \tag{3.228}$$

for $\log\frac{1}{\delta_k} < |x| < \delta_k^{-1}$. Inequalities (3.225), (3.226), and (3.228) imply

$$\int_{r_0|x|<|y|<\delta_k^{-1}} \log\frac{|y|}{|x-y|}\cdot\tilde{V}_k(y)e^{u_k(y)} = O((\log\frac{1}{\delta_k})^{-2+4\varepsilon}),\quad \log\frac{1}{\delta_k} < |x| < \delta_k^{-1},$$

and therefore,

$$u_k(x) = \frac{1}{2\pi}\int_{|y|<r_0|x|} \log\frac{|y|}{|x-y|}\cdot \tilde{V}_k(y)e^{u_k(y)} + O(1)$$
$$= -\frac{\tilde{\beta}_k(x)}{2\pi}\cdot \log|x| + \frac{1}{2\pi}\int_{|y|<r_0|x|} \log\left|\frac{|x-y|}{|x||y|}\right|^{-1}\cdot \tilde{V}_k(y)e^{u_k(y)} + O(1) \tag{3.229}$$

by (3.211).

Now (3.224) implies

$$\left|-\frac{\tilde{\beta}_k(x)}{2\pi}\log|x| + \frac{\beta_k}{2\pi}\log|x|\right| \le C|\log\frac{1}{\delta_k}|^{-1+4\varepsilon} = O(1) \tag{3.230}$$

for $\log\frac{1}{\delta_k} < |x| < \delta_k^{-1}$, while we obtain

$$|y| < r_0|x| \quad \Rightarrow \quad \frac{1-r_0}{|y|} < \frac{1}{|y|} - \frac{1}{|x|} \le \frac{|x-y|}{|x||y|} = \frac{1}{|y|} + \frac{1}{|x|} < \frac{1+r_0}{|y|}$$

by $0 < r_0 < 1$. From (3.206) and (3.222) it follows that

$$\frac{1}{2\pi}\int_{|y|<r_0|x|} \left|\log\left|\frac{|x-y|}{|x||y|}\right|^{-1}\right| \tilde{V}_k(y)e^{u_k(y)}$$
$$\le \frac{1}{2\pi}\int_{|y|<\delta_k^{-1}} \left(|\log|y|| + \log\max\left\{(1-r_0)^{-1}, (1+r_0)^{-1}\right\}\right) \tilde{V}_k(y)e^{u_k(y)}$$
$$\le C, \quad x\in B_{\delta_k^{-1}}. \tag{3.231}$$

Then (3.229), (3.230), and (3.231) imply

$$u_k(x) = \frac{\beta_k}{2\pi}\log\frac{1}{|x|} + O(1) \quad \text{uniformly in } \log\frac{1}{\delta_k} < |x| < \delta_k^{-1},$$

and hence (3.221). □

Lemma 3.34 *There is $R \gg 1$ such that*

$$\left|\nabla u_k(x) + \frac{\beta_k}{2\pi}\frac{x}{|x|^2}\right| \le C\left\{\delta_k + |x|^{-2}\right\}, \quad R < |x| < \delta_k^{-1} \tag{3.232}$$

as $k \to \infty$.

Proof By (3.76) we have

$$\nabla v_k(x) = -\frac{1}{2\pi}\int_{|y|<1} \frac{x-y}{|x-y|^2}\cdot V_k(y)e^{v_k(y)} + O(1)$$

uniformly in $x \in \overline{B}$, which means

$$\begin{aligned}\nabla u_k(x) &= \delta_k \nabla v_k(\delta_k x)\\ &= -\frac{\delta_k}{2\pi}\int_{|y|<1} \frac{\delta_k x - y}{|\delta_k x - y|^2}\cdot V_k(y)e^{v_k(y)} + O(\delta_k)\\ &= -\frac{1}{2\pi}\int_{|y|<\delta_k^{-1}} \frac{x-y}{|x-y|^2}\cdot \tilde{V}_k(y)e^{u_k(y)} + O(\delta_k)\end{aligned}$$

uniformly in $x \in B_{\delta_k^{-1}}$ by (3.203). Hence we obtain

$$\nabla u_k(x) + \frac{\beta_k}{2\pi}\cdot\frac{x}{|x|^2} = \frac{1}{2\pi}\int_{|y|<\delta_k^{-1}} \left(\frac{x}{|x|^2} - \frac{x-y}{|x-y|^2}\right)\cdot \tilde{V}_k(y)e^{u_k(y)} + O(\delta_k).$$

Since

$$|y-x| > \frac{|x|}{2},\ 0<\theta<1 \quad\Rightarrow\quad \left|\frac{\partial}{\partial\theta}\frac{x-\theta y}{|x-\theta y|^2}\right| \le \frac{2|x|\,|y|}{|x-\theta y|^3} \le \frac{4|y|}{|x|^2}$$

it holds that

$$|y-x| > \frac{|x|}{2} \quad\Rightarrow\quad \left|\frac{x}{|x|^2} - \frac{x-y}{|x-y|^2}\right| \le \frac{4|y|}{|x|^2},$$

while there arises

$$|y-x| < \frac{|x|}{2} \quad\Rightarrow\quad \left|\frac{x}{|x|^2} - \frac{x-y}{|x-y|^2}\right| \le \frac{1}{|x|} + \frac{1}{|x-y|} < \frac{2}{|x-y|}.$$

Hence we obtain

$$\begin{aligned}\left|\nabla u_k(x) + \frac{\beta_k}{2\pi}\cdot\frac{x}{|x|^2}\right| &\le \frac{2}{\pi|x|^2}\int_{|y|<\delta_k^{-1},\ |y-x|>|x|/2} |y|\tilde{V}_k(y)e^{u_k(y)}\\ &+ \frac{1}{\pi}\int_{B(x,|x|/2)} \frac{1}{|x-y|}\cdot \tilde{V}_k(y)e^{u_k(y)} + O(\delta_k)\end{aligned} \tag{3.233}$$

uniformly in $x \in B_{\delta_k^{-1}}$.

Inequality (3.222) is applicable to the first term on the right-hand side of (3.233). It holds that

$$\frac{2}{\pi |x|^2} \int_{|y|<\delta_k^{-1},\ |y-x|>|x|/2} |y| \tilde{V}_k(y) e^{u_k(y)} \leq \frac{C}{|x|^2} \tag{3.234}$$

uniformly in $x \in B_{\delta_k^{-1}}$. Lemmas 3.31 and 3.32, on the other hand, are valid for the second term, and there arises

$$\tilde{V}_k(x) e^{u_k(x)} \leq |x|^{-7/2}, \quad |x| > R \tag{3.235}$$

for $R \gg 1$ as $k \to \infty$. Since

$$|y - x| < \frac{|x|}{2} \quad \Rightarrow \quad \frac{|x|}{2} < |y| < \frac{3|x|}{2}$$

it follows that

$$\frac{1}{\pi} \int_{B(x,|x|/2)} \frac{1}{|x-y|} \cdot \tilde{V}_k(y) e^{u_k(y)} \leq C|x|^{-7/2} \cdot \int_0^{|x|/2} \frac{1}{r} \cdot r dr = C'|x|^{-5/2}, \tag{3.236}$$

and therefore, we get (3.232) by (3.233), (3.234), and (3.236). □

Regarding (3.204)–(3.205), inequality (3.195) is (3.221) with β_k replaced by 8π, and with the efficient region extended to $R < |x| < \delta_k^{-1}$.

Lemma 3.35 *Relation (3.206) is refined as*

$$\beta_k = 8\pi + O((\log \frac{1}{\delta_k})^{-1}), \quad k \to \infty. \tag{3.237}$$

Proof By Lemmas 3.33 and 3.34 it holds that

$$u_k(x) = v_k(\delta_k x) + 2 \log \delta_k = \frac{\beta_k}{2\pi} \log \frac{1}{|x|} + O(1)$$

$$\nabla u_k(x) = \delta_k \nabla v_k(\delta_k x) = -\frac{\beta_k}{2\pi} \cdot \frac{x}{|x|^2} + O(\delta_k + \frac{1}{|x|^2})$$

uniformly in $\log \frac{1}{\delta_k} < |x| < \delta_k^{-1}$, which means

$$v_k(x) = \frac{\beta_k}{2\pi} \log \frac{1}{|x|} - (\frac{\beta_k}{2\pi} - 2) \log \frac{1}{\delta_k} + O(1)$$

$$\nabla v_k(x) = -\frac{\beta_k}{2\pi} \cdot \frac{x}{|x|^2} + O(1) + O(\frac{\delta_k}{|x|^2}) \tag{3.238}$$

uniformly in $\delta_k \log \frac{1}{\delta_k} < |x| < 1$. Here we apply the Pohozaev inequality (3.200) to obtain

$$\int_{B_k} (2V_k(x) + x \cdot \nabla V_k(x))\, e^{v_k}\, dx$$
$$= r_k \int_{\partial B_k} V_k(x)e^{v_k} - \frac{1}{2}|\nabla v_k|^2 + (\frac{\partial v_k}{\partial \nu})^2\, d\sigma \tag{3.239}$$

for $B_k = B(0, \delta_k \log \frac{1}{\delta_k})$.

For the first term on the left-hand side of (3.239), the first equality of (3.238) is applicable. It holds that

$$\begin{aligned}\int_{B_k} 2V_k(x)e^{v_k} &= 2\beta_k - \int_{B\setminus B_k} 2V_k(x)e^{v_k} = 2\beta_k + O(1)\int_{B\setminus B_k} e^{v_k} \\ &= 2\beta_k + O(e^{-(\frac{\beta_k}{2\pi}-2)\log\frac{1}{\delta_k}}) \cdot \int_{B\setminus B_k} |x|^{-\frac{\beta_k}{2\pi}} \\ &= 2\beta_k + O(e^{-(\frac{\beta_k}{2\pi}-2)\log\frac{1}{\delta_k}} \cdot (\delta_k \log\frac{1}{\delta_k})^{-\frac{\beta_k}{2\pi}+2}) \\ &= 2\beta_k + O((\log\frac{1}{\delta_k})^{-1}).\end{aligned} \tag{3.240}$$

Concerning the second term, on the other hand, we use $\|\nabla V_k\|_\infty \leq C$ and (3.202), to deduce

$$\int_{B_k} x \cdot \nabla V_k(x)e^{v_k} = O(\delta_k \log\frac{1}{\delta_k}) \cdot \int_{B_k} V_k(x)e^{v_k} = O((\log\frac{1}{\delta_k})^{-1}). \tag{3.241}$$

For the first term on the right-hand side of (3.239), the first equality of (3.238) is applicable. It holds that

$$\begin{aligned}r_k \int_{\partial B_k} V_k(x)e^{v_k}\, d\sigma &= O(\delta_k \log\frac{1}{\delta_k}) \int_{\partial B_k} e^{v_k}\, d\sigma \\ &= O(\delta_k \log\frac{1}{\delta_k} \cdot e^{-(\frac{\beta_k}{2\pi}-2)\log\frac{1}{\delta_k}} \cdot (\delta_k \log\frac{1}{\delta_k})^{-\frac{\beta_k}{2\pi}+1}) \\ &= O((\log\frac{1}{\delta_k})^{-1}).\end{aligned} \tag{3.242}$$

For the second and the third terms on the right-hand side of (3.239), we apply the second equality of (3.238). Since

$$\frac{\delta_k}{r_k^2} = \frac{1}{\delta_k(\log\delta_k)^2} \to +\infty,$$

we obtain

$$\nabla v_k(x) = -\frac{\beta_k}{2\pi} \cdot \frac{x}{|x|^2} + O(\frac{\delta_k}{r_k^2}), \quad x \in \partial B_k. \tag{3.243}$$

By

$$r_k^2 \cdot \frac{\delta_k^2}{r_k^4} = \frac{\delta_k^2}{r_k^2} = (\log \frac{1}{\delta_k})^{-2}$$

we have

$$\begin{aligned} r_k \int_{\partial B_k} -\frac{1}{2}|\nabla v_k|^2 + (\frac{\partial v_k}{\partial \nu})^2 \, d\sigma &= \frac{r_k}{2} \cdot \left(\frac{\beta_k}{2\pi}\right)^2 \cdot \frac{1}{r_k^2} \cdot 2\pi r_k + r_k^2 \cdot O(\frac{\delta_k}{r_k^3}) \\ &= \frac{\beta_k^2}{4\pi} + O(\frac{\delta_k}{r_k}) = \frac{\beta_k^2}{4\pi} + O((\log \frac{1}{\delta_k})^{-1}). \end{aligned} \tag{3.244}$$

Then, equalities (3.239), (3.240), (3.241), (3.242), and (3.244) imply

$$2\beta_k = \frac{\beta_k^2}{4\pi} + O((\log \frac{1}{\delta_k})^{-1}), \quad k \to \infty$$

and hence (3.237). □

Proof It holds that

$$|u_k(x) + 4 \log |x|| \le C, \quad \log \frac{1}{\delta_k} < |x| < \delta_k^{-1} \tag{3.245}$$

by Lemmas 3.33 and 3.35, while

$$|u_k(x) + 4 \log |x|| \le C, \quad |x| = R \tag{3.246}$$

is valid as $k \to \infty$, for $R \gg 1$ fixed, by (3.204)–(3.205).

Let $c_1 \gg 1$ and put

$$w_\pm(x) = 4 \log \frac{1}{|x|} \pm c_1(1 - |x|^{-1/2}), \quad |x| \ge R.$$

It holds that

$$\Delta w_\pm = \mp \frac{1}{4} c_1 |x|^{-5/2},$$

while (3.235) implies

$$0 \le -\Delta u_k = \tilde{V}_k(x)e^{u_k} \le |x|^{-7/2}, \quad |x| \ge R.$$

Regarding (3.245), $|x| = \log \frac{1}{\delta_k}$, and (3.246), we apply the comparison theorem to obtain

$$w_-(x) \le u_k(x) \le w_+(x), \quad R < |x| < \log \frac{1}{\delta_k},$$

and hence

$$|u_k(x) + 4\log|x|| \le C, \quad R < |x| < \delta_k^{-1}. \tag{3.247}$$

Then, (3.195) is obtained by (3.204)–(3.205) and (3.247). □

3.4.9 Ma–Wei Theorem

Theorems 2.6–2.7 concerning (2.65) are valid for

$$-\Delta v = \lambda V(x)e^v \text{ in } \Omega, \quad v|_{\partial\Omega} = 0 \tag{3.248}$$

for $0 < V = V(x) \in C^1(\overline{\Omega})$ if the Hamiltonian is replaced by

$$H_\ell(x_1, \cdots, x_m) = \frac{1}{2}\sum_{j=1}^m R(x_j) + \sum_{1\le i<j\le m} G(x_i, x_j) + \frac{1}{8\pi}\sum_{j=1}^m \log V(x_j).$$

The proof, however, requires an essentially different approach.

First, the Brezis–Merle theorem guarantees the finiteness of the blowup set. Second, a scaling limit is used for realization of the quantized mass and exclusion of the boundary blowup. Finally, the location of blowup points is prescribed by the Pohozaev identity or the duality argument [26].

Theorem 2.5 ensures the invariance of the topological degree of the solution set to (2.92) for each connected component of $[0, +\infty)\setminus 8\pi\mathbf{N}$. By the quantized blowup mechanism of (3.248) this property is extended to

$$-\Delta u = \frac{\lambda K(x)e^u}{\int_\Omega K(x)e^u} \text{ in } \Omega, \quad u|_{\partial\Omega} = 0. \tag{3.249}$$

This value is actually determined by the genus of Ω [7]. We show the above result in the following form [25], where $K(x)$ may vary as $K_k \to K$ in $C^1(\overline{\Omega})$.

Theorem 3.38 (Ma–Wei) *Given a bounded domain $\Omega \subset \mathbf{R}^2$ with smooth boundary $\partial\Omega$ and $0 < K = K(x) \in C^1(\overline{\Omega})$, let (λ_k, u_k), $k = 1, 2, \cdots$, be a sequence of solutions to*

$$-\Delta u_k = \frac{\lambda_k K(x)e^{u_k}}{\int_\Omega K(x)e^{u_k}} \text{ in } \Omega, \quad u_k|_{\partial\Omega} = 0, \tag{3.250}$$

satisfying

$$\lim_{k\to\infty} \lambda_k = \lambda_0 \in (0, +\infty), \quad \lim_{k\to\infty} \|v_k\|_\infty = +\infty.$$

Then it holds that $\lambda_0 = 8\pi\ell$, $\ell \in \mathbf{N}$. *Passing to a subsequence, furthermore, there are distinct interior points* $x_1^*, \cdots, x_\ell^*$ *such that*

$$\nabla_{x_j} H_\ell(x_1^*, \cdots, x_\ell^*) = 0, \quad j = 1, \cdots, \ell, \tag{3.251}$$

where

$$H_\ell(x_1, \cdots, x_\ell) = \frac{1}{2}\sum_{j=1}^{\ell} R(x_j) + \sum_{1\le i<j\le \ell} G(x_i, x_j) + \frac{1}{8\pi}\sum_{j=1}^{\ell} \log K(x_j) \tag{3.252}$$

and $\mathscr{S} = \{x_1^*, \cdots, x_\ell^*\}$ *is the blowup set of* $\{u_k\}$. *It holds also that (2.100), i.e.,*

$$u_k \to 8\pi\sum_{j=1}^{\ell} G(\cdot, x_j^*) \quad \text{locally uniformly in } \overline{\Omega}\setminus\mathscr{S}. \tag{3.253}$$

The following identity [20] is used to derive (3.251). Recall that ν denotes the outer unit normal vector.

Lemma 3.36 (Kazdan–Warner) *If*

$$-\Delta u = W(x)e^u \quad \text{in } B \tag{3.254}$$

for $B = B(0, 1) \subset \mathbf{R}^2$ *and* $W \in C^1(\overline{\Omega})$, *it holds that*

$$\int_B [\nabla W(x)]e^u \, dx = \int_{\partial B} \frac{\partial u}{\partial \nu}\nabla u - \frac{1}{2}|\nabla u|^2\nu + W(x)e^u \, dS. \tag{3.255}$$

Proof By (3.254) it holds that

$$-\int_B \Delta u \, \nabla u = \int_B W(x)\nabla e^u = \int_{\partial B} W(x)e^u\nu - \int_B [\nabla W(x)]e^u. \tag{3.256}$$

Here, we obtain

$$I_{ij} = \int_B u_{ii}u_j = \int_{\partial B} \nu_i u_i u_j - \int_B u_i u_{ij} = \int_{\partial B} \nu_i u_i u_j - \nu_i u u_{ij} \, dS + \int_B u u_{iij}$$

$$= \int_{\partial B} \nu_i u_i u_j - \nu_i u u_{ij} + \nu_j u u_{ii} \, dS - I_{ij}$$

and hence

$$I_{ij} = \frac{1}{2} \int_{\partial B} \nu_i u_i u_j - \nu_i u u_{ij} + \nu_j u u_{ii} \, dS. \tag{3.257}$$

Then (3.257) implies

$$\begin{aligned}
\int_B \Delta u \, \nabla u &= \frac{1}{2} \int_{\partial B} \frac{\partial u}{\partial \nu} \nabla u - u \frac{\partial}{\partial \nu}(\nabla u) + \nu u \Delta u \, dS \\
&= \int_{\partial B} \frac{\partial u}{\partial \nu} \nabla u - \frac{1}{2} \frac{\partial}{\partial \nu}(u \nabla u) + \frac{1}{2} \nu u \Delta u \, dS \\
&= \int_{\partial B} \frac{\partial u}{\partial \nu} \nabla u - \frac{1}{2} \int_B \Delta(u \nabla u) - \nabla(u \Delta u) \, dx \\
&= \int_{\partial B} \frac{\partial u}{\partial \nu} \nabla u - \frac{1}{2} \int_B \nabla(|\nabla u|^2) \\
&= \int_{\partial B} \frac{\partial u}{\partial \nu} \nabla u - \frac{1}{2} |\nabla u|^2 \nu \, dS.
\end{aligned} \tag{3.258}$$

From (3.256) and (3.258) it follows that (3.255). □

By (3.249) we obtain

$$-\Delta v = V(x) e^v \text{ in } \Omega, \quad v = u - \log \int_\Omega K(x) e^u, \quad V(x) = \lambda K(x).$$

According to (3.250), there arises

$$-\Delta v_k = V_k(x) e^{v_k} \quad \text{in } \Omega$$

and

$$a \leq V_k(x) \leq b, \ \|\nabla V_k\|_\infty \leq C, \ \int_\Omega V_k(x) e^{v_k} \leq C, \quad k = 1, 2, \cdots$$

with $a, b > 0$. Then the Brezis–Merle theorem and the Li–Shafrir theorem ensure the following alternatives up to a subsequence.

1. $\{v_k\}$ is locally uniformly bounded in Ω.
2. $v_k \to -\infty$ locally uniformly in Ω.
3. There arise $\ell \in \mathbf{N}$, $x_1^* \cdots, x_\ell^* \in \Omega$, $n_1 \cdots, n_\ell \in \mathbf{N}$, and $x_k^j \to x_j^*$, $1 \leq k \leq \ell$, such that

$$v_k(x) \to -\infty \quad \text{locally uniformly in } \Omega \setminus \{x_1^* \cdots, x_\ell^*\}$$
$$v_k(x_k^j) \to +\infty, \quad 1 \leq j \leq \ell$$
$$V_k(x)e^{v_k}dx \rightharpoonup 8\pi \sum_{j=1}^{\ell} n_j \delta_{x_j^*}(dx) \quad \text{in } \mathcal{M}(\Omega). \tag{3.259}$$

Taking the preliminaries for the proof, we note

$$\|u_k\|_{W^{1,q}} \leq C(q), \quad 1 \leq q < 2 \tag{3.260}$$

by the L^1 elliptic estimate to (3.250). Since

$$-\Delta u_k = V_k(x)e^{v_k}$$

is locally uniformly bounded in $\Omega \setminus \mathcal{S}$, therefore, the interior blowup points of $\{u_k\}$ belong to $\mathcal{S} = \{x_1^*, \cdots, x_\ell^*\}$. Hence each $B = B(x_j^*, r)$, $0 < r \ll 1$, admits

$$0 \leq \max_{\partial B} u_k \leq C_4,$$

which implies

$$\max_{\partial B} v_k - \min_{\partial B} v_k = \max_{\partial B} u_k - \min_{\partial B} u_k \leq C'.$$

The Y.Y. Li estimate is applicable now, and in particular, it follows that $n_j = 1$ for $j = 1, \cdots, \ell$, in the third equality of (3.259).

Uniform monotonicity of u_k near $\partial\Omega$ holds by the following lemma [15], and then the proof of Theorem 2.5 excludes its boundary blowup. Then, there arises the third alternative, $\mathcal{S} \neq \emptyset$, and (3.259) is valid in $\mathcal{M}(\overline{\Omega})$,

$$-\Delta v_k \, dx = V_k(x)e^{v_k}dx \rightharpoonup 8\pi \sum_{j=1}^{\ell} \delta_{x_j^*}(dx) \quad \text{in } \mathcal{M}(\overline{\Omega}). \tag{3.261}$$

Then, (3.261) implies (3.253).

Lemma 3.37 *There is $0 < \gamma, \delta \ll 1$ determined by the domain $\Omega \subset \mathbf{R}^2$, such that any solution $u = u(x)$ to (3.249) satisfies*

$$\frac{d}{dt} u(x_0 + t\xi) < 0, \quad -\gamma < t < 0,$$

where ν_{x_0} denotes the outer unit normal vector at $x_0 \in \partial\Omega$ and $\xi \in \mathbf{R}^2$ is the normal vector satisfying $|\theta| < \delta$ for $\nu_{x_0} \cdot \xi = \cos\theta$.

Proof It holds that

$$- \Delta u = W(x)e^u \text{ in } \Omega, \quad u|_{\partial\Omega} = 0 \tag{3.262}$$

for

$$W(x) = \lambda K(x) / \int_\Omega K(x)e^u.$$

Since $\partial\Omega$ is smooth, each $x_0 \in \partial\Omega$ admits $B(x_1, r) \subset \Omega^c$ such that $\overline{B(x_1, r)} \cap \overline{\Omega} = \{x_0\}$. The Kelvin transformation

$$y = r^2 \frac{x - x_1}{|x - x_1|^2}, \quad w(y) = u(x),$$

implies

$$\begin{aligned} -\Delta u &= f(y, w) \text{ in } \Omega' \subset B(0, r), \quad w|_{\partial\Omega'} = 0 \\ f(y, w) &= \frac{r^4}{|y|^4} W(x_1 + r^2 \frac{y}{|y|^2})e^w \end{aligned}$$

by (3.262). The method of moving planes at $x_0 \in \partial\Omega'$ works if $f(y, w)$ is non-increasing in the direction $|y|$.

We confirm this property, putting

$$y = \frac{\rho(x_0 - x_1)}{r}, \quad 0 < \rho < r.$$

Since $(x_1 - x_0)/r = \nu_{x_0}$ it follows that

$$\begin{aligned} \frac{\partial}{\partial\rho} f(y, w)\Big|_{\rho=r} &= \frac{\partial}{\partial\rho}\{\frac{r^4}{\rho^4} W(x_1 + \frac{r}{\rho}(x_0 - x_1))\}\Big|_{\rho=r} \cdot e^w \\ &= \{-4r^{-1} W(x_0) + \nabla W(x_0) \cdot \frac{x_1 - x_0}{r}\}e^w \\ &= W(x_0)\{-4r^{-1} + \frac{\partial \log W}{\partial\nu_{x_0}}\}e^w \\ &= W(x_0)\{-4r^{-1} + \frac{\partial \log K}{\partial\nu_{x_0}}\}e^w. \end{aligned} \tag{3.263}$$

Hence there is $0 < r_0 \ll 1$ such that the right-hand side of (3.263) is negative for any $x_0 \in \partial\Omega$, provided $0 < r < r_0$. Thus we obtain the result. □

Proof *(Theorem 3.38)* The proof of Theorem 2.5 assures an open set $\hat{\omega}$ satisfying $\partial\Omega \subset \hat{\omega}$ such that

$$\|u_k\|_{L^\infty(\omega)} \leq C, \quad \omega = \Omega \cap \hat{\omega},$$

because of (3.260) and Lemma 3.37.

To show (3.251)–(3.252), we apply Lemma 3.36 to

$$W_k(x) = \frac{\lambda_k K(x)}{\int_\Omega K(x) e^{u_k}}, \quad B = B(x_i^*, \varepsilon), \ 0 < \varepsilon \ll 1$$

for $i = 1, \cdots, \ell$. Since

$$-\Delta u_k = W_k(x) e^{u_k} \text{ in } B(x_i^*, \varepsilon)$$

it holds that

$$\int_{B(x_i^*,\varepsilon)} [\nabla W_k(x)] e^{u_k} = \int_{\partial B(x_i^*,\varepsilon)} \frac{\partial u_k}{\partial \nu} \nabla u_k - \frac{1}{2} |\nabla u_k|^2 \nu + W_k(x) e^{u_k} \, dS$$

and hence

$$\int_{B(x_i^*,\varepsilon)} [\nabla \log K] W_k(x) e^{u_k}$$
$$= \int_{\partial B(x_i^*,\varepsilon)} \frac{\partial u_k}{\partial \nu} \nabla u_k - \frac{1}{2} |\nabla u_k|^2 \nu + W_k(x) e^{u_k} \, dS. \tag{3.264}$$

Then, (3.261) implies

$$\lim_{k\to\infty} \int_{B(x_i^*,\varepsilon)} [\nabla \log K] W_k(x) e^{u_k} = 8\pi \nabla \log K(x_i^*), \tag{3.265}$$

while we obtain

$$\lim_{k\to\infty} \int_{\partial B(x_i^*,\varepsilon)} \frac{\partial u_k}{\partial \nu} \nabla u_k - \frac{1}{2} |\nabla u_k|^2 \nu + W_k(x) e^{u_k} \, dS$$
$$= 64\pi^2 \int_{\partial B(x_i^*,\varepsilon)} \frac{\partial \tilde{G}}{\partial \nu} \nabla \tilde{G} - \frac{1}{2} |\nabla \tilde{G}|^2 \nu \, dS \tag{3.266}$$

by (3.253), where

$$\tilde{G} = \sum_{j=1}^{\ell} G(\cdot, x_j^*).$$

Since equalities (3.264), (3.265), and (3.266) imply

$$\frac{1}{8\pi}\nabla \log K(x_i^*) = \int_{\partial B(x_i^*,\varepsilon)} \frac{\partial \tilde{G}}{\partial \nu}\nabla \tilde{G} - \frac{1}{2}|\nabla \tilde{G}|^2 \nu \, dS,$$

the desired (3.251)–(3.252) is reduced to

$$\int_{\partial B(x_i^*,\varepsilon)} \frac{\partial \tilde{G}}{\partial \nu}\nabla \tilde{G} - \frac{1}{2}|\nabla \tilde{G}|^2 \nu \, d\sigma = -\nabla \tilde{H}(x_i^*) + o(1), \quad \varepsilon \downarrow 0 \tag{3.267}$$

for

$$\tilde{H}(x) = K(x, x_i^*) + \sum_{j\neq i} G(x, x_j^*), \quad K(x, x_i^*) = G(x, x_i^*) + \frac{1}{2\pi}\log|x - x_i^*|.$$

In fact, $\tilde{H} = \tilde{H}(x)$ is smooth at $x = x_i^*$ and there arises

$$\nabla \tilde{G} = -\frac{\nu}{2\pi\varepsilon} + \nabla \tilde{H} \quad \text{on } \partial B(x_i^*, \varepsilon)$$

by

$$\tilde{G}(x) = \frac{1}{2\pi}\log\frac{1}{|x - x_i^*|} + \tilde{H}(x).$$

Since

$$\int_{\partial B(x_i^*,\varepsilon)} \nu \, d\sigma = 0,$$

the left-hand side on (3.267) is equal to

$$\begin{aligned}
&\int_{\partial B(x_i^*,\varepsilon)} \left(-\frac{1}{2\pi\varepsilon} + \frac{\partial \tilde{H}}{\partial \nu}\right)\left(-\frac{\nu}{2\pi\varepsilon} + \nabla \tilde{H}\right) - \frac{1}{2}\left|-\frac{\nu}{2\pi\varepsilon} + \nabla \tilde{H}\right|^2 \nu \, d\sigma \\
&= \frac{1}{2\pi\varepsilon}\int_{\partial B(x_i^*,\varepsilon)} -\frac{\partial \tilde{H}}{\partial \nu}\nu - \nabla \tilde{H} + (\nu\cdot\nabla \tilde{H})\nu \, d\sigma + o(1) \\
&= -\nabla \tilde{H}(x_i^*) + o(1),
\end{aligned}$$

which means that (3.267) holds. □

Remark 3.10 Theorems 3.31 and 3.37 imply the correspondence of Morse indices between the linearized operator $-\Delta - \lambda_k e^{u_k}$ under the Dirichlet boundary condition for $k \gg 1$ and the critical point $(x_1^*, \cdots, x_m^*)$ of $H(x_1, \cdots, x_m)$ defined by (2.10), where (λ_k, u_k), $k = 1, 2, \cdots$, is a family of solutions to (2.6) for $(\lambda, u) = (\lambda_k, u_k)$ satisfying

$$\lambda_k \to 0, \quad u_k \to 8\pi \sum_{j=1}^{m} G(\cdot, x_j^*) \text{ locally uniformly in } \overline{\Omega} \setminus \mathscr{S},$$

where $\mathscr{S} = \{x_1^*, \cdots, x_m^*\} \subset \Omega$. See [18, 19, 31].

References

1. P. Aviles, Local behavior of solutions of some elliptic equations. Commun. Math. Phys. **108**, 177–192 (1987)
2. C. Bandle, *Isoperimetric Inequalities and Applications* (Pitman, Boston, 1980)
3. S. Baraket, F. Pacard, Construction of singular limits for a semilinear elliptic equation in dimension 2. Calc. Vari. Partial Differ. Equations **6**, 1–38 (1998)
4. H. Brezis, Y.Y. Li, I. Shafrir, A sup + inf inequality for some nonlinear elliptic equations involving exponential nonlinearities. J. Funct. Anal. **115**, 344–358 (1993)
5. H. Brezis, F. Merle, Uniform estimates and blow-up behavior for solutions of $-\Delta u = V(x)e^u$ in two dimensions. Commun. Partial Differ. Equations **16**, 1223–1253 (1991)
6. L.A. Caffarelli, B. Gidas, J. Spruck, Asymptotic symmetry and local behavior of semilinear elliptic equations with critical Sobolev growth. Commun. Pure Appl. Math. **42**, 271–297 (1989)
7. C.-C. Chen, C.-S. Lin, Topological degree for a mean field equation on Riemann surfaces. Commun. Pure Appl. Math. **56**, 1667–1727 (2003)
8. W. Chen, C. Li, Classification of solutions of some nonlinear elliptic equations. Duke Math. J. **63**, 615–622 (1991)
9. X. Chen, Remarks on the existence of branch bubbles on the blowup analysis of equation $-\Delta u = e^{2u}$ in dimension two. Commun. Anal. Geom. **7**, 295–302 (1999)
10. E.A. Coddington, N. Levinson, *Theory of Ordinary Differential Equations* (McGraw Hill Publishing, New York, 1955)
11. E.N. Dancer, Some notes on the method of moving planes. Bull. Austral. Math. Soc. **46**, 425–434 (1992)
12. D.G. De Figueiredo, P.L. Lions, R.D. Nussbaum, A priori estimates and existence of positive solutions of semilinear elliptic equations. J. Math. Pures Appl. **61**, 41–63 (1882)
13. K. Deimling, *Nonlinear Functional Analysis* (Springer, Berlin, 1985)
14. W. Ding, J. Jost, J. Li, G. Wang, Existence results for mean field equations. Ann. Inst. H. Poincaré Analyse non linèaire **16**, 653–666 (1999)
15. B. Gidas, W.-M. Ni, L. Nirenberg, Symmetry and related properties via the maximum principle. Commun. Math. Phys. **68**, 209–243 (1979)
16. B. Gidas, J. Spruck, Global and local behavior of positive solutions of nonlinear elliptic equations. Commun. Pure Appl. Math. **34**, 525–598 (1981)
17. B. Gidas, J. Spruck, A priori bounds for positive solutions of nonlinear elliptic equations. Commun. Partial Differ. Equations **6**, 883–901 (1981)
18. F. Gladiali, M. Grossi, On the spectrum of a nonlinear planar problem. Ann. Inst. Henri Poincaré Anal. Non Linéaire **26**, 191–222 (2009)
19. F. Gladiali, M. Grossi, H. Ohtsuka, T. Suzuki, Morse index of multiple blow-up solution to the two-dimensional Gel'fand problem. Commun. Partial Differ. Equations **39**, 2028–2063 (2014)
20. J.L. Kazdan, F.W. Warner, Integrability conditions for $\Delta u = k - Ke^{\alpha u}$ with applications to Riemannian geometry. Bull. Amer. Math. Soc. **77**, 819–823 (1971)
21. Y.Y. Li, Harnack type inequality: the method of moving planes. Commun. Math. Phys. **200**, 421–444 (1999)
22. Y.Y. Li, I. Shafrir, Blow up analysis for solutions of $-\Delta u = Ve^u$ in dimension two. Indiana Univ. Math. J. **43**, 1255–1270 (1994)

23. Y. Li, W.-M. Ni, Radial symmetry of positive solutions of nonlinear elliptic equations in$\mathbf{R}^n$. Commun. Partial Differ. Equations **18**, 1043–1054 (1993)
24. C.S. Lin, An expository survey on recent development of mean field equation. Discrete Contin. Dyn. Syst. Ser. A **19**, 387–410 (2007)
25. L. Ma, J.C. Wei, Convergence of Liouville equation. Comment. Math. Helv. **76**, 506–514 (2001)
26. H. Ohtsuka, T. Suzuki, Palais-Smale sequence relative to the Trudinger-Moser inequality. Calc. Var. Partial Differ. Equations **17**, 235–255 (2003)
27. Q.H. Phan, P. Souplet, Liouville type theorems and bounds of solutions of Hardy-Hénon equations. J. Differ. Equations **252**, 2544–2562 (2012)
28. P. Polačik, P. Quittner, P. Souplet, Singularities and decay estimates in superlinear problems via Liouville-type theorems, Part I: elliptic equations and systems. Duke Math. J. **139**, 555–579 (2007)
29. S. Sakaguchi, T. Suzuki, Nonexistence of solutions for a degenerate parabolic equation describing imperfect ignition. Nonlinear Anal. **31**, 665–669 (1998)
30. S. Sakaguchi, T. Suzuki, On a class of L^1-illposed quasilinear parabolic equations. Adv. Differ. Equations **4**, 671–696 (1999)
31. T. Sato, T. Suzuki, Morse indices of the solutions to the Liouville-Gel'fand problem with variable coefficients. Funkc. Ekvacioj. **61**, 229–265 (2018)
32. T. Sato, T. Suzuki, F. Takahashi, p-capacity of singular set of p-harmonic function is zero. Electron J. Differ. Equations **67**, 1–15 (2011)
33. I. Shafrir, A $\sup + \inf$ inequality for the equation $-\Delta u = Ve^u$. C. R. Acad. Sci. Paris Série I **315**, 159–164 (1992)
34. M. Struwe, G. Tarantello, On multivortex solutions in Chern-Simons-Higgs gauge theory. Boll. Unione Math. Ital. Sez. B(8) **1**, 109–121 (1998)
35. T. Suzuki, *Mean Field Theories and Dual Variation, Mathematical Structures of the Mesoscopic Model*, 2nd edn. (Atlantis Press, Paris, 2015)
36. T. Suzki, *Interface vanishing for nonstationary Maxwell equation*, preprint, 2019
37. T. Suzuki, *Semilinear Elliptic Equations, Modern and Classical Theories*, 2nd edn. (De Gruyter, Berlin, 2020)
38. T. Suzuki, F. Takahashi, Capacity estimate for the blow-up set of parabolic equations. Math. Z. **259**, 867–878 (2008)
39. D. Ye, Une remarque sur le comportement asymptotique des solutions de $-\Delta u = \lambda f(u)$. C. R. Acad. Sci. Paris, Série I **325**, 1279–1282 (1997)

Chapter 4
Transformation Theory

Abstract This chapter describes the fourth of Liouville's theorems, which concerns volume derivatives on deforming domains. This transformation theory provides a mathematical formulation of several physical phenomena involving material transport. Its applications, however, spread, beyond mathematical physics, into various areas in analysis and engineering. The first section is a summary. In the second section, we confirm a fundamental elliptic H^1 regularity on Lipschitz domains, and then turn to H^2 regularity on convex domains in accordance with the second fundamental form. In the third section, we describe the above mentioned Liouville's theorem and application to the free boundary problem. In the fourth section we turn to the problem of fluid mechanics, associated with both coordinates of Lagrange and Euler, to reach the blowup of the solution to irrotational Euler flow. In the fifth section, we describe an abstract theory of shape optimization. The final section is devoted to Hadamard's variational formulae for Green's function of Laplacians. The first volume derivative is used for modeling of fundamental equations in physics. It is to be a basic tool to approach the principle of self-organization in the following chapter, besides the scaling described in the previous chapter. The second volume derivative, on the other hand, is more associated with geometric objects, the second fundamental form of the boundary.

4.1 Introduction

4.1.1 Transformation of Variables

Mathematical modeling and analysis of physical laws are based on the study of a family of transformations, T_t, $|t| \ll 1$, on a bounded domain $\Omega \subset \mathbf{R}^N$, satisfying

$$T_t = I + tS + \frac{t^2}{2}R + o(t^2) \tag{4.1}$$

T. Suzuki, *Liouville's Theory in Linear and Nonlinear Partial Differential Equations*, Springer Asia Pacific Mathematics Series 8,
https://doi.org/10.1007/978-981-95-1975-0_4

as $t \to 0$ uniformly on Ω, where $I : \Omega \to \Omega$ denotes the identity mapping. A dynamical system, for example, induces such a deformation. If Ω is a Lipschitz domain and $v = v(x)$ is a Lipschitz continuous vector field on $\overline{\Omega}$, thus, we can define $T_t : \Omega \to \Omega_t = T_t(\Omega)$ by $T_t x_0 = x(t)$, where $x = x(t)$ is the solution to

$$\frac{dx}{dt} = v(x), \quad x|_{t=0} = x_0.$$

Then (4.1) holds for

$$Sx = v(x), \quad Rx = [(v \cdot \nabla)v](x).$$

Liouville's formulae are concerned with the asymptotics of several quantities under this family of transformations. Using the transformation of variables they are reduced to the derivatives in t of the Jacobian of $T_t : \Omega \to \Omega_t = T_t(\Omega)$, denoted by $J(T_t)$. Theorem 4.8 of this chapter says that

$$J(T_t) = 1 + t\nabla \cdot S + \frac{t^2}{2}\left(\nabla \cdot R + (\nabla \cdot S)^2 - J(S) : {}^tJ(S)\right) + o(t^2), \tag{4.2}$$

where ${}^tJ(S)$ is the transpose matrix of $J(S)$ and

$$A : B = \sum_{i,j=1}^{N} a_{ij}b_{ij}, \quad A = (a_{ij}), \ B = (b_{ij}).$$

4.1.2 Volume Derivatives and Applications

Equation (4.2) induces the first and the second variational formulae of the volume integral

$$\left.\frac{d}{dt}\int_{\Omega_t} c(x,t)\, dx\right|_{t=0}, \quad \left.\frac{d^2}{dt^2}\int_{\Omega_t} c(x,t)\, dx\right|_{t=0} \tag{4.3}$$

under the transformation (4.1) as in Theorems 4.9 and 4.10 below. The first derivative is used to derive several fundamental equations in mathematical physics. Here we pick up latent heat and ideal flow, to reach the notions of *Stefan condition* and flux. The second derivative, on the other hand, involves the second fundamental form of the hypersurface $\partial\Omega$ with its mean curvature $\nabla \cdot n$, where n denotes the outer normal derivatives. This formula is a base of H^2 regularity of solutions to the Poisson equation on convex domains. The most direct application of these formulae, however, lies in the study of free boundary problems and shape optimization in engineering.

4.1.3 Organization of Chap. 4

Section 4.2 is concerned with the non-standard elliptic regularity. We confirm the notion of Lipschitz domain and its fundamental properties: trace and integration by parts in Theorems 4.1 and 4.3 below, respectively. Study of the sectional curvature induces a trace identity, Lemma 4.2, which assures non-standard elliptic regularity for an H^1 solution to the Poisson equations

$$-\Delta u = f \text{ in } \Omega, \quad u|_{\partial\Omega} = 0 \tag{4.4}$$

and

$$-\Delta u + u = f \text{ in } \Omega, \quad \left.\frac{\partial u}{\partial n}\right|_{\partial\Omega} = 0. \tag{4.5}$$

Theorems 4.5 and 4.6 in Sect. 4.2.4 thus confirm $u \in H^2(\Omega)$ if $f \in L^2(\Omega)$ and Ω is convex.

Free boundary problems are treated in Sect. 4.3. Showing (4.2), we calculate the derivatives in (4.3). Lemma 4.2 in the previous section is used to reduce the second volume derivative there, associated with the *second fundamental form* of $\partial\Omega$. A classical free boundary problem is then introduced. Reformulating it as a variational formula, we illustrate first derivatives of associated variational functionals (Theorem 4.13). Finally, we derive the Stefan condition to describe latent heat, applying the first volume derivative.

Section 4.4 is a continuation of the application of the first volume derivative to mathematical physics. The basic notion of flux is formulated by the comparison of the first volume derivative and the divergence formula of Gauss. The second moment under the Lagrange coordinate then implies a decay of the total pressure in time of a *compressible Euler flow*. The other formulation of a Hamilton system, finally, induces the irrotational blowup of the solution.

In shape optimization, one defines a cost function of shapes. It is defined in accordance with the solution to the fundamental equation, which is to be finite dimensional under discritization. Then the domain variation of Hadamard type is used to execute to solve this optimization problem effectively. This concept of optimum design is described in Sect. 4.5 in an abstract manner.

The final section, Sect. 4.6 is devoted to Hadamard's variational formula for Green's function for $-\Delta$ under the Dirichlet boundary condition. These formulae, Theorems 4.20 and 4.21, are described directly from the study of *sectional curvatures*. Then the second derivative for the Neumann boundary condition is recovered by the second fundamental form on $\partial\Omega$ [12].

4.2 Non Standard Elliptic Regularity

4.2.1 Lipschitz Domains

Given an open set $\Omega \subset \mathbf{R}^N$, we say that its boundary $\Gamma = \partial\Omega$ is Lipschitz continuous if the following conditions are satisfied:

1. Each $x \in \Gamma$ takes an open rectangle O with $x \in O$ and new orthogonal coordinates $y = (y', y_N)$, $y' = (y_1, \cdots, y_{N-1})$ such that
$$O = \{y \mid |y_j| < a_j,\ 1 \leq j \leq N\},$$
where $a_j > 0$, $1 \leq j \leq N$.
2. There is a Lipschitz continuous function $\varphi : O' \to \mathbf{R}$,
$$O' = \{y' \mid |y_j| < a_j,\ 1 \leq j \leq N-1\}$$
such that $|\varphi(y')| \leq a_N/2$ for $y' \in O'$ and
$$\Omega \cap O = \{y \mid y_N > \varphi(y'),\ y' \in O\}, \quad \Gamma \cap O = \{y \mid y_N = \varphi(y'),\ y' \in O\}.$$

A bounded domain $\Omega \subset \mathbf{R}^N$ is called a Lipschitz domain if its boundary is Lipschitz continuous. In the case of $N = 2$, for example, such a domain admits corners except for cusps on the boundary.

We study the elliptic boundary value problem, that is, the Dirichlet problem for the Laplace equation
$$\Delta u = 0 \text{ in } \Omega, \quad u = g \text{ on } \partial\Omega \tag{4.6}$$
and that for the Poisson equation
$$-\Delta u = f \text{ in } \Omega, \quad u = 0 \text{ on } \partial\Omega \tag{4.7}$$
using the Laplacian $\Delta = \sum_{i=1}^N \partial^2/\partial x_i^2$. The well-posedness of such problems is classical if $\partial\Omega$ is sufficiently smooth. At least three approaches have been tried for (4.6), since Perron, Fredholm, and Riemann [13, 38].

By Perron's method in Sect. 3.2, given $g \in C(\partial\Omega)$ there is a unique solution $u \in C^2(\Omega) \cap C(\overline{\Omega})$ to (4.6) if and only if any boundary point is regular in the sense of potential theory. A well-known sufficient condition for this regularity is the outscribing ball condition, which means that any $x_0 \in \partial\Omega$ takes an open ball B such that $\overline{\Omega} \cap \overline{B} = \{x_0\}$. Note that not all Lipschitz domains satisfy this condition. A sufficient condition of this outscribing ball condition is that $\partial\Omega$ is C^1, which means that the above $\varphi : O' \to \mathbf{R}$ is C^1. The other is the convexity of Ω.

Concerning (4.7), the *Schauder theory* is classical. It requires that $\partial\Omega$ is $C^{2,\theta}$, $0 < \theta < 1$, which means again that the above $\varphi : O' \to \mathbf{R}$ is provided with this regularity. If this condition is satisfied, there is a unique solution $u \in C^{2,\theta}(\overline{\Omega})$ for given $f \in C^{\theta}(\overline{\Omega})$. The L^p, $1 < p < \infty$, *theory* is also classical. Here we require that $\partial\Omega$ is $C^{1,1}$, which is defined similarly by $\varphi : O' \to \mathbf{R}$. In this case, given $f \in L^p(\Omega)$, there is a unique solution $u \in W^{2,p}(\Omega)$ to (4.7) with the boundary value $u|_{\partial\Omega} \in W^{2-1/p,p}(\partial\Omega)$ taken in the sense of *trace*. Here, to define the Sobolev space on the boundary with the fractional exponent we use a norm associated with boundary integrals [1, 26, 27], recalling that the Sobolev space $W^{s,p}(\Omega)$, $s = m+\sigma$, $0 < \sigma < 1$, $1 \leq p < \infty$, is the set of $v \in W^{m,p}(\Omega)$ satisfying

$$\int\int_{\Omega\times\Omega} \frac{|\partial^\alpha u(x) - \partial^\alpha u(x')|^p}{|x-x'|^{N+\sigma p}}\, dxdx' < +\infty, \quad |\alpha| = m.$$

If $p = 2$ this space is defined equivalently using the local chart $\varphi : O' \to \mathbf{R}$ and Fourier transformation [25]. The higher regularity is also available for $m = 2, 3, \cdots$. Thus if $\partial\Omega$ is $C^{m,1}$ and $f \in W^{m-1,p}(\Omega)$ then it holds that $u \in W^{m+1,p}(\Omega)$. Similarly, if $\partial\Omega$ is $C^{m+1,\theta}$, $0 < \theta < 1$, and $f \in C^{m-1,\theta}(\overline{\Omega})$ then it holds that $u \in C^{m+1,\theta}(\overline{\Omega})$.

These regularity theories are, however, invalid for general Lipschitz domains. Typical examples are observed for polygons on the plane [5, 10]. In fact, if $f(z)$ is holomorphic in $z = x_1 + \imath x_2$ then its real part is harmonic in $x = (x_1, x_2)$. We take the harmonic function

$$u = r^m \sin m\theta = \mathrm{Im}\,(x_1 + \imath x_2)^m, \quad x = re^{\imath\theta}, \; m > 0, \tag{4.8}$$

on the sector $0 < \theta < \alpha$. This function satisfies the Dirichlet boundary condition on $\theta = 0$ and $\theta = \alpha$ if and only if $m\alpha = \pi$. Then

$$u_{rr} \sim r^{m-2}, \quad r \downarrow 0$$

unless $m = 1$, and therefore, $u_{rr} \in L^2$ near $r = 0$ if and only if $m > 1$, which means $\alpha < \pi$.

This example indicates the role of the angular α in the L^p theory, $1 < p < \infty$, for (4.7). We recall that if $\partial\Omega$ is C^2 then this elliptic regularity is always satisfied. By this example, we notice the role of convexity of the domain again, that is, for the H^2-regularity of the Poisson equation (4.7).

In fact, if Ω is convex and $f \in L^2(\Omega)$, there is a unique solution $u \in H^2(\Omega) \cap H_0^1(\Omega)$ to (4.7). The Neumann case is similar; if Ω is convex and $f \in L^2(\Omega)$ satisfies $\int_\Omega f\, dx = 0$, there is a unique $u \in H^2(\Omega)$ up to an additive constant such that

$$-\Delta u = f \quad \text{in } \Omega, \qquad \frac{\partial u}{\partial n} = 0 \quad \text{on } \partial\Omega, \tag{4.9}$$

where the normal derivative $\partial u/\partial n \in H^{1/2}(\partial\Omega) = W^{1/2,2}(\partial\Omega)$ of u is taken in the sense of trace. It may be useful to take

$$u = \text{Re } z^m = r^m \cos m\theta$$

in the sector $0 < \theta < \alpha$ to understand this property of H^2-regularity in accordance with the convexity of Ω.

The mixed problem is more delicate. This problem is formulated as

$$-\Delta u = f \text{ in } \Omega, \quad u = 0 \text{ on } \gamma_0, \quad \frac{\partial u}{\partial n} = 0 \text{ on } \gamma_1, \tag{4.10}$$

using γ_i, $i = 0, 1$, closed sets with non-void interiors γ_i°, $i = 0, 1$, such that

$$\gamma_0 \cup \gamma_1 = \partial\Omega, \quad \gamma_0^\circ \cap \gamma_1^\circ = \emptyset.$$

We can use again the above example (4.8), $u = r^m \sin m\theta$ in the sector $0 < \theta < \alpha$. Here we take $\gamma_1 : \theta = 0$ and $\gamma_2 : \theta = \alpha$. Actually, this harmonic function satisfies the second boundary condition of (4.10) if and only if $m\alpha = \pi/2$. Hence the requirement $u \in H^2(\Omega)$ near $r = 0$ is reduced to $\theta \leq \pi/2$.

As we have described, H^2-regularity actually holds in both (4.7) and (4.9) if either $\partial\Omega$ is C^2 or Ω is convex. Since these results are localized, this H^2-regularity holds also to (4.10), provided that $\gamma_1 \cap \gamma_2 = \emptyset$, that is, the portions where the Dirichlet and the Neumann boundary conditions are imposed are disconnected. Thus we obtain the H^2-regularity in (4.10) in this case, if either one of γ_i, $i = 1, 2$, is a boundary of a convex domain containing Ω and the other is C^2, or both γ_i, $i = 1, 2$, are C^2.

The elliptic regularity for corner domains has been studied in detail since [21]. Roughly speaking, the seeds of irregularity of the solution are only of finite dimension, which derives the Fredholm property of the elliptic operator [9]. For a convex domain on the plane we have $W^{1,p}$-regularity for $|p-2| \ll 1$ under the Dirichlet boundary condition [23]. This $W^{1,p}$-regularity of the Dirichlet problem holds for any bounded domain with C^1-boundary of arbitrary dimension [33]. From the above example we need $m > 1 - \frac{2}{p}$ for the property on the sector. Hence $W^{1,p}$-regularity on the planar domain for any $1 < p < \infty$ is valid only for the convex case.

4.2.2 H^1-Solution

Fundamental properties of Lipschitz domains are described in [14, 15, 22, 26] in accordance with the unique solvability of the boundary value problems for Laplace and Poisson equations, say, (4.6) and (4.7), respectively.

First, Poincaré's inequality

$$\|v\|_2 \leq C_1 \|\nabla v\|_2, \quad v \in H_0^1(\Omega), \tag{4.11}$$

guarantees that $-\Delta : H_0^1(\Omega) \to H^{-1}(\Omega) = H_0^1(\Omega)'$ is an isomorphism for any bounded domain Ω. Hence each $f \in H^{-1}(\Omega)$ takes a unique $u \in H_0^1(\Omega)$ such that

$$(\nabla u, \nabla v) = \langle v, f \rangle, \quad \forall v \in H_0^1(\Omega).$$

Here and henceforth, $(\ , \ )$ and $\langle \ , \ \rangle$ denote the standard L^2 norm and the duality pairing between $H_0^1(\Omega)$ and $H^{-1}(\Omega) = H_0^1(\Omega)'$, respectively. As far as the existence of the H^1-solution is concerned, therefore, the Dirichlet problem for the Laplace equation (4.6) is reduced to that of the Poisson equation (4.7) if the boundary value g is represented as $h|_{\partial\Omega}$ in the sense of trace for some $h \in H^1(\Omega)$. The following theorem actually guarantees this procedure.

Theorem 4.1 *If $\Omega \subset \mathbf{R}^N$ is a Lipschitz domain, the mapping*

$$v \in H^1(\Omega)/H_0^1(\Omega) \mapsto v|_{\partial\Omega} \in H^{1/2}(\partial\Omega)$$

is well-defined and an isomorphism. The integration by parts

$$(v, \frac{\partial w}{\partial x_i}) = -(\frac{\partial v}{\partial x_i}, w) + \int_{\partial\Omega} w n^i v, \quad 1 \leq i \leq N \tag{4.12}$$

is valid for $v, w \in H^1(\Omega)$, noting $n \in L^\infty(\partial\Omega, \mathbf{R}^N)$.

The following property also assures several fundamental properties of $H^1(\Omega)$ for the Lipschitz domain Ω such as the Sobolev embedding

$$H^1(\Omega) \hookrightarrow L^{2N/(N-2)}(\Omega)$$

for $N \geq 3$.

Theorem 4.2 *If $\Omega \subset \mathbf{R}^N$ is a Lipschitz domain then $C^\infty(\overline{\Omega})$ is dense in $H^1(\Omega)$ where*

$$C^\infty(\overline{\Omega}) = \{v : \overline{\Omega} \to \mathbf{R} \mid \exists \tilde{v} \in C_0^\infty(\mathbf{R}^N), \ \tilde{v}|_{\overline{\Omega}} = v\}.$$

In the following theorem the trace $\varphi|_{\partial\Omega} \in H^{1/2}(\partial\Omega)$ is taken for $\varphi \in H^1(\Omega)$. Also Δv of $v \in H^1(\Omega)$ is taken in the sense of distributions, recalling that

$$H_0^1(\Omega) \subset H^1(\Omega) \quad \Rightarrow \quad H^1(\Omega)' \subset H^{-1}(\Omega) = H_0^1(\Omega)'.$$

Finally, the inclusion

$$H^{1/2}(\partial\Omega) \hookrightarrow L^2(\partial\Omega)$$

is dense if $\partial\Omega$ is Lipschitz continuous, which realizes the Gel'fand triple

$$H^{1/2}(\partial\Omega) \hookrightarrow L^2(\partial\Omega) \cong L^2(\partial\Omega)' \hookrightarrow H^{-1/2}(\partial\Omega) = H^{1/2}(\partial\Omega)'.$$

Theorem 4.3 *If $\Omega \subset \mathbf{R}^N$ is a Lipschitz domain and $v \in H^1(\Omega)$ admits $\Delta v \in H^1(\Omega)'$ then there is a unique element in $H^{-1/2}(\partial\Omega)$ denoted by $\partial v/\partial n$, satisfying*

$$(\nabla v, \nabla\varphi) = \langle\varphi, -\Delta v\rangle + \langle\varphi, \frac{\partial v}{\partial n}\rangle, \quad \forall\varphi \in H^1(\Omega). \tag{4.13}$$

Proof We have only to show that the element $\partial v/\partial n \in H^{-1/2}(\partial\Omega)$ is defined by

$$\langle\varphi, \frac{\partial v}{\partial n}\rangle = (\nabla v, \nabla\varphi) + \langle\varphi, \Delta v\rangle\,, \quad \varphi \in H^1(\Omega),$$

that is, the right-hand side is independent of the choice of the extension of $\varphi \in H^{1/2}(\partial\Omega)$ to $H^1(\Omega)$. This property means

$$(\nabla v, \nabla\varphi) = \langle\varphi, -\Delta v\rangle\,, \quad \forall\varphi \in H_0^1(\Omega), \tag{4.14}$$

which follows from the definition of Δv, that is,

$$(\nabla v, \nabla\varphi) = \langle\varphi, -\Delta v\rangle\,, \quad \forall\varphi \in C_0^\infty(\Omega). \tag{4.15}$$

In fact (4.14) follows from (4.15) because $H_0^1(\Omega)$ is the closure of $C_0^\infty(\Omega)$ in $H^1(\Omega)$. □

The above $\partial v/\partial n \in H^{-1/2}(\partial\Omega)$ is not regarded as a function on $\partial\Omega$, which, however, is identified with the normal derivative of v in the Neumann boundary value problem such as (4.9), if it is Lipschitz continuous on $\overline{\Omega}$. Actually, the weak form of (4.9) is formulated as finding $u \in H^1(\Omega)$ such that

$$(\nabla u, \nabla\varphi) = \langle\varphi, f\rangle, \quad \forall\varphi \in H^1(\Omega), \tag{4.16}$$

using this notion of $\partial u/\partial n$. Then, given

$$f \in V = H^1(\Omega)', \quad \langle 1, f\rangle = 0, \tag{4.17}$$

there is a unique solution $u \in H^1(\Omega)$ to (4.16) satisfying

$$\int_\Omega u\, dx = 0.$$

To prove this fact, first, we note the Poincaré-Wirtinger inequality,

$$\|v\|_2 \le C_2\|\nabla v\|_2, \quad v \in V_0 \tag{4.18}$$

valid for

$$V_0 = \{v \in H^1(\Omega) \mid \int_\Omega v\, dx = 0\}.$$

Given $f \in H^1(\Omega)'$, next, we regard $\varphi \in V_0 \mapsto \langle \varphi, f\rangle$ as a bounded linear functional. Then the representation theorem of Riesz guarantees the existence of a unique $u \in V_0$ satisfying

$$(\nabla u, \nabla \varphi) = \langle \varphi, f\rangle, \quad \varphi \in V_0.$$

Finally, this u satisfies (4.16), if the second relation of (4.17) holds.

We have confirmed that if Ω is a Lipschitz domain and $v \in H^1(\Omega)$, then $v \in H^{1/2}(\partial\Omega)$ holds in the sense of trace. If $\Delta v \in H^1(\Omega)'$, furthermore, then

$$\frac{\partial v}{\partial n} \in H^{-1/2}(\partial\Omega) = H^{1/2}(\partial\Omega)'$$

is defined by (4.13). This mapping is continuous in the sense that

$$\left\|\frac{\partial v}{\partial n}\right\|_{H^{-1/2}(\partial\Omega)} \le C_3(\|\nabla v\|_{L^2(\Omega)} + \|\Delta v\|_{H^1(\Omega)'}).$$

We cannot, however, define

$$v \in H^{-1/2}(\partial\Omega)$$

for general $v \in L^2(\Omega)$, nor

$$\frac{\partial v}{\partial n} \in H^{-1/2}(\partial\Omega)$$

for general $v \in H^1(\Omega)$.

If $v \in H^2(\Omega)$, on the contrary, we have $\nabla v \in H^1(\Omega, \mathbf{R}^N)$. Using $n \in L^\infty(\partial\Omega, \mathbf{R}^N)$, we can thus define

$$n \cdot \nabla v \in L^2(\partial\Omega).$$

We have also $\Delta v \in L^2(\Omega) \hookrightarrow H^1(\Omega)'$ and hence $\partial v/\partial n \in H^{-1/2}(\partial\Omega)$ is defined by (4.13). These two terms are actually consistent.

Theorem 4.4 *Let $\Omega \subset \mathbf{R}^N$ be a Lipschitz domain and $v \in H^2(\Omega)$. Then the above $n \cdot \nabla v \in L^2(\partial\Omega)$ is identified with $\partial v/\partial n \in H^{-1/2}(\partial\Omega)$ in Theorem 4.3 as a distribution on $\partial\Omega$.*

Proof Let $\omega \subset \mathbf{R}^N$ be an open set containing $\partial\Omega$. Take $\varphi \in C_0^\infty(\omega)$ and its zero extension which is regarded as an element in $H^1(\Omega)$.

Given $v \in H^2(\Omega)$, then, we have

$$\left\langle \varphi, \frac{\partial v}{\partial n} \right\rangle = (\nabla v, \nabla \varphi) + (\varphi, \Delta v)$$

by (4.13). The right-hand side is now equal to $\langle \nabla v, n\varphi \rangle$ by (4.12). Hence we obtain the conclusion by Theorem 4.3. □

4.2.3 Sectional Curvatures

If $\Omega \subset \mathbf{R}^N$ is a Lipschitz domain, we can define the standard measure on $\Gamma = \partial\Omega$ using the local chart. Also any Lipschitz continuous function on Γ takes a Lipschitz continuous extension to its neighborhood through a flow to be used in Sect. 4.3.1. The tangential space $T_\xi(\Gamma)$, furthermore, is defined for almost every $\xi \in \Gamma$ with the outer unit normal vector denoted by n. Using an orthonormal basis $\{s_i \mid i = 1, \cdots, N-1\}$ of $T_\xi(\Gamma)$, we thus obtain the frame

$$\{s_1, \cdots, s_{N-1}, n\} \tag{4.19}$$

almost everywhere on Γ.

If $\Omega \subset \mathbf{R}^N$ is a $C^{1,1}$ domain we can take a Lipschitz continuous frame (4.19) on Γ. Then we cut Γ by the plane made by $\{s_i, n\}$, $1 \leq i \leq N-1$, denoted by Π, to obtain a curve $\gamma = \Pi \cap \Gamma$. Taking the parametrization $x = x(s) \in \gamma$ with the arc-length s, we have

$$s_i = \dot{x} \equiv \frac{\partial x}{\partial s}, \quad |s_i| = 1,$$

which implies

$$\dot{s}_i \cdot s_i = 0.$$

Since $\dot{s}_i \in \Pi$, we define the *sectional curvature* κ_i by

$$\dot{s}_i = -\kappa_i n, \tag{4.20}$$

recalling n is outer with respect to Ω. On the other hand, we have

$$n \cdot s_i = 0, \quad |n| = 1,$$

which results in $\dot{n} \cdot s_i = \kappa_i$ and $\dot{n} \cdot n = 0$ along this parametrization. Hence it holds that

$$\dot{n} = \kappa_i s_i \tag{4.21}$$

by $\dot{n} \in \Pi$.

Concerning s_j for $j \neq i$, we use

$$s_j \cdot s_j = s_j \cdot n = 0, \quad |s_j| = 1.$$

Since $\dot{s}_j$ defined on γ belongs to the space generated by s_i, s_j, n, it follows that

$$\dot{s}_j = 0 \tag{4.22}$$

similarly. We write the relations (4.20) and (4.22) as

$$\frac{\partial s_i}{\partial s_i} = -\kappa_i n, \quad \frac{\partial s_j}{\partial s_i} = 0, \ j \neq i, \tag{4.23}$$

parametrizing γ by s_i. Hence the vector s_j on γ is parametrized by the line elements s_i, $1 \leq i \leq N-1$, in (4.23).

Next, we use a tubular neighborhood $\mathcal{O}$ and extend the above frame (4.19) to a neighborhood of Γ. Hence we rewrite γ locally as a graph of $z = f(y)$ satisfying $f(0) = f'(y) = 0$ to take

$$\mathcal{O} = \{(y, f(y) + t) \mid |y|, \ |t| < \varepsilon\}, \quad 0 < \varepsilon \ll 1.$$

We thus have a fiber of γ on Π with the length parameter t, which results in the parametrization $n = n(t, s)$ and $s_i = s_i(t, s)$ satisfying

$$n' \equiv \frac{\partial n}{\partial t} = 0 \quad \text{at } t = s = 0.$$

Then we obtain $s_i' = 0$ there, by

$$n \cdot s_i = 0, \quad |s_i| = 1.$$

Using notations similar to (4.23), we end up with (4.21), the Frenet-Serret formula

$$\frac{\partial n}{\partial s_i} = \kappa_i s_i, \quad 1 \leq i \leq N-1 \tag{4.24}$$

and also

$$\frac{\partial n}{\partial n} = 0, \tag{4.25}$$

where $\partial/\partial n$ denotes the directional derivative toward $n \in S^{N-1}$.

Henceforth, we put

$$a \otimes b = (a_i b_j) = a\,{}^t b$$

for the vectors $a = (a_i)$ and $b = (b_i)$, and also

$$\nabla n = {}^t(\nabla \otimes n) = \left(\frac{\partial n^j}{\partial x_k}\right)_{jk} \tag{4.26}$$

for $n = (n^j)$. Using the frame (4.19) as an orthonormal system in $\mathbf{R}^N$, we have

$${}^t(\nabla n) = \sum_{i=1}^{N-1} \frac{\partial n}{\partial s_i} \otimes s_i + \frac{\partial n}{\partial n} \otimes n = \sum_{i=1}^{N-1} \kappa_i s_i \otimes s_i \tag{4.27}$$

by (4.24) and (4.25). It holds that

$$\nabla \cdot n = \mathrm{tr}(\nabla n) = \sum_{i=1}^{N-1} \kappa_i. \tag{4.28}$$

This value called the mean curvature is independent of the choice of frame.

Remark 4.1 By the notation of (4.26) above, it holds that

$$\nabla \otimes (f\boldsymbol{a}) = (\nabla f) \otimes \boldsymbol{a} + f(\nabla \otimes \boldsymbol{a}) = (\nabla f) \otimes \boldsymbol{a} + f^t(\nabla \boldsymbol{a}),$$

where f and $\boldsymbol{a}$ are scalar and vector fields, respectively.

Lemma 4.1 *If $\Omega \subset \mathbf{R}^N$ is a $C^{1,1}$-domain and $f \in C^{1,1}(\overline{\Omega})$ then it holds that*

$$\nabla^2 f = \sum_{i=1}^{N-1} {}^t(\nabla s_i)\frac{\partial f}{\partial s_i} + {}^t(\nabla n)\frac{\partial f}{\partial n} + \sum_{i,j=1}^{N-1} s_i \otimes s_j \frac{\partial^2 f}{\partial s_i \partial s_j}$$

$$+ \sum_{i=1}^{N-1} (s_i \otimes n + n \otimes s_i)\frac{\partial^2 f}{\partial s_i \partial n} + n \otimes n \frac{\partial^2 f}{\partial n^2} \tag{4.29}$$

a.e. on $\Gamma = \partial\Omega$, where $\nabla^2 f = \left(\frac{\partial^2 f}{\partial x_i \partial x_j}\right)$ denotes the Hesse matrix of f.

Proof Using the orthonormal coordinate (4.19), we have

$$\begin{aligned}\nabla^2 f &= \nabla \otimes \nabla f = \nabla \otimes \left(\sum_{i=1}^{N-1} s_i \frac{\partial f}{\partial s_i} + n \frac{\partial f}{\partial n} \right) \\ &= \sum_{i=1}^{N-1} {}^t(\nabla s_i) \frac{\partial f}{\partial s_i} + {}^t(\nabla n) \frac{\partial f}{\partial n} + \sum_{i=1}^{N-1} \nabla \frac{\partial f}{\partial s_i} \otimes s_i + \nabla \frac{\partial f}{\partial n} \otimes n.\end{aligned}$$

Here the third and the fourth terms on the right-hand side are equal to

$$\begin{aligned}\sum_{i=1}^{N-1} \nabla \left(\frac{\partial f}{\partial s_i} \right) \otimes s_i &= \sum_{i=1}^{N-1} \left(\sum_{j=1}^{N-1} s_j \frac{\partial^2 f}{\partial s_j \partial s_i} + n \frac{\partial^2 f}{\partial n \partial s_i} \right) \otimes s_i \\ &= \sum_{i,j=1}^{N-1} [s_j \otimes s_i] \frac{\partial^2 f}{\partial s_i \partial s_j} + \sum_{i=1}^{N} [n \otimes s_i] \frac{\partial^2 f}{\partial s_i \partial n}\end{aligned}$$

and

$$\begin{aligned}\nabla \left(\frac{\partial f}{\partial n} \right) \otimes n &= \left(\sum_{i=1}^{N-1} s_i \frac{\partial^2 f}{\partial s_i \partial n} + n \frac{\partial^2 f}{\partial n^2} \right) \otimes n \\ &= \sum_{i=1}^{N-1} (s_i \otimes s_j) \frac{\partial^2 f}{\partial s_i \partial n} + (n \otimes n) \frac{\partial^2 f}{\partial n^2},\end{aligned}$$

respectively. Hence it follows that

$$\begin{aligned}\nabla^2 f = &\sum_{i=1}^{N-1} {}^t(\nabla s_i) \frac{\partial f}{\partial s_i} + {}^t(\nabla n) \frac{\partial f}{\partial n} + \sum_{i,j=1}^{N-1} s_i \otimes s_j \frac{\partial^2 f}{\partial s_i \partial s_j} \\ &+ \sum_{i=1}^{N-1} (s_i \otimes n + n \otimes s_i) \frac{\partial^2 f}{\partial s_i \partial n} + n \otimes n \frac{\partial^2 f}{\partial n^2}.\end{aligned}$$

The proof is complete. □

The above lemma implies the following result.

Corollary 4.1 *If $\Omega \subset \mathbf{R}^N$ is a $C^{1,1}$-domain and $f \in C^{1,1}(\overline{\Omega})$ then it holds that*

$$\Delta f = (\nabla \cdot n) \frac{\partial f}{\partial n} + \frac{\partial^2 f}{\partial n^2} + \sum_{i=1}^{N-1} \frac{\partial^2 f}{\partial s_i^2} \quad \text{a.e. on } \Gamma. \tag{4.30}$$

Proof In (4.29) we have

$$s_k \otimes n + n \otimes s_k = (s_k^i n^j + n^i s_k^j)_{ij}$$

and hence

$$\mathrm{tr}(s_k \otimes n + n \otimes s_k) = \sum_i (s_k^i n^i + n^i s_k^i) = 2n \cdot s_k = 0.$$

We have also

$$\mathrm{tr}[{}^t(\nabla n)] = \nabla \cdot n, \quad \mathrm{tr}(n \otimes n) = |n|^2 = 1, \quad \mathrm{tr}[{}^t(\nabla s_i)] = 0$$

by

$$[{}^t(\nabla s_i)]s_k \cdot s_\ell = \frac{\partial s_i}{\partial s_k} \cdot s_\ell = -\kappa_i \delta_{ik} n \cdot s_\ell = 0, \quad [{}^t(\nabla s_i)]n \cdot n = \frac{\partial s_i}{\partial n} \cdot n = 0.$$

Then (4.30) follows. □

4.2.4 Convex Domains

The standard elliptic regularity is concerned with a bounded domain $\Omega \subset \mathbf{R}^N$ with C^2 boundary $\partial\Omega$. Given $f \in L^2(\Omega)$, thus, the solution $u = u(x) \in H_0^1(\Omega)$ to

$$-\Delta u = f \text{ in } \Omega, \quad u|_{\partial\Omega} = 0$$

belongs to $H^2(\Omega)$. Convexity of the domain, however, compensates the lack of smoothness of C^2 for this H^2 regularity of the solution [16, 19].

Theorem 4.5 (Kadlec) *If $\Omega \subset \mathbf{R}^N$ is a convex bounded open set and $f \in L^2(\Omega)$, there is a unique $u \in H^2(\Omega)$ such that*

$$-\Delta u = f \text{ in } \Omega, \quad u|_{\partial\Omega} = 0. \tag{4.31}$$

Theorem 4.6 (Grisvard–Iooss) *If $\Omega \subset \mathbf{R}^N$ is a convex bounded open set and $f \in L^2(\Omega)$, there is a unique $u \in H^2(\Omega)$ such that*

$$-\Delta u + u = f \text{ in } \Omega, \quad \left.\frac{\partial u}{\partial n}\right|_{\partial\Omega} = 0, \tag{4.32}$$

where n denotes the outer normal unit vector.

Remark 4.2 Equations (4.31) and (4.32 mean

$$u \in H_0^1(\Omega), \quad (\nabla u, \nabla \varphi) = (f, \varphi), \ \forall \varphi \in H_0^1(\Omega)$$

and

$$u \in H^1(\Omega), \quad (\nabla u, \nabla \varphi) + (u, \varphi) = (f, \varphi), \ \forall \varphi \in H^1(\Omega),$$

respectively. The unique existence of these solutions is a consequence of the representation theorem of Riesz. Hence the above theorems are concerned with the regularity of the solution.

These theorems are based on an identity valid for Lipschitz Ω, where the outer unit normal vector n belongs to $L^\infty(\partial\Omega)$. Henceforth,

$$\gamma : H^1(\Omega) \to H^{1/2}(\partial\Omega)$$

denotes the trace operator assured by the Lipschitz continuity of $\partial\Omega$. Henceforth we put $\{e_1, \cdots, e_{N-1}, n\}$ for the frame of $\mathscr{M}$, and s_i, $1 \leq i \leq N-1$, for the arc-length parameter along $\mathscr{C}_i$, the curve made by the cross section of $\mathscr{M} = \partial\Omega$ and the plane generated by $\{e_i, n\}$.

The second fundamental form may be identified with

$$\mathscr{B}(\xi, \eta) = -\frac{\partial n}{\partial \xi} \cdot \eta$$

defined for

$$\xi = \sum_{i=1}^{N-1} \xi_i e_i, \quad \eta = \sum_{i-1}^{N-1} \eta_i e_i \in \mathscr{M}.$$

It holds that

$$\mathscr{B}(\xi, \eta) = -\sum_{j,k} \frac{\partial n}{\partial s_j} \cdot e_k \xi_j \eta_k = -\sum_{i=1}^{N-1} \kappa_i \xi_i \eta_i, \tag{4.33}$$

where κ_i, $1 \leq i \leq N-1$, are the sectional curvatures satisfying

$$\frac{\partial \nu}{\partial s_j} \cdot e_k = \delta_{jk} \kappa_j.$$

The trace of the bilinear form, therefore, is equal to the mean curvature,

$$\mathrm{tr}\, \mathscr{B} = -\sum_{i=1}^{N-1} \kappa_i = -\nabla \cdot n.$$

Finally, the tangential part and the tangential derivative of the vector field ζ and the scalar field z on $\mathscr{M}$, denoted by ζ_T and $\nabla_T z$, respectively, are defined by

$$\zeta_T = \zeta - (\zeta \cdot n)n, \quad \nabla_T z = \nabla z - (n \cdot \nabla z)n.$$

Theorem 4.7 *If $\Omega \subset \mathbf{R}^N$ is a bounded Lipschitz domain and $v \in H^1(\Omega)^N$, it holds that*

$$\int_\Omega (\nabla \cdot v)^2 - \sum_{i,j=1}^N \frac{\partial v_i}{\partial x_i}\frac{\partial v_j}{\partial x_i}\, dx = -2\,\langle(\gamma v)_T, \nabla_T(\gamma v \cdot n)\rangle$$
$$- \int_{\partial\Omega} \mathscr{B}((\gamma v)_T, (\gamma v)_T) + (tr\, \mathscr{B})[(\gamma v \cdot n)^2]\, dS.$$

The following result provides a starting point of the proof of Theorem 4.5.

Corollary 4.2 *If $\Omega \subset \mathbf{R}^N$ is a bounded convex Lipschitz domain and $u \in H^2 \cap H_0^1(\Omega)$, it holds that*

$$\sum_{i,j=1}^N \int_\Omega (\frac{\partial^2 u}{\partial x_i \partial x_j})^2\, dx \leq \int_\Omega (\Delta u)^2\, dx. \tag{4.34}$$

Proof Given $u \in H^2 \cap H_0^1(\Omega)$, put

$$v = \nabla u \in H^1(\Omega)^N.$$

It holds that

$$(\gamma v)_T = 0.$$

From the assumption on Ω, we have $\kappa_i \geq 0$ for $1 \leq i \leq N-1$. Then it follows that

$$\sum_{i,j=1}^N \int_\Omega (\frac{\partial^2 u}{\partial x_i \partial x_j})^2\, dx = \sum_{i,j=1}^N \int_\Omega \frac{\partial v_i}{\partial x_j}\frac{\partial v_j}{\partial x_i}\, dx$$
$$\leq \int_\Omega (\nabla \cdot v)^2\, dx = \int_\Omega (\Delta u)^2\, dx$$

from Theorem 4.7, and the proof is complete. □

Lemma 4.2 *Under the assumption of Theorem 4.7 it holds that*

$$
\begin{aligned}
&(v \cdot n)\nabla \cdot v - [(v \cdot \nabla)v] \cdot n \\
&\quad = \nabla_T \cdot ((v \cdot n)v_T) - (tr\, \mathscr{B})(v \cdot n)^2 - \mathscr{B}(v_T, v_T) - 2v_T \cdot \nabla_T(v \cdot n)
\end{aligned}
$$

on $\mathscr{M} = \partial\Omega$.

Proof We have (4.33),

$$
\mathscr{B}(\xi, \eta) = -\sum_{j,k} \frac{\partial n}{\partial s_j} \cdot e_k \xi_j \eta_k, \quad \xi = \sum_j \xi_j e_j, \ \eta = \sum_j \eta_j e_j,
$$

and

$$
\nabla u = \nabla_T u + \frac{\partial u}{\partial n} n, \quad v = v_T + v_n n, \ v_n = v \cdot n \quad \text{in } \mathbf{R}^N,
$$

where u and v are scalar and vector fields near $\mathscr{M} = \partial\Omega$, respectively. Using

$$
v = \sum_k v_k e_k + v_n n,
$$

we obtain

$$
\begin{aligned}
\nabla \cdot v &= \sum_j \frac{\partial v}{\partial s_j} \cdot e_j + \frac{\partial v}{\partial n} \cdot n \\
&= \sum_{j,k} (\frac{\partial v_k}{\partial s_j} e_k + v_k \frac{\partial e_k}{\partial s_j}) \cdot e_j + \sum_j (\frac{\partial v_n}{\partial s_j} n + v_n \frac{\partial n}{\partial s_j}) \cdot e_j \\
&\quad + \sum_k (\frac{\partial v_k}{\partial n} e_k + v_k \frac{\partial e_k}{\partial n}) \cdot n + (\frac{\partial v_n}{\partial n} n + v_n \frac{\partial n}{\partial n}) \cdot n
\end{aligned}
$$

with

$$
\frac{\partial n}{\partial n} \cdot n = 0.
$$

Here we use a frame, called the *Fermi coordinate*, satisfying

$$
\frac{\partial e_k}{\partial n} \cdot n = 0, \tag{4.35}
$$

to obtain

$$
\nabla \cdot v = \sum_j \frac{\partial v_j}{\partial s_j} + \sum_{j,k} v_k \frac{\partial e_k}{\partial s_j} \cdot e_j + v_n \sum_j \frac{\partial n}{\partial s_j} \cdot e_j + \frac{\partial v_n}{\partial n}. \tag{4.36}
$$

It holds also that

$$
\begin{aligned}
(v \cdot \nabla)v &= \sum_j v_j \frac{\partial v}{\partial s_j} + v_n \frac{\partial v}{\partial n} \\
&= \sum_{j,k} v_j (\frac{\partial v_k}{\partial s_j} e_k + v_k \frac{\partial e_k}{\partial s_j}) + \sum_j v_j (\frac{\partial v_n}{\partial s_j} n + v_n \frac{\partial n}{\partial s_j}) \\
&\quad + \sum_j v_n (\frac{\partial v_k}{\partial n} e_k + v_k \frac{\partial e_k}{\partial n}) + v_n (\frac{\partial v_n}{\partial n} n + v_k \frac{\partial n}{\partial n})
\end{aligned}
$$

and hence

$$
((v \cdot \nabla)v) \cdot n = \sum_{j,k} v_j v_k \frac{\partial e_k}{\partial s_j} \cdot n + \sum_j v_j \frac{\partial v_n}{\partial s_j} + v_n \frac{\partial v_n}{\partial n} \tag{4.37}
$$

by (4.35) and

$$
\frac{\partial n}{\partial s_j} \cdot n = 0, \quad \frac{\partial n}{\partial n} = 0.
$$

Then we obtain

$$
\begin{aligned}
&v_n \nabla \cdot v - ((v \cdot \nabla)v) \cdot n \\
&= v_n \sum_j \frac{\partial v_j}{\partial s_j} + v_n \sum_{j,k} v_k \frac{\partial e_k}{\partial s_j} \cdot e_j + v_n^2 \sum_j \frac{\partial n}{\partial s_j} \cdot e_j \\
&\quad - \sum_{j,k} v_j v_k \frac{\partial e_k}{\partial s_j} \cdot n - \sum_j v_j \frac{\partial v_n}{\partial s_j} \\
&= v_n \sum_j \frac{\partial v_j}{\partial s_j} + v_n \sum_{j,k} v_k \frac{\partial e_k}{\partial s_j} \cdot e_j - (\mathrm{tr}\, \mathscr{B}) v_n^2 \\
&\quad - \mathscr{B}(v_T, v_T) - \sum_j v_j \frac{\partial v_n}{\partial s_j}.
\end{aligned} \tag{4.38}
$$

Finally,

$$
\begin{aligned}
\nabla_T \cdot (v_n v_T) &= \sum_j \frac{\partial (v_n v_T)}{\partial s_j} \cdot e_j = \sum_{j,k} \frac{\partial (v_n v_k e_k)}{\partial s_j} \cdot e_j \\
&= \sum_{j,k} \{ \frac{\partial v_n}{\partial s_j} v_k e_k + v_n \frac{\partial v_k}{\partial s_j} e_k + v_n v_k \frac{\partial e_k}{\partial s_j} \} \cdot e_j
\end{aligned}
$$

$$= \sum_{j,k} \frac{\partial v_n}{\partial s_j} v_j + v_n \sum_j \frac{\partial v_j}{\partial s_j} + v_n \sum_{j,k} v_k \frac{\partial e_k}{\partial s_j} \cdot e_j. \tag{4.39}$$

By (4.37)–(4.39) we obtain

$$\begin{aligned} &v_n \nabla \cdot v - ((v \cdot \nabla) v) n \\ &= \nabla_T \cdot (v_n v_T) - (\operatorname{tr} \mathscr{B}) v_n^2 - \mathscr{B}(v_T, v_T) - 2 \sum_j v_j \frac{\partial v_n}{\partial s_j} \\ &= \nabla_T \cdot (v_n v_T) - (\operatorname{tr} \mathscr{B}) v_n^2 - \mathscr{B}(v_T, v_T) - 2 v_T \cdot \nabla_T v_n, \end{aligned}$$

and the proof is complete. □

Proof *(Theorem 4.7)* We assume $v \in H^2(\Omega)^N$, relying on a standard argument of approximation. By Green's formula we have

$$\begin{aligned} \int_\Omega (\nabla \cdot v)^2 \, dx &= \sum_{i,j=1}^N \int_\Omega \frac{\partial v_i}{\partial x_i} \frac{\partial v_j}{\partial x_j} \, dx \\ &= - \sum_{i,j=1}^N \int_\Omega v_i \frac{\partial^2 v_i}{\partial x_i \partial x_j} \, dx + \sum_{i,j=1}^N \int_{\partial\Omega} v_i \frac{\partial v_j}{\partial x_j} n_j \, dS \\ &= \sum_{i,j=1}^N \int_\Omega \frac{\partial v_i}{\partial x_j} \frac{\partial v_j}{\partial x_i} \, dx + \sum_{i,j=1}^N \int_{\partial\Omega} v_i \frac{\partial v_j}{\partial x_j} n_i - v_i \frac{\partial v_j}{\partial x_i} n_j \, dS, \end{aligned}$$

which implies

$$\begin{aligned} I(v) &\equiv \int_\Omega (\nabla \cdot v)^2 - \sum_{i,j=1}^N \frac{\partial v_i}{\partial x_j} \frac{\partial v_j}{\partial x_i} \, dx \\ &= \int_{\partial\Omega} (v \cdot n) \nabla \cdot v - [(v \cdot \nabla) v] \cdot n \, dS. \end{aligned}$$

Then Lemma 4.2 is applicable to the right-hand side.

If ζ is a vector field on $\mathscr{M} = \partial\Omega$, we obtain

$$\nabla_T \cdot \zeta = \sum_{j=1}^{N-1} \frac{\partial \zeta}{\partial s_j} \cdot s_j = \sum_{j,k=1}^{N-1} (\frac{\partial}{\partial s_j} [(\zeta \cdot s_k) s_k]) \cdot s_j$$

and hence

$$\int_{\partial\Omega} \nabla_T \cdot ((v \cdot n) v_T) \, dS = 0.$$

Then the result follows. □

Applying Theorem 4.7 to

$$v = \nabla u, \qquad \left. \frac{\partial u}{\partial n} \right|_{\partial\Omega} = 0,$$

we obtain

$$\int_{\Omega} (\Delta u)^2 - \sum_{i,j=1} (\frac{\partial^2 u}{\partial x_i \partial x_j})^2 \, dx = -\int_{\partial\Omega} \mathscr{B}(\nabla_T u, \nabla_T u) \, dS \tag{4.40}$$

by $\gamma v \cdot n = 0$. If Ω is convex, the bilinear form $\mathscr{B}(\cdot, \cdot)$ is non-positive by (4.33). Hence there arises (4.34), which is a base for the proof of Theorem 4.6, similarly to Corollary 4.2 for Theorem 4.5.

4.3 Free Boundary Problems

4.3.1 Domain Perturbations

We continue to assume that $\Omega \subset \mathbf{R}^N$ is a Lipschitz domain. Let

$$T_t : \overline{\Omega} \to \overline{\Omega_t}, \ \Omega_t = T_t \Omega, \ |t| < \varepsilon, \quad 0 < \varepsilon \ll 1$$

be a family of bi-Lipschitz homeomorphisms, with $T_t x$ being once or twice differentiable in t for each $x \in \overline{\Omega}$. Henceforth, we say that $\{T_t\}$ is once or twice differentiable in t for short, if

$$T_t = I + tS + o(t), \quad T_t = I + tS + \frac{1}{2} t^2 R + o(1)$$

holds uniformly on $\overline{\Omega}$, respectively, with Lipschitz continuous T, S, and R. Then we put

$$\delta\rho = \left. \frac{\partial T_t}{\partial t} \right|_{t=0} \cdot n = S \cdot n, \quad \delta^2 \rho = \left. \frac{\partial^2 T_t}{\partial t^2} \right|_{t=0} \cdot n = R \cdot n.$$

To perform the boundary trial method for free boundary problems we need to construct such domain perturbations numerically [42].

Example 4.1 (Dynamical Perturbations) Let $v = v(x)$ be a Lipschitz vector field defined in a domain $\tilde{\Omega}$ containing $\overline{\Omega}$. Given $x \in \tilde{\Omega}$, we take the ordinary differential equation

$$\frac{dc(t)}{dt} = v(c(t)), \quad c(0) = x$$

to define the integral curve $c : (-\varepsilon, \varepsilon) \to \mathbf{R}^n$, $|t| < \varepsilon$. Writing $c(t) = T_t x$, we obtain a family of bi-Lipschitz homeomorphisms $T_t : \overline{\Omega} \to \overline{\Omega_t}$, $|t| \ll 1$. Then it holds that

$$Sx = \left.\frac{\partial}{\partial t} T_t x\right|_{t=0} = v(T_t x)|_{t=0} = v(x)$$

$$Rx = \left.\frac{\partial^2}{\partial t^2} T_t x\right|_{t=0} = \left.\frac{\partial}{\partial t} v(T_t x)\right|_{t=0} = [(v \cdot \nabla)v](x).$$

This $\{T_t\}$ is a differentiable deformation and is twice-differentiable if ∇v is furthermore Lipschitz continuous in $\tilde{\Omega}$. Then it holds that

$$T_t(x) = x + tv(x) + \frac{t^2}{2}[(v \cdot \nabla)v](x) + o(t^2)$$

uniformly in $x \in \overline{\Omega}$.

Example 4.2 (Normal Perturbations) If $\partial\Omega$ is $C^{1,1}$, the unit normal vector $n = n_x$ is Lispschitz continuous on $\Gamma = \partial\Omega$. Then we can take a bi-Lipschitz deformation of Γ by

$$\Gamma_t : x + t \cdot \delta\rho(x) n_x, \quad x \in \Gamma$$

for $|t| \ll 1$, where $\delta\rho = \delta\rho(x)$, $x \in \Gamma$, is a Lipschitz continuous function. Then there is a domain $\Omega_t \subset \mathbf{R}^N$ such that $\Gamma_t = \partial\Omega_t$. There is, also, a bi-Lipschitz mapping $T_t : \overline{\Omega} \to \overline{\Omega_t}$, $|t| \ll 1$, which satisfies

$$\delta\rho = \left.\frac{\partial T_t}{\partial t}\right|_{t=0} \cdot n, \quad \delta^2\rho = \left.\frac{\partial^2 T_t}{\partial t^2}\right|_{t=0} \cdot n = 0 \qquad \text{on } \Gamma = \partial\Omega.$$

Although this deformation has been used in the classical theory [12], it does not work for the general Lipschitz domain, for example, if $\partial\Omega$ has a corner.

Definition 4.1 Given $\gamma \subset \Gamma = \partial\Omega$, we say that the non-peeling off (NPO) condition on γ is achieved if $S \cdot n = 0$ on γ.

An NPO. deformation is used in the filtration problem [29]. It is realized if $\delta\rho = 0$ on γ in the normal perturbation. Dynamical perturbation achieves it when $v \cdot n = 0$ on γ. An advantage of the dynamical perturbation is the group property of $\{T_t\}$ indicated by $T_t \circ T_s = T_{t+s}$ for $|t|, |s| \ll 1$, $T_0 = I$.

4.3.2 *Variational Jacobian Formulae*

We follow the standard notation of matrices $A = (a_{ij})_{i:1\downarrow N, j:1\rightarrow N}$. The transpose of the matrix A is denoted by ${}^tA$. Given $A = (a_{ij})$ and $B = (b_{ij})$, we put

$$A : B = \sum_{i,j} a_{ij} b_{ij}.$$

For the Lipschitz continuous vector field T on $\overline{\Omega}$, let

$$J(T) = {}^t\left(\frac{\partial y_\ell}{\partial x_k}\right)_{k,\ell}$$

be its Jacobi matrix, where $y = Tx$.

Theorem 4.8 ([44]) *Let $\Omega \subset \mathbf{R}^N$ be a Lipschitz domain and $\{T_t\}$ be a differentiable deformation. Then it holds that*

$$\frac{\partial}{\partial t} \det J(T_t)\Big|_{t=0} = \nabla \cdot S. \tag{4.41}$$

If $\{T_t\}$ is twice differentiable, furthermore, we have

$$\frac{\partial^2}{\partial t^2} \det J(T_t)\Big|_{t=0} = \nabla \cdot R + (\nabla \cdot S)^2 - J(S) : {}^tJ(S). \tag{4.42}$$

Proof For simplicity we take the case that $\{T_t\}$ is twice differentiable. Then both $S = {}^t(S^1, \ldots, S^N)$ and $R = {}^t(R^1, \cdots, R^N)$ are Lipschitz continuous. Put

$$S^i_j = \frac{\partial S^i}{\partial x_j}, \quad R^i_j = \frac{\partial R^i}{\partial x_j}. \tag{4.43}$$

It holds that

$$J(T_t) = \begin{pmatrix} 1 + tS_1^1 + \frac{1}{2}t^2R_1^1 & \cdots & tS_N^1 + \frac{1}{2}t^2R_N^1 \\ \vdots & \cdots & \vdots \\ tS_1^N + \frac{1}{2}t^2R_1^N & \cdots & 1 + tS_N^N + \frac{1}{2}t^2R_N^N \end{pmatrix} + o(t^2)$$

and hence

$$\det J(T_t) = 1 + t\nabla \cdot S + t^2 \sum_{i<j} (S_i^i S_j^j - S_j^i S_i^j) + \frac{t^2}{2} \sum_i R_i^i + o(t^2).$$

Then we obtain (4.41) and (4.42) by

$$\begin{aligned} \frac{\partial^2}{\partial t^2} \det J(T_t)\bigg|_{t=0} &= 2\sum_{i<j} (S_i^i S_j^j - S_j^i S_i^j) + \nabla \cdot R \\ &= \sum_{i \neq j} (S_i^i S_j^j - S_j^i S_i^j) + \nabla \cdot R = \sum_{i,j} (S_i^i S_j^j - S_j^i S_i^j) + \nabla \cdot R \\ &= \nabla \cdot R + (\nabla \cdot S)^2 - J(S) : {}^t J(S) \end{aligned}$$

and the proof is complete. □

4.3.3 *Volume Derivatives*

Liouville's transformation theory is a set of variational formulae concerning volume and surface integrals described in Lagrange coordinates. These formulas are used to derive a representation of boundary integrals of the first variation of Green's function. Recall the notations in Sect. 4.3.1, and put

$$\dot{c} = c_t|_{t=0}, \quad \ddot{c} = c_{tt}|_{t=0}.$$

Theorem 4.9 (First Volume Derivative) *Let $\Omega \subset \mathbf{R}^N$ be a Lipschitz domain and $\{T_t\}$ be a differentiable deformation. Let, furthermore, $c = c(x,t)$ be a Lipschitz continuous function in $\bigcup_t \Omega_t \times \{t\}$ for $\Omega_t = T_t\Omega$. Then it holds that*

$$\frac{d}{dt} \int_{\Omega_t} c \bigg|_{t=0} = \int_\Omega \dot{c} + \nabla \cdot (cS)|_{t=0} \, dx = \int_\Omega \dot{c} \, dx + \int_{\partial\Omega} c|_{t=0} \, \delta\rho \, dS. \tag{4.44}$$

Proof Letting $J_t(x) = J(T_t x)$, we have

$$\int_{\Omega_t} c(y,t)\,dy = \int_\Omega c(T_t x, t) \det J_t(x)\,dx$$

and hence

$$\frac{d}{dt}\int_{\Omega_t} c(y,t)\,dy\bigg|_{t=0}$$
$$= \int_\Omega \left(\dot{c}(T_t x, t) + \nabla c(T_t x, t) \cdot \frac{\partial T_t x}{\partial t} \right) \det J_t(x) + c(T_t x, t)\frac{\partial}{\partial t}\det J_t(x)\bigg|_{t=0} dx$$
$$= \int_\Omega \dot{c} + \nabla c \cdot S + c\nabla \cdot S|_{t=0}\,dx$$

by (4.41). Then (4.44) follows from the divergence formula of Gauss. □

Theorem 4.10 (Second Volume Derivative) *Let $\{T_t\}$ be twice differentiable and $c = c(x,t)$ be $C^{1,1}$ in $\bigcup_t \Omega_t \times \{t\}$. Then it holds that*

$$\frac{d^2}{dt^2}\int_{\Omega_t} c\bigg|_{t=0} = \int_\Omega \ddot{c}$$
$$+ \int_{\partial\Omega} (2\dot{c} + [(S\cdot\nabla)c])\delta\rho + c(\delta^2\rho - S_T \cdot \nabla_T \delta\rho + \delta\rho\nabla_T \cdot S_T)\bigg|_{t=0} dS$$
$$- \int_{\partial\Omega} c|_{t=0} \left\{ (tr\,\mathscr{B})(\delta\rho)^2 + \mathscr{B}(S_T, S_T) \right\} dS, \tag{4.45}$$

recalling $\delta\rho = S\cdot n$ and $\delta^2\rho = R\cdot n$.

Proof First, we have

$$\frac{d^2}{dt^2}\int_{\Omega_t} c\bigg|_{t=0} = \int_\Omega \frac{\partial^2}{\partial t^2} c(T_t x, t) \det J_t x\bigg|_{t=0} = \int_\Omega \ddot{c} + 2\nabla\dot{c}\cdot S + (\nabla^2 c)[S,S]$$
$$+ \nabla c \cdot R + c\frac{\partial^2}{\partial t^2}\det J_t + 2\left[\dot{c} + \nabla c \cdot S\right]\frac{\partial}{\partial t}\det J_t\bigg|_{t=0} dx,$$

recalling

$$\frac{\partial}{\partial t}c(T_t x, t) = (c_t + \nabla c \cdot \frac{\partial T_t}{\partial t})(T_t x, t).$$

Then equalities (4.41) and (4.42) imply

$$\frac{d^2}{dt^2}\int_{\Omega_t} c\bigg|_{t=0} = \int_\Omega \ddot{c} + 2\nabla\dot{c}\cdot S + (\nabla^2 c)[S,S] + \nabla c \cdot R$$

$$+2(\dot{c}+\nabla c\cdot S)\nabla\cdot S+c(\nabla\cdot R+(\nabla\cdot S)^2-J(S):{}^tJ(S)\Big|_{t=0}\,dx$$

$$=\int_{\Omega}\ddot{c}+2\nabla\cdot(\dot{c}S)+(\nabla^2c)[S,S]+\nabla\cdot(cR)+2(\nabla c\cdot S)(\nabla\cdot S)$$

$$+c((\nabla\cdot S)^2-J(S):{}^tJ(S))\Big|_{t=0}\,dx$$

$$=\int_{\Omega}\ddot{c}\,dx+\int_{\partial\Omega}2\dot{c}\delta\rho+c\delta^2\rho\Big|_{t=0}\,dS$$

$$+\int_{\Omega}(\nabla^2c)[S,S]+2(\nabla c\cdot S)\nabla\cdot S+c((\nabla\cdot S)^2-J(S):{}^tJ(S))\Big|_{t=0}\,dx.$$

We have

$$(\nabla\cdot S)^2-J(S):{}^tJ(S)=\sum_{i,j}(S_i^iS_j^j-S_i^jS_j^i),$$

recalling (4.43), and hence

$$\frac{d^2}{dt^2}\int_{\Omega_t}c\,dx\Big|_{t=0}-\int_{\Omega}\ddot{c}\,dx-\int_{\partial\Omega}2\dot{c}\delta\rho+c\delta^2\rho\Big|_{t=0}\,dS$$
$$=\sum_{i,j}\int_{\Omega}c_{ij}S^iS^j+2c_iS^iS_j^j+c(S_i^iS_j^j-S_i^jS_j^i)\Big|_{t=0}\,dx.$$

Then we use

$$\sum_{i,j}\int_{\Omega}c_{ij}S^iS^j+2c_iS^iS_j^j\Big|_{t=0}\,dx$$
$$=\int_{\partial\Omega}(\nabla c\cdot S)(S\cdot n)|_{t=0}\,dS+\sum_{i,j}\int_{\Omega}c_i\left(-\frac{\partial}{\partial x_j}(S^iS^j)+2S^iS_j^j\right)\Big|_{t=0}\,dx$$
$$=\int_{\partial\Omega}(\nabla c\cdot S)\delta\rho|_{t=0}\,dS+\sum_{i,j}\int_{\Omega}c_i(S^iS_j^j-S_j^iS^j)\Big|_{t=0}\,dx$$

to obtain

$$\frac{d^2}{dt^2}\int_{\Omega_t}c\,dx\Big|_{t=0}-\int_{\Omega}\ddot{c}\,dx-\int_{\partial\Omega}(2\dot{c}+\nabla c\cdot S)\delta\rho+c\delta^2\rho\Big|_{t=0}\,dS$$
$$=\sum_{i,j}\int_{\Omega}c_i(S^iS_j^j-S_j^iS^j)+c(S_i^iS_j^j-S_i^jS_j^i)\Big|_{t=0}\,dx$$

$$= \sum_{i,j} \int_{\Omega} S^j_j \frac{\partial}{\partial x_i}(cS^i) - S^i_j \frac{\partial}{\partial x_i}(cS^j)\Big|_{t=0} dx \equiv I.$$

This integral is treated similarly as in Sect. 4.2.4, and therefore,

$$I = \sum_{i,j} \int_{\partial\Omega} S^j_j c S^i n_i - S^i_j c S^j n_i\Big|_{t=0} dS + \sum_{i,j} \int_{\Omega} -cS^i S^j_{ij} + cS^j S^i_{ij}\Big|_{t=0} dx$$
$$= \int_{\partial\Omega} c|_{t=0} [(\nabla \cdot S)\delta\rho - (S \cdot \nabla)S \cdot n]\, dS.$$

Then we apply Lemma 4.2 to conclude

$$\begin{aligned} I &= \int_{\partial\Omega} c|_{t=0} ((\nabla \cdot S)(S \cdot n) - [(S \cdot \nabla)S] \cdot n)\, dS \\ &= \int_{\partial\Omega} c|_{t=0} (\nabla_T \cdot (S \cdot n) S_T - (\operatorname{tr} \mathscr{B})(S \cdot n)^2 \\ &\quad - \mathscr{B}(S_T, S_T) - 2S_T \cdot \nabla_T (S \cdot n))\, dS \\ &= -\int_{\partial\Omega} c|_{t=0} \{(\operatorname{tr} \mathscr{B})(\delta\rho)^2 + \mathscr{B}(S_T, S_T) + S_T \cdot \nabla_T(\delta\rho) - \delta\rho \nabla_T \cdot S_T\}\, dS, \end{aligned}$$

recalling $S \cdot n = \delta\rho$, and hence the conclusion. □

In the above proof we use

$$\sum_{i,j} \int_{\Omega} c_i (S^i S^i_j - S^i_j S^j) + c(S^i_i S^j_j - S^j_i S^i_j)\, dx$$
$$= -\int_{\partial\Omega} (\nabla_T c \cdot S_T)\delta\rho + c[(\operatorname{tr} \mathscr{B})(\delta\rho)^2 + \mathscr{B}(S_T, S_T) + 2S_T \cdot \nabla_T(\delta\rho)]\, dS$$

for $\delta\rho = S \cdot n$. If $S^i = \frac{\partial u}{\partial x_i}$ this identity means

$$\sum_{i,j} \int_{\Omega} c_i \left(\frac{\partial u}{\partial x_i} \frac{\partial^2 u}{\partial x_i \partial x_j} - \frac{\partial^2 u}{\partial x_i \partial x_j} \frac{\partial u}{\partial x_j}) + c(\frac{\partial^2 u}{\partial x_i^2} \frac{\partial^2 u}{\partial x_j^2} - (\frac{\partial^2 u}{\partial x_i \partial x_j})^2 \right) dx$$
$$= -\int_{\partial\Omega} (\nabla_T c \cdot S_T)\delta\rho + c[(\operatorname{tr} \mathscr{B})(\delta\rho)^2 + \mathscr{B}(S_T, S_T) + 2S_T \cdot \nabla_T(\delta\rho)]\, dS.$$

If $u|_{\partial\Omega} = 0$, we have $S_T = 0$ and hence

$$II + \int_{\Omega} [(\Delta u)^2 - \sum_{i,j} (\frac{\partial^2 u}{\partial x_i \partial x_j})^2] c\, dx = \int_{\partial\Omega} (\nabla \cdot n)(\frac{\partial u}{\partial n})^2 c\, dS$$

by $\nabla \cdot n = -\text{tr}\,\mathscr{B}$, where

$$
\begin{aligned}
II &\equiv \sum_{i,j} \int_\Omega c_i \left(\frac{\partial u}{\partial x_i} \frac{\partial^2 u}{\partial x_i \partial x_j} - \frac{\partial^2 u}{\partial x_i \partial x_j} \frac{\partial u}{\partial x_j} \right) dx \\
&= \sum_{i,j} \int_\Omega (c_i - c_j) \frac{\partial u}{\partial x_i} \frac{\partial^2 u}{\partial x_i \partial x_j}\, dx \\
&= \frac{1}{2} \sum_{i,j} \int_\Omega (c_i - c_j) \left(\frac{\partial u}{\partial x_i} - \frac{\partial u}{\partial x_j} \right) \frac{\partial^2 u}{\partial x_i \partial x_j}\, dx.
\end{aligned}
$$

Note that Corollary 4.2 in the previous section is the case of $c \equiv 1$.

If

$$
\left. \frac{\partial u}{\partial n} \right|_{\partial \Omega} = 0,
$$

similarly, we have $S \cdot n = \delta \rho = 0$ for $S = \nabla u$. Then it follows that

$$
II + \int_\Omega [(\Delta u)^2 - \sum_{i,j} (\frac{\partial^2 u}{\partial x_i \partial x_j})^2] c\, dx = -\int_{\partial \Omega} c \mathscr{B}(S_T, S_T)\, dS. \tag{4.46}
$$

Equality (4.40) is the case of $c \equiv 1$ in (4.46).

4.3.4 *Filtration*

Filtration is a free boundary problem concerning

$$
u(x) = p(x) + x_2
$$

for $x = (x_1, x_2)$, where $p(x)$ denotes the pressure of water. The boundary of the domain Ω is composed of Γ_i, $i = 1, 2, 3, 4$, where Γ_1 is the base, Γ_3 is composed of disjoint sides according to up and down waves, respectively, Γ_4 is the main free boundary, and Γ_4 arises in the down wave with Γ_3. Furthermore, h_1 and h_2 are the height of water up and down waves, respectively.

Hence

$$
v = -\kappa \nabla u
$$

is the velocity, and the problem is formulated by the quasi-variational problem

$$\int_{\Omega} \nabla u \cdot \nabla \zeta \le 0, \quad \forall \zeta \in H^1(\mathscr{D}), \ \zeta \ge 0 \text{ on } S_2, \ \zeta = 0 \text{ on } S_3,$$

where $\mathscr{D} \subset \mathbf{R}^2$ indicates the dam containing Ω with the boundary composed of S_i, $i = 1, 2, 3$. Here, $S_1 = \Gamma_1$ is a base curve with left and right end points denoted by ζ_1 and ζ_2, respectively. It is contained in $\{x_2 < h_1\}$. We thus obtain

$$S_2 = \partial\mathscr{D} \cap \{(x_1, x_2) \mid x_1 > a, \ x_2 > h_2\}, \quad S_3 = \partial\mathscr{D} \setminus (S_1 \cup S_2),$$

where a is the value of x_1 coordinate at the left point of $\partial\mathscr{D} \cap \{x_2 = h_1\}$, denoted by ζ_3. A constraint on $u \in H^1(\mathscr{D})$ is the condition

$$u = u^0 \quad \text{on } \Gamma_2 \cup \Gamma_3 \cup \Gamma_4,$$

where

$$u^0(x) = \begin{cases} h_1 \text{ in } \mathscr{D}_0 \\ x_2 \text{ in } \mathscr{D}_1 \\ h_2 \text{ in } \mathscr{D}_3. \end{cases}$$

Hence $\mathscr{D}_1$ is the upper portion of $\mathscr{D}$ higher than Γ_2 and

$$\mathscr{D}_0 = \mathscr{D} \cap \{x_1 < a\}, \quad \mathscr{D}_3 = (\mathscr{D} \cap \{x_2 < h_2\}) \setminus \mathscr{D}_0.$$

The Euler-Lagrange equation of this variational problem is given by

$$\Delta u = 0 \text{ in } \Omega, \quad \frac{\partial u}{\partial \nu} = 0 \text{ on } \Gamma_1, \quad u = u^0, \ \frac{\partial u}{\partial \nu} = 0 \text{ on } \Gamma_2$$

$$u = u^0 \text{ on } \Gamma_3, \quad u = u^0, \ \frac{\partial u}{\partial \nu} \le 0 \text{ on } \Gamma_4.$$

To introduce a variational reformulation we define the set of admissible domains denoted by $\mathscr{A}(\mathscr{D})$. Hence we say $\Omega \in \mathscr{A}(\mathscr{D})$ if and only if $\Omega \subset \mathscr{D}$ is a Lipschitz simply-connected domain, $S_1 \cup S_3 \subset \partial\Omega$, $\partial\Omega(S_1 \cup S_3)$ is a Lipschitz monotone-decreasing graph in the direction x_2. Then we define the closed cone

$$\mathscr{K} = \{v \in H^1(\Omega) \mid v = 0 \text{ on } \Gamma_1 \cup \Gamma_2 \cup \Gamma_3, \ v \ge 0 \text{ on } \Gamma_4\}$$

and its dual cone

$$\mathscr{K}^* = \{(\nabla v, \nabla \chi) \le 0, \ \forall \chi \in \mathscr{K}(\Omega)\}.$$

Given $\chi \in H^2(\Omega)$, we have $\chi \in K^*(\Omega)$ if and only if $\frac{\partial \chi}{\partial \nu} \le 0$ on Γ_4.

Let

$$A(\Omega) = \{v \in K^*(\Omega) \mid v = u^0 \text{ on } \Gamma_2 \cup \Gamma_3 \cup \Gamma_4\}$$
$$a(\Omega) = \inf_{v \in A(\Omega)} D_\Omega(v), \quad D_\Omega(u) = \frac{1}{2}\int_\Omega |\nabla v|^2.$$

This $a(\Omega)$ is attained by some $u = u_\Omega \in A(\Omega)$ satisfying

$$\Delta u = 0 \text{ in } \Omega, \quad u = u^0 \text{ on } \Gamma_2 \cup \Gamma_3 \cup \Gamma_4, \quad \left.\frac{\partial u}{\partial \nu}\right|_{\Gamma_4} \le 0, \quad \left.\frac{\partial u}{\partial \nu}\right|_{\Gamma_1} = 0.$$

Let, next,

$$B(\Omega) = \{v \in H^1(\Omega) \mid v = u^0 \text{ on } \Gamma_3 \cup \Gamma_4\}$$
$$b(\Omega) = \inf_{v \in B(\Omega)} D_\Omega(v).$$

This $b(\Omega)$ is attained by some $w = w_\Omega \in B(\Omega)$ satisfying

$$\Delta w = 0 \text{ in } \Omega, \quad w = u^0 \text{ on } \Gamma_3 \cup \Gamma_4, \quad \left.\frac{\partial w}{\partial \nu}\right|_{\Gamma_1 \cup \Gamma_2} = 0.$$

The following fact is easy to see.

Theorem 4.11 *Let $J(\Omega) = a(\Omega) - b(\Omega)$. Then*

$$\inf_{\Omega \in \mathscr{A}(\mathscr{D})} J(\Omega) = 0$$

is attained by $\Omega \in \mathscr{A}(\mathscr{D})$ if and only if it is a solution to the filtration problem.

Each admissible domain $\Omega \in \mathscr{A}(\mathscr{D})$ is associated with the mapping

$$\varphi = \varphi_\Omega : B = B(0,1) \subset \mathbf{R}^2 \to \Omega,$$

which has a bi-Lipschitz continuous extension $\overline{B} \to \overline{\Omega}$ satisfying the three-point condition, $\varphi(z_i) = \zeta_i$, $i = 1, 2, 3$, where

$$z_1 = 1, \ z_2 = e^{\imath 2\pi/3}, \ z_3 = e^{-\imath 2\pi/3}.$$

Theorem 4.12 ([42]) *The set $\mathscr{A}(\mathscr{D}) \neq \emptyset$ is closed with respect to the uniform convergence and is sequentially compact. If*

$$\Omega_k \to \Omega \text{ uniformly in } \mathscr{A}(\mathscr{D}), \quad \lim_{k\to\infty} J(\Omega_k) = 0,$$

it holds that $J(\Omega) = 0$.

To realize a minimizing sequence for

$$\inf_{\mathscr{A}(\mathscr{D})} J,$$

bi-Lipschitz domain perturbation $T_t : \Omega \to \Omega_t$ is useful. Let

$$T_t x = x + tSx + \frac{t^2}{2}Rx + o(t^2) \quad \text{uniformly in } x \in \Omega$$

as $t \to 0$. This perturbation satisfies the NPO condition if $v \in V_1(\Omega_t)$ is equivalent to $v \circ T_t \in V_1(\Omega)$ for $|t| \ll 1$, where

$$V_1(\Omega_t) = \{v \in H^1(\Omega_t) \mid v|_{\Gamma_3 \cup \Gamma_4^t} = 0\}, \quad \Omega_0 = \Omega.$$

Theorem 4.13 ([43]) *The first variation of* $a(\Omega_t)$ *exists as*

$$\delta a(\Omega) \equiv \frac{d}{dt} a(\Omega_t)\Big|_{t=0} = \frac{1}{2}\left\langle 1 - \left(\frac{\partial P_\Omega}{\partial \nu}\right)^2, \delta\rho \right\rangle_{\Gamma_2 \cup \Gamma_4},$$

where $\langle\ ,\ \rangle$ *denotes the pairing and*

$$\delta\rho = S \cdot \nu, \quad p_\Omega = u_\Omega - x_2$$

for $u = u_\Omega$ *satisfying*

$$\Delta u = 0 \text{ in } \Omega, \quad u = u_0 \text{ on } \Gamma_2 \cup \Gamma_3 \cup \Gamma_4, \quad \frac{\partial u}{\partial \nu}\Big|_{\Gamma_4} \le 0, \quad \frac{\partial u}{\partial \nu}\Big|_{\Gamma_1} = 0.$$

Hence $u = u_\Omega$ *attains*

$$a(\Omega) = \inf_{v \in \mathscr{A}(\Omega)} D_\Omega(v).$$

If $\{T_t\}$ *satisfies the NPO condition, furthermore, the first variation of* $b(\Omega_t)$ *exists as*

$$\delta b(\Omega) \equiv \frac{d}{dt} b(\Omega_t)\Big|_{t=0} = \frac{1}{2}\left\langle \left(\frac{\partial w_\Omega}{\partial \tau}\right)^2, \delta\rho \right\rangle_{\Gamma_2},$$

where τ *denotes the unit tangential vector and* $w = w_\Omega$ *attains* $b(\Omega)$. *Hence it holds that*

$$\Delta w = 0 \; in \; \Omega, \quad w = u^0 \; in \; \Gamma_3 \cup \Gamma_4, \quad \left.\frac{\partial w}{\partial \nu}\right|_{\Gamma_4 \cup \Gamma_2} = 0.$$

The first variation of the functional J is applied to design iterative schemes for the filtration problem. Given $\mathscr{D} \subset \mathbf{R}^2$ and h_1, h_2, we seek the flow region Ω and the potential function u by an iterative scheme. Let Ω^k and $\Gamma_2^k \subset \partial\Omega_k$ be the k-th guess of the flow region and the free boundary, respectively. Since the first variation of the functional $J : \mathscr{A}(\mathscr{D}) \to \mathbf{R}$ is

$$\delta J(\Omega^k) = \left\langle 1 - \left(\frac{\partial p_\Omega^k}{\partial \nu}\right)^2 - \left(\frac{\partial w_\Omega^k}{\partial \tau}\right)^2, \delta\rho \right\rangle_{\Gamma_2^k},$$

an intuitive iterative scheme is defined by

$$\Gamma_2^{k+1} = \{x + \varepsilon F V(x)\nu(x) \mid x \in \Gamma_2^k\}$$

for

$$FV(x) = 1 - \left(\frac{\partial p_\Omega^k}{\partial \nu}\right)^2 - \left(\frac{\partial w_\Omega^k}{\partial \tau}\right)^2, \quad x \in \Gamma_2^k$$

with $0 < \varepsilon \ll 1$. This scheme of the steepest descent method does not fit numerical experiments well. After several iterations, Γ^k becomes very jagged and computation cannot be carried out anymore.

The effectiveness of the H^1-gradient method [3] is confirmed. It is given by

$$\Gamma_2^{k+1} = \{x - z^k(x)\nu(x) \mid x \in \Gamma_2^k\},$$

where $z^k \in H^1(\Omega^k)$ denotes the solution to the boundary value problem

$$\Delta z^k = 0 \text{ in } \Omega^k, \quad \left. z^k \right|_{\Gamma^3 \cup \Gamma_4^k} = 0, \quad \left.\frac{\partial z^k}{\partial \nu}\right|_{\Gamma^1} = 0, \quad \frac{\partial z^k}{\partial \nu} = FV \text{ on } \Gamma^k.$$

We conclude this subsection with the formulation of the Toland duality to

$$\tilde{J}(\Omega) = \frac{1}{b(\Omega)} - \frac{1}{a(\Omega)}$$

which is valid in the region where both $a(\Omega)^{-1}$ and $b(\Omega)^{-1}$ are convex in $\Omega \in \mathscr{A}(\mathscr{D})$. Although this formulation is formal it may be worth noting that the second eigenvalue to $-\Delta$ with Neumann boundary condition, denoted by $\lambda_2 = \lambda_2(\Omega)$ is harmonic convex, that is, λ_2^{-1} is convex with respect to the domain perturbation [12].

Let X be a Banach space over $\mathbf{R}$ and X^* be its dual space. The pairing is denoted by $\langle\ ,\ \rangle$. The Toland duality is based on the Fenchel-Morreaux duality. First, given $F : X \to [-\infty, +\infty]$, its Legendre transformation is defined by

$$F^*(p) = \sup_{x \in X}\{\langle x, p\rangle - F(x)\}, \quad p \in X^*.$$

Second, if $F : X \to (-\infty, +\infty]$ is proper, convex, and lower semi-continuous, this $F^* : X^* \to (-\infty, +\infty]$ is also, and it holds that $F^{**} = F$, where

$$F^{**}(x) = \sup_{p \in X^*}\{\langle x, p\rangle - F^*(x)\}.$$

The Toland duality is concerned with

$$\tilde{J}(x) = G(x) - F(x), \quad \tilde{J}^*(x) = F^*(x) - G^*(x),$$

where $F, G : X \to (-\infty, +\infty]$ are proper, convex, and lower semi-continuous [37]. Using the Lagrange function

$$L(x, p) = F^*(p) + G(x) - \langle x, p\rangle,$$

we have

$$\inf_{(x,p) \in X \times X^*} L(x, p) = \inf_{p \in X^*} \tilde{J}^*(p) = \inf_{x \in X} \tilde{J}(x)$$

and

$$L|_{p \in \partial G(x)} = \tilde{J}(x) = \inf_{p \in X^*} L(x, p), \quad L|_{x \in \partial F^*(x)} = \tilde{J}^*(p) = \inf_{x \in X} L(x, p). \tag{4.47}$$

Then local minima of $\tilde{J}$ and $\tilde{J}^*$ are infinitesimally stable if the F and G are analytic [39–41].

Remark 4.3 Equalities (4.47) are called the *unfolding-minimality* [37]. Several models in non-equilibrium thermodynamics are provides with the semi-structure. The stationary problem is then realized as a nonlinear eigenvalue problem involving the conservative quantity. A local minimum of the associated variational functional is dynamically stable because of this semi-structure. Examples are in phase transition [17], phase separation [18], and shape memory [35].

4.3.5 Stefan Condition as Heat Transfer

Here we treat the two phase Stefan problem and its mathematical structures, from the viewpoint of Liouville's theory. Let θ be the relative temperature, and assume that the heat conductor is water and ice if $\theta > 0$ and $\theta < 0$, respectively. Then we obtain the heat equation

$$c\rho\theta_t = \nabla \cdot (\kappa \nabla \theta) \qquad \text{in } \{\theta \neq 0\}\,. \tag{4.48}$$

The region $\{x \in \Omega \mid \theta(x,t) = 0\}$ is composed of an interface, denoted by Γ_t, where the phase transition occurs and the latent heat is exchanged. Usually, the density $\rho = \rho(\theta)$ depends continuously on $\theta \in \mathbf{R}$, but the specific heat $c = c(\theta)$ and the conductivity $\kappa = \kappa(\theta)$ have the discontinuity of the first kind at $\theta = 0$.

To describe the motion of Γ_t, we take the level-set approach and introduce the C^1 function $\Phi = \Phi(x,t)$ satisfying

$$\Gamma_t : \Phi(\cdot, t) = 0. \tag{4.49}$$

Let ν be the outer unit normal vector of Γ_t from $\{\theta > 0\}$ at $x \in \Gamma_t$. If x moves $\nu \Delta h$ during the small time interval Δt, then it holds that

$$\Phi(x + \nu \Delta h, t + \Delta t) = 0.$$

Taking the infinitesimal approximation of this relation, we obtain

$$(\nu \cdot \nabla \Phi)\Delta h + \Phi_t \Delta t = 0 \qquad \text{on } \Gamma_t. \tag{4.50}$$

Meanwhile, $\ell \Delta h$ is radiated from the unit area on Γ_t as the latent heat, where $\ell = \lambda \rho$ with λ standing for the latent heat per unit weight, and, therefore, the Newton-Fourier-Fick heat energy balance law guarantees the relation

$$\ell \Delta h = -\left[\kappa \frac{\partial \theta}{\partial \nu}\right]_-^+ \Delta t, \tag{4.51}$$

where

$$[A]_-^+ = A_+ - A_-, \qquad A_\pm(x) = \lim_{y \in \{\pm\theta > 0\},\ y \to x} A(y).$$

Combining (4.50) with (4.51), thus, we obtain the Stefan condition

$$\ell \Phi_t = \frac{\partial \Phi}{\partial \nu} \cdot \left[\kappa \frac{\partial \theta}{\partial \nu}\right]_-^+ \qquad \text{on } \Gamma_t, \tag{4.52}$$

which comprises the *Stefan problem* with (4.48) and (4.49).

Then we use the *Kirchhoff transformation*

$$u = \int_0^\theta \kappa(\theta')d\theta'$$

and the enthalpy $H = H(u)$ defined by

$$H(u) = \begin{cases} \int_0^\theta \rho(\theta')c(\theta')d\theta' - \ell, & u < 0 \\ \int_0^\theta \rho(\theta')c(\theta')d\theta', & u > 0, \end{cases}$$

which satisfies

$$H'(u) = \frac{\rho(\theta)c(\theta)}{\kappa(\theta)}, \ u \neq 0, \quad H(+0) - H(-0) = \ell.$$

Concerning $u = u(x, t)$, we obtain

$$\nabla u = \kappa \nabla \theta, \quad H(u)_t = \frac{\rho c}{\kappa} \cdot \theta_t \cdot \kappa$$

and, therefore,

$$H(u)_t = \Delta u \qquad \text{in} \bigcup_{0<t<T} (\Omega \setminus \Gamma_t) \times \{t\} = Q \setminus \Gamma, \quad u|_{t=0} = u_0(x)$$

$$\ell \Phi_t = [\nabla u \cdot \nabla \Phi]_-^+ \quad \text{on} \bigcup_{0<t<T} \Gamma_t \times \{t\} \equiv \Gamma \tag{4.53}$$

by (4.48) and (4.51), where $Q = \Omega \times (0, T)$.

Since $\Gamma = \{\Phi = 0\}$, there arises

$$\Phi_t + v \cdot \nabla \Phi = 0,$$

where v stands for the velocity of the movement of Γ. Then we obtain

$$\Phi_t = -(v \cdot \nu)\frac{\partial \Phi}{\partial \nu}$$

by $v = (v \cdot \nu)\nu$. The Stefan condition (4.53), on the other hand, implies

$$\ell \Phi_t = \left[\frac{\partial u}{\partial \nu}\right]_-^+ \frac{\partial \Phi}{\partial \nu}.$$

We thus reach the system

$$H(u)_t = \Delta u \text{ in } Q \setminus \Gamma, \quad u|_{\partial\Omega} = 0, \quad u|_{t=0} = u_0(x)$$

$$[u]_-^+ = 0, \ [H(u)]_-^+ = \ell \text{ on } \Gamma \equiv \{\Phi = 0\}$$
$$v \cdot \nu = -\frac{1}{\ell}\left[\frac{\partial u}{\partial \nu}\right]_-^+ \quad \text{on } \Gamma. \tag{4.54}$$

Now we show that (4.54) implies

$$H(u)_t = \Delta u \text{ in } Q = \Omega \times (0,T), \quad u|_{\partial\Omega} = 0, \quad u|_{t=0} = u_0(x) \tag{4.55}$$

in the sense of distributions. To this end, take $\varphi \in C_0^\infty(Q)$ and put $\Omega_t^\pm = \{\pm\Phi(\cdot,t) > 0\}$. Then, Liouville's theorem assures

$$\begin{aligned}
\frac{d}{dt}\int_{\Omega_t^\pm} H(u)\varphi &= \int_{\Omega_t^\pm} H(u)_t\varphi + H(u)\varphi_t \, dx \pm \int_{\partial\Omega_t^\pm} H(u)\varphi v \cdot \nu \\
&= \pm\int_{\partial\Omega_t^\pm} \varphi\frac{\partial u}{\partial \nu} - \frac{\partial \varphi}{\partial \nu}u + H(u)v \cdot \nu\varphi \, dS \\
&\quad + \int_{\Omega_t^\pm} u\Delta\varphi + H(u)\varphi_t \, dx.
\end{aligned}$$

It follows that

$$\frac{d}{dt}\int_\Omega H(u)\varphi = \int_\Omega u\Delta\varphi + H(u)\varphi_t \, dx$$

and hence

$$\int\int_Q u\Delta\varphi + H(u)\varphi_t \, dxdt = 0$$

from $\varphi|_{t=0,T} = 0$, which means that (4.55) holds as $\mathscr{D}'(Q)$.

Here, $H = H(u)$ takes the discontinuity of the first kind at $u = 0$. If ρ is constant and c and κ are piecewise constant, then it holds that

$$H'(u) = \begin{cases} c_+\kappa_+, \ u > 0 \\ c_-\kappa_-, \ u < 0 \end{cases}, \qquad H(+0) - H(-0) = \ell.$$

In any case, we define the *maximum monotone graph*, still denoted by $H = H(u)$ in $\mathbf{R} \times \mathbf{R}$, putting $H(0) = [H(-0), H(+0)]$. Then we obtain the unique existence of the weak solution

$$u = u(x,t) \in L^\infty(\Omega \times (0,T))$$

to (4.55) global in time [20, 28]. More precisely, there is $h = h(x,t)$ such that $h \in H(u)$ a.e. and

$$\int\int_Q u\Delta\varphi + h\varphi_t \, dxdt + \int_\Omega u_0\varphi(\cdot, 0)dx = 0$$

for any $\varphi = \varphi(x,t) \in C^{2,1}(\overline{\Omega} \times [0,T])$ with $\varphi|_{t=T} = 0$, where $Q = \Omega \times (0,T)$ and $\Gamma = \partial\Omega \times (0,T)$.

Using the single-valued maximal monotone graph $f(v) = H^{-1}(v)$, we can derive

$$v_t = \Delta f(v)$$

from (4.55), and then the nonlinear semigroup theory [4, 24] is applicable. In more detail, if $f : \mathbf{R} \to \mathbf{R}$ is a non-decreasing continuous function satisfying $f(0) = 0$, then the operator $Av = -\Delta f(v)$ with the domain

$$D(A) = \left\{ v \in L^1(\Omega) \mid f(v) \in W_0^{1,1}(\Omega),\ \ \Delta f(v) \in L^1(\Omega) \right\}$$

is maximum monotone [8], and hence generates a contraction semigroup [8] in $X = L^1(\Omega)$, denoted by $\{T_t\}_{t \geq 0}$. Then $v(\cdot, t) = T_t v_0$ is regarded as a solution to

$$v_t = \Delta f(v), \quad f(v)|_{\partial\Omega} = 0, \quad v|_{t=0} = v_0, \tag{4.56}$$

and in this sense the problem (4.56) is well-posed in $X = L^1(\Omega)$ global in time.

4.4 Conservation Laws

4.4.1 Ideal Fluids

A solid can be put on a desk with which some part of it is not in contact. This phenomenon is due to the tangential force. Fluid is a continuum, for which tangential force is absent at equilibrium. The ideal fluid, on the other hand, is a continuum for which tangential force is absent even when it is moving. Liouville's transformation formulae take a fundamental role in the theory of continuum mechanics both in modeling and analysis.

If $v = v(x,t)$ is the velocity of a fluid, the position $x = x(t) \in \mathbf{R}^3$ of a particle in this fluid is subject to

$$\frac{dx}{dt} = v(x,t), \quad x(s) = \xi, \tag{4.57}$$

where $t = s$ and $x = \xi$ stand for the initial time and the initial position, respectively. Then we define the propagator $\{T(t,s)\}$ by

$$x(t) = T(t, s)\xi.$$

The acceleration vector of this particle is given by

$$\frac{d}{dt}v(x(t), t) = (v_t + (v \cdot \nabla)v)(x(t), t) \equiv \frac{Dv}{Dt}(x(t), t)$$

and hence the Euler equation arises as the equation of motion

$$\rho\frac{Dv}{Dt} = \rho F - \nabla p, \tag{4.58}$$

where ρ, p, and F denote the density, the pressure, and the outer force, respectively. Here, the left-hand side of (4.58) indicates the mass times acceleration vector, while ρF on the right-hand side is compensated by the gradient of the pressure, $-\nabla p$.

The other derivation of (4.58) uses the equation of continuity

$$\rho_t + \nabla \cdot \rho v = 0, \tag{4.59}$$

which describes the mass conservation,

$$\frac{d}{dt}\int_\omega \rho\, dx = -\int_{\partial\omega} \nu \cdot j\, ds. \tag{4.60}$$

Here,

$$j = \rho v \tag{4.61}$$

is the flux of ρ and ω is an arbitrary subdomain with smooth boundary.

To confirm (4.58) from this point of view, we assume that the rate of change of momentum of a moving piece of ideal fluid is equal to the total force acting on it. This total force is composed of the surface and the volume forces, and, hence it follows that

$$\frac{d}{dt}\int_{T(t,s)\omega} \rho v\, dx = -\int_{T(t,s)\omega} p\nu\, dS + \int_{T(t,s)\omega} \rho F\, dx \tag{4.62}$$

where ν denotes the outer unit normal vector. Regarding $x = T(t, s)\xi$ as a transformation of variables

$$\xi \in \mathbf{R}^3 \quad \mapsto \quad x \in \mathbf{R}^3, \tag{4.63}$$

we have

$$J_t(\xi) = 1 + (t - s)\nabla \cdot v(\xi, s) + o(t - s), \quad t \to s \tag{4.64}$$

by Liouville's formula, where

$$J_t = \det(\frac{\partial x_i}{\partial \xi_j})$$

denotes the Jacobian.

At $t = s$, therefore, the left-hand side of (4.62) is equal to

$$\begin{aligned}
&\frac{d}{dt}\int_\omega (\rho v)(x(\xi,t),t)|J_t(\xi)|\,d\xi\Big|_{t=s} \\
&= \int_\omega \frac{D}{Dt}(\rho v) + \rho v(\nabla\cdot v)\,dx \\
&= \int_\omega (\frac{D\rho}{Dt} + \rho\nabla\cdot v)v + \rho\frac{Dv}{Dt}\,dx. \qquad (4.65)
\end{aligned}$$

Using the equation of continuity (4.59), we obtain

$$\int_\omega \rho\frac{Dv}{Dt}\,dx = \int_\omega -\nabla p + \rho F\,dx$$

by the divergence formula. Then (4.58) follows because the sub-domain ω is arbitrary.

The equation of continuity (4.60) can be derived from

$$0 = \frac{d}{dt}\int_{T(t,s)\omega} \rho(x,t)\,dx\Big|_{t=s} \qquad (4.66)$$

which indicates the mass conservation in the Lagrange coordinate. In fact, under the transformation (4.63), equality (4.66) reads as

$$\begin{aligned}
0 &= \frac{d}{dt}\int_\omega \rho(x(\xi,t),t)|J_t(\xi)|\,d\xi\Big|_{t=s} = \int_\omega \frac{D\rho}{Dt} + \rho\nabla\cdot v\,dx \\
&= \int_\omega \rho_t + \nabla\cdot\rho v\,dx \qquad (4.67)
\end{aligned}$$

similarly to (4.65). Then (4.59) follows because the subdomain ω is arbitrary.

Regarding (4.60) as the definition of the flux j, we apply the divergence of Gauss to get

$$\int_\omega \rho_t = -\int_\omega \nabla\cdot j. \qquad (4.68)$$

Comparing (4.67) and (4.68), we agree the relation (4.61).

In the barotropic fluid the density is a function of the pressure, represented by the state equation $\rho = \rho(p)$. In this case, using

$$P(p) = \int^p \frac{d\xi}{\rho(\xi)}$$

in (4.58), we obtain

$$v_t + (v \cdot \nabla)v = -\nabla P + F. \tag{4.69}$$

From the thermodynamical law, it follows that

$$Td_eS = dU + pdV, \tag{4.70}$$

where d_eS, T, U, p, and V denote entropy variation, temperature, inner energy, pressure, and volume, respectively. The fluid is said to be isentropic if $d_eS = 0$. Since $V = 1/\rho$ we have

$$dU = \frac{p}{\rho^2}d\rho$$

by (4.70) in this case, which means

$$dw = \frac{1}{\rho}dp, \quad w = U + \frac{p}{\rho}. \tag{4.71}$$

This w, called the specific enthalpy, is a thermodynamical quantity. Hence it holds that $w = w(p)$ by (4.71), and also a functional relation between ρ and p. This relation is indicated as the state equation $p = p(\rho)$. Isentropic fluid is thus barotropic.

If the outer force is a potential, we have $F = \nabla\varphi$ with a scalar field φ in (4.69). Then it holds that

$$\frac{Dv}{Dt} = -\nabla Q, \quad Q = P - \varphi \tag{4.72}$$

which implies

$$\omega_t + \nabla \times (v \cdot \nabla)v = 0, \quad \omega = \nabla \times v.$$

Since

$$(v \cdot \nabla)v = \frac{1}{2}\nabla|v|^2 - v \times \omega \tag{4.73}$$

and

$$\nabla \times (v \times \omega) = (\omega \cdot \nabla)v - (v \cdot \nabla)\omega + v\nabla \cdot \omega - \omega\nabla \cdot v$$
$$= -(v \cdot \nabla)\omega + (\omega \cdot \nabla)v - \omega\nabla \cdot v$$

it holds that

$$\frac{D\omega}{Dt} = (\omega \cdot \nabla)v - (\nabla \cdot v)\omega. \tag{4.74}$$

Equation (4.59) means

$$\frac{D\rho}{Dt} = -(\nabla \cdot v)\rho,$$

and, therefore, we obtain

$$\frac{D}{Dt}(\frac{\omega}{\rho}) = (\frac{\omega}{\rho} \cdot \nabla)v \tag{4.75}$$

by (4.74).

From (4.57) we have

$$\frac{DX}{Dt} = (X \cdot \nabla)v, \quad X|_{t=s} = I$$

for the matrix

$$X(t,s) = \frac{\partial}{\partial \xi}T(t,s)\xi,$$

which implies

$$(\frac{\omega}{\rho})(x(t),t) = X(t,s)(\frac{\omega}{\rho})(\xi,s), \quad x(t) = T(t,s)\xi \tag{4.76}$$

by (4.75). Equality (4.76), called the Cauchy integral, ensures the Lagrange vortex theorem,

$$\nabla \times v|_{t=0} = 0 \quad \Rightarrow \quad \nabla \times v = 0, \ \forall t. \tag{4.77}$$

Thus, barotropic ideal fluids under the potential force are classified into with or without vorticities.

Circulation along the oriented Jordan curve $C \subset \mathbf{R}^3$ is denoted by

$$\Gamma(C) = \int_C v \cdot dx = \int\int_{\mathscr{M}} \omega \cdot dS,$$

where $\mathscr{M}$ is a closed surface satisfying $\partial\mathscr{M} = C$ and dS stands for the vector surface element. A tube composed of the stream lines of $\omega = \nabla \times v$, denoted by $\mathscr{T} \subset \mathbf{R}^3$, is called a vortex tube. Then the Stokes formula implies Helmholtz's circulation theorem, that is if C_i, $i = 1, 2$, are oriented Jordan curves surrounding $\mathscr{T}$ it holds that $\Gamma(C_1) = \Gamma(C_2)$.

If $\mathscr{M} \subset \mathbf{R}^3$ is a closed surface such that $\partial\mathscr{M} = C$, then it holds that

$$\frac{d}{dt}\int_{C(t,s)} v \cdot dx\bigg|_{t=s} = \int_C \frac{D}{Dt}(v \cdot dx)\bigg|_{t=s} = \int_C \frac{Dv}{Dt} \cdot dx\bigg|_{t=s} + \int_C v \cdot \frac{D}{Dt}(dx)\bigg|_{t=s}$$

for $C(t,s) = T(t,s)C$. From (4.72) it follows that

$$\int_C \frac{Dv}{Dt} \cdot dx = -\int_C \nabla Q \cdot dx = 0,$$

while

$$v \cdot \frac{D}{Dt}dx = v \cdot d(\frac{Dx}{Dt}) = v \cdot dv = \frac{1}{2}d|v|^2$$

implies

$$\int_C v \cdot \frac{D}{Dt}(dx) = 0.$$

We thus end up with the circulation theorem of Kelvin,

$$\frac{d}{dt}\int_{C(t,s)} v \cdot dx\bigg|_{t=s} = 0. \tag{4.78}$$

If we put $\mathscr{M}(t,s) = T(t,s)\mathscr{M}$ then (4.78) reads as the Helmholtz vortex theorem,

$$\frac{d}{dt}\int_{\mathscr{M}(t,s)} \nabla \times v \cdot dS\bigg|_{t=s} = 0.$$

Then it follows that (4.77) again.

4.4.2 Gas Dynamics

Here we assume rapid decay at infinity in space and also sufficient regularity of physical quantities. First, (4.59) implies $\rho \geq 0$ from $\rho|_{t=0} \geq 0$. The total mass conservation,

$$M = \int_{\mathbf{R}^3} \rho \, dx,$$

also follows from this equation. Assuming (4.69) with $F = 0$, we now derive the total energy conservation

$$E = \int_{\mathbf{R}^3} \frac{\rho}{2} |v|^2 + Q(\rho) \, dx, \quad Q''(\rho) = \frac{p'(\rho)}{\rho}. \tag{4.79}$$

In fact, in this case, Eq. (4.69) is reduced to

$$(\rho v)_t + \nabla \cdot \rho v \otimes v + \nabla p = 0 \tag{4.80}$$

by (4.59), where $v \otimes v = (v^i v^j)$ for $v = (v^j)$. Here we have

$$\begin{aligned}\int_{\mathbf{R}^3} [\nabla \cdot \rho v \otimes v] \cdot v &= \int_{\mathbf{R}^3} \left[\partial_j(\rho v^i v^j)\right] v^i = -\int_{\mathbf{R}^3} \rho v^j v^i \partial_j v^i \\ &= -\frac{1}{2} \int_{\mathbf{R}^3} \rho v^j \partial_j |v|^2 = \frac{1}{2} \int_{\mathbf{R}^3} |v|^2 \nabla \cdot (\rho v) = -\frac{1}{2} \int_{\mathbf{R}^3} |v|^2 \rho_t\end{aligned}$$

and

$$\int_{\mathbf{R}^3} (\rho v)_t \cdot v = \int_{\mathbf{R}^3} \rho_t |v|^2 + \rho v_t \cdot v \, dx = \int_{\mathbf{R}^3} \rho_t |v|^2 + \frac{1}{2} \rho \partial_t |v|^2 \, dx,$$

which results in

$$\begin{aligned}\int_{\mathbf{R}^3} [(\rho v)_t + \nabla \cdot \rho v \otimes v] \cdot v &= \frac{1}{2} \int_{\mathbf{R}^3} |v|^2 \rho_t + \rho \partial_t |v|^2 \, dx \\ &= \frac{1}{2} \frac{d}{dt} \int_{\mathbf{R}^3} |v|^2 \rho.\end{aligned} \tag{4.81}$$

Next, we have

$$\int_{\mathbf{R}^3} \nabla p \cdot v = \int_{\mathbf{R}^3} \nabla Q(\rho) \cdot \rho v = \int_{\mathbf{R}^3} Q'(\rho) \rho_t = \frac{d}{dt} \int_{\mathbf{R}^3} Q(\rho). \tag{4.82}$$

Then (4.79) follows from (4.80)–(4.82).

Without vortices, the barotropic fluid is called irrotational. If $F = 0$, Eq. (4.69) is reduced to

$$v_t = -\nabla(P + \frac{1}{2}|v|^2)$$

by (4.73) and $\omega = \nabla \times v = 0$. Then we obtain Bernoulli's law,

$$\nabla(-\psi_t + P + \frac{1}{2}|v|^2) = 0,$$

using the vector potential, $v = -\nabla\psi$. Hence we may suppose

$$\psi_t = P + \frac{1}{2}|v|^2$$

together with $P(p) = Q'(\rho)$. The total energy is then written as

$$E(\rho, \psi) = \int_{\mathbf{R}^3} Q(\rho) + \frac{\rho}{2}|\nabla\psi|^2 \, dx$$

and it holds that

$$\frac{\delta E}{\delta \psi} = -\nabla \cdot \rho\nabla\psi = \nabla \cdot \rho v, \quad \frac{\delta E}{\delta \rho} = P + \frac{1}{2}|v|^2.$$

The irrotational barotropic fluid without outer force is thus formulated by the Hamilton system

$$\psi_t = \frac{\delta E}{\delta \rho}, \quad \rho_t = -\frac{\delta E}{\delta \psi}. \tag{4.83}$$

The total energy conservation, $\frac{dE}{dt} = 0$, now follows from (4.83).

Turning to the kinetic formulation [30], we use the second moment in the Lagrange coordinate, $|x - tv|^2$. Since

$$(|x - tv|^2\rho)_t = |x - tv|^2\rho_t - 2(x - tv) \cdot (v + tv_t)\rho$$
$$\nabla \cdot |x - tv|^2\rho v = |x - tv|^2\nabla \cdot \rho v + \sum_{i,j} \rho v^j \partial_j (x_i - tv^i)^2$$

with

$$\rho v^j \partial_j (x_i - tv^i)^2 = 2\rho v^j (x_i - tv^i)(\delta_{ij} - t\partial_j v^i),$$

it follows that

$$\sum_{i,j} \rho v^j \partial_j (x_i - tv^i)^2 = 2(x - tv) \cdot (v - t(v \cdot \nabla)v)\rho.$$

By (4.59) and (4.58) with $F = 0$, we thus obtain

$$(|x-tv|^2\rho)_t+\nabla\cdot|x-tv|^2\rho v=-2t(x-tv)\cdot\rho(v_t+(v\cdot\nabla)v)$$
$$=2t(x-tv)\cdot\nabla p. \tag{4.84}$$

In (4.84) we use

$$x\cdot\nabla p=\nabla\cdot xp-np,\quad n=3 \tag{4.85}$$

and also

$$v\cdot\nabla p=p'(\rho)v\cdot\nabla\rho=p'(\rho)(\nabla\cdot\rho v-\rho\nabla\cdot v)$$
$$=-p_t-\nabla\cdot(p'(\rho)\rho v)+\frac{p''(\rho)\rho+p'(\rho)}{p'(\rho)}v\cdot\nabla p,$$

which results in

$$\frac{p''(\rho)\rho}{p'(\rho)}v\cdot\nabla p=p_t+\nabla\cdot(p'(\rho)\rho v).$$

If the state equation is described by

$$p=A\rho^\gamma,\quad A>0,\ 1<\gamma<2 \tag{4.86}$$

it follows that

$$v\cdot\nabla p=\frac{p_t}{\gamma-1}+\frac{\gamma}{\gamma-1}\nabla\cdot pv \tag{4.87}$$

and

$$Q(\rho)=\frac{A\rho^\gamma}{\gamma-1},\quad E=\int_{\mathbf{R}^3}\frac{\rho}{2}|v|^2+\frac{A\rho^\gamma}{\gamma-1}\,dx.$$

In (4.86), $A>0$ is a constant determined by the total entropy of the system, and $1<\gamma<2$ is the adiabatic constant. The isentropic fluid is assumed here, while $\gamma=5/3,7/5,\ldots$ when the gas is mono-atomic, bi-atomic, and so on, and $\gamma=4/3$ in the case of the excessive radiational pressure [6].

Equalities (4.84), (4.85), and (4.87) are summarized as

$$(|x-tv|^2\rho)_t+\nabla\cdot|x-tv|^2\rho v$$
$$=\nabla\cdot(2txp-\frac{2\gamma t^2}{\gamma-1}vp)-(\frac{2t^2p}{\gamma-1})_t-2(n-\frac{2}{\gamma-1})tp, \tag{4.88}$$

which implies

$$\frac{d}{dt}\int_{\mathbf{R}^3}|x-tv|^2\rho+\frac{2t^2p}{\gamma-1}\,dx+\int_{\mathbf{R}^3}\frac{n(\gamma-1)-2}{t}\cdot\frac{2t^2p}{\gamma-1}=0. \tag{4.89}$$

If $\gamma\geq 1+\frac{2}{n}$, therefore, it follows that

$$\frac{d}{dt}\int_{\mathbf{R}^3}\frac{2t^2p}{\gamma-1}\leq -\frac{d}{dt}\int_{\mathbf{R}^3}|x-tv|^2\rho$$

and hence $\int_{\mathbf{R}^3}p=O(t^{-2})$. If $1<\gamma<1+\frac{2}{n}$, on the other hand, we have

$$\frac{d}{dt}\int_{\mathbf{R}^3}\frac{2t^2p}{\gamma-1}-\frac{\alpha}{t}\int_{\mathbf{R}^3}\frac{2t^2p}{\gamma-1}=-\frac{d}{dt}\int_{\mathbf{R}^3}|x-tv|^2\rho$$

for $\alpha=2-n(\gamma-1)\in(0,2)$, and, therefore, $\int_{\mathbf{R}^3}p=O(t^{-2+\alpha})$. In any case it holds that

$$\int_{\mathbf{R}^3}p=o(1),\quad t\uparrow+\infty. \tag{4.90}$$

4.4.3 Irrotational Blowup

Using (4.83) and (4.90), we can show that the irrotational flow cannot survive in the energy class. For this purpose we write (4.83) as

$$\rho_t=-\nabla\cdot\rho v,\quad \psi_t=\frac{1}{2}|v|^2+\frac{A\gamma\rho^{\gamma-1}}{\gamma-1},\quad v=-\nabla\psi \tag{4.91}$$

and assume $\psi_0=\psi|_{t=0}\geq 0$ without loss of generality. Then we obtain $\rho,\psi\geq 0$ with

$$\begin{aligned}\frac{d}{dt}\int_{\mathbf{R}^3}\rho\psi&=\int\rho_t\psi+\rho\psi_t\,dx=\int_{\mathbf{R}^3}-\frac{\rho}{2}|v|^2+\frac{A\gamma\rho^\gamma}{\gamma-1}\,dx\\&=-E+\frac{\gamma+1}{\gamma-1}\int_{\mathbf{R}^3}A\rho^\gamma=-E+o(1),\end{aligned}$$

which is impossible by $E>0$.

A similar phenomenon arises if the gas has a positive mass and zero velocity at ∞. Thus we have the following theorem of blowup. See [32, 45] for the other cases.

Theorem 4.14 ([36]) *There is no irrotational global-in-time classical solution to (4.58)–(4.59) with (4.86) if the initial value* (ρ_0, v_0) *satisfies* $\rho_0 > 0$ *in* $\mathbf{R}^3$,

$$v_0 = 0, \ \ \rho_0 = \overline{\rho}, \quad |x| > R, \tag{4.92}$$

and $E < 0$*, where* $\overline{\rho} > 0$ *and* $R > 0$ *are constants and*

$$E = \int_{\mathbf{R}^3} \frac{\rho_0}{2}|v_0|^2 + \frac{A\rho_0^\gamma}{\gamma-1} - \frac{A\overline{\rho}^\gamma}{\gamma-1}\,dx$$

stands for the defect energy.

Proof Since system (4.58)–(4.59) is hyperbolic, it is provided with the property of finite propagation [31, 32]. It holds that

$$v(x,t) = 0, \ \ \rho(x,t) = \overline{\rho}, \quad |x| > R + \sigma t \tag{4.93}$$

by (4.92), where $\sigma = (\gamma\overline{\rho}^{\gamma-1})^{1/2}$. Hence the defect energy and the defect mass defined by

$$E = \int_{\mathbf{R}^3} \frac{\rho}{2}|v|^2 + \frac{A\rho^\gamma}{\gamma-1} - \frac{A\overline{\rho}^\gamma}{\gamma-1}\,dx, \quad M = \int_{\mathbf{R}^3} \rho - \overline{\rho}\,dx \tag{4.94}$$

converge and are independent of t.

Note

$$(|x-tv|^2\rho)_t + \nabla\cdot|x-tv|^2\rho v = \nabla\cdot(2tx(p-\overline{p}) - \frac{2\gamma t^2}{\gamma-1}vp)$$

$$-(\frac{2t^2(p-\overline{p})}{\gamma-1})_t - 2(n-\frac{2}{\gamma-1})t(p-\overline{p}), \quad \overline{p} = A\overline{\rho}^\gamma$$

derived from (4.88). By an argument analogous to the above, then we obtain

$$\int_{\mathbf{R}^3} p - \overline{p}\,dx \le O(t^{-\min(2,2-\alpha)}), \tag{4.95}$$

recalling $\alpha = 2 - n(\gamma-1)$, $n = 3$. In particular, it holds that

$$E \le \int_{\mathbf{R}^3} \frac{\rho}{2}|v|^2\,dx + o(1) \tag{4.96}$$

by (4.93) and (4.94).

Turning to (4.91), we may assume $\psi_0 = 0$, $|x| > R$, for $\psi_0 = \psi|_{t=0}$. Then it follows that

$$\psi(x,t) = \frac{A\gamma\overline{\rho}^{\gamma-1}}{\gamma-1}t, \quad |x| > R + \sigma t. \tag{4.97}$$

Equality (4.97) justifies

$$\begin{aligned}\frac{d}{dt}\int_{\mathbf{R}^3} \rho(\psi - \frac{A\gamma\overline{\rho}^{\gamma-1}}{\gamma-1}t) &= \int_{\mathbf{R}^3} -\frac{\rho}{2}|v|^2 + \frac{A\gamma\rho}{\gamma-1}(\rho^{\gamma-1} - \overline{\rho}^{\gamma-1})\,dx \\ &\le -E + \int_{\mathbf{R}^3} \frac{A\gamma\overline{\rho}^{\gamma-1}}{\gamma-1}(\overline{\rho} - \rho) + o(1) \\ &= -E - \frac{A\gamma\overline{\rho}^{\gamma-1}}{\gamma-1}M + o(1)\end{aligned}$$

together with (4.95)–(4.96). Hence we obtain

$$\begin{aligned}-Et + o(t) &\ge \int_{\mathbf{R}^3} \rho(\psi - \frac{A\gamma\overline{\rho}^{\gamma-1}}{\gamma-1}t)\,dx + \frac{A\gamma\overline{\rho}^{\gamma-1}}{\gamma-1}Mt \\ &= \int_{\mathbf{R}^3} \rho(\psi - \frac{A\gamma\overline{\rho}^{\gamma-1}}{\gamma-1}t) + \frac{A\gamma\overline{\rho}^{\gamma-1}}{\gamma-1}(\rho - \overline{\rho})t\,dx \\ &= \int_{\mathbf{R}^3} \rho\psi - \frac{A\gamma\overline{\rho}^{\gamma}}{\gamma-1}t\,dx.\end{aligned} \tag{4.98}$$

It follows, on the other hand, that

$$\psi_t\rho t \ge \frac{A\gamma\rho^\gamma}{\gamma-1}t$$

from (4.91), and hence

$$\begin{aligned}\rho\psi - \frac{A\gamma\rho^\gamma}{\gamma-1}t &\ge \rho\psi - \psi_t\rho t = 2\rho\psi - \frac{\partial}{\partial t}(\psi\rho t) + \psi\rho_t t \\ &= 2\rho\psi - \frac{\partial}{\partial t}\{\psi(\rho - \overline{\rho})t\} - \overline{\rho}\frac{\partial}{\partial t}(\psi t) + \psi\rho_t t \\ &= 2\rho\psi - \frac{\partial}{\partial t}\{\psi(\rho - \overline{\rho})t\} - \overline{\rho}\frac{\partial}{\partial t}\{(\psi - \frac{A\gamma\overline{\rho}^{\gamma-1}}{\gamma-1}t)t\} - 2t\frac{A\gamma\overline{\rho}^{\gamma}}{\gamma-1} + \psi\rho_t t.\end{aligned}$$

Then we have

$$\begin{aligned}\int_{|x|<R+\sigma t} \rho\psi - \frac{A\gamma\rho^\gamma}{\gamma-1}t\,dx &\ge 2\int_{|x|<R+\sigma t} \rho\psi - \frac{A\gamma\overline{\rho}^{\gamma}}{\gamma-1}t\,dx \\ &+ \int_{\mathbf{R}^3} -\frac{\partial}{\partial t}\{\psi(\rho - \overline{\rho})t + \overline{\rho}(\psi - \frac{A\gamma\overline{\rho}^{\gamma-1}}{\gamma-1}t)t\} + \psi\rho_t t\,dx\end{aligned}$$

$$= 2\int_{|x|<R+\sigma t} \rho\psi - \frac{A\gamma\rho^\gamma}{\gamma-1}t\,dx + o(t)$$
$$+ \int_{\mathbf{R}^3} -\frac{\partial}{\partial t}\{\psi(\rho-\overline{\rho})t + \overline{\rho}(\psi - \frac{A\gamma\overline{\rho}^{\gamma-1}}{\gamma-1}t)t\} + \psi\rho_t t\,dx$$

by (4.95), which means

$$\int_{|x|<R+\sigma t} \rho\psi - \frac{A\gamma\rho^\gamma}{\gamma-1}t\,dx$$
$$\leq \int_{\mathbf{R}^3} \frac{\partial}{\partial t}\{\psi(\rho-\overline{\rho})t + \overline{\rho}(\psi - \frac{A\gamma\overline{\rho}^{\gamma-1}}{\gamma-1}t)t\} - \psi\rho_t t\,dx + o(t)$$
$$= \frac{d}{dt}\int_{\mathbf{R}^3} \psi(\rho-\overline{\rho})t + \overline{\rho}(\psi - \frac{A\gamma\overline{\rho}^{\gamma-1}}{\gamma-1}t)t\,dx + t\int_{\mathbf{R}^3} \rho|v|^2\,dx + o(t)$$
$$= \frac{d}{dt}\int_{\mathbf{R}^3} (\psi\rho - \frac{A\gamma\overline{\rho}^{\gamma}}{\gamma-1}t)t\,dx + t\int_{\mathbf{R}^3} \rho|v|^2\,dx + o(t) \tag{4.99}$$

by (4.91). Since the right-hand side of (4.99) is equal to

$$\int_{|x|<R+\sigma t} \psi\rho - \frac{A\gamma\rho^\gamma}{\gamma-1}t\,dx + t\frac{d}{dt}\int_{\mathbf{R}^3} \psi\rho - \frac{A\gamma\overline{\rho}^\gamma}{\gamma-1}t\,dx + 2tE + o(t)$$

by (4.95) and (4.96), it follows that

$$-2E + o(1) \leq \frac{d}{dt}\int_{\mathbf{R}^3} \psi\rho - \frac{A\gamma\overline{\rho}^\gamma}{\gamma-1}t\,dx. \tag{4.100}$$

Combining (3.135) and (4.100), we obtain

$$-2Et + o(t) \leq \int_{\mathbf{R}^3} \psi\rho - \frac{A\gamma\overline{\rho}^\gamma}{\gamma-1}t\,dx \leq -Et + o(t)$$

and hence $E \geq 0$ in the case of $T = +\infty$. □

4.5 Shape Optimizations

4.5.1 Theory of Optimizations

In free boundary problems in mathematical science, over-determined boundary conditions are set following physical laws, to prescribe the domain. Shape optimization in engineering, on the other hand, provides a domain maximizing income or minimizing cost. Usually, it is composed of the process of discretization and the theory of finite dimensional optimization. The finite element method is applicable for the former [10]. Here we describe the latter in an abstract manner. See [2] for applications of this abstract theory.

We begin with the Karush–Kuhn–Tucker (KKT) theory, to describe the fundamental theory of optimization. It is, in fact, concerned with

$$\min_{x\in\mathbf{R}^d}\{f_0(x) \mid f_j(x) \leq 0,\ 1 \leq j \leq m\}, \tag{4.101}$$

where $f_j = f_j(x) : \mathbf{R}^d \to \mathbf{R}$, $0 \leq j \leq m$, are C^1 functions. Put

$$S = \{x \in \mathbf{R}^d \mid f_j(x) \leq 0,\ 1 \leq j \leq m\}$$

and $I(x) = \{1 \leq j \leq m \mid f_j(x) = 0\}$ for $x \in S$. Let

$$C_s(x) = \{e \in S^{d-1} \mid \exists y_k \in S \to x,\ e = \lim_{k\to\infty} \frac{y_k - x}{|y_k - x|}\}$$

and

$$T_s(x) = \{e \in S^{d-1} \mid e \cdot \nabla f_j(x) \leq 0,\ \forall j \in I(x)\}$$

be the sets of admissible directions and linearly admissible directions, respectively. It holds that $C_s(x) < T_s(x)$ for any $x \in S$.

Constraint qualification of the first order is a requirement of $C_s(x) = T_s(X)$, while the Lagrangian is defined as

$$\mathscr{L}(x,\lambda) = f_0(x) + \sum_{j=1}^m \lambda_j f_j(x).$$

Then we obtain the following theorem.

Theorem 4.15 (KKT) *Assume that $x \in S$ is a solution to (4.101), provided with the constraint qualification of the first order, and*

$$\{\nabla f_j(x) \mid j \in I(x)\}$$

are linearly independent. Then there is $\lambda = (\lambda_1, \cdots, \lambda_m)$ satisfying

$$f_j(x) \leq 0,\ \lambda_j \geq 0,\ f_j(x) = 0,\ \lambda_j f_j(x) = 0, \quad 1 \leq j \leq m$$

and

$$\frac{\partial \mathscr{L}}{\partial x}(x,\lambda) = 0. \tag{4.102}$$

Corollary 4.3 *Let* $S_0 = \{x \in \mathbf{R}^d \mid f_j(x = 0),\ 1 \leq j \leq m\}$ *and* $x \in S_0$ *be a solution to* $\{f_0(x) \mid x \in S_0\}$. *Assume that*

$$\{\nabla f_j(x) \mid 1 \leq j \leq m\}$$

are linearly independent. Then there is $\lambda = (\lambda_1, \cdots, \lambda_m)$ *such that (4.102) holds.*

Proof This problem is equivalent to

$$\min_{x \in \mathbf{R}^d} \{f_0(x) \mid \pm f_j(x) \leq 0,\ 1 \leq j \leq m\}.$$

Then the result follows from the previous theorem. □

Remark 4.4 It holds that

$$\frac{\partial \mathscr{L}}{\partial x}(x, \lambda) = \nabla f_0(x) + {}^t(\nabla h(x))[\lambda]$$

for

$$h(x) = \begin{pmatrix} f_1(x) \\ \vdots \\ f_m(x) \end{pmatrix}, \quad \nabla h(x) = \begin{pmatrix} \frac{\partial f_1}{\partial x_1} & \cdots & \frac{\partial f_1}{\partial x_d} \\ \vdots & \ddots & \vdots \\ \frac{\partial f_m}{\partial x_1} & \cdots & \frac{\partial f_m}{\partial x_d} \end{pmatrix},$$

by $x = (x_1, \cdots, x_d)$ and

$$\nabla f = \begin{pmatrix} \frac{\partial f}{\partial x_1} \\ \vdots \\ \frac{\partial f}{\partial x_d} \end{pmatrix}.$$

Hence (4.102) means $x \in \mathbf{R}^d$ and

$$\nabla f_0(x) + {}^t(\nabla h(x))[\lambda] = 0, \quad \lambda = (\lambda_1, \cdots, \lambda_m).$$

4.5.2 Abstract Theory of Optimum Design

Typical optimum design, after the reduction of discretization, is composed of a cost function, state equation, and synthesis parameters, denoted by

$$f_0 = f_0(\xi, u) : \mathbf{R}^\ell \times \mathbf{R}^n \to \mathbf{R}, \quad h = h(\xi, u) : \mathbf{R}^\ell \times \mathbf{R}^n \to \mathbf{R}^n,$$

and $\xi \in \mathbf{R}^\ell$, respectively,

$$\min_{(\xi,u)\in\mathbf{R}^\ell\times\mathbf{R}^n} \{f_0(\xi,u) \mid h(\xi,u)=0\}. \tag{4.103}$$

To understand the formulation, let $h(\xi,u)=0$ be solved as $u=u(\xi)$ locally, for the moment. Then problem (4.103) is reduced to

$$\min_{\xi\in\mathbf{R}^\ell} \tilde{f}_0(\xi), \quad \tilde{f}_0(\xi) = f_0(\xi, u(\xi)).$$

If ξ is a solution to (4.103), therefore, it holds that

$$0 = \frac{\partial \tilde{f}_0}{\partial \xi}(\xi) = \frac{\partial f_0}{\partial \xi}(\xi, u(\xi)) + {}^t(\frac{\partial f_0}{\partial u})(\xi, u(\xi))\frac{\partial u}{\partial^t \xi}, \tag{4.104}$$

under the agreement of

$$\frac{\partial f}{\partial \xi} = \begin{pmatrix} \frac{\partial f}{\partial \xi_1} \\ \vdots \\ \frac{\partial f}{\partial \xi_\ell} \end{pmatrix}, \quad \frac{\partial f}{\partial u} = \begin{pmatrix} \frac{\partial f}{\partial u_1} \\ \vdots \\ \frac{\partial f}{\partial u_n} \end{pmatrix}, \quad \frac{\partial u}{\partial^t \xi} = \begin{pmatrix} \frac{\partial u_1}{\partial \xi_1} & \cdots & \frac{\partial u_1}{\partial \xi_\ell} \\ \vdots & \vdots & \vdots \\ \frac{\partial u_n}{\partial \xi_1} & \cdots & \frac{\partial u_n}{\partial \xi_\ell} \end{pmatrix}.$$

Since $h(\xi, u(\xi)) = 0$ locally, on the other hand, we obtain

$$0 = \frac{\partial h}{\partial \xi}(\xi, u(\xi)) + {}^t(\frac{\partial h}{\partial u})(\xi, u(\xi))\frac{\partial u}{\partial^t \xi}. \tag{4.105}$$

Following the above observation, we now define the adjoint problem by

$${}^t(\frac{\partial h}{\partial u})(\xi, u)[v] = \frac{\partial f_0}{\partial u}(\xi, u), \quad v \in \mathbf{R}^n, \tag{4.106}$$

using

$$\frac{\partial h}{\partial u} = \begin{pmatrix} \frac{\partial f_1}{\partial u_1} & \cdots & \frac{\partial f_1}{\partial u_n} \\ \vdots & \ddots & \vdots \\ \frac{\partial f_n}{\partial u_1} & \cdots & \frac{\partial f_n}{\partial u_n} \end{pmatrix}.$$

It follows that

$${}^t(\frac{\partial f_0}{\partial u})(\xi, u(\xi))\frac{\partial u}{\partial^t \xi} = {}^t v \frac{\partial h}{\partial u}(\xi, u(\xi))\frac{\partial u}{\partial^t \xi} = -v \cdot \frac{\partial h}{\partial \xi}(\xi, u(\xi))$$

from (4.105). Hence we obtain

$$\frac{\partial f_0}{\partial \xi}(\xi, u(\xi)) = v \cdot \frac{\partial h}{\partial \xi}(\xi, u(\xi)) \tag{4.107}$$

by (4.104). Equalities (4.104), (4.106), and (4.107) induce the adjoint method without solving $h(\xi, u) = 0$ directly as $u = u(\xi)$.

Given (ξ, u), thus, we define $v = v(\xi, u)$ by (4.106), and then solve

$$\frac{\partial h}{\partial \xi}(\xi, u) + {}^t(\frac{\partial h}{\partial u})(\xi, u)\frac{\partial u}{\partial^t \xi} = 0$$
$$v(\xi, u) \cdot \frac{\partial h}{\partial \xi}(\xi, u) + {}^t(\frac{\partial f_0}{\partial u})(\xi, u)\frac{\partial u}{\partial^t \xi} = 0$$

with respect to (ξ, u). Note that if h is linear in u, Eq. (4.106) provides an explicit form of $v = v(\xi, u)$.

This adjoint problem has a formulation of Lagrangian multiplier. Letting

$$\mathscr{L}(\xi, u, v) = f_0(\xi, u) - v \cdot h(\xi, u),$$

thus, we seek its critical point (ξ, u, v). In fact, there arises

$$\frac{\partial \mathscr{L}}{\partial u}(\xi, u, v) \equiv \frac{\partial f_0}{\partial u}(\xi, u) - v \cdot \frac{\partial h}{\partial u}(\xi, u) = 0$$
$$\frac{\partial \mathscr{L}}{\partial v}(\xi, u, v) \equiv -h(\xi, u) = 0$$
$$\frac{\partial \mathscr{L}}{\partial \xi}(\xi, u, v) \equiv \frac{\partial f_0}{\partial \xi}(\xi, u) - v \cdot \frac{\partial h}{\partial \xi}(\xi, u) = 0$$

in this case.

4.5.3 *Optimum Design with Constraint*

A discrete optimum design with constraint is formulated by the C^1 functions

$$f_j = f_j(\xi, u) : \mathbf{R}^\ell \times \mathbf{R}^n \to \mathbf{R}, \quad 0 \le j \le m$$
$$h = h(\xi, u) : \mathbf{R}^\ell \times \mathbf{R}^n \to \mathbf{R}^n$$

as

$$\min_{(\xi,u)\in\mathbf{R}^\ell\times\mathbf{R}^n} \{f_0(\xi, u) \mid h(\xi, u) = 0, \quad f_j(\xi, u) \le 0, \ 1 \le j \le m\}. \tag{4.108}$$

To formulate the adjoint system and Lagrangian multiplier methods, assume, for the moment that the state equation $h(\xi, u) = 0$ is solved as $u = u(\xi)$ locally, and put

$$\tilde{f}_0(\xi) = f_0(\xi, u(\xi)).$$

Then, (4.108) is equivalent to

$$\min_{\xi \in \mathbf{R}^\ell} \{\tilde{f}_0(\xi) \mid \tilde{f}_j(\xi) \leq 0, \ 1 \leq j \leq m\}.$$

If $\xi \in \mathbf{R}^\ell$ solves this problem with constraint qualification of the first kind, therefore, there is $\lambda = (\lambda_1, \cdots, \lambda_m)$ such that

$$\lambda_j \geq 0, \ f_j(\xi, u) \leq 0, \ \lambda_j f_j(\xi, u) = 0, \quad 1 \leq j \leq m, \tag{4.109}$$

and

$$\begin{aligned} 0 = \frac{\partial \tilde{f}_0}{\partial \xi}(\xi) + \sum_{j=1}^m \lambda_j \frac{\partial \tilde{f}_j}{\partial \xi}(\xi) = \frac{\partial f_0}{\partial \xi}(\xi, u) + {}^t(\frac{\partial f_0}{\partial u})(\xi, u)\frac{\partial u}{\partial^t \xi} \\ + \sum_{j=1}^m \lambda_j \{\frac{\partial f_j}{\partial \xi}(\xi, u) + {}^t(\frac{\partial f_j}{\partial \xi})(\xi, u)\frac{\partial u}{\partial^t \xi}\} \end{aligned} \tag{4.110}$$

for $u = u(\xi)$. It holds also that

$$\frac{\partial h}{\partial \xi}(\xi, u(\xi)) + {}^t(\frac{\partial h}{\partial u})(\xi, u(\xi))\frac{\partial u}{\partial^t \xi} = 0 \tag{4.111}$$

by $h(\xi, u(\xi)) = 0$.

The adjoint problem is now formulated as

$${}^t(\frac{\partial h}{\partial u})(\xi, u)[v_j] = \frac{\partial f_j}{\partial u}(\xi, u), \quad v_j \in \mathbf{R}^n, \ 0 \leq j \leq m. \tag{4.112}$$

Then (4.111) implies

$${}^t(\frac{\partial f_j}{\partial u})(\xi, u(\xi))\frac{\partial u}{\partial^t \xi} = {}^t v_j \frac{\partial h}{\partial u}(\xi, u(\xi))\frac{\partial u}{\partial^t \xi} = -v_j \cdot \frac{\partial h}{\partial \xi}(\xi, u(\xi)),$$

and therefore,

$$0 = \frac{\partial f_0}{\partial \xi}(\xi, u) - v_j \cdot \frac{\partial h}{\partial \xi}(\xi, u) + \sum_{j=1}^m \lambda_j \{\frac{\partial f_j}{\partial \xi}(\xi, u) - v_j \cdot \frac{\partial h}{\partial \xi}(\xi, u)\} \tag{4.113}$$

for $u = u(\xi)$ by (4.110).

In the adjoint method, therefore, we solve (4.112) for

$$v_j = v_j(\xi, u), \quad 0 \le j \le m,$$

given (ξ, u). Then seek (λ, ξ, u) satisfying (4.109), (4.110), and (4.113). For the Lagrangian multiplier method we take

$$\mathscr{L}(\xi, u, v_0, \cdots, v_m, \lambda) = \mathscr{L}_0(\xi, u, v_0) + \sum_{j=1}^{m} \mathscr{L}_j(\xi, u, v_j)$$

for

$$\lambda = (\lambda_1, \cdots, \lambda_m), \quad \mathscr{L}_j(\xi, u, v_j) = f_j(\xi, u) - v_j \cdot h(\xi, u), \ 0 \le j \le m.$$

Under the constraint

$$\lambda_j \ge 0, \ \lambda_j \mathscr{L}_j(\xi, u, v_j) = 0, \ \mathscr{L}_j(\xi, u, v_j) \le 0, \quad 1 \le j \le m$$

we find a critical point of this $\mathscr{L}$, regarding

$$\frac{\partial \mathscr{L}}{\partial u} = \frac{\partial f_0}{\partial u}(\xi, u) - v_0 \cdot \frac{\partial h}{\partial u}(\xi, u) + \sum_{j=1}^{m} \lambda_j \{\frac{\partial f_j}{\partial u}(\xi, u) - v_j \cdot \frac{\partial h}{\partial u}(\xi, u)\}$$

$$\frac{\partial L}{\partial v_0} = -h(\xi, u)$$

$$\frac{\partial \mathscr{L}}{\partial \xi} = \frac{\partial f_0}{\partial \xi}(\xi, u) - v_0 \cdot \frac{\partial h}{\partial \xi}(\xi, u) + \sum_{j=1}^{m} \lambda_j \{\frac{\partial f_j}{\partial \xi}(\xi, u) - v_j \cdot \frac{\partial h}{\partial \xi}(\xi, u)\}.$$

4.6 Hadamard's Variational Formula

4.6.1 An Abstract Theorem

Creating new numerical schemes for free boundary problems [43], we have several analytic problems such as the compactness of the sequence of two-dimensional domains defined by conformal mappings [7] and also clarifying the advantage of the H^1 gradient method [3].

Generalization of Hadamard's variational formula is the other issue. First, domain perturbation must be realized. Thus we have to construct a scheme to create such perturbations. Second, conical domains are more desirable in practice. Here, Lipschitz domains are in a suitable category from both numerical and theoretical points of view. Third, variational formulas for mixed boundary conditions are necessary because over-determined boundary conditions are imposed on the free boundary. Then we find several technical difficulties because of the lack of regularity of the solution. Finally, the second variational formula is needed to make the computation be adaptive because its instability of the scheme is relaxed by Newton's method.

This section is devoted to analytic foundations of these problems, selecting Green's function as a target. This project is not always necessary in our original attempt of improving the visualization of free boundaries. It, however, contains significant suggestions from the theories of function spaces, elliptic regularity, tensor analysis, and differential geometry. Here we develop a general theory, expanding classical descriptions in two or three dimensions [12].

The derivation of variational formulas is hard even formally, but a fundamental theoretical obstruction is the change of domain where Green's function is defined. Its differentiability, on the other hand, is taken only in the interior, while Green's function is not free from the shape of the boundary. This discrepancy is the origin of the difficulty if one uses the Euler coordinate. Then we use the Lagrange derivative to ensure the differentiability where an abstract theorem is applicable.

Let V be a Hilbert space over $\mathbf{R}$, and

$$a_t : V \times V \to \mathbf{R}, \quad t \in (-\varepsilon, \varepsilon) \equiv I,$$

be a family of symmetric bilinear forms. We assume their uniform boundedness and coercivity. Hence there are $\delta > 0$ and $M > 0$ such that

$$a_t(u,u) \geq \delta \|u\|_V^2, \quad |a_t(u,v)| \leq M\|u\|_V \|v\|_V \tag{4.114}$$

for any $u, v \in V$ and $t \in I$. Given $f : I \to V'$, we have a unique $u = u(t) \in V$, $t \in I$, such that

$$a_t(u(t), v) = \langle v, f(t) \rangle, \quad \forall v \in V, \tag{4.115}$$

by the representation theorem of Riesz.

Theorem 4.16 ([34]) *Let the above $a_t(\cdot,\cdot)$ and $f(t)$ be strongly differentiable in t. Namely, first, there is a symmetric bilinear form denoted by $\dot{a}_t : V \times V \to \mathbf{R}$ such that*

$$\lim_{h\to 0} \sup_{\|u\|\leq 1,\ \|v\|\leq 1} \left| \frac{a_{t+h}(u,v) - a_t(u,v)}{h} - \dot{a}_t(u,v) \right| = 0, \quad \forall t \in I.$$

Next, there is $\dot{f} = \dot{f}(t) \in V'$ such that

$$\lim_{h\to 0} \sup_{\|v\|\leq 1} \left| \frac{\langle v, f(t+h)\rangle - \langle v, f(t)\rangle}{h} - \langle v, \dot{f}(t)\rangle \right| = 0, \quad \forall t \in I.$$

Then $u = u(t) \in V$ defined by (4.115) is strongly differentiable in t, that is,

$$\exists \dot{u}(t) \in V, \quad \lim_{h\to 0} \left\| \frac{u(t+h)-u(t)}{h} - \dot{u}(t) \right\|_V = 0, \quad \forall t \in I, \tag{4.116}$$

and it holds that

$$a_t(\dot{u}(t), v) + \dot{a}_t(u(t), v)) = \langle v, \dot{f}(t)\rangle, \quad \forall v \in V,\ t \in I. \tag{4.117}$$

The second strong differentiability of $u = u(t)$ arises similarly if

$$a_t : V \times V \to \mathbf{R}$$

and $f = f(t) : I \to V'$ are twice differentiable. Then it holds that

$$a_t(\ddot{u}(t), v) + 2\dot{a}_t(\dot{u}(t), v) + \ddot{a}_t(u(t), v) = \langle v, \ddot{f}(t)\rangle, \quad \forall v \in V,\ t \in I \tag{4.118}$$

where

$$\lim_{h\to 0} \sup_{\|u\|\le 1,\ \|v\|\le 1} \left| \frac{\dot{a}_{t+h}(u, v) - \dot{a}_t(u, v)}{h} - \ddot{a}_t(u, v) \right| = 0$$

$$\lim_{h\to 0} \sup_{\|v\|\le 1} \left| \frac{\langle v, \dot{f}(t+h)\rangle - \langle v, \dot{f}(t)\rangle}{h} - \langle v, \ddot{f}(t)\rangle \right| = 0$$

$$\lim_{h\to 0} \left\| \frac{\dot{u}(t+h) - \dot{u}(t)}{h} - \ddot{u}(t) \right\|_V = 0.$$

To prove strong differentiability, for example, first, we define $\dot{u}$ by (4.117) and then derive (4.116) using (4.114).

4.6.2 Green's Function Revisited

Let $\Omega \subset \mathbf{R}^N$ be a Lipschitz domain. Recall that Green's function for $-\Delta$ with $\cdot|_{\partial\Omega} = 0$, denoted by $G = G(x, y)$, $(x, y) \in \overline{\Omega} \times \Omega$, is defined by

$$-\Delta G(\cdot, y) = \delta(\cdot - y) \quad \text{in } \Omega, \quad G(\cdot, y)|_{\partial\Omega} = 0,$$

where $\delta = \delta(x)$ is the delta function. The fundamental solution $\Gamma(x) = \gamma(|x|)$ of $-\Delta$,

$$\gamma(r) = \begin{cases} \frac{1}{2\pi} \log \frac{1}{r}, & N = 2 \\ \frac{1}{(N-2)\omega_N} r^{2-N}, & N \geq 3 \end{cases}$$

takes the property

$$-\Delta \Gamma(\cdot - y) = \delta(\cdot - y),$$

where ω_N denotes the $(N-1)$-dimensional volume of the N-dimensional unit ball. Hence the function $u = u(x)$ defined by

$$G(\cdot, y) = \Gamma(\cdot - y) + u \tag{4.119}$$

satisfies

$$\Delta u = 0 \quad \text{in } \Omega, \qquad u = -\Gamma(\cdot - y) \quad \text{on } \partial\Omega. \tag{4.120}$$

We reduce this boundary value problem of the Laplace equation to that of the Poisson equation, taking open sets $\omega \subset\subset \hat{\omega}$ satisfying $\partial\Omega \subset \omega$ and $y \notin \hat{\omega}$ and $\varphi = \varphi(x) \in C_0^\infty(\mathbf{R}^N)$ such that $0 \leq \varphi \leq 1$ and

$$\varphi(x) = \begin{cases} 1, & x \in \omega \\ 0, & x \notin \hat{\omega}. \end{cases}$$

In fact, (4.120) is reduced to

$$-\Delta w = g \quad \text{in } \Omega, \qquad w = 0 \quad \text{on } \partial\Omega \tag{4.121}$$

concerning $w = u - \varphi\Gamma(\cdot - y)$ for $g = \Delta(\varphi\Gamma(\cdot - y))$. Then this $g = g(x) \in C_0^\infty(\mathbf{R}^N)$ is regarded as an element of $H^{-1}(\Omega)$ by

$$\langle v, g \rangle = -(\nabla v, \nabla(\varphi\Gamma(\cdot - y))), \quad v \in H_0^1(\Omega).$$

Hence the boundary value problem (4.121) admits a unique solution $w \in H_0^1(\Omega)$. Then the solution to (4.120) is obtained by $u = w + \varphi\Gamma(\cdot - y) \in H^1(\Omega)$ because $\Gamma(\cdot - y)$ is smooth in $\hat{\omega}$, which implies the well-definedness of

$$0 = G(\cdot, y) \in H^{1/2}(\partial\Omega), \qquad \frac{\partial G(\cdot, y)}{\partial n} \in H^{-1/2}(\partial\Omega) \tag{4.122}$$

by (4.119). In the following theorem, the pairing $\langle\ ,\ \rangle$ is taken between $H^{1/2}(\partial\Omega)$ and $H^{-1/2}(\partial\Omega)$.

Theorem 4.17 *If $\Omega \subset \mathbf{R}^N$ is a Lipschitz domain and $f \in H^1(\Omega)$ is harmonic in Ω, then it holds that*

$$f(y) = -\left\langle f, \frac{\partial G(\cdot, y)}{\partial n} \right\rangle, \quad y \in \Omega. \tag{4.123}$$

Proof Since $\Delta f = 0 \in H^1(\Omega)'$ we have

$$(\nabla f, \nabla \varphi) = \left\langle \varphi, \frac{\partial f}{\partial n} \right\rangle, \quad \varphi \in H^1(\Omega) \tag{4.124}$$

by Theorem 4.3. Given $y \in \Omega$, we take

$$\varphi_\varepsilon(x) = \begin{cases} \gamma(|x-y|) & |x-y| \geq \varepsilon \\ \gamma(\varepsilon) & |x-y| < \varepsilon \end{cases}$$

for $0 < \varepsilon \ll 1$, which belongs to $H^1(\Omega)$. Then we obtain

$$\begin{aligned}(\nabla f, \nabla \varphi_\varepsilon) &= \int_{\Omega \setminus B(y,\varepsilon)} \nabla f(x) \cdot \nabla \Gamma(x-y)\, dx \\ &= \int_{\partial\Omega} f(x) \frac{\partial \Gamma(x-y)}{\partial n_x}\, ds_x - \int_{\partial B(y,\varepsilon)} f(x) \frac{\partial \Gamma(x-y)}{\partial n_x}\, ds_x\end{aligned}$$

using the traces $f \in H^{1/2}(\partial\Omega)$ and $f \in H^{1/2}(\partial B(y, \varepsilon))$. By Weyl's lemma, $f = f(x)$ is smooth in Ω. Hence by the classical argument, the left-hand side of (4.124) for $\varphi = \varphi_\varepsilon$, that is, $(\nabla f, \nabla \varphi_\varepsilon)$, converges to

$$\int_{\partial\Omega} f(x) \frac{\partial \Gamma(x-y)}{\partial n_x}\, ds_x + f(y)$$

as $\varepsilon \downarrow 0$. Since its right-hand side, $\left\langle \varphi_\varepsilon, \frac{\partial f}{\partial n} \right\rangle$, is independent of $0 < \varepsilon \ll 1$, we end up with

$$\left\langle \Gamma(\cdot - y), \frac{\partial f}{\partial n} \right\rangle = \left\langle f, \frac{\partial \Gamma(\cdot - y)}{\partial n} \right\rangle + f(y). \tag{4.125}$$

We define $u = u(x)$ by (4.119), and note $\Delta f = \Delta u = 0 \in H^1(\Omega)'$. Then it follows that

$$(\nabla u, \nabla f) = \left\langle u, \frac{\partial f}{\partial n} \right\rangle = \left\langle f, \frac{\partial u}{\partial n} \right\rangle \tag{4.126}$$

by Theorem 4.3. Since (4.122) implies

$$\left\langle u + \Gamma(\cdot - y), \frac{\partial f}{\partial n} \right\rangle = \left\langle G(\cdot, y), \frac{\partial f}{\partial n} \right\rangle = 0$$

it follows that

$$\left\langle \Gamma(\cdot - y), \frac{\partial f}{\partial n} \right\rangle = -\left\langle u, \frac{\partial f}{\partial n} \right\rangle = -\left\langle f, \frac{\partial u}{\partial n} \right\rangle$$

from (4.126). Hence we obtain

$$f(y) = -\left\langle f, \frac{\partial \Gamma(\cdot - y)}{\partial n} + \frac{\partial u}{\partial n} \right\rangle = -\left\langle f, \frac{\partial G(\cdot, y)}{\partial n} \right\rangle$$

by (4.125). □

4.6.3 Lagrange Derivative

We have two formulations of variational formulae, using Euler and Lagrange coordinates. In the Lagrange coordinate, we do not have any derivative losses in space variables. The variational formulae, however, are not clear compared with the ones in Euler coordinates.

We recall that $\Omega \subset \mathbf{R}^N$ is a Lipschitz domain and $\{T_t\}$ is either a once or twice differentiable deformation. Green's function for $-\Delta$ in Ω_t with $\cdot|_{\partial\Omega_t} = 0$ is denoted by $G_t = G_t(x, y)$. We fix $y \in \Omega$ to define $u = u(\cdot, t)$ by (4.119):

$$G_t(x, y) = \Gamma(x - y) + u(x, t), \tag{4.127}$$

which satisfies

$$\Delta u(\cdot, t) = 0 \quad \text{in } \Omega_t, \qquad u(\cdot, t) = -\Gamma(\cdot - y) \quad \text{on } \partial\Omega_t. \tag{4.128}$$

As in Sect. 4.6.2 this problem is reduced to the boundary value problem of the Poisson equation

$$-\Delta w(\cdot, t) = g \quad \text{in } \Omega_t, \qquad w(\cdot, t) = 0 \quad \text{on } \partial\Omega_t \tag{4.129}$$

where

$$g = \Delta(\varphi\Gamma(\cdot - y))$$

for $\varphi \in C_0^\infty(\mathbf{R}^N)$. Given $y \in \Omega$, this $\varphi = \varphi(x)$ is taken uniformly in $|t| \ll 1$.

Theorem 4.18 ([44]) *Let $\Omega \subset \mathbf{R}^N$ be a Lipschitz domain and $\{T_t\}$, $t \in I = (-\varepsilon, \varepsilon)$, be a once or twice differentiable deformation. Then, according to its differentiabilities we have the existence of*

$$\frac{\partial u}{\partial t}(T_t x, t), \quad \frac{\partial^2 u}{\partial t^2}(T_t x, t)$$

strongly in $H^1(\Omega)$.

Proof The weak form of (4.129) is

$$\int_{\Omega_t} \nabla w(\cdot, t) \cdot \nabla \varphi \, dy = \int_{\Omega_t} \varphi g \, dy, \quad \forall \varphi \in H_0^1(\Omega_t). \tag{4.130}$$

Given $\psi \in H_0^1(\Omega)$, we define $v = v(\cdot, t)$ by

$$\varphi(y, t) = \psi(T_t^{-1} y), \quad y \in \Omega_t, \quad |t| \ll 1.$$

It holds that $v(\cdot, t) \in H_0^1(\Omega_t)$ because $T_t : \overline{\Omega} \to \overline{\Omega_t}$ is bi-Lipschitz. Then (4.130) implies

$$\int_{\Omega} [J_t(x)^{-1} \nabla v(x, t) \cdot J_t(x)^{-1} \nabla \psi(x)] \det J_t(x) \, dx$$
$$= \int_{\Omega} \psi(x) g(T_t x) \det J_t(x) \, dx$$

for

$$J_t = J(T_t), \quad v(x, t) = w(T_t x, t).$$

Now we apply Theorem 4.16 for $V = H_0^1(\Omega)$,

$$f(t) = h(\cdot, t) \det J_t(\cdot), \quad h(x, t) = g(T_t x),$$

and

$$a_t(v, \psi) = \int_{\Omega} [J_t^{-1} \nabla v \cdot J_t^{-1} \nabla \psi] \det J_t \, dx.$$

From the dominated convergence theorem the strong differentiability of $v(\cdot, t) \in H_0^1(\Omega)$ once or twice in t follows, which implies those of $u(T_t x, t)$ and $G_t(T_t x, y)$. □

To conclude the subsection, we show concrete forms of the Lagrange derivative of $G(\cdot, y)$. The complicated formulae below will be reduced to simpler ones in the

Euler coordinate under the cost of the regularity of u. Let $\dot{v} = \dot{v}(0) \in H_0^1(\Omega)$ and $\ddot{v} = \ddot{v}(0) \in H_0^1(\Omega)$ satisfy (4.117) and (4.118), respectively. Put

$$\dot{f} = \dot{f}(0), \quad \ddot{f} = \ddot{f}(0),$$

recalling Theorem 4.8.

First,

$$f(x,t) = g(T_t x) \det J_t(x)$$

satisfies

$$\frac{\partial f}{\partial t}(x,t) = \frac{\partial T_t x}{\partial t} \cdot \nabla g(T_t x) + g(T_t x) \frac{\partial}{\partial t} \det J_t(x)$$

and hence

$$\dot{f} = S\nabla g + g\nabla \cdot S = \nabla \cdot (Sg).$$

We have also

$$\begin{aligned} \ddot{f} &= R \cdot \nabla g + [\nabla^2 g]S \cdot S + (S \cdot \nabla g)\nabla \cdot S + g\left[\nabla \cdot R + (\nabla \cdot S)^2 - J(S) : {}^t J(S)\right] \\ &= \nabla \cdot (gR) + [\nabla^2 g]S \cdot S + (S \cdot \nabla g)\nabla \cdot S + g\left[(\nabla \cdot S)^2 - J(S) : {}^t J(S)\right]. \end{aligned}$$

Second, the strong derivatives

$$\dot{a}_t = \dot{a}_t(v, \psi), \; \ddot{a}_t = \ddot{a}_t(v, \psi), \quad v, \psi \in H_0^1(\Omega), \; |t| \ll 1$$

of $a_t = a_t(v, \psi)$ are given by

$$\begin{aligned} \dot{a}_t(v, \psi) = \int_\Omega & \left[\frac{\partial J_t^{-1}}{\partial t} \nabla v \cdot J_t^{-1} \nabla \psi + J_t^{-1} \nabla v \cdot \frac{\partial J_t^{-1}}{\partial t} \nabla \psi \right] \det J_t \\ & + \left[J_t^{-1} \nabla v \cdot J_t^{-1} \nabla \psi \right] \frac{\partial}{\partial t} \det J_t \, dx \end{aligned}$$

and

$$\begin{aligned} & \ddot{a}_t(v, \psi) \\ & = \int_\Omega \left[\frac{\partial^2 J_t^{-1}}{\partial t^2} \nabla v \cdot J_t^{-1} \nabla \psi + 2 \frac{\partial J_t^{-1}}{\partial t} \nabla v \cdot \frac{\partial J_t^{-1}}{\partial t} \nabla \psi \right. \end{aligned}$$

$$+ J_t^{-1}\nabla v \cdot \frac{\partial^2 J_t^{-1}}{\partial t^2}\nabla\psi\Bigg] \det J_t$$

$$+ 2\left[\frac{\partial J_t^{-1}}{\partial t}\nabla v \cdot J_t^{-1}\nabla\psi + J_t^{-1}\nabla v \cdot \frac{\partial J_t^{-1}}{\partial t}\nabla\psi\right]\frac{\partial}{\partial t}\det J_t$$

$$+ \left[J_t^{-1}\nabla v \cdot J_t^{-1}\nabla\psi\right]\frac{\partial^2}{\partial t^2}\det J_t\, dx,$$

respectively. Here we have

$$\left.\frac{\partial J_t^{-1}}{\partial t}\right|_{t=0} = -J(S), \quad \left.\frac{\partial^2 J_t^{-1}}{\partial t^2}\right|_{t=0} = -J(R) + 2J(S)^2$$

by

$$J_0 = I, \quad \left.\frac{\partial J_t}{\partial t}\right|_{t=0} = J(S), \quad \left.\frac{\partial^2 J_t}{\partial t^2}\right|_{t=0} = J(R).$$

In fact, the relation $J_t^{-1}J_t = I$ implies

$$\frac{\partial J_t^{-1}}{\partial t}J_t + J_t^{-1}\frac{\partial J_t}{\partial t} = 0, \quad \frac{\partial^2 J^{-1}}{\partial t^2}J_t + 2\frac{\partial J_t^{-1}}{\partial t}\frac{\partial J_t}{\partial t} + J_t^{-1}\frac{\partial^2 J_t}{\partial t^2} = 0.$$

Hence it follows that

$$\dot{a}_0(v,\psi) = -\int_\Omega \left(J(S) + {}^tJ(S) - (\nabla\cdot S)I\right)\nabla v \cdot \nabla\psi\, dx$$

and

$$\begin{aligned}
\ddot{a}_0(v,\psi) &= \int_\Omega \left[-J(R) + 2J(S)^2\right]\nabla v \cdot \nabla\psi + 2J(S)\nabla v \cdot J(S)\nabla\psi \\
&\quad + \nabla v \cdot \left[-J(R) + 2J(S)^2\right]\nabla\psi \\
&\quad -2\left[J(S)\nabla v \cdot \nabla\psi + \nabla v \cdot J(S)\nabla\psi\right]\nabla\cdot S \\
&\quad + \nabla v \cdot \nabla\psi\left[\nabla\cdot R + (\nabla\cdot S)^2 - J(S) : {}^tJ(S)\right] dx \\
&= \int_\Omega \left(\left[-J(R) + 2J(S)^2\right] + {}^t\left[-J(R) + 2J(S)^2\right]\right)\nabla v \cdot \nabla\psi \\
&\quad + 2\left({}^tJ(S)\cdot J(S) - \left[J(S) + {}^tJ(S)\right]\nabla\cdot S\right)\nabla v \cdot \nabla\psi \\
&\quad + \left(\nabla\cdot R + (\nabla\cdot S)^2 - J(S) : {}^tJ(S)\right)\nabla v \cdot \nabla\psi\, dx.
\end{aligned}$$

Then the strong derivatives $\dot{v}$ and $\ddot{v}$ in H_0^1 are defined by

$$a_0(\dot{v},\psi)+\dot{a}_0(v,\psi)=\langle\psi,\dot{f}\rangle,\qquad \forall\psi\in H_0^1(\Omega)$$
$$a_0(\ddot{v},\psi)+2\dot{a}_0(\dot{v},\psi)+\ddot{a}_0(v,\psi)=\langle\psi,\ddot{f}\rangle,\quad \forall\psi\in H_0^1(\Omega).$$

4.6.4 Euler Derivative

Theorem 4.18 guarantees the differentiability of $u = u(\cdot,t)$ in $\mathscr{D}'(\Omega)$ for any Lipschitz domain Ω. To prescribe the boundary value of $\dot{u}$, however, we require an additional regularity for $u = u(x)$.

Theorem 4.19 (**[44]**) *Let $\Omega\subset\mathbf{R}^N$ be a Lipschitz domain and $G = G(\cdot,y)$, $y\in\Omega$, be Green's function for $-\Delta$ provided with the Dirichlet boundary condition. Define $u = u(\cdot,t)$ by (4.127). Then, if $\{T_t\}$ is a differentiable deformation, the first variation*

$$\dot{u}=\delta G(\cdot,y)=\left.\frac{\partial G_t}{\partial t}(\cdot,y)\right|_{t=0}$$

of $G_t = G_t(x,y)$ exists in the sense of distributions in Ω. It belongs to $H^1(\Omega)$ if and only if $S\cdot\nabla u\in H^1(\Omega)$. If this condition is satisfied then it holds that $S\cdot\nabla G(\cdot,y)\in H^{1/2}(\partial\Omega)$ and

$$\Delta\dot{u}=0 \text{ in } \Omega,\quad \dot{u}=-S\cdot\nabla G(\cdot,y) \text{ on } \partial\Omega. \tag{4.131}$$

If $\{T_t\}$ is a twice differentiable deformation then the second variation

$$\ddot{u}=\delta^2 G(\cdot,y)=\left.\frac{\partial^2}{\partial t^2}G_t(\cdot,y)\right|_{t=0}$$

of $G_t = G_t(x,y)$ exists in the sense of distributions in Ω. It belongs to $H^1(\Omega)$ if and only if

$$2S\cdot\nabla\dot{u}+R\cdot\nabla u+[\nabla^2 u]S\cdot S\in H^1(\Omega).$$

If this condition is satisfied then it holds that $H\in H^{1/2}(\partial\Omega)$ for

$$H=2S\cdot\nabla\dot{u}+R\cdot\nabla G(\cdot,y)+[\nabla^2 G(\cdot,y)]S\cdot S, \tag{4.132}$$

and furthermore,

$$\Delta\ddot{u}=0 \quad \text{in } \Omega,\qquad \ddot{u}=-H \quad \text{on } \partial\Omega.$$

Proof We use the method of Sect. 4.3.3. Let $J_t(x) = J(T_t x)$. Given $\varphi \in C_0^\infty(\tilde{\Omega})$, we have

$$\int_{\Omega_t} u(z,t)\varphi(z)\,dz = \int_{\Omega} u(T_t x, t)\varphi(T_t x)\det J_t(x)\,dx$$

for $|t| \ll 1$, where $\tilde{\Omega}$ is a domain satisfying $\overline{\Omega} \subset \tilde{\Omega}$. Here, Theorem 4.18 guarantees the existence of $\frac{\partial u}{\partial t}(T_t x, t)$ strongly in $L^2(\Omega)$, in particular. Since that of

$$\frac{\partial}{\partial t}\left[\varphi(T_t x)\det J_t(x)\right]$$

is immediate we obtain the existence of

$$\begin{aligned}
&\frac{d}{dt}\int_{\Omega} u(z,t)\varphi(z)\,dz\Big|_{t=0} \\
&= \int_{\Omega} \left[\frac{\partial u}{\partial t}(T_t x, t)\cdot\varphi(T_t x) + u(T_t x, t)\frac{\partial T_t x}{\partial t}\cdot\nabla\varphi(T_t x)\right]\det J_t(x) \\
&\quad + u(T_t x, t)\varphi(T_t x)\frac{\partial}{\partial t}\det J_t(x)\Big|_{t=0}\,dx \\
&= \int_{\Omega} \frac{\partial u}{\partial t}(T_t x, t)\Big|_{t=0}\varphi + u(S\cdot\nabla\varphi + \varphi\nabla\cdot S)\,dx \\
&= \int_{\Omega} \frac{\partial u}{\partial t}(T_t x, t)\Big|_{t=0}\varphi + u\nabla\cdot(S\varphi)\,dx
\end{aligned}$$

by Theorem 4.8, recalling that S is a Lipschitz continuous vector field on $\overline{\Omega}$. Then it holds that

$$\frac{d}{dt}\int_{\Omega} u(z,t)\varphi(z)\,dz\Big|_{t=0} = \int_{\Omega}\left[\frac{\partial u}{\partial t}(T_t x, t)\Big|_{t=0} - S\nabla\cdot u\right]\varphi\,dx$$

by $u \in H^1(\Omega)$, which means the existence of $\dot{u} = \frac{\partial u}{\partial t}\big|_{t=0}$ in $\mathscr{D}'(\Omega)$ with

$$\dot{u} = \frac{\partial u}{\partial t}(T_t\cdot, t)\Big|_{t=0} - S\cdot\nabla u.$$

The formal calculation

$$\frac{\partial u}{\partial t}(T_t x, t)\Big|_{t=0} = \dot{u}(x) + S\cdot\nabla u \tag{4.133}$$

is thus justified in the context of distributions with the right-hand side belonging to $L^2(\Omega)$.

The strong differentiability of

$$u(T_t\cdot, t) \in H^1(\Omega), \ t \in I,$$

implies that in $H^{1/2}(\partial\Omega)$. Since $0 = G_t(T_t\cdot, y) \in H^{1/2}(\partial\Omega)$ it holds that

$$\frac{\partial u}{\partial t}(T_t\cdot, t)\bigg|_{t=0} = -\frac{\partial \Gamma}{\partial t}(T_t\cdot - y)\bigg|_{t=0} = -S\cdot\nabla\Gamma(\cdot - y) \quad \text{in } H^{1/2}(\partial\Omega).$$

Therefore, if $S\cdot\nabla u \in H^1(\Omega)$ the above $\dot{u}$ is in $H^1(\Omega)$ and takes the boundary value

$$\dot{u} = -S\cdot\nabla(\Gamma(\cdot - y) + u) = -S\cdot\nabla G(\cdot, y) \in H^{1/2}(\partial\Omega).$$

We thus obtain the first part of the theorem because $\Delta\dot{u} = 0$ in $\mathscr{D}'(\Omega)$ is obvious.

The existence of

$$\ddot{u} = \frac{\partial^2 u}{\partial t^2}\bigg|_{t=0} \qquad \text{in } \mathscr{D}'(\Omega)$$

is similarly proven if $\{T_t\}$ is twice differentiable. First, the formal equality

$$\frac{\partial^2 u}{\partial t^2}(T_t\cdot, t)\bigg|_{t=0} = \ddot{u} + 2S\cdot\nabla\dot{u} + R\cdot\nabla u + [\nabla^2 u]S\cdot S \tag{4.134}$$

arises with the symmetric tensor

$$\nabla^2 u = \left(\frac{\partial^2 u}{\partial x_i \partial x_j}\right)$$

indicating the Hesse matrix of u. In fact, we have, formally,

$$\frac{\partial^2 u}{\partial t^2}(T_t x, t) = \frac{\partial}{\partial t}\left\{\dot{u}(T_t x, t) + \frac{\partial T_t x}{\partial t}\cdot\nabla u(T_t x, t)\right\}$$

with

$$\frac{\partial \dot{u}}{\partial t}(T_t x, t) = \ddot{u}(T_t x, t) + \frac{\partial T_t x}{\partial t}\cdot\nabla\dot{u}(T_t x, t)$$

and

$$\frac{\partial}{\partial t}\left[\frac{\partial T_t x}{\partial t}\cdot\nabla u(T_t x, t)\right] = \frac{\partial^2 T_t x}{\partial t^2}\cdot\nabla u(T_t x, t) + \frac{\partial T_t x}{\partial t}\cdot\frac{\partial}{\partial t}\nabla u(T_t x, t)$$

$$= \frac{\partial^2 T_t x}{\partial t^2} \cdot \nabla u(T_t x, t) + \frac{\partial T_t x}{\partial t} \cdot \left\{ \nabla \dot{u}(T_t x, t) + \frac{\partial T_t x}{\partial t} \nabla^2 u(T_t x, t) \right\}$$

$$= \frac{\partial^2 T_t x}{\partial t^2} \cdot \nabla u(T_t x, t) + \frac{\partial T_t}{\partial t} \cdot \nabla \dot{u}(T_t x, t) + [\nabla^2 u(T_t x, t)] \frac{\partial T_t x}{\partial t} \cdot \frac{\partial T_t x}{\partial t}.$$

Putting $t = 0$, we thus obtain (4.134).

Next, we confirm that the last three terms on the right-hand side of (4.134) are regarded as distributions in Ω. In fact, first, S and R are Lipschitz continuous. Then $u \in H^1(\Omega)$ implies $R \cdot \nabla u \in L^2(\Omega)$. Second, the distributions $S \cdot \nabla \dot{u}$ and $[\nabla^2 u] S \cdot S$ are defined by

$$\langle \varphi, S \cdot \nabla \dot{u} \rangle = -(\dot{u}, \nabla \cdot (S\varphi))$$

$$\left\langle \varphi, \frac{\partial^2 u}{\partial x_i \partial x_j} S^i S^j \right\rangle = -\left(\frac{\partial u}{\partial x_i}, \frac{\partial}{\partial x_j} (S^i S^j \varphi) \right), \quad \varphi \in C_0^\infty(\Omega),$$

using $\dot{u} \in L^2(\Omega)$ and $u \in H^1(\Omega)$. Hence the last three terms on the right-hand side of (4.134) are actually distributions.

The final part is the justification of the derivation of (4.134) in the sense of distributions. This process is the same as in the case of the first variation. Thus, taking $\varphi \in C_0^\infty(\Omega)$, we can show the existence of

$$\frac{d^2}{dt^2} \langle \varphi, u(\cdot, t) \rangle \Big|_{t=0}$$

with the equality

$$\frac{d^2}{dt^2} \int_\Omega u(z, t) \varphi(z) \, dz \Big|_{t=0} = \int_\Omega \frac{\partial^2 u}{\partial t^2} (T_t x, t) \Big|_{t=0} \varphi$$

$$- 2\dot{u} \nabla \cdot (S\varphi) - \sum_{i,j} \frac{\partial u}{\partial x_i} \frac{\partial}{\partial x_j} (S^i S^j \varphi) \, dx, \quad z = T_t x.$$

To complete the proof, we regard $u(T_t x, t) = -\Gamma(T_t x - y)$ as a function of $x \in \partial\Omega$. Obviously, this function is twice differentiable in t strongly in $H^{1/2}(\partial\Omega)$, and it holds that

$$\frac{\partial \Gamma}{\partial t}(T_t \cdot -y) \Big|_{t=0} = S \cdot \nabla \Gamma(\cdot - y)$$

$$\frac{\partial^2 \Gamma}{\partial t^2}(T_t \cdot -y) \Big|_{t=0} = R \cdot \nabla \Gamma(\cdot - y) + [\nabla^2 \Gamma(\cdot - y)] S \cdot S.$$

Since it is obvious that $\ddot{u}$ is harmonic in Ω we obtain the second part of the theorem similarly. □

Remark 4.5 The differentiability of $G_t(\cdot, y)$, $y \in \Omega$ is assured in the weak topology of $L^2(\Omega)$, by using its zero extension outside Ω_t. The zero extension of the test function $\varphi \in L^2(\Omega)$ outside Ω is also necessary.

4.6.5 $C^{1,1}$-Domains

If $\Omega \subset \mathbf{R}^N$ is a $C^{1,1}$-domain there is a Lipschitz continuous frame (4.19) on $\partial\Omega$. Here the vector field n on $\partial\Omega$ has a Lipschitz extension on $\overline{\Omega}$. In fact, first, we extend it near $\partial\Omega$ using the flow described in Sect. 4.3.1 and then take a cut-off function to extend it on $\overline{\Omega}$.

Given $v \in H^2(\Omega)$, the function $n \cdot \nabla v$ belongs to $H^{1/2}(\partial\Omega)$. By the identification given in Theorem 4.4 it thus holds that $\partial v/\partial n \in H^{1/2}(\partial\Omega)$. We have also

$$\frac{\partial v}{\partial s_i} = s_i \cdot \nabla v \in H^{1/2}(\partial\Omega)$$

because s_i, $1 \leq i \leq N-1$, are Lipschitz continuous. If $v \in H_0^1(\Omega)$, furthermore, we obtain

$$0 = \int_{\partial\Omega} (s_i \cdot \nabla)(v\varphi)\, ds = \int_{\partial\Omega} [(s_i \cdot \nabla)v]\varphi\, ds$$

for any $\varphi \in C_0^\infty(\omega)$ by $v|_{\partial\Omega} = 0$, using the local chart and a neighborhood ω of $\partial\Omega$. Hence it follows that

$$\frac{\partial v}{\partial s_i} = 0, \quad i = 1, \cdots, N-1,$$

which implies

$$\begin{aligned} S \cdot \nabla v &= [(S \cdot n)n + \sum_{i=1}^{N-1} (S \cdot s_i)s_i)] \cdot \nabla v = (S \cdot n)(n \cdot \nabla v) \\ &= \delta\rho \frac{\partial v}{\partial n} \in H^{1/2}(\partial\Omega), \quad \forall v \in H^2(\Omega) \cap H_0^1(\Omega), \end{aligned}$$

recalling $S \cdot \nabla v \in H^1(\Omega)$ by the Lipschitz continuity of S. This category of Ω is thus efficient to represent the first variational formula in Euler coordinates in $H^1(\Omega)$.

In the following theorem, the pairing $\langle\ ,\ \rangle$ is identified with the L^2 inner product on $\partial\Omega$.

Theorem 4.20 (Hadamard) *If $\Omega \subset \mathbf{R}^N$ is a $C^{1,1}$-domain then it holds that*

$$\dot{u} = \delta G(\cdot, y) \in H^1(\Omega).$$

This function is harmonic in Ω and we have

$$\delta G(x, y) = \left\langle \delta\rho \frac{\partial G(\cdot, y)}{\partial n}, \frac{\partial G(\cdot, x)}{\partial n} \right\rangle, \quad x, y \in \Omega. \tag{4.135}$$

Proof Since Ω is $C^{1,1}$ we have

$$u = G(\cdot, y) - \Gamma(\cdot - y) \in H^2(\Omega)$$

by the elliptic regularity, recalling $y \in \Omega$. Then it follows that

$$n \cdot \nabla G(\cdot, y) = \frac{\partial G(\cdot, y)}{\partial n} \in H^{1/2}(\partial\Omega) \tag{4.136}$$

and hence

$$S \cdot \nabla G(\cdot, y) = \delta\rho \frac{\partial G(\cdot, y)}{\partial n} \in H^{1/2}(\partial\Omega). \tag{4.137}$$

Then we obtain (4.135) by Theorems 4.17 and 4.19. □

Remark 4.6 The notations of the above theorem are the same as those of [11], except for the direction of n.

The second variational formula in Theorem 4.19 contains $\dot{u}$ in the first term of H defined by (4.132). The following lemma is used to reduce this term to an integration on Ω. Here the pairing $\langle \cdot, \cdot \rangle$ on the right-hand side is taken between $H^{1/2}(\partial\Omega)$ and $H^{-1/2}(\partial\Omega)$.

Lemma 4.3 *If $\Omega \subset \mathbf{R}^N$ is a $C^{1,1}$-domain, then it holds that*

$$(\nabla \delta G(\cdot, x), \nabla \delta G(\cdot, y)) = -\left\langle \delta\rho \frac{\partial G(\cdot, x)}{\partial n}, \frac{\partial \delta G(\cdot, y)}{\partial n} \right\rangle, \quad x, y \in \Omega. \tag{4.138}$$

Proof For $C^{1,1}$-domain Ω, we have $u \in H^2(\Omega)$, (4.136), (4.137), and

$$\dot{u} = \delta G(\cdot, y) \in H^1(\Omega).$$

Given $\varphi \in H^1(\Omega)$ with $\Delta\varphi = 0$, therefore, we have

$$(\nabla \delta G(\cdot, y), \nabla \varphi) = (\nabla \dot{u}, \nabla \varphi) = \left\langle \dot{u}, \frac{\partial \varphi}{\partial n} \right\rangle = -\left\langle \delta\rho \frac{\partial G(\cdot, y)}{\partial n}, \frac{\partial \varphi}{\partial n} \right\rangle.$$

Putting $\varphi = \delta G(\cdot, x) \in H^1(\Omega)$, $x \in \Omega$, we obtain (4.138) with x and y changed. □

4.6.6 $C^{2,\theta}$-Domains

For more regular domains, we have $C^{2,\theta}$ regularity of (4.7) for $0 < \theta < 1$. This theory is valid for the $C^{2,\theta}$-domain, because $\Gamma(\cdot - y)$ is C^∞ in ω, a neighborhood of $\partial\Omega$ such that $y \notin \omega$.

Then $u \in C^{2,\theta}(\overline{\Omega})$ follows, which implies the same regularity of $G(\cdot, y)$ near $\partial\Omega$. In particular, we obtain

$$\frac{\partial^2 G(\cdot, y)}{\partial n^2} = -(\nabla \cdot n)\frac{\partial G(\cdot, y)}{\partial n} \text{ on } \partial\Omega \tag{4.139}$$

by Corollary 4.1 and $\Delta u = 0$ in Ω. Here we use the method of Sect. 4.2.3.

Lemma 4.4 *If $\Omega \subset \mathbf{R}^N$ is a $C^{2,\theta}$-domain and $\{T_t\}$ is a twice differentiable deformation then it holds that*

$$[\nabla^2 G(\cdot, y)]S\cdot S = 2\delta\rho \sum_{i=1}^{N-1} \mu_i \frac{\partial^2 G(\cdot, y)}{\partial s_i \partial n} + \sum_{i=1}^{N-1} \kappa_i(\mu_i^2 - (\delta\rho)^2)\frac{\partial G(\cdot, y)}{\partial n} \tag{4.140}$$

on $\partial\Omega$, where $\mu_i = S \cdot s_i$.

Proof We have (4.139) and

$$G(\cdot, y) = \frac{\partial G(\cdot, y)}{\partial s_i} = \frac{\partial^2 G(\cdot, y)}{\partial s_i \partial s_j} = 0 \quad \text{on } \partial\Omega, \quad i, j = 1, 2, \cdots, N-1.$$

Hence it holds that

$$\begin{aligned}
&\nabla^2 G(\cdot, y) \\
&= \frac{\partial G(\cdot, y)}{\partial n} {}^t(\nabla n) + \sum_{i=1}^{N-1} (s_i \otimes n + n \otimes s_i)\frac{\partial^2 G(\cdot, y)}{\partial s_i \partial n} + n \otimes n \frac{\partial^2 G(\cdot, y)}{\partial n^2} \\
&= \frac{\partial G(\cdot, y)}{\partial n}({}^t(\nabla n) - (\nabla \cdot n) n \otimes n) \\
&\quad + \sum_{i=1}^{N-1} \frac{\partial^2 G(\cdot, y)}{\partial s_i \partial n}(s_i \otimes n + n \otimes s_i).
\end{aligned} \tag{4.141}$$

Here we have

$$S = \sum_{i=1}^{N-1} \mu_i s_i + (\delta\rho) n. \tag{4.142}$$

First, the relations

$$[n \otimes n] n \cdot n = \sum_{i,j} n^i n^j n^j n^i = |n|^2 \cdot |n|^2 = 1$$

$$[n \otimes n] s_k \cdot n = \sum_{i,j} n^i n^j s_k^j n^i = (n \cdot s_k)|n|^2 = 0$$

$$[n \otimes n] s_k \cdot s_\ell = \sum_{i,j} n^i n^j s_k^i s_\ell^j = (n \cdot s_k)(n \cdot s_\ell) = 0$$

imply

$$[n \otimes n] S \cdot S = (\delta\rho)^2.$$

Second, it holds that

$$[{}^t(\nabla n)] n \cdot n = \sum_{i,j} \frac{\partial n^i}{\partial x_j} n^j n^i = \frac{1}{2} \sum_{i,j} \frac{\partial n^{i2}}{\partial x_j} n^j = \frac{1}{2}(n \cdot \nabla)|n|^2 = 0$$

$$[{}^t(\nabla n)] s_k \cdot n = \sum_{i,j} \frac{\partial n^i}{\partial x_j} s_k^j n^i = \frac{1}{2} \sum_{i,j} \frac{\partial n^{i2}}{\partial x_k} s_k^j = \frac{1}{2}(s_k \cdot \nabla)|n|^2 = 0$$

by (4.27). Then, since

$$[s_m \otimes s_m](s_k \cdot s_\ell) = \sum_{i,j} s_m^i s_m^j s_k^j s_\ell^i = (s_m \cdot s_k)(s_m \cdot s_\ell) = \delta_{mk}\delta_{m\ell}$$

we have

$$[{}^t(\nabla n)] s_k \cdot s_\ell = \sum_{m=1}^{N-1} \kappa_m \delta_{mk} \delta_{m\ell} = \kappa_k \delta_{k\ell},$$

which implies

$$[{}^t(\nabla n)] S \cdot S = \sum_{k,\ell=1}^{N-1} \mu_k \mu_\ell [{}^t(\nabla n)] s_k \cdot s_\ell = \sum_{k,\ell=1}^{N-1} \mu_k \mu_\ell \kappa_k \delta_{k\ell} = \sum_{\ell=1}^{N-1} \mu_\ell^2 \kappa_\ell.$$

For the second term of the right-hand side of (4.141), we use

$$[s_k \otimes n]n \cdot n = \sum_{i,j} s_k^i n^j n^j n^i = (s_k \cdot n)|n|^2 = 0$$

$$[s_k \otimes n]s_\ell \cdot n = \sum_{i,j} s_k^i n^j s_\ell^j n^i = (n \cdot s_\ell)(s_k \cdot n) = 0$$

$$[s_k \otimes n]s_\ell \cdot s_m = \sum_{i,j} s_k^i n^j s_\ell^j s_m^i = (n \cdot s_\ell)(s_k \cdot s_m) = 0$$

$$[s_k \otimes n]n \cdot s_\ell = \sum_{i,j} s_k^i n^j n^j s_\ell^i = (s_k \cdot s_\ell)|n|^2 = \delta_{k\ell}$$

and also

$$[n \otimes s_k]n \cdot n = [n \otimes s_k]s_\ell \cdot s_m = [n \otimes s_k]n \cdot s_\ell = 0$$

$$[n \otimes s_k]s_\ell \cdot n = \delta_{k\ell}.$$

Then it follows that

$$[s_i \otimes n + n \otimes s_i]S \cdot S = 2(\delta\rho)\mu_i.$$

We thus end up with (4.140). □

The right-hand side of (4.140) has a description independent of the choice of $\{s_1, \cdots, s_{N-1}\}$. First, we have

$$\sum_{i=1}^{N-1} \mu_i \frac{\partial^2 G(\cdot, y)}{\partial s_i \partial n} = (S \cdot \nabla)\frac{\partial G(\cdot, y)}{\partial n}.$$

Next, equality (4.24) implies

$$S \cdot \frac{\partial n}{\partial s_i} = \kappa_i \mu_i.$$

Therefore, there arises

$$\begin{aligned}\sum_{i=1}^{N-1} \kappa_i \mu_i^2 &= \sum_{i=1}^{N-1} (S \cdot \frac{\partial n}{\partial s_i}) s_i \cdot S = \left[\sum_{i=1}^{N-1} s_i \frac{\partial n}{\partial s_i}\right] S \cdot S \\ &= \left[(\nabla - \frac{\partial}{\partial n})n\right] S \cdot S = (\nabla n)S \cdot S.\end{aligned}$$

We thus obtain

$$[\nabla^2 G(\cdot, y)]S \cdot S = 2(\delta\rho)(S \cdot \nabla)\frac{\partial G(\cdot, y)}{\partial n} + (\nabla n)S \cdot S - (\nabla \cdot n)(\delta\rho)^2 \frac{\partial G(\cdot, y)}{\partial n}$$

on $\partial\Omega$.

4.6.7 $C^{2,1}$-Domains

Henceforth, we assume that $\Omega \subset \mathbf{R}^N$ is a $C^{2,1}$-domain and

$$S = \left.\frac{\partial T_t}{\partial t}\right|_{t=0} \in C^{2,1}(\overline{\Omega}, \mathbf{R}^N). \tag{4.143}$$

In this case it holds that $u \in W^{3,p}(\Omega)$, $1 < p < \infty$, and hence $u \in C^2(\overline{\Omega})$. Thus Lemma 4.1 is applicable to $f = u$.

Theorem 4.21 ([44]) *Let $\Omega \subset \mathbf{R}^N$ be a $C^{2,1}$ domain and $\{T_t\}$ be a twice differentiable deformation satisfying (4.143). Then we have $\ddot{u} = \delta^2 G(\cdot, y) \in H^1(\Omega)$. This function is harmonic in Ω and it holds that*

$$\delta^2 G(x, y) = -2(\nabla \delta G(\cdot, x), \nabla \delta G(\cdot, y)) + \left\langle \chi \frac{\partial G(\cdot, x)}{\partial n}, \frac{\partial G(\cdot, y)}{\partial n} \right\rangle \tag{4.144}$$

for $x, y \in \Omega$ with Lipschitz continuous χ defined by

$$\chi = \delta^2\rho + \sum_{i=1}^{N-1} \left\{ \kappa_i(\mu_i^2 - (\delta\rho)^2) - 2\mu_i \frac{\partial \delta\rho}{\partial s_i} \right\}, \tag{4.145}$$

recalling $\mu_i = S \cdot s_i$.

Proof From the assumption we have $u \in H^3(\Omega)$ and also $S \cdot \nabla u \in H^2(\Omega)$ by (4.143). Hence it holds that

$$S \cdot \nabla G(\cdot, y) = \delta\rho \frac{\partial G(\cdot, y)}{\partial n} \in H^{3/2}(\partial\Omega),$$

which means the existence of $g \in H^2(\Omega)$ such that $g = S \cdot \nabla G(\cdot, y)$ on $\partial\Omega$. Then the elliptic regularity applied to (4.131) implies

$$\dot{u} = \delta G(\cdot, y) \in H^2(\Omega).$$

Thus we obtain $S \cdot \nabla \dot{u} \in H^1(\Omega)$, and therefore,

$$S \cdot \nabla \delta G(\cdot, y) = \sum_{i=1}^{N-1} \mu_i \frac{\partial \delta G(\cdot, y)}{\partial s_i} + \delta\rho \frac{\partial \delta G(\cdot, y)}{\partial n}$$

in $H^{1/2}(\partial\Omega)$.

We have also $S \cdot \nabla \dot{u} \in H^1(\Omega)$ and hence

$$R \cdot \nabla G(\cdot, y) = \delta^2 \rho \, \frac{\partial G(\cdot, y)}{\partial n} \quad \text{on } \partial\Omega.$$

Then it follows that

$$\begin{aligned} \ddot{u} = -H = &-2 \sum_{i=1}^{N-1} \mu_i \left(\frac{\partial \delta G(\cdot, y)}{\partial s_i} + \delta\rho \frac{\partial^2 G(\cdot, y)}{\partial s_i \partial n} \right) - 2\delta\rho \frac{\partial \delta G(\cdot, y)}{\partial n} \\ &- \left[\delta^2 \rho + \sum_{i=1}^{N-1} \kappa_i (\mu_i^2 - (\delta\rho)^2) \right] \frac{\partial G(\cdot, y)}{\partial n} \quad \text{on } \partial\Omega \end{aligned}$$

from Theorem 4.19. Since

$$\delta G(\cdot, y) = \dot{u} = -\delta\rho \frac{\partial G(\cdot, y)}{\partial n} \quad \text{on } \partial\Omega$$

we have

$$\begin{aligned} \frac{\partial \delta G(\cdot, y)}{\partial s_i} + \delta\rho \frac{\partial^2 G(\cdot, y)}{\partial s_i \partial n} &= -\frac{\partial}{\partial s_i} \left(\delta\rho \frac{\partial G(\cdot, y)}{\partial n} \right) + \delta\rho \frac{\partial^2 G(\cdot, y)}{\partial s_i \partial n} \\ &= -\frac{\partial \delta\rho}{\partial s_i} \frac{\partial G(\cdot, y)}{\partial n} \quad \text{on } \partial\Omega. \end{aligned}$$

Thus we end up with

$$\ddot{u} = -2\delta\rho \frac{\partial \dot{u}}{\partial n} - \chi \frac{\partial G(\cdot, y)}{\partial n} \quad \text{on } \partial\Omega$$

with (4.145) and the proof is complete by Lemma 4.3. □

An immediate consequence [12] is the following corollary.

Corollary 4.4 (Garabedian–Schiffer) *Let $\{T_t\}$ be a family of normal perturbations associated with $\delta\rho = S \cdot n$. Then it holds that*

$$\begin{aligned} \delta^2 G(x, y) = &-2(\nabla \delta G(\cdot, x), \nabla \delta G(\cdot, y)) \\ &- \left\langle (\nabla \cdot n)(\delta\rho)^2 \frac{\partial G(\cdot, x)}{\partial n}, \frac{\partial G(\cdot, y)}{\partial n} \right\rangle, \quad x, y \in \Omega \quad (4.146) \end{aligned}$$

under the assumption of Theorem 4.21.

Proof The formula follows from the previous theorem because in this case it holds that $\delta^2\rho = 0$ and $\mu_i = S \cdot s_i = 0$, $i = 1, 2, \cdots, N-1$. □

To treat the dynamical perturbation, we represent χ independent of the choice of the frame.

Lemma 4.5 *It holds that*

$$\chi = \delta^2\rho - (\nabla \cdot n)(\delta\rho)^2 + \frac{\partial(\delta\rho)^2}{\partial n} - (S \cdot \nabla S) \cdot n - S \cdot \nabla\delta\rho. \tag{4.147}$$

Proof Since (4.28) it holds that

$$\chi = \delta^2\rho - (\nabla \cdot n)(\delta\rho)^2 + \sum_{i=1}^{N-1}\left(\kappa_i\mu_i^2 - 2\mu_i\frac{\partial\delta\rho}{\partial s_i}\right).$$

Then equality (4.28) implies

$$\kappa_i\mu_i^2 = \kappa_i(S \cdot s_i)^2 = (S \cdot s_i)(S \cdot \frac{\partial n}{\partial s_i}),$$

where

$$S \cdot \frac{\partial n}{\partial s_i} = \frac{\partial}{\partial s_i}(S \cdot n) - \frac{\partial S}{\partial s_i} \cdot n = \frac{\partial\delta\rho}{\partial s_i} - \frac{\partial S}{\partial s_i} \cdot n.$$

Then we obtain

$$\chi = \delta^2\rho - (\nabla \cdot n)(\delta\rho)^2 - \sum_{i=1}^{N-1}(S \cdot s_i)\left\{\frac{\partial\delta\rho}{\partial s_i} + \frac{\partial S}{\partial s_i} \cdot n\right\} \tag{4.148}$$

by $\mu_i = S \cdot s_i$.

The third term of the right-hand side of (4.148) is treated as follows. First, we have

$$\sum_{i=1}^{N-1}(S \cdot s_i)(\frac{\partial S}{\partial s_i} \cdot n) = \left[\sum_{i=1}^{N-1}(S \cdot s_i)\frac{\partial S}{\partial s_i}\right] \cdot n = \left[\sum_{i=1}^{N-1}(S \cdot s_i)s_i \cdot \nabla S\right] \cdot n$$
$$= [(S - (\delta\rho)n) \cdot \nabla S] \cdot n = (S \cdot \nabla S) \cdot n - \delta\rho\frac{\partial S}{\partial n} \cdot n.$$

Here, we have (4.27) so that

$$\frac{\partial S}{\partial n} \cdot n = \frac{\partial}{\partial n}(S \cdot n) = \frac{\partial\delta\rho}{\partial n}.$$

Hence it follows that

$$\sum_{i=1}^{N-1}(S\cdot s_i)(\frac{\partial S}{\partial s_i}\cdot n)=(S\cdot\nabla S)\cdot n-\frac{1}{2}\frac{\partial(\delta\rho)^2}{\partial n}.$$

We have, similarly,

$$\sum_{i=1}^{N-1}(S\cdot s_i)\frac{\partial\delta\rho}{\partial s_i}=\sum_{i=1}^{N-1}(S\cdot s_i)s_i\cdot\nabla\delta\rho=(S-(\delta\rho)n)\cdot\nabla\delta\rho$$
$$=S\cdot\nabla\delta\rho-\frac{1}{2}\frac{\partial}{\partial n}(\delta\rho)^2$$

and then (4.147) follows. □

Corollary 4.5 *Let $\{T_t\}$ be a family of dynamical perturbations associated with the vector field v. Then it holds that*

$$\delta^2G(x,y)=-2(\nabla\delta G(\cdot,x),\nabla\delta G(\cdot,y))-\left\langle\sigma\frac{\partial G(\cdot,x)}{\partial n},\frac{\partial G(\cdot,y)}{\partial n}\right\rangle$$
$$\sigma=(\nabla\cdot n)(v\cdot n)^2-\frac{\partial(v\cdot n)^2}{\partial n}+(v\cdot\nabla)(v\cdot n),\quad x,y\in\Omega \tag{4.149}$$

under the assumption of Theorem 4.21.

Proof In this case we have $S\cdot\nabla S=R$ so that

$$\chi=-(\nabla\cdot n)(\delta\rho)^2+\frac{\partial(\delta\rho)^2}{\partial n}-S\cdot\nabla\delta\rho.$$

Then (4.149) follows from $S=v$ and $\delta\rho=v\cdot n$. □

4.6.8 Examples

If $N=2$ and $\Omega=B(0,1)$, then we obtain

$$G(x,y)=\frac{1}{2\pi}\log\left|\frac{1-\bar{z}w}{z-w}\right|,$$

where $z=x_1+\imath x_2$, $w=y_1+\imath y_2$ for $x=(x_1,x_2)$, $y=(y_1,y_2)$.

Dynamical Perturbations
Using $v(x)=x$, we have $T_tx=e^tx$. Take $y=0$:

$$G_t(x,0) = \frac{1}{2\pi}\log\frac{e^t}{|x|}.$$

Then it holds that

$$u(x,t) = G_t(x,0) - \Gamma(x) = \frac{t}{2\pi}$$

$$\delta G(x,0) = \dot{u}(x,0) = \frac{1}{2\pi}$$

$$\delta^2 G(x,0) = \ddot{u}(x,0) = 0, \quad x \in \Omega.$$

Since $n = \frac{x}{|x|}$ and $S = v = x$, we obtain

$$\delta\rho = n\cdot x = |x| = 1, \quad \frac{\partial(\delta\rho)^2}{\partial n} = 2, \quad \nabla\cdot n = 1, \quad S\cdot\nabla\delta\rho = 1.$$

Then, we get

$$(\nabla\cdot n)(\delta\rho)^2 - \frac{\partial(\delta\rho)^2}{\partial n} + S\cdot\nabla\delta\rho = 0,$$

which implies

$$(\nabla\delta G(\cdot,x), \nabla\delta G(\cdot,0)) = 0 \tag{4.150}$$

by (4.149).

In fact, since

$$G(x,y) = \frac{1}{4\pi}\log\frac{1-2x\cdot y+|x|^2|y|^2}{|x-y|^2} \tag{4.151}$$

it holds that

$$G_t(x,y) = G(e^{-t}x, e^{-t}y) = \frac{1}{4\pi}\log\frac{e^{2t}-2x\cdot y+e^{-2t}|x|^2|y|^2}{|x-y|^2}.$$

Then we obtain

$$\delta G(x,y) = \left.\frac{\partial G_t(x,y)}{\partial t}\right|_{t=0} = \frac{1}{2\pi}\cdot\frac{1-|x|^2|y|^2}{1-2x\cdot y+|x|^2|y|^2},$$

which implies

$$\nabla_x \delta G(x, y) = \frac{1}{\pi} \cdot \frac{2|y|^2(-1 + x \cdot y)x + (1 - |x|^2|y|^2)y}{(1 - 2x \cdot y + |x|^2|y|^2)^2}.$$

We have

$$(\nabla \delta G(\cdot, x), \nabla \delta G(\cdot, y)) = \frac{1}{\pi^2} \int_\Omega \frac{2|x|^2(-1 + z \cdot x)z + (1 - |z|^2|x|^2)x}{(1 - 2z \cdot x + |z|^2|x|^2)^2}$$
$$\cdot \frac{2|y|^2(-1 + z \cdot y)z + (1 - |z|^2|y|^2)y}{(1 - 2z \cdot y + |z|^2|y|^2)^2}\, dz$$

and hence (4.150).

Normal Perturbations

We have

$$\Gamma_t : x + t\delta\rho(x)n_x, \quad x \in \Gamma = \partial\Omega, \quad n = \frac{x}{|x|} = x.$$

If $\delta\rho(x) = 1$, it holds that $\Gamma_t = (1 + t)\Gamma$. Then we obtain

$$G_t(x, y) = G\left((1 + t)^{-1}x, (1 + t)^{-1})y\right), \quad G_t(x, 0) = \frac{1}{2\pi} \log \frac{1 + t}{|x|}$$

and hence

$$u(x, t) = G_t(x, 0) - \Gamma(x) = \frac{1}{2\pi} \log(1 + t)$$
$$\delta G(x, 0) = \dot{u}(x) = \frac{1}{2\pi}$$
$$\delta^2 G(x, 0) = -\frac{1}{2\pi}.$$

Now it holds that

$$G_t(x, y) = \frac{1}{4\pi} \log \frac{(1 + t)^2 1 - 2x \cdot y + (1 + t)^{-2}|x|^2|y|^2}{|x - y|^2}$$

and hence

$$\delta G(x, y) = \frac{1}{2\pi} \cdot \frac{1 - |x|^2|y|^2}{1 - 2x \cdot y + |x|^2|y|^2}.$$

This form is the same as in the Lie perturbation which guarantees (4.150).

Next, by (4.151) we obtain

$$\nabla_x G(x, y) = \frac{1}{4\pi} \cdot \frac{-2y + 2|y|^2 x}{1 - 2x \cdot y + |x|^2|y|^2} - \frac{1}{4\pi} \cdot \frac{2(x - y)}{|x - y|^2}$$

$$x \cdot \nabla G_x(x, y) = \frac{1}{2\pi} \cdot \frac{|x|^2|y|^2 - x \cdot y}{1 - 2x \cdot y|x|^2|y|^2} - \frac{1}{2\pi} \cdot \frac{|x|^2 - x \cdot y}{|x - y|^2}.$$

Then it holds that

$$x \cdot \nabla_x G(x, y)|_{|x|=1} = \left.\frac{\partial G(x, y)}{\partial n_x}\right|_{|x|=1} = \frac{1}{2\pi} \cdot \frac{|y|^2 - x \cdot y}{1 - 2x \cdot y + |y|^2} - \frac{1}{2\pi} \cdot \frac{1 - x \cdot y}{|x - y|^2}$$
$$= \frac{1}{2\pi} \cdot \frac{|y|^2 - 1}{|x - y|^2}$$

and hence

$$\left\langle \frac{\partial G(\cdot, x)}{\partial n}, \frac{\partial G(\cdot, y)}{\partial n} \right\rangle = \frac{1}{4\pi^2} \int_{|z|=1} \frac{(1 - |x|^2)(1 - |y|^2)}{|z - x|^2 |z - y|^2} \, dS_z.$$

We get

$$\left\langle \frac{\partial G(\cdot, x)}{\partial n}, \frac{\partial G(\cdot, 0)}{\partial n} \right\rangle = \frac{1}{2\pi}$$

by the Poisson formula, which agrees with the Garabedian–Schiffer formula (4.146).

References

1. R.A. Adams, *Sobolev Spaces* (Academic Press, Boston, 1978)
2. H. Azegami, *Shape Optimization Problems* (Springer, Singapore, 2020)
3. H. Azegami, K. Takeuchi, A smoothing method for shape optimization: traction method using the Robin condition. Int. J. Comput. Methods **3**, 21–33 (2006)
4. V. Barbu, *Nonlinear Semigroups and Differential Equations in Banach Spaces* (Nordhoff, Leiden, 1976)
5. S.C. Brenner, L.R. Scott, *The Mathematical Theory of Finite Element Methods*, 3rd edn. (Springer, New York, 2007)
6. S. Chandrasekhar, *An Introduction to the Study of Stellar Structure* (University of Chicago Press, Chicago, 1939)
7. R. Courant, *Dirichlet's Principle, Conformal Mappings, and Minimal Surfaces* (Springer, New York, 1970)
8. M.G. Crandall, T. Liggett, Generation of semi-groups of nonlinear transformations on general Banach spaces. Am. J. Math. **93**, 265–293 (1971)
9. M. Dauge, *Elliptic Boundary Value Problems on Corner Domains*. L.N. in Math., vol. 1341 (Springer, Berlin, 1988)

10. H. Fujita, N. Saito, T. Suzuki, *Operator Theory and Numerical Methods* (North-Holland, Amsterdam, 2001)
11. P.R. Garabedian, *Partial Differential Equations*, 2nd edn. (Chelsea Publishing Comp., New York, 1986)
12. P.R. Garabedian, M. Schiffer, Convexity of domain functionals. J. Anal. Math. **2**, 281–368 (1952–1953)
13. D. Gilbarg, N.S. Trudinger, *Elliptic Partial Differential Equations of Second Order*, 2nd edn. (Springer, New York, 1983)
14. V. Girault, P.-A. Raviart, *Finite Element Methods for Navier-Stokes Equations* (Springer, Berlin, 1986)
15. P. Grisvard, *Elliptic Problems in Nonsmooth Domains* (Pitman, Boston, 1985)
16. P. Grisvard, G. Iooss, Problémes aux limites unilatéraux dans des domaines non réguliers. Publ. Séminaries Mathématiques, Univ. Rennes **9**, 1–26 (1975)
17. A. Ito, T. Suzuki, Asymptotic behavior of the solution to the non-isothermal phase field equation. Nonlinear Anal. **64**, 2454–2479 (2006)
18. A. Ito, T. Suzuki, Asymptotic behavior of the solution to the non-isothermal phase separation. Nonlinear Anal. **68**, 1825–1843 (2008)
19. J. Kadlec, On the regularity of the solution of the Poisson problem on a domain with boundary locally similar to the boundary of a convex open set. Czechoslovak Math. J. **14**, 386–393 (1964)
20. S.L. Kamenomostskaja, On the Stefan problem (in Russian). Mat. Sbornik **53**, 488–514 (1961)
21. V.A. Kondrat'ev, Boundary problems for elliptic equations in domains with conical or angular points. Trans. Mosc. Math. Soc. **16**, 227–313 (1967)
22. A. Kufner, O. John, S. Fučík, *Function Spaces* (Academia, Prague, 1977)
23. N.G. Meyers, An L^p-estimate for the gradient of solutions of second order elliptic divergence equations. Ann. Scoul. Norm. Sup. Pisa, III **17**, 189–206 (1963)
24. I. Miyadera, *Nonlinear Semigroups* (American Mathematical Society, Providence, 1992)
25. S. Mizohata, *The Theory of Partial Differential Equations* (Cambridge University Press, Cambridge, 1973)
26. J. Nečas, *Le Méthodes Directes en Théorie des Equations Elliptiques* (Masson, Paris, 1967)
27. J. Nečas, *Introduction to the Theory of Nonlinear Elliptic Equations* (Teubner, Leipzig, 1983)
28. O.A. Oleinik, A method of solution of the general Stefan problem (in Russian). Soviet Math. Dokl. **1**, 1350–1354 (1960)
29. N.L. Othman, T. Suzuki, T. Tsuchiya, A combined scheme for computing numerical solutions of a free boundary problem. Int. J. Math. Comput. Simul. **1**, 53–60 (2011)
30. B. Pertham, *Kinetic Formulation of Conservation Laws* (Oxford University Press, Oxford, 2002)
31. T.C. Sideris, Formation of singularities in solutions to nonlinear hyperbolic equations. Arch. Ration. Mech. Anal. **86**, 369–381 (1984)
32. T.C. Sideris, Formation of singularities in three-dimensional compressible fluids. Commun. Math. Phys. **101**, 475–485 (1985)
33. C.G. Simader, *On Dirichlet's Boundary Value Problem*. Lecture Notes in Math., vol. 268 (Springer, Berlin, 1972)
34. T. Suzuki, Full-discrete finite element approximation of evolution equation $u_t + A(t)u = 0$ of parabolic type. J. Fac. Sci. Univ. Tokyo, Sec IA **29**, 195–240 (1982)
35. T. Suzuki, Thermomechanical system equations on shape memory materials, in *Current Advances in Nonlinear Analysis and Related Topics*, ed. by T. Aiki, N. Kenmochi, M. Niezgodka, M. Otani. Gakuto International Series on Mathematical Science and Applications, vol. 32 (2010), pp. 189–213
36. T. Suzuki, Irrotational blowup of the solution to compressible Euler equation. J. Math. Fluid Mech. **15**, 617–633 (2013)
37. T. Suzuki, *Mean Field Theories and Dual Variation, Mathematical Structures of the Mesoscopic Model*, 2nd edn. (Atlantis Press, Paris, 2015)
38. T. Suzuki, T. Senba, *Applied Analysis, Mathematical Methods for Natural Science*, 2nd edn. (Imperial College Press, London, 2011)

39. T. Suzuki, S. Tasaki, Stationary Fix-Caginalp equation with non-local term. Nonlinear Anal. **71**, 1329–1349 (2009)
40. T. Suzuki, S. Tasaki, Stationary solutions to the Falk system on shape memory materials. Math. Methods Appl. Sci. **33**, 994–1011 (2010)
41. T. Suzuki, S. Tasaki, Stationary solutions to a thermoelastic system on shape memory materials. Nonlinearity **23**, 2623–2656 (2010)
42. T. Suzuki, T. Tsuchiya, Convergence analysis of trial free boundary methods for the two-dimensional filtration problem. Numer. Math. **100**, 537–564 (2005)
43. T. Suzuki, T. Tsuchiya, Weak formulation of Hadamard variation applied to the filtration problem. Jpn. J. Ind. Appl. Math. **28**, 327–350 (2011)
44. T. Suzuki, T. Tsuchiya, First and second Hadamard variational formulae of the Green function for general domain perturbations. J. Math. Soc. Jpn. **68**, 1389–1419 (2016)
45. Z. Xin, Blowup of smooth solutions to the compressible Navier-Stokes equation with compact density. Commun. Pure Appl. Math. **51**, 229–240 (1998)

Chapter 5
Potentials of Self-Organization: A Fusion

Abstract The last of Liouville's theorems treated in this monograph arises in accordance with classical mechanics. It is an integrability of the Hamilton system under the transformation of sympletic variables. The geometric structure hidden behind it is a suggestion to understand the principle of relaxation of singularities, caused by the skew-symmetric interaction of species and also the symmetry achieved by the action–reaction law in the duality between particle density and field distributions. The first section is a summary. In the second section we observe this fact of singularity cancellation in classical mechanics to reach the notion of a Poisson manifold. In the third section we confirm that the relaxation of the singularity, in the models of multi-species in mathematical biology, emerges from this Poisson structure, particularly in dissipative Lotka–Volterra systems. Motivated by this observation, we treat the reaction–diffusion systems with super-critical growth rate in the fourth section. The fifth section is devoted to the quantized blowup mechanism realized in the 2D Smoluchowski–Poisson equation at three levels of time scale; stationary, in finite time, and in infinite time. This model is a simplified system of chemotaxis, but is concerned, rather more, with the motion of the mean field of many point vorticities in relaxation time, that is, from quasi-equilibrium to equilibrium. This physical background is the origin of recursive hierarchy observed in this model, realized as a collapse and sub-collapse dynamics controlled by a Hamiltonian. There Liouville's theory of transformation acts as a basic tool. In the final section we turn to a problem in mathematical oncology, the chemotactic paradox. We thus conclude this chapter with a fusion of Liouville's theory in several areas, that is, the method of Lagrange coordinates and classification of entire solutions, which results in quantized blowup mechanism, duality between field and particle, recursive hierarchy, nonlinear spectral mechanics, space homogenization of isolated systems, emergence from reaction network, and cancellation of singularities in multi-component systems.

T. Suzuki, *Liouville's Theory in Linear and Nonlinear Partial Differential Equations*,
Springer Asia Pacific Mathematics Series 8,
https://doi.org/10.1007/978-981-95-1975-0_5

5.1 Introduction

5.1.1 Mathematics of Interaction

Besides the four Liouville theorems used in the previous chapters, at least one important one is left. It is in the field of classical mechanics, obtained by converting the Hamilton system

$$\dot{q} = \frac{\partial H}{\partial p}, \quad \dot{p} = -\frac{\partial H}{\partial q} \tag{5.1}$$

into

$$\dot{F} = \{F, H\}, \quad \forall F = F(q, p),$$

where

$$\{F, G\} = \frac{\partial F}{\partial q} \cdot \frac{\partial G}{\partial p} - \frac{\partial F}{\partial p} \cdot \frac{\partial G}{\partial q}.$$

Hence $F = F(q, p)$ is a first integral of (5.1) if it holds that $\{F, H\} = 0$. The notion of a *sympletic manifold* has emerged from this theorem, which is generalized to the *Poisson manifold* with possible odd dimensions.

These Hamilton, sympletic, and Poisson structures are observed in several biological models concerning the interaction of multiple species. Among them are the prey–predator system (5.77), Gierer–Meinhardt system (5.89) with (5.90), and three component system (5.95) in Sect. 5.3.3, which are transformed into (5.81) with (5.82), (5.93) with (5.94), and (5.97) with (5.98), respectively. Conservation laws provided with these models bind the components of the solution on a manifold.

These properties even assure the existence of a global-in-time solution to the reaction diffusion system

$$\begin{aligned} &\tau_j \frac{\partial u_j}{\partial t} - d_j \Delta u_j = f_j(u) \quad \text{in } \Omega \times (0, T) \\ &\left.\frac{\partial u_j}{\partial \nu}\right|_{\partial\Omega} = 0, \quad \left.u_j\right|_{t=0} = u_{j0}(x) > 0, \end{aligned} \tag{5.2}$$

where $\Omega \subset \mathbf{R}^n$ is a bounded domain with smooth boundary $\partial\Omega$, $\tau = (\tau_j) > 0$, $d = (d_j) > 0$, $1 \leq j \leq N$, are constants, and $u_{j0} = u_{j0}(x) \in C(\overline{\Omega})$. The nonlinearity $f = (f_j(u))$ is locally Lipschitz continuous on $\overline{\mathbf{R}}_+$, satisfying quasi-positivity (5.167) and total mass dissipation (5.168) below. A recent result [17] has revealed that if the growth rate is up to quadratic the solution is uniformly bounded and global in time for any space dimension n (Theorem 5.12 in Sect. 5.4.5).

5.1.2 Non-equilibrium Thermodynamics

Quasi-positivity of the nonlinearity in (5.2) assures non-negativity of the solution $u = (u_j(x,t)) \geq 0$ from the theory of invariant regions [15]. Total mass dissipation implies

$$\|\tau \cdot u(\cdot,t)\|_1 \leq \|\tau \cdot u_0\|_1 \tag{5.3}$$

and then $n = 2$ has been thought to be a critical dimension if the growth rate of nonlinearity is up to quadratic [62]. More precisely, a standard combination of the Gagliardo–Nirenberg inequality and semigroup estimate in $n = 2$ ensures local-in-space control of the solution, and hence ε regularity as in Theorem 5.7 in Sect. 5.4.1.

The Smoluchowski–Poisson equation

$$\begin{aligned} & u_t = \nabla \cdot (\nabla u - u\nabla v) \text{ in } \Omega \times (0,T) \\ & \left.\frac{\partial u}{\partial \nu} - u\frac{\partial v}{\partial \nu}\right|_{\partial\Omega} = 0, \quad u|_{t=0} = u_0(x) \geq 0 \\ & -\Delta v = u, \quad v|_{\partial\Omega} = 0 \end{aligned} \tag{5.4}$$

is the other example of such a structure, where $\Omega \subset \mathbf{R}^n$ is a bounded domain with smooth boundary $\partial\Omega$. This system is invariant under the self-similar transformation

$$u_\mu(x,t) = \mu^2 u(\mu x, \mu^2 t), \ v_\mu(x,t) = v(\mu x, \mu^2 t), \quad \mu > 0,$$

except for the initial boundary condition, as a consequence of the quadratic nonlinearity. There arise also positivity preserving $u = u(x,t) \geq 0$ and total mass conservation

$$\|u(\cdot,t)\|_1 = \|u_0\|_1 \equiv \lambda, \tag{5.5}$$

and from this structure, $n = 2$ is regarded as a critical dimension of this system for the existence of a solution global in time.

A reaction–diffusion system with dissipation prevents formation of measures in finite or infinite time because of the L^1 compactness of the principal part, Lemma 5.13 in Sect. 5.4.2. There may be, however, transit to measures in the Smoluchowski–Poisson equation, which means blowup of the classical solution. The measure formed in this process, furthermore, is a finite sum of delta functions called *collapses* in the case of two space dimensions. This property is (5.12) and (5.47)–(5.48) described below.

System (5.4) is a fundamental equation in non-equilibrium statistical mechanics, proposed in the context of astrophysics [70]. Hence there is decreasing of the *Helmholtz free energy*,

$$\mathscr{F}(u) = \int_{\Omega} u(\log u - 1) - \frac{1}{2}\langle(-\Delta)^{-1}u, u\rangle, \tag{5.6}$$

besides the total mass conservation (5.5), where $v = (-\Delta)^{-1}u$ indicates

$$-\Delta v = u \text{ in } \Omega, \quad v|_{\partial\Omega} = 0. \tag{5.7}$$

This property is described by

$$\frac{d}{dt}\mathscr{F}(u) = -\int_{\Omega} u|\nabla(\log u - v)|^2, \tag{5.8}$$

which may be derived by writing (5.4) as a *model (B) equation* [77] associated with (5.6),

$$u_t = \nabla \cdot u\nabla\delta\mathscr{F}(u), \ u > 0 \ \text{ in } \Omega \times (0, T), \quad \left.\frac{\partial}{\partial\nu}\delta\mathscr{F}(u)\right|_{\partial\Omega} = 0$$

$$\delta\mathscr{F}(u) = \log u - (-\Delta)^{-1}u.$$

The other model (B) equation derived from this $\mathscr{F}(u)$ is the equation of *logarithmic diffusion*

$$u_t = \Delta\delta\mathscr{F}(u) \text{ in } \Omega \times (0, T), \quad \left.\frac{\partial}{\partial\nu}\delta\mathscr{F}(u)\right|_{\partial\Omega} = 0, \tag{5.9}$$

which coincides with Hamilton's *normalized Ricci flow* [25] if $\lambda = 8\pi$ and $\Omega = S^2$. Hence study of (5.9) provides an analytic approach to this geometric flow [37].

The *Trudinger–Moser inequality* is a fundamental feature of this $\mathscr{F}(u)$ from the viewpoint of real analysis, that is,

$$\inf\{\mathscr{F}(u) \mid u \geq 0, \ \|u\|_1 = 8\pi\} > -\infty.$$

As $n = 2$ is the critical dimension, $\lambda = 8\pi$ is thus a critical mass in (5.4). By (5.5), (5.7), and (5.8), actually, the stationary state of (5.4) is formulated as

$$u > 0, \ \log u - v = \text{constant}, \ -\Delta v = u \ \text{ in } \Omega, \quad v|_{\partial\Omega} = 0, \ \|u\|_1 = \lambda,$$

that is, the Boltzmann–Poisson equation

$$-\Delta v = \frac{\lambda e^v}{\int_{\Omega} e^v} \text{ in } \Omega, \quad v|_{\partial\Omega} = 0, \tag{5.10}$$

of which the quantized blowup mechanism and Hamiltonian control, referred to as the recursive hierarchy, are noticed in Theorem 2.5 of Chap. 2. We have, in particular, $\lambda_0 \in 8\pi\mathbf{N}$ for $\mathbf{N} = \{1, 2, \cdots\}$, if

$$\|u_k\|_\infty \to +\infty, \quad \lambda_k \to \lambda_0,$$

arises, where (λ_k, v_k), $k = 1, 2, \cdots$, denotes a family of solutions to (5.10).

Here we observe these properties in the dynamics (5.4). Based on the study of free energy and second moment, this value $\lambda = 8\pi$ is realized as the threshold for the blowup in finite time of the solution [5, 6, 19, 54, 55, 66, 75]. Hence $\lambda = \|u_0\|_1 > 8\pi$ with a concentration at some interior point of $u_0 = u_0(x)$ implies $T < +\infty$, while $\lambda < 8\pi$ ensures

$$T = +\infty, \quad \|u(\cdot, t)\|_\infty \leq C, \tag{5.11}$$

differently from the reaction–diffusion system (5.2) with total mass dissipation and quadratic growth rate, which induces (5.11) for any n and λ.

Concerning the quantized blowup mechanism and Hamiltonian control of (5.4), first, we mention that blowup in finite time exhibits formation of *collapses* with quantized mass. Hence $T < +\infty$ in (5.4) implies $\mathscr{S} \subset \Omega$, $\sharp\mathscr{S} < +\infty$, and

$$u(x, t)dx \rightharpoonup \sum_{x_0 \in \mathscr{S}} m(x_0)\delta_{x_0}(dx) + f(x)dx \quad \text{in } \mathscr{M}(\overline{\Omega}) = C(\overline{\Omega})' \tag{5.12}$$

as $t \uparrow T$, where $m(x_0) \in 8\pi\mathbf{N}$, $0 < f = f(x) \in L^1(\Omega) \cap C(\overline{\Omega} \setminus \mathscr{S})$, and

$$\mathscr{S} = \{x_0 \in \overline{\Omega} \mid \exists x_k \to x_0, \ \exists t_k \uparrow T, \ u(x_k, t_k) \to +\infty\} \tag{5.13}$$

denotes the blowup set.

So far, the method of formal asymptotic expansion has been applied to pick up several profiles of blowup solution in finite time; that is, a radially symmetric blowup pattern with bounded free energy [27] and collision of sub-collapses [49]. These profiles are actually included in the illustration of the behavior of general solutions as in Theorems 5.15 and 5.16 of Sect. 5.5.7. Significance of these theorems lies in *collapse mass quantization*,

$$m(x_0) \in 8\pi\mathbf{N} \tag{5.14}$$

in (5.12). The underlying driving force which induces (5.14) from the quantized blowup mechanism of the family of stationary solutions is described in Chapter 1 of [75]. The quantization (5.14), however, is different from that of the family of solutions to the Boltzmann–Poisson equation (5.10), where each collapse takes the normalized mass 8π and location of these collapses is controlled by the Hamiltonian defined by (5.16) below.

The collapse formed at each blowup point of (5.4), actually, is composed of a finite sum of *sub-collapses*, the rescaled, translated solutions to the stationary problem on the whole space, called the Liouville equation in elliptic theory. A collapse composed of one sub-collapse is said to be simple. Otherwise, it is multiple, provided with the collision of sub-collapses. In this terminology, a family of blowing-up stationary solutions exhibits always simple collapses; collision of sub-collapses indicated by $m(x_0) = 8\pi m$, $m \geq 2$, in (5.4), actually, was missed in the first approach of the author [73–75].

The above-stated mechanism of formation of collapses in (5.4) is different also from that in the Gel'fand–Liouville equation (3.74) without boundary condition. In fact, Theorem 3.30 in Chap. 3 clarifies formation of sub-collapses at the origin, with possibly different concentration rates. In (5.4), on the other hand, movement of these sub-collapses is subject to the free Hamiltonian, (5.44)–(5.45) in the following subsection.

The profile of blowup in infinite time in (5.4), second, is rather more similar to that of the family of stationary solutions formulated by (5.10) as is described in Theorem 2.5 of Chap. 2. This profile is actually observed in the collapse dynamics. Theorem 5.19 in Sect. 5.5.9 thus says that

$$T = +\infty, \quad \limsup_{t \uparrow +\infty} \|u(\cdot, t)\|_\infty = +\infty \tag{5.15}$$

occurs only if

$$\lambda = \|u_0\|_1 \in 8\pi \mathbf{N}$$

and there is a critical point $x_* = (x_1^*, \cdots, x_m^*)$ of $H = H(x)$, $x = (x_1, \cdots, x_m)$, defined by

$$H(x_1, \cdots, x_m) = \frac{1}{2} \sum_{j=1}^{m} R(x_j) + \sum_{1 \leq i < j \leq m} G(x_i, x_j) \tag{5.16}$$

for $\lambda = 8\pi m$. These two fundamental profiles of blowup of the solution in finite and in infinite time govern dynamics of (5.4) essentially in accordance with the shape of domain and also the concentration of initial value [37]. See Remarks 5.28–5.34 in Sects. 5.5.9–5.5.11.

In the Smoluchowski–Poisson equation (5.4), the quantized blowup mechanism is thus realized at three levels of time evolution; stationary, in finite time, and in infinite time, while recursive hierarchy is realized, both in the family of stationary solutions and non-stationary solutions blowing up in infinite time. *Recursive hierarchy*, furthermore, is detected even in the blowup in finite time, under the scaling limit of the solution, that is, (5.44)–(5.45) and (5.46) in Sect. 5.1.3 for multiple and simple blowup points, respectively.

Totally, singular limits emerged from the total set of stationary solutions control blowup mechanisms not only in infinite time but also finite time, as collapse and sub-collapse quantization and recursive hierarchy in (5.4). This property is called *nonlinear spectral mechanics*, because the stationary state takes the form of a nonlinear eigenvalue problem (5.10) [75]. In the context of theoretical sciences, this dynamics may be called the *potential of self-organization*, because blowup with a conservative quantity, realized as an eigenvalue in a stationary state, exhibits the profile of *bottom up self-organization* in closed thermodynamical systems [77].

5.1.3 Quantized Blowup Mechanism in Dynamics

Here we describe the story behind collapse mass quantization for solutions to (5.4) blowing-up in finite time.

First, formation of collapse, (5.12) for $T < +\infty$, is a consequence of ε regularity, which means the existence of $\varepsilon_0 > 0$ such that

$$\lim_{R\downarrow 0}\limsup_{t\uparrow T} \|u(\cdot,t)\|_{L^1(\Omega\cap B(x_0,R))} < \varepsilon_0 \quad \Rightarrow \quad x_0 \notin \mathscr{S}, \tag{5.17}$$

and the *monotonicity formula*

$$\left|\frac{d}{dt}\int_\Omega u\varphi\right| \leq C(\lambda)\|\nabla\varphi\|_{C^1} \tag{5.18}$$

valid for

$$\varphi \in C^2(\overline{\Omega}), \ \left.\frac{\partial\varphi}{\partial\nu}\right|_{\partial\Omega} = 0.$$

In fact, inequality (5.18) improves (5.17) to

$$x_0 \in \mathscr{S} \quad \Rightarrow \quad \lim_{R\downarrow 0}\liminf_{t\uparrow T} \|u(\cdot,t)\|_{L^1(\Omega\cap B(x_0,R)} \geq \varepsilon_0, \tag{5.19}$$

which implies $\sharp\mathscr{S} < +\infty$ and also (5.12),

$$u(x,t)dx \rightharpoonup \sum_{x_0\in\mathscr{S}} m(x_0)\delta_{x_0}(dx) + f(x)dx \quad \text{in } \mathscr{M}(\overline{\Omega}), \tag{5.20}$$

for $\mathscr{M}(\overline{\Omega}) = C(\overline{\Omega})'$ with

$$m(x_0) \geq \varepsilon_0, \quad 0 \leq f = f(x) \in L^1(\Omega) \cap C(\overline{\Omega}\setminus\mathscr{S}) \tag{5.21}$$

[66].

Property (5.19) is a standard consequence of the combination of Gagliardo–Nirenberg inequality in two space dimensions, semigroup estimate, and total mass control (5.5) as is noticed in the previous subsection, while the monotonicity formula (5.18) is a consequence of the *weak form* of (5.4),

$$\frac{d}{dt}\int_{\Omega} u\varphi = \int_{\Omega} u\Delta\varphi + \frac{1}{2}\iint_{\Omega\times\Omega} \rho_{\varphi}(x,x')u\otimes u, \tag{5.22}$$

where $u \otimes u = u(x,t)u(x',t)dxdx'$ and

$$\rho_{\varphi}(x,x') = \nabla\varphi(x)\cdot\nabla_x G(x,x') + \nabla\varphi(x')\cdot\nabla_{x'}G(x,x'). \tag{5.23}$$

Deriving (5.22)–(5.23) is called the method of symmetrization, because it is based on the symmetry of Green's function $G(x',x) = G(x,x')$. This profile, concerned with the Poisson part of (5.4), arises in accordance with self-adjointness in operator theory [18] and that with the action–reaction law in classical mechanics [77]. There is thus a structure comparable to the spectral theory treated in Chap. 1 in the Poisson part of (5.4); symmetry of operators and duality between boundary conditions and that between particle distribution u and the created potential v.

This symmetry of Green's function, however, induces a cancellation of singularities caused by the self-interaction of many particles, where u stands for their distribution; a similar but different feature is observed in the reaction–diffusion system (5.2) with total mass dissipation.

More precisely, since behavior of $G = G(x,x')$ on $(\Omega \cap B(x_0,R)) \times (\Omega \cap B(x_0,R))$ for $0 < R \ll 1$ and $x_0 \in \partial\Omega$, is controlled by the fundamental solution

$$\Gamma(x) = \frac{1}{2\pi}\log\frac{1}{|x|} \tag{5.24}$$

and the conformal mapping $X : \Omega \cap B(x_0,R) \to \mathbf{R}^2_+$, we obtain

$$\rho_{\varphi} \in L^{\infty}(\Omega\times\Omega) \tag{5.25}$$

and hence (5.18) [75, 76]; careful treatment is required for this $\rho_{\varphi}(x,x')$ because it is not continuous at $x = x'$.

A *weak solution* is then defined by [67], based on the weak form (5.22). Let

$$\mathscr{Y} = \{\varphi \in C^2(\overline{\Omega}) \mid \left.\frac{\partial\varphi}{\partial\nu}\right|_{\partial\Omega} = 0\}$$

$$\mathscr{X}_0 = \{\rho_{\varphi} + \psi \mid \varphi \in \mathscr{Y},\ \psi \in C(\overline{\Omega}\times\overline{\Omega})\}$$

and $\mathscr{X}$ be the closure in $L^{\infty}(\Omega\times\Omega)$ of the topological vector space $\mathscr{X}_0$. We say that

$$0 \leq \mu = \mu(dx,t) \in C_*([0,T], \mathscr{M}(\overline{\Omega}))$$

is a weak solution to (5.4) if there is $\mathscr{N} = \mathscr{N}(\cdot, t) \in L^\infty_*([0, T], \mathscr{X}')$, called the *multiplicated operator* satisfying the following properties:

1. The mapping $t \in [0, T] \mapsto \langle \varphi, \mu(dx, t) \rangle$ is absolutely continuous for $\varphi \in \mathscr{Y}$.
2. It holds that

$$\frac{d}{dt}\langle \varphi, \mu \rangle = \langle \Delta\varphi, \mu \rangle + \frac{1}{2}\langle \rho_\varphi, \mathscr{N}(\cdot, t) \rangle \quad \text{a.e. } t \in [0, T].$$

3. There arises that $\mathscr{N} \geq 0$ and $\mathscr{N}|_{C(\overline{\Omega}\times\overline{\Omega})} = \mu \otimes \mu$.

Here are several direct consequences. First, it holds that

$$\mu(\overline{\Omega}, t) = \mu(\overline{\Omega}, 0) \equiv \lambda, \quad 0 \leq t \leq T.$$

Second, we obtain

$$\left| \frac{d}{dt}\langle \varphi, \mu(dx, t) \rangle \right| \leq C(\lambda) \|\nabla\varphi\|_{C^1} \tag{5.26}$$

for $\varphi \in \mathscr{Y}$ similarly to (5.18). Third, since the above $\mathscr{X}$ is *separable*, there is a generation of weak solutions with the aid of (5.26).

More precisely, if $\mu_k(dx, t) \in C_*([0, T], \mathscr{M}(\overline{\Omega}))$, $k = 1, 2, \cdots$, is a sequence of weak solutions associated with the multiplicated operators $\mathscr{N}_k \in L^\infty_*([0, T], \mathscr{X}')$, $k = 1, 2, \cdots$, satisfying

$$0 \leq \mu_k(\overline{\Omega}, t) \leq C, \ \|\mathscr{N}_k(\cdot, t)\|_{\mathscr{X}'} \leq C, \quad 0 \leq t \leq T, \ k = 1, 2, \cdots, \tag{5.27}$$

then there is a subsequence, denoted by the same symbol, such that

$$\begin{aligned} \mu_k(dx, t) &\rightharpoonup \mu(dx, t) && \text{in } C_*([0, T], \mathscr{M}(\overline{\Omega})) \\ \mathscr{N}_k(\cdot, t) &\rightharpoonup \mathscr{N}(\cdot, t) && \text{in } L^\infty_*([0, T], \mathscr{X}'), \end{aligned}$$

where $\mu(dx, t)$ a weak solution to (5.4) with the multiplicated operator $\mathscr{N}(\cdot, t)$.

Any classical solution $u = u(x, t)$, fourth, is identified with the weak solution $\mu(dx, t) = u(x, t)dx$, provided with the associated multiplicated operator

$$\mathscr{N}(\cdot, t) = u(x, t)u(x', t)\, dxdx'.$$

Given a family of classical solutions $u_k = u_k(x, t)$, $k = 1, 2, \cdots$, defined on $\overline{\Omega} \times [0, T]$, then, the condition (5.27) is satisfied if

$$\|u_{0k}\|_1 \leq C, \quad k = 1, 2, \cdots,$$

where $u_{0k} = u_k|_{t=0} \geq 0$.

Finally, the blowup criterion for classical solutions derived from the *second moment* [6, 54] is also efficient; there is thus no weak solution

$$\mu(dx, t) \in C_*([0, T], \mathcal{M}(\overline{\Omega}))$$

for any $T > 0$, if, for example,

$$\mu(\{x_0\}, 0) > 8\pi, \quad \lim_{R \downarrow 0} \frac{1}{R^2} \langle |x - x_0|^2 \varphi_{x_0,R}, \mu(dx, 0) \rangle = 0 \tag{5.28}$$

is the case for some $x_0 \in \Omega$, where $\varphi = \varphi_{x_0,R} \in \mathcal{Y}$ is a cut-off function:

$$0 \leq \varphi \leq 1, \quad \varphi = 0 \text{ on } \overline{\Omega} \cap B(x_0, R)^c, \quad \varphi = 1 \text{ on } \overline{\Omega} \cap B(x_0, R/2).$$

Besides the above properties of the weak solution to (5.4) noted in [67], *weak Liouville property* is valid for its full orbit on the whole plane. In fact, the proof of this property for classical solution [42] is valid for the weak solution. More precisely, the local second moment and scaling invariance $u_\mu(x, t) = \mu^2 u(\mu x, \mu^2 t)$, $\mu > 0$, of

$$u_t = \Delta u - \nabla \cdot u \nabla \Gamma * u, \; u \geq 0 \quad \text{in } \mathbf{R}^2 \times (-\infty, +\infty), \tag{5.29}$$

assures either $M = 0$ or $M = 8\pi$, for any weak solution

$$0 \leq a(dx, t) \in C_*(-\infty, +\infty; \mathcal{M}(\mathbf{R}^2))$$

to (5.29) satisfying $a(\mathbf{R}^2, 0) = M < +\infty$.

Note that $\Gamma(x)$ is the fundamental solution to $-\Delta$ given by (5.24), $\mathcal{M}(\mathbf{R}^2) = C_\infty(\mathbf{R}^2)'$, and $C_\infty(\mathbf{R}^2)$ denotes the set of continuous functions on $\mathbf{R}^2 \cup \{\infty\}$, the one point compactification of $\mathbf{R}^2$, taking the value zero at ∞. We have also total mass conservation,

$$a(\mathbf{R}^2, t) = M, \quad -\infty < t < +\infty$$

for the weak solution $a = a(dy, t)$ to (5.29).

Turning back to the blowup in finite time of the classical solution to (5.4), take $x_0 \in \mathcal{S}$ for the blowup set $\mathcal{S}$ defined by (5.13). Then we introduce the *backward self-similar transformation*

$$z(y, s) = (T - t) u(x, t), \quad y = (x - x_0)/(T - t)^{1/2}, \; s = -\log(T - t),$$

to obtain

$$z_s = \Delta z - \nabla \cdot z \nabla (w + \frac{|y|^2}{4}) \qquad \text{in } \bigcup_{s > -\log T} \Omega_s \times \{s\}$$

$$\frac{\partial z}{\partial \nu} - z \frac{\partial}{\partial \nu}(w + \frac{|y|^2}{4}) = 0 \qquad \text{on } \bigcup_{s > -\log T} \partial \Omega_s \times \{s\}, \tag{5.30}$$

where $\Omega_s = (T-t)^{-1/2}(\Omega - \{x_0\})$ and

$$w(y, s) = \int_{\Omega_s} G_s(y, y') z(y', s)\, dy', \quad G_s(y, y') = G(x, x').$$

The above-described notion of weak solution and the principle of its generation are extended and valid for (5.30). From the profile of Green's function $G(x, x')$ mentioned above in connection with (5.25), any $s_k \uparrow +\infty$ admits a subsequence, denoted by the same symbol, such that $z(dy, s + s_k)dy$ admits $*$-weak convergence to produce a weak solution to the full orbit on the whole and on a half space according to $x_0 \in \Omega$ and $x_0 \in \partial\Omega$, respectively. Henceforth, $z(y, s)$ is set to zero where it is not defined [75].

More precisely, if $x_0 \in \Omega$, we have

$$z(dy, s + s_k)dy \rightharpoonup \zeta(dy, s) \quad \text{in } C_*(-\infty, +\infty; \mathscr{M}(\mathbf{R}^2)) \tag{5.31}$$

as $k \to \infty$ up to a subsequence, where $\zeta = \zeta(dy, s)$ is a weak solution to

$$z_s = \Delta z - \nabla \cdot z \nabla (\Gamma * z + \frac{|y|^2}{4}) \quad \text{in } \mathbf{R}^2 \times (-\infty, +\infty) \tag{5.32}$$

for $\Gamma = \Gamma(x)$ defined by (5.24). If $x_0 \in \partial\Omega$, on the other hand, it holds that (5.31) with supp $\zeta(\cdot, s) \subset \overline{\mathbf{R}^2_+}$ for $\mathbf{R}^2_+ = \{(y_1, y_2) \mid y_2 > 0\}$ under suitable rotation of y. This $\zeta = \zeta(dy, s)$ is a weak solution to

$$z_s = \Delta z - \nabla \cdot z \nabla (E * z + \frac{|y|^2}{4}) \quad \text{in } \mathbf{R}^2_+ \times (-\infty, +\infty)$$

$$\frac{\partial z}{\partial \nu} - z \frac{\partial E * z + \frac{|y|^2}{4}}{\partial \nu} = 0 \qquad \text{on } \partial \mathbf{R}^2_+ \times (-\infty, +\infty), \tag{5.33}$$

where

$$E(y, y') = \Gamma(y - y') - \Gamma(y - y'_*) \tag{5.34}$$

and $y_* = (y_1, -y_2)$ for $y = (y_1, y_2)$.

Under this process of rescaling, some information of $u(x, t)$ as $t \uparrow T$ is lost. It is usual in the blowup analysis, developed for elliptic problem in Sect. 3 for example, as the tail moves to infinity in taking the limit. Monotonicity formula (5.18) then ensures the following fundamental properties in this argument, called the *parabolic envelope* [64, 75].

It is indicated as (5.306)–(5.307) in Sect. 5.5.5, that is,

$$\zeta(\mathbf{R}^2, s) = m(x_0), \quad \langle |y|^2, \zeta(dy, s)\rangle \leq C, \quad -\infty < s < +\infty. \tag{5.35}$$

Then $\mathscr{S} \subset \Omega$ follows first, because there is no weak solution $\zeta = \zeta(dy, s)$ to (5.33) satisfying $\zeta(\mathbf{R}^2, 0) > 0$ [76]. At this stage we are concentrating on the study of (5.32) with (5.35) to complete the proof of $m(x_0) \in 8\pi\mathbf{N}$.

An ε regularity is valid for this equation [69]. Hence the singular part of $\zeta(dy, s)$, denoted by $\zeta^s(dy, s)$, is composed of a finite sum of delta functions of uniformly bounded positions of centers by the second inequality of (5.35):

$$\zeta^s(dy, s) = \sum_{j=1}^{m(s)} \tilde{m}_j(s)\delta_{y_j(s)}(dy)$$
$$m(s) \leq m(x_0)/\varepsilon_0, \quad |y_j(s)| \leq C, \quad \tilde{m}_j(s) \geq \varepsilon_0. \tag{5.36}$$

There is, furthermore, $R > 0$ such that $\zeta(dy, s)$ is composed of uniformly bounded regular parts $\zeta^{ac}(dy, s)$ in $B_R^c \times (-\infty, +\infty)$; it thus holds that $\zeta(dy, s) = \zeta^{ac}(dy, s)$ in $(\mathbf{R}^2 \setminus B_R) \times (-\infty, +\infty)$ and

$$\|\zeta(\cdot, s)\|_{L^\infty(|y|\geq R)} \leq C, \quad -\infty < s < +\infty. \tag{5.37}$$

To establish this outer uniform estimate, we apply *scaling-back* to $\zeta(dy, s)$ used in [75],

$$\zeta(dy, s) = e^{-s}A(dy', s') \quad y' = e^{-s/2}y, \ s' = -e^{-s}, \tag{5.38}$$

to obtain a weak solution $A = A(dy', s') \in C_*(-\infty, 0; \mathscr{M}(\mathbf{R}^2))$ to

$$A_{s'} = \Delta A - \nabla \cdot A\nabla\Gamma * A, \ A \geq 0 \quad \text{in } \mathbf{R}^2 \times (-\infty, 0). \tag{5.39}$$

Scaling invariance of (5.39) makes ε regularity for the classical solution scale free. Regularity Lemma 5.31 is applicable to the weak solution $A(dy', s')$ generated by a family of classical solutions. Through (5.38) it produces a similar property to $\zeta(dy, s)$ indicated by (5.312) below, that is,

$$\zeta(B(y_0, 2r), s) < \varepsilon_0 \quad \Rightarrow \quad \|\zeta(\cdot, s)\|_{L^\infty(B(y_0,r))} \leq Cr^{-2} \tag{5.40}$$

valid for any $y_0 \in \mathbf{R}^2$, $r > 0$, and $-\infty < s < +\infty$. Then (5.37) is obtained for $R \gg 1$ again by the second inequality of (5.35). This (5.40) implies also the profile of the singular part of $\zeta(dy, s)$ indicated by (5.36).

Inequality (5.37) and $\zeta(\mathbf{R}^2, s) = m(x_0)$ now control the chemotaxis term $\Gamma * z$ in (5.32) for $|y| > R$, while the $|y|^2/4$ term there brings the mass of ζ far away as $s \uparrow +\infty$. This property, combined with the second inequality of (5.35) again, induces the *residual vanishing* indicated by

$$\zeta^{ac}(dy, s) = 0 \quad \text{in } (\mathbf{R}^2 \setminus B_R) \times (-\infty, +\infty),$$

and hence $\zeta(dy, s) = \zeta^s(dy, s)$ in $\mathbf{R}^2 \times (-\infty, +\infty)$ by the unique continuation theorem for parabolic equations.

An argument using the second moment in an outer region is also efficient here (Remark 5.14 in Sect. 5.5.6). This *outer second moment* is actually non-decreasing because of $|y|^2/4$ term in (5.32), and then a contradiction to the second parabolic envelope arises with $s \uparrow +\infty$, if residual vanishing is not achieved. Meanwhile we obtain the type II blowup rate (Remark 5.18 in Sect. 5.5.6).

If one uses the result of [69],

$$\tilde{m}_j(s) = 8\pi \tag{5.41}$$

in (5.36) at this stage, it follows that

$$\zeta(dy, s) = 8\pi \sum_{j=1}^{m} \delta_{y_j(s)}(dy) \tag{5.42}$$

and hence collapse mass quantization $m(x_0) \in 8\pi\mathbf{N}$ with $m = m(x_0)/(8\pi)$. Scaling-back (5.38), however, makes this process possible to detect sub-collapse dynamics.

In fact, we have readily shown that $A = A(dy', s') \in C_*(-\infty, 0; \mathcal{M}(\mathbf{R}^2))$, the weak solution to (5.39), is composed of a finite sum of delta functions. Given $s_0 \in \mathbf{R}$, therefore, we can take the *translation limit* as $s'_k \uparrow +\infty$, denoted by $\tilde{A}(dy, s)$ in $\mathbf{R}^2 \times (-\infty, +\infty)$, of $\hat{A}(dy', s_0 + s'_k)$, where

$$\hat{A}(dy, s) = A(dy', s'), \quad y = y' - y_j(s_0).$$

Then it follows that (5.41) for $s = s_0$, from the *dilation limit* of $\hat{A}_{\beta_k}(dy, s)$ as $\beta_k \downarrow 0$ and the weak Liouville property to (5.29), where

$$\hat{A}_\beta(dy', s') = \beta^2 \hat{A}(dy, s), \quad y = \beta y', \ s = \beta^2 s'.$$

Having (5.42), we apply (5.38) to obtain

$$A(dy', s') = 8\pi \sum_{j=1}^{m} \delta_{y'_j(s')}(dy'), \ s' < 0, \quad m = m(x_0)/(8\pi). \tag{5.43}$$

Then, local second moment applied to (5.39) assures the *sub-collapse dynamics*

$$\frac{dy'_j(s')}{ds'} = 8\pi \nabla_{y_j} H^0_m(y'_1, \cdots, y'_m) \quad \text{a.e. } s' < 0$$
$$\lim_{s' \uparrow 0} y'_j(s') = 0, \ 1 \le j \le m, \tag{5.44}$$

with

$$\Gamma^0(y_1, \cdots, y_m) = \sum_{1 \le i < j \le m} \Gamma(y_i - y_j) \tag{5.45}$$

for $m \ge 2$, and

$$A(dy', s') = 8\pi \delta_0(dy'), \quad s' < 0 \tag{5.46}$$

for $m = 1$ (Remark 5.17 in Sect. 5.5.6). The conclusion $f > 0$ on $\overline{\Omega} \setminus \mathscr{S}$ in (5.12), finally, is derived from the strong maximum principle for parabolic equations.

5.1.4 Recursive Hierarchy in Dynamics

Here we describe the profile of the solution to (5.4) blowing-up in infinite time, that is, total mass quantization and recursive hierarchy.

Assuming (5.15) in (5.4), we have

$$\exists t_k \uparrow +\infty, \quad \lim_{k \to \infty} \|u(\cdot, t_k)\|_\infty = +\infty.$$

There is a subsequence, denoted by the same symbol, such that

$$u(x, t + t_k)dx \rightharpoonup \mu(dx, t) \quad \text{in } C_*(-\infty, +\infty; \mathscr{M}(\overline{\Omega})), \tag{5.47}$$

where $\mu(dx, t)$ is a weak solution to (5.4). Then improved regularity due to [68], Lemma 5.31 in Sect. 5.5.4, ensures $\sharp \mathscr{S}_t < +\infty$ for

$$\mathscr{S}_t = \{x_0 \in \overline{\Omega} \mid \exists x_k \to x_0, \ u(x_k, t + t_k) \to +\infty\}$$

and

$$\mu^s(dx, t) = \sum_{x_0 \in \mathscr{S}_t} m_t(x_0)\delta_{x_0}(dx), \quad m_t(x_0) \ge \varepsilon_0. \tag{5.48}$$

It holds also that $\mathscr{S}_0 \neq \emptyset$ by (5.15).

To confirm $\mathscr{S}_T \subset \Omega$ for any $T \in \mathbf{R}$, we take $x_0 \in \mathscr{S}_T \cap \partial\Omega$, the backward self-similar transformation

$$\zeta(dy, s) = (T-t)\mu(dx, t), \quad y = (x - x_0)/(T-t), \ s = -\log(T-t),$$

and $s_k \uparrow +\infty$. Then, up to a subsequence $\zeta(dy, s + s_k)$, $k = 1, 2, \cdots$, generates a weak solution

$$\tilde{\zeta}(dy, s) \in C_*(-\infty, +\infty; \mathscr{M}(\overline{\mathbf{R}}^2_+))$$

to (5.33) with $\tilde{\zeta}(\overline{\mathbf{R}}^2_+, 0) = m(x_0) > 0$, a contradiction. Thus we obtain $\mathscr{S}_t \subset \Omega$ for any $-\infty < t < +\infty$.

Given $x_0 \in \mathscr{S}_{t_0}$, next, we take

$$\mu_\beta(dx, t) = \beta^2 \mu(dx', dt'), \quad x' = \beta x + x_0, \ t' = \beta^2 t + t_0$$

for $\beta > 0$. Any $\beta_k \downarrow 0$ admits a subsequence, denoted by the same symbol, such that

$$\mu_{\beta_k}(dx, t) \rightharpoonup \tilde{\mu}(dx, t) \quad \text{in } C_*(-\infty, +\infty; \mathscr{M}(\mathbf{R}^2)).$$

This $\tilde{\mu}(dx, t)$ is a weak solution to (5.29) satisfying $\tilde{\mu}(\mathbf{R}^2, 0) = m_{t_0}(x_0)$ and hence it follows that $m_{t_0}(x_0) = 8\pi$.

We now use the local second moment to follow collapse dynamics in (5.48). In Lemma 5.34 in Sect. 5.5.10, equality (5.48) is refined as

$$\mu^s(dx, t) = 8\pi \sum_{j=1}^{m} \delta_{x_j(t)}(dx), \quad -\infty < t < +\infty \tag{5.49}$$

with $m \leq \lambda/(8\pi)$ independent of t and $x_j = x_j(t)$, $1 \leq j \leq m$, satisfying

$$\frac{dx_j}{dt} = 8\pi \nabla_{x_j} H(x_1, \cdots, x_m), \quad -\infty < t < +\infty, \quad 1 \leq j \leq m, \tag{5.50}$$

where $H = H_m(x)$, $x = (x_1, \cdots, x_m)$, is the point vortex Hamiltonian defined in (5.16).

The argument used for exclusion of boundary blowup and the instant blowup criterion (5.28) to $\mu(dx, t)$ guarantees that the closure of orbit $\mathscr{O} = \{x(t)\}$ made by the solution $x(t) = (x_1(t), \cdots, x_m(t))$, $-\infty < t < +\infty$, to (5.50) is contained in $\Omega^m \setminus D$, where

$$D = \{(x_1, \cdots, x_m) \mid x_i = x_j, \ \exists i \neq j\}.$$

Since (5.50) is an anti-gradient flow satisfying

$$\frac{dH}{dt} = 8\pi|\nabla H|^2 \geq 0, \quad H = H(x(t)),$$

the theory of gradient inequality [29] assures that there are critical points x_*, x^*, which may be equal, of $H(x)$ that realize $\mathscr{O}$ as their connecting orbit:

$$\lim_{t\downarrow-\infty} x(t) = x_*, \quad \lim_{t\uparrow+\infty} x(t) = x^*. \tag{5.51}$$

In particular, $H_m(x)$ for $m = \lambda/(8\pi)$ must have at least one critical point. This property is sensitive to the shape of domain, where the elliptic theory commits in [82].

Total mass quantization $\lambda = \|u_0\|_1 \in 8\pi\mathbf{N}$ is now a consequence of the residual vanishing,

$$\mu^{ac}(dx,t) = 0. \tag{5.52}$$

To establish this property, we note that above compactness of $\overline{\mathscr{O}}$ ensures $0 < r \ll 1$ such that

$$B(x_j(t), r) \cap \mathscr{S}_t = \{x_j(t)\}, \quad -\infty < t < +\infty, \ 1 \leq j \leq m.$$

Extending the argument for radially symmetric solutions [58, 78], we use the local second moment of defect mesure. Subject to (5.50), we derive (5.423) below, that is,

$$\frac{d}{dt}\int_{B(x_j(t),r)} (|x - x_j(t)|^2 - r^2)u_k$$

$$\leq 4\int_{B(x_j(t),r)} u_k - \frac{1}{2\pi}\left(\int_{B(x_j(t),r)} u_k\right)^2 + C\int_{B(x_j(t),r)} |x - x_j(t)|u_k$$

by the *first volume derivative* shown in Chap. 4.

Making $k \to \infty$ in (5.47), we arrive at

$$\frac{dI}{dt} \leq CI, \quad -\infty < t < +\infty$$

for $0 < r \ll 1$, where

$$I(t) = \int_{B(x_j(t),r)} (|x - x_j(t)|^2 - r^2)f \leq 0$$

for $f = f(x,t) \geq 0$ defined by

$$f(x,t)dx = \mu^{ac}(dx,t).$$

Then it follows that $I \equiv 0$, and hence (5.52) (Theorem 5.38).

5.1.5 Mathematical Oncology

Reaction, diffusion, and chemotaxis are formulated by Liouville's first volume derivative in Sect. 5.3.2. Multi-scale models are then constructed for several problems in mathematical oncology, where an appropriate formulation of chemotaxis is required. Detailed mass balance is achieved in a numerical scheme provided with positivity and total mass preservation, which realizes the transient probability of particles. Then we define the particle velocity and discrete flux to execute hybrid simulations: continuous–discrete, and deterministic–stochastic. There are several applications in clinical and theoretical studies.

5.1.6 Organization of Chap. 5

Section 5.2 is concerned with fundamental profiles of geometry of classical mechanics. We describe the notion of manifold, Riemann and informatic geometries, and sympletic and Poisson manifolds. Meanwhile Liouville's theorem on the first integral of a Hamilton system is confirmed.

The effect of interaction of multiple species is studied in Sect. 5.3. We begin with fundamentals of mathematical modeling using ordinary differential equations, distinguishing reactions of the zero-th, the first, and the second orders. A diffusion equation is then derived from the notions of flux and gradient of particle distribution. Here we mention the role of Einstein's formula to take the mean field limit of the master equation. The semigroup estimate, a fundamental tool in mathematical analysis of reaction–diffusion equation, is also described. Then we turn to several biological models provided with Poisson structure, which controls their global-in-time dynamics.

This section ends with a class of skew-symmetric Lotka–Volterra N-systems, provided with $(N-2)$ entropies, which induces foliations composed of one dimensional stationary states and nested periodic orbits. This result of the cancellation of singularities of interaction in multi-species is stated in Theorems 5.4 and 5.5.

Section 5.4 deals with the reaction–diffusion system (5.2). We observe that the critical dimension for the existence of a uniformly bounded global-in-time solution is $n = 2$, approaching from the total mass control (5.3). Evidence of this criticality is presented in both space and time variables. This critical dimension, however, is exceeded by the other approach; duality, Alexandroff–Bakelman–Pucci and Krylov–Safonov inequalities, and scaling. We finally illustrate the method of *interpolation regularity* [16, 17] to reach Theorem 5.12 without restriction of space dimension for the quadratic growth rate of the nonlinearity.

The Smoluchowski–Poisson equation (5.4) has its origin in cell biology and non-equilibrium thermodynamics. In Sect. 5.5 we follow the Keller–Segel model to reduce a form of the Smoluchowski–Poisson equation. Then we turn to non-equilibrium statistical mechanics, particularly, the Kramers–Moyal expansion, to reach the Smoluchowski part. Modeling in non-equilibrium thermodynamics is also described, using maximal entropy production. The viral theorem concerning the second moment of mass density is a fundamental scheme to control dynamics of models in physics, as in the case of the Euler equation in the previous chapter, besides total mass conservation and decreasing free energy.

Then we show a quantized blowup mechanism and Hamiltonian control in blowup in finite time and infinite time of (5.4) in two space dimensions, where the critical dimension and critical mass are detected in accordance with total mass conservation and free energy decreasing, respectively, under the presence of scaling invariance. We begin with the model on the whole space, to show formation of collapses, generation of scaling limit accompanied with the backward self-similar transformation, residual vanishing leading to the collapse mass quantization, and the sub-collapse dynamics observed in the scaling-back of the limit measure.

We turn to the study on the bounded domain provided with Dirichlet and Neumann boundary conditions for the Poisson part. Having distinguished the simplicity of collapses and collision of sub-collapses, we show type II blowup rate and free energy transmission in accordance with the boundedness of the free energy. Then, we show the exclusion of boundary blowup for the model described in this section, (5.4). Finally, we study blowup in infinite time. We show the formation of collapses with quantized mass, Hamiltonian control upon collapse dynamics, and initial mass quantization in turn, taking account of the effect of boundary conditions of the Poisson part.

The final section, Sect. 5.6, is devoted to mathematical oncology. First, we describe a tumor growth model using hyperbolic equations. Second, a fundamental system on angiogenesis is derived, using the theories of chemical reaction and transport. A mathematical basis on hybrid simulation is then prepared in the context of numerical analysis. Finally, a new theory for chemotactic paradox is presented using a Turing paradigm concerning RAS pathways.

5.2 Geometry of Classical Mechanics

5.2.1 Manifolds

Most mathematical models are concerned with the interaction of multiple components. If these components are involved by conservation laws, for example, their existence region is reduced to a lower-dimensional manifold, and a comparable profile is observed even for gradient systems. Formation of singularities to the solution is avoided by this structure, or restricted to striking patterns recognized as self-organization. The final chapter begins with a review of several concepts of Riemannian geometry, to access an unexpected gift from Liouville's several theories, the understanding of competitive and cooperative interactions in the presence of geometric structure of the component of the solution.

First, we generalize the theories of surfaces and hypersurfaces in Sects. 2.2 and 4.2, respectively. A manifold M is thus a Hausdorff space covered by a family of open sets, $M = \bigcup_{\alpha \in \mathscr{A}} U_\alpha$, provided with the homeomorphism

$$\psi_\alpha : U_\alpha \to \psi_\alpha(U_\alpha) = \{x^1, \cdots, x^n\} \subset \mathbf{R}^n, \ \alpha \in \mathscr{A},$$

where

$$\psi_\beta \circ \psi_\alpha^{-1} : \psi_\alpha(U_\alpha \cap U_\beta) \to \psi_\beta(U_\alpha \cap U_\beta)$$

is C^∞ if $U_\alpha \cap U_\beta \neq \emptyset$. The tangent space T_pM at $p \in M$ is defined as a linear space formed by infinitesimal directions of curves on M at p, the basis of which is denoted by

$$\{\frac{\partial}{\partial x^1}, \cdots, \frac{\partial}{\partial x^n}\} \tag{5.53}$$

according to the local chart $(U_\alpha; x^1, \cdots, x^n)$ such that $p \in U_\alpha$. Then the tangent bundle is induced by

$$TM = \bigcup_{p \in M} T_pM.$$

The cotangent space T_p^*M is defined by the dual space of the tangent space, $(T_pM)^*$. The dual basis of (5.53) is denoted by

$$\{dx^1, \cdots, dx^n\}.$$

Then the cotangent bundle is formed as

$$T^*M = \bigcup_{p \in M} T_p^*M.$$

The smooth mappings

$$p \in M \mapsto X_p \in T_pM, \quad p \in M \mapsto \omega_p \in T_p^*M$$

are called vector field and 1-form, respectively. The set of vector fields and 1-forms are denoted by $\mathscr{X}(M)$ and $\mathscr{D}^1(M)$, respectively.

The *co-variant derivative* ∇ is a mapping

$$(X, Y) \in \mathscr{X}(M) \times \mathscr{X}(M) \to \nabla_X Y \in \mathscr{X}(M)$$

satisfying

$$\nabla_X(Y + Z) = \nabla_X Y + \nabla_X Z$$
$$\nabla_X(fY) = (Xf)Y + f\nabla_X Y$$
$$\nabla_{X+Y} Z = \nabla_X Z + \nabla_Y Z$$
$$\nabla_{fX} Y = f\nabla_X Y$$

for any $X, Y, Z \in \mathscr{X}(M)$ and $f \in C^\infty(M)$. This axiom determines the connection coefficient Γ_{ij}^k by

$$\nabla_{\frac{\partial}{\partial x^i}} \frac{\partial}{\partial x^j} = \Gamma_{ij}^k \frac{\partial}{\partial x^k}$$

through the local chart under the Einstein convention. Thus ∇ is represented by Γ_{ij}^k with appropriate rules on the transformation of local charts, and for this reason ∇ is sometimes called the (affine) connection.

The curvature tensor is a mapping $R :\in \mathscr{X}(M) \times \mathscr{X}(M) \times \mathscr{X}(M) \to \mathscr{X}(M)$ defined by

$$R(X, Y, Z) = \nabla_X(\nabla_Y Z) - \nabla_Y(\nabla_X Z) - \nabla_{[X,Y]} Z,$$

where $[X, Y] = XY - YX$ denotes the commutator. The torsion tensor, on the other hand, is a mapping $T : \mathscr{X}(M) \times \mathscr{X}(M) \to \mathscr{X}(M)$ defined by

$$T(X, Y) = \nabla_X Y - \nabla_Y X - [X, Y].$$

A manifold M provided with the connection ∇ is said to be flat if curvature and torsion tensors are free, that is, $R \equiv 0$ and $T \equiv 0$. Then there is a local chart satisfying $\Gamma_{ij}^k = 0$ for any k, i, j. Connection free, thus, induces a parallel translation.

More precisely, if M is a manifold provided with the connection ∇, and

$$C : p(t) \in M, \quad a \leq t \leq b$$

is a curve on M, we say that $Z \in \mathscr{X}(M)$ is parallel to ∇ along C if $\nabla_{\dot{p}} Z = 0$ for $a \leq t \leq b$. If

$$\nabla_{\dot{p}} \dot{p} = 0, \quad a \leq t \leq b,$$

this C is called geodesic. A geodesic curve is a minimum length curve connecting any two close points on it, if this M is a Riemann manifold and ∇ is the Levi–Chivita connection formulated below.

5.2.2 Riemann Manifolds and Informatic Geometry

Given a vector space V, let V^* be its dual space. A multi-linear form

$$F : \overbrace{V^* \times \cdots \times V^*}^{r} \times \underbrace{V \times \cdots V}_{s} \to \mathbf{R}$$

is called an (r, s) tensor on V. If F and G are (r, s) and (r', s') tensors, respectively, their tensor product $F \otimes G$ is an $(r + r', s + s')$ tensor defined by

$$(F \otimes G)(x, x' : y, y') = F(x, x') \cdot G(y, y'),$$

where $(x, x', y, y') \in V^r \times V^{r'} \times (V^*)^r \times (V^*)^{r'}$.

A Riemann metric g on a manifold M is a smooth mapping $p \in M \mapsto g_p$ that is a (0, 2) tensor on T_pM, symmetric and positive definite. This metric is thus represented by $g = g_{ij}dx^i \otimes dx^j$ under the local chart, where (g_{ij}) defined for

$$g_{ij} = g(\frac{\partial}{\partial x^i}, \frac{\partial}{\partial x^j})$$

is a positive definite symmetric matrix. A manifold provided with a Riemann metric is called a Riemann manifold.

The Riemann metric $g(\; , \;)$ induces a connection ∇, called the Levi–Chivita connection , satisfying

$$Xg(Y, Z) = g(\nabla_X Y, z) + g(Y, \nabla_X Z), \quad X, Y, Z \in \mathscr{X}(M). \tag{5.54}$$

Such a connection is called metric in general. A connection ∇ on a Riemann manifold (M, g), furthermore, is Levi–Chivita if and only if it is metric and torsion free.

Let (M, g) be a Riemann manifold and ∇, ∇^* be connections. We say that ∇^* is a dual of ∇ if

$$Xg(Y, Z) = g(\nabla_X Y, Z) + g(Y, \nabla^*_X Z), \quad X, Y, Z \in \mathscr{X}(M),$$

and that (∇, ∇^*) is a dual affine structure to (M, g). By (5.54), therefore, the Levi–Chivita connection $\nabla = \nabla^*$ provides a trivial dual affine structure to (M, g). We say that a manifold M is dual flat if there is a dual affine structure (∇, ∇^*) of (M, g)

such that $R = R^* = 0$ and $T = T^* = 0$, where R and R^* are curvatures with respect to ∇ and ∇^*, respectively, and T and T^* are torsions with respect to ∇ and ∇^*, respectively. The dual flat property is a fundamental structure of informatic geometry.

5.2.3 *Analytic Mechanics*

If mass particles $x_i = x_i(t) \in \mathbf{R}^3$, $i = 1, 2, \cdots, f$, are subject to the potential energy $U = U(x_1, x_2, \cdots, x_f)$, then Newton's equation of motion takes the form

$$\frac{dp_i}{dt} = -\frac{\partial U}{\partial x_i}, \qquad i = 1, 2, \cdots, f. \tag{5.55}$$

Kinetic energy then arises,

$$K(\dot{x}_1, \dot{x}_2, \cdots, \dot{x}_f) = \frac{1}{2}\sum_{i=1}^{f} m_i \dot{x}_i^2,$$

and it holds that

$$p_i = \frac{\partial K}{\partial \dot{x}_i}$$

where $p_i = m_i \dot{x}_i$ denotes the momentum.

If $(q_1, \cdots, q_f)$ denotes the generalized coordinate then it holds that

$$x_1 = x_1(q_1, \cdots, q_f; t), \quad \ldots \quad , \quad x_f = x_f(q_1, \cdots, q_f; t)$$

and hence

$$\dot{x}_i = \sum_j \frac{\partial x_i}{\partial q_j}\dot{q}_j + \frac{\partial x_i}{\partial t}. \tag{5.56}$$

Regarding $(q, \dot{q}, t)$ as independent variables, we differentiate (5.56) with respect to $\dot{q}_j$ to obtain

$$\frac{\partial \dot{x}_i}{\partial \dot{q}_j} = \frac{\partial x_i}{\partial q_j}.$$

Hence we have

$$\frac{\partial K}{\partial \dot{q}_j} = \sum_k \frac{\partial K}{\partial \dot{x}_k}\frac{\partial \dot{x}_k}{\partial \dot{q}_j} = \sum_k p_k \frac{\partial x_k}{\partial q_j},$$

and then it follows that

$$\begin{aligned}\frac{d}{dt}\frac{\partial K}{\partial \dot{q}_j} &= \sum_k \left(\frac{dp_k}{dt}\frac{\partial x_k}{\partial q_j} + p_k \frac{d}{dt}\frac{\partial x_k}{\partial q_j}\right) \\ &= \sum_k \left(-\frac{\partial U}{\partial x_k}\frac{\partial x_k}{\partial q_j} + \frac{\partial K}{\partial \dot{x}_k}\frac{\partial \dot{x}_k}{\partial q_j}\right) = -\frac{\partial U}{\partial q_j} + \frac{\partial K}{\partial q_j}\end{aligned}$$

by

$$\frac{d}{dt}\frac{\partial x_k}{\partial q_j} = \frac{\partial \dot{x}_k}{\partial q_j}.$$

If we regard K as a function of $(q, \dot{q}, t)$ by (5.56) and take the Lagrangian,

$$L(q, \dot{q}, t) = K(q, \dot{q}, t) - U(q, t), \tag{5.57}$$

it follows that

$$\frac{d}{dt}\frac{\partial L}{\partial \dot{q}_j} = \frac{\partial L}{\partial q_j}, \qquad j = 1, \cdots, f. \tag{5.58}$$

Equation (5.58), called Lagrange's equation of motion, is nothing but the Euler equation of the variational problem $\delta S = 0$ under the constraint $\delta q(t_1) = \delta q(t_2) = 0$ where

$$S = \int_{t_1}^{t_2} L\left(q(t), \dot{q}(t), t\right) dt$$

denotes the action integral defined for $q = q(t) \in \mathbf{R}^{3f}$ with $t \in [t_1, t_2]$. This fact is called Hamilton's principle of least action.

The *Legendre transformation* of $\dot{q} \mapsto L = L(q, \dot{q}, t)$ is given by

$$L^*(p) = \sup_{\dot{q}}\{p \cdot \dot{q} - L(\dot{q})\} = p \cdot \dot{q} - L, \quad p = \frac{\partial L}{\partial \dot{q}}.$$

Writing this L^* as

$$H = H(p, q, t) = \sum_j p_j \dot{q}_j - L, \quad p_j = \frac{\partial L}{\partial \dot{q}_j}, \tag{5.59}$$

we call it the *Hamiltonian* provided with the generalized momentum p and the generalized coordinate q.

From (5.58) we have

$$\begin{aligned} dL &= \frac{\partial L}{\partial q}dq + \frac{\partial L}{\partial \dot{q}}d\dot{q} + \frac{\partial L}{\partial t}dt \\ &= \frac{d}{dt}\left(\frac{\partial L}{\partial \dot{q}}\right)dq + \frac{\partial L}{\partial \dot{q}}d\dot{q} + \frac{\partial L}{\partial t}dt = \dot{p}dq + pd\dot{q} + \frac{\partial L}{\partial t}dt, \end{aligned}$$

which implies

$$\begin{aligned} dH &= \dot{q}dp + pd\dot{q} - dL = \dot{q}dp - \dot{p}dq - \frac{\partial L}{\partial t}dt \\ &= \frac{\partial H}{\partial p}dp + \frac{\partial H}{\partial q}dq + \frac{\partial H}{\partial t}dt, \end{aligned}$$

or Hamilton's canonical equation

$$\dot{q} = \frac{\partial H}{\partial p}, \quad \dot{p} = -\frac{\partial H}{\partial q} \tag{5.60}$$

which makes up a Hamilton system. Then we obtain

$$\dot{H} \equiv \frac{d}{dt}H(p(t), q(t)) = \frac{\partial H}{\partial q} \cdot \dot{q} + \frac{\partial H}{\partial p} \cdot \dot{p} = \frac{\partial H}{\partial q} \cdot \frac{\partial H}{\partial p} - \frac{\partial H}{\partial p} \cdot \frac{\partial H}{\partial q} = 0.$$

Let

$$\{F, G\} = \frac{\partial F}{\partial q} \cdot \frac{\partial G}{\partial p} - \frac{\partial F}{\partial p} \cdot \frac{\partial G}{\partial q}$$

for $F = F(q, p)$ and $G = G(q, p)$. Then (5.60) is equivalent to

$$\dot{F} = \{F, H\}, \quad \forall F = F(q, p), \tag{5.61}$$

which leads to the other Liouville theorem that F is a first integral or a conservative quantity of (5.60) if and only if $\{F, H\} = 0$.

The mapping $(F, G) \mapsto \{F, G\}$ is bilinear and skew-symmetric. It holds also that the Jacobi identity

$$\{F, \{G, H\}\} + \{H, \{F, G\}\} + \{G, \{H, F\}\} = 0$$

and the Leibniz identity

$$\{FG, H\} = F\{G, H\} + \{F, H\}G.$$

Such a $\{\cdot, \cdot\}$ is called a *Poisson bracket*.

If $Q = \mathbf{R}^{3f}$ denotes the configuration manifold, then the Hamiltonian is realized as a mapping $H : (q, p) \in T^*Q \mapsto H(q, p) \in \mathbf{R}$. The Hamilton vector field is defined by

$$X_H = \begin{pmatrix} \frac{\partial H}{\partial p} \\ -\frac{\partial H}{\partial q} \end{pmatrix}$$

and then (5.60) is equivalent to

$$\dot{z} = X_H z, \quad z = \begin{pmatrix} q \\ p \end{pmatrix}.$$

5.2.4 Sympletic and Poisson Manifolds

Recall that $\mathscr{D}^1(M)$ denotes the set of 1-forms on M. Each element in $\mathscr{D}^1(M)$ induces a 2-form by exterior product $\wedge$ and exterior derivative d. Henceforth the set of two forms on M is denoted by $\Lambda^2 T^* M$. By the local chart each $\omega \in \Lambda^2 T^* M$ is represented by a smooth function valued in skew-symmetric matrices. Then it forms a bilinear, skew-symmetric mapping $\omega : \mathscr{X}(M) \times \mathscr{X}(M) \to \mathbf{R}$; for example it follows that

$$\langle (u, v), dx^1 \wedge dx^2 \rangle = u^1 v^2 - u^2 v^1$$

for

$$u = u^1 \frac{\partial}{\partial x_1} + u^2 \frac{\partial}{\partial x_2}, \quad v = v^1 \frac{\partial}{\partial x^1} + v^2 \frac{\partial}{\partial x^2}.$$

A 2-form ω is said to be non-degenerate if the associated skew-symmetric matrix represented by local chart is non-degenerate. It is said to be closed if $d\omega = 0$. A sympletic manifold is a manifold M provided with a non-degenerate, closed 2-form denoted by ω. Then Darboux's theorem ensures that this (M, ω) is locally equivalent to the canonical sympletic form $(\mathbf{R}^{2n}, \omega_0)$, where

$$\omega_0 = dx \wedge dy \equiv dx^1 \wedge dy^1 + \cdots + dx^n \wedge dy^n, \quad (x, y) \in \mathbf{R}^n \times \mathbf{R}^n.$$

Liouville's 1-form $\theta = q_i dp^i$ is introduced in accordance with analytic mechanics, which results in the sympletic form $\omega = -d\theta$ for $p^i = x^i, q^i = y^i$. Recall that $\mathscr{X}(M)$ denotes the set of vector fields on M.

Theorem 5.1 *Let (M, ω) be a sympletic manifold and $H \in C^\infty(M)$. Define $X_H \in \mathscr{X}(M)$, called the Hamilton vector field, by*

$$\omega(X_H, \cdot) = dH \tag{5.62}$$

as elements in $\mathscr{D}^1(M)$. *Then* $\{F, G\} = \omega(X_F, X_G)$ *for* $F, G \in C^\infty(M)$ *is a Poisson bracket, that is, skew-symmetric bilinear form provided with the Jacobi and Leibniz identities.*

Let M be a manifold, and $X \in \mathscr{X}(M)$ and $f \in C^\infty(M)$ be given. Then the Lie derivative $\mathscr{L}_X f$ is defined by

$$\mathscr{L}_X f(x) = \left.\frac{d}{dt} f \circ \Phi_{-t}(x)\right|_{t=0}, \quad x \in M,$$

where $\{\Phi_t\}$ is the local dynamical system generated by X:

$$\frac{d}{dt}\Phi_t(x) = X(\Phi_t(x)), \quad \Phi_t(x)|_{t=0} = x.$$

A manifold P provided with a Poisson bracket $\{\ ,\ \} : C^\infty(M) \times C^\infty(M) \to C^\infty(M)$ is called a Poisson manifold. A sympletic manifold is Poisson, under the bracket induced in the previous theorem. Let $(P, \{\ ,\ \})$ be a Poisson manifold and $H \in C^\infty(P)$. Define $X_H \in \mathscr{X}(M)$ simply from the Poisson structure, differently as in (5.62) provided with the sympletic structure, by

$$\mathscr{L}_{X_H} F = \{F, H\}, \quad \forall F \in C^\infty(P).$$

Then we call

$$\dot{x} = X_H(x)$$

a Hamilton system on P.

Theorem 5.2 *A sympletic manifold* (M, ω) *is a Poisson manifold with* $\{\ ,\ \}$ *defined in the previous theorem. Both definitions on the Hamilton vector field* X_H *for* $H \in \mathscr{X}(M)$ *above, using the sympletic form and the Poisson bracket, coincide.*

An example of an odd dimensional Poisson manifold is $P = \mathbf{R}^3$ and

$$\{F, G\} = -\nabla C \cdot \nabla F \times \nabla G$$

for fixed $C \in C^\infty(P, \mathbf{R}^3)$. Then the Hamilton system induced by the Hamiltonian $H \in C^\infty(P)$ becomes

$$\dot{x} = \nabla C \times \nabla H.$$

This structure is observed in the Lotka–Volterra system described in the following section. See [28] for the proof of all the facts above and Theorems 5.1–5.2.

5.3 Interaction of Multi Species

5.3.1 *Mathematics of Reaction*

Diffusion is formulated as a material transport, where the flux is realized as a product of mass and velocity of particles by the formulae of Liouville and Gauss. Interaction, on the other hand, results in reaction, where thermodynamical laws bound the components of the solution on a manifold. Fusion of two concepts of Liouville's theory in mathematical physics, that is, material transport and analytic mechanics, thus induces the description of self-organization emerging from interaction of multiple species and transportation of materials.

To describe the balance of the amount, first, let $u = u(t)$ be a quantity of some material. Compare two models of ordinary differential equations,

$$\frac{du}{dt} = a, \quad u(0) = u_0 > 0 \tag{5.63}$$

and

$$\frac{du}{dt} = au, \quad u(0) = u_0 > 0 \tag{5.64}$$

with a constant a, which are concerned with different laws of variation in time. The first model (5.63) stands for supply or consumption according to $a > 0$ and $a < 0$, respectively.

In the latter case the solution

$$u(t) = at + u_0$$

becomes eventually negative in spite of $u_0 > 0$. This feature has a strong contrast with the case of $a < 0$ in (5.64), where the solution

$$u(t) = u_0 e^{at}$$

stays positive even if it decays fast as $t \uparrow +\infty$.

The logistic equation

$$\frac{du}{dt} = u(1-u), \quad u(0) = u_0 > 0$$

admits a global-in-time solution $u = u(t) >$ converging to 1 as $t \uparrow +\infty$. This feature is associated with the stability of two equilibria, $u = 0$ and $u = 1$. There may arise the blowup of the solution in accordance with the nonlinearity of the right-hand side. A typical example is

$$\frac{du}{dt} = u^2, \quad u(0) = u_0 > 0,$$

of which the solution

$$u(t) = (T - t)^{-1}, \quad T = u_0^{-1},$$

is usually regarded as expiring at the blowup time $t = T$. Activation and inhibition may be modeled by

$$u_t = (\alpha v)u, \quad v_t = -(\beta u)v,$$

using positive constants α and β.

The general *autonomous system*

$$\frac{du}{dt} = f(u), \quad u(0) = u_0 \in \mathbf{R}^N_+$$

can generate a local-in-time orbit, provided that $f = f(u)$ is locally Lipschitz continuous, where

$$\mathbf{R}^N_+ = \{u = (u_i) \in \mathbf{R}^N \mid u_i > 0, \ 1 \leq i \leq N\}$$

denotes the *positive cone*. The solutions keeps the positivity $u = u(t) \in \mathbf{R}^N_+$, if this $f = f(u)$ is *quasi-positive*,

$$f_i(\hat{u}_i) \geq 0, \quad u \in \mathbf{R}^N_+, \ 1 \leq i \leq N, \tag{5.65}$$

where $f = (f_i)$ and $\hat{u}_i = (u_1, \cdots, u_{i-1}, 0, u_{i+1}, \cdots, u_N)$ for $u = (u_i) \in \mathbf{R}^N_+$.

The fundamental process of the chemical reaction

$$A + B \rightarrow AB \ (k), \quad AB \rightarrow A + B \ (\ell) \tag{5.66}$$

is formulated by the quadratic process using mass action law. Here we assume, first, that the collision of two particles causes a reaction with a definite rate. Second, the collision rate is proportional to the product of concentrations of particles. Hence k and ℓ in (5.66) denote the attachment and detachment reaction rates, and if $[X]$ stands for the concentration of X-molecules there arises

$$\frac{d}{dt}[A] = -k[A][B] + \ell[AB]$$
$$\frac{d}{dt}[B] = -k[A][B] + \ell[AB]$$
$$\frac{d}{dt}[AB] = k[A][B] - \ell[AB].$$

This system implies the total mass conservations of A and B molecules,

$$\frac{d}{dt}([A]+[AB])=0, \quad \frac{d}{dt}([B]+[AB])=0.$$

If B is always homo-dymerized as in BB, reaction law (5.66) implies

$$A+BB \rightarrow ABB\ (2k), \quad ABB \rightarrow A+BB\ (\ell)$$

and then it follows that

$$\begin{aligned}
\frac{d}{dt}[A] &= -2k[A][BB]+\ell[ABB] \\
\frac{d}{dt}[BB] &= -2k[A][BB]+\ell[ABB] \\
\frac{d}{dt}[ABB] &= 2k[A][BB]-\ell[ABB]
\end{aligned}$$

from the above-described mass action law. Similarly, if B is always hetero-dymerized as in AB, we obtain

$$A+AB \rightarrow ABA\ (k), \quad ABA \rightarrow A+AB\ (2\ell)$$

which results in

$$\begin{aligned}
\frac{d}{dt}[A] &= -k[A][AB]+2\ell[ABA] \\
\frac{d}{dt}[AB] &= -k[A][AB]+2\ell[ABA] \\
\frac{d}{dt}[ABA] &= k[A][AB]-2\ell[ABA].
\end{aligned}$$

To formulate the polymerization $A+A \rightarrow AA$, we take the case that this A is masked as in AB and define the reaction rates k_* and ℓ_* by

$$A+AB \rightarrow AAB\ (k_*), \quad AAB \rightarrow A+AB\ (\ell_*).$$

If N_A denotes the number of the particle A, then the number of the collision of two A-particles is $\frac{1}{2}N_A(N_A-1) \approx \frac{N_A^2}{2}$. Hence the above mass action law induces

$$A+A \rightarrow AA\ (k_*/2), \quad AA \rightarrow A+A\ (\ell_*).$$

Regarding the symmetry of the process, we reach

$$\frac{d}{dt}[A] = 2(-\frac{k_*}{2}[A]^2 + \ell_*[AA])$$
$$\frac{d}{dt}[AA] = \frac{k_*}{2}[A]^2 - \ell_*[AA].$$

We can construct a reaction network, piling these fundamental processes up. In some cases the application of dimension analysis is useful to detect unknown coefficients [33].

5.3.2 *Mathematics of Diffusion*

Partial differential equations are used to realize the averaged movement of particles interacting with the environment continuously distributed in space and time. The gradient of the scalar field is the fundamental notion to describe the action of individuals inside a medium.

As is confirmed in Chap. 4, if Ω_t denotes a domain in $\mathbf{R}^N$ moving with the velocity $v = v(\cdot, t)$, and $\rho = \rho(x, t)$ is the smooth scalar field varying in t, then Liouville's theorem guarantees

$$\frac{d}{dt}\int_{\Omega_t} \rho \, dx = \int_{\Omega_t} \rho_t + \nabla \cdot \rho v \, dx. \tag{5.67}$$

Mass conservation has two forms. The differential equation

$$\rho_t = -\nabla \cdot j \tag{5.68}$$

comes from the Euler coordinate, that is,

$$\frac{d}{dt}\int_{\omega} \rho \, dx = -\int_{\partial\omega} \nu \cdot j \, dS$$

with $\omega \subset \mathbf{R}^n$ standing for an arbitrary fixed domain, and hence $j = j(x, t)$ stands for the flux of this flow. The integral equation

$$\frac{d}{dt}\int_{\Omega_t} \rho \, dx = 0,$$

on the other hand, is treated by the Lagrange coordinate. Hence there arises

$$\rho_t + \nabla \cdot \rho v = 0 \tag{5.69}$$

by (5.67). Recall that (4.61),

$$j = \rho v, \tag{5.70}$$

is obtained by (5.68) and (5.69). If the mass density varies under this flow, (5.68) is replaced by

$$\rho_t + \nabla \cdot \rho v = \rho X, \tag{5.71}$$

where X stands for the changing rate of mass density. Equations (5.70) and (5.71) take the fundamental role in transport theory described in Sect. 4.4.1 in the previous chapter.

The diffusion equation then emerges as

$$\rho_t = \Delta \rho, \tag{5.72}$$

if the above j is proportional to $\nabla \rho$ and the diffusion coefficient D is put to be one:

$$j = -\nabla \rho.$$

Equation (5.72), however, arises also as a limit of the distribution function of a randomly moving particle subject to the master equation. The diffusion coefficient D is then associated with the mean jump length Δx and the mean waiting time τ through the Einstein formula

$$D = \frac{(\Delta x)^2}{2n\tau},$$

where n denotes the space dimension [83]. See Sect. 5.6.2 below.

The semilinear parabolic equation

$$\frac{\partial u}{\partial t} - \Delta u = f(u), \quad \left.\frac{\partial u}{\partial \nu}\right|_{\partial \Omega} = 0, \quad u|_{t=0} = u_0(x) \tag{5.73}$$

provides a model in many fields. The Neumann boundary condition in (5.73) can be replaced by the Dirichlet boundary condition and others. If $f = f(u)$ is locally Lipschitz continuous it is transformed into the integral equation in a function space, say $X = L^\infty(\Omega)$,

$$u(t) = e^{t\Delta} u_0 + \int_0^t e^{(t-s)\Delta} f(u(s))\, ds, \tag{5.74}$$

where $-\Delta$ is provided with the Neumann boundary condition. Equation (5.74) is solved by the fixed point argument in the space $C([0, T]; X)$. Since the existence time T is estimated below by $\|u_0\|_\infty$, it holds that

$$T < +\infty \quad \Rightarrow \quad \lim_{t \uparrow T} \|u(\cdot, t)\|_{\infty} = +\infty. \tag{5.75}$$

Thus the orbit of (5.73) generated by the smooth initial value u_0 is classified into compact global in time, blowup in finite time, blowup in infinite time indicated by

$$T = +\infty, \quad \limsup_{t \uparrow +\infty} \|u(\cdot, t)\|_{\infty} = +\infty.$$

Using the semigroup estimate

$$\|e^{t\Delta} u\|_r \leq C \max\{1, t^{-\frac{n}{2}(\frac{1}{q} - \frac{1}{r})}\} \|u\|_q, \quad 1 \leq q \leq r \leq \infty, \tag{5.76}$$

this theory is extended to $X = L^p(\Omega)$, $1 \leq p \leq \infty$ according to the growth rate of $f = f(u)$.

Similarly to the spatially homogeneous autonomous case, the reaction diffusion system

$$\tau_j \frac{\partial u_j}{dt} - d_j \Delta u_j = f_j(u) \text{ in } \Omega \times (0, T), \quad u_j\big|_{t=0} = u_{j0}(x), \quad \left.\frac{\partial u_j}{\partial \nu}\right|_{\partial\Omega} = 0$$

for $1 \leq j \leq N$ generates a local-in-time orbit, provided that $f = f(u)$ is locally Lipschitz continuous, where $\Omega \subset \mathbf{R}^n$ is a bounded domain with smooth boundary $\partial\Omega$ and $0 \leq u_{j0} = u_{j0}(x) \in C(\overline{\Omega})$. The solutions keeps the non-negativity

$$u = u(x, t) \in \overline{\mathbf{R}}_+^N,$$

if this $f = f(u)$ is quasi-positive, (5.65).

5.3.3 *Poisson Structure of Biological Model*

The reaction–diffusion equation in two species is formulated by

$$u_t = \varepsilon^2 \Delta u + f(u, v), \quad \tau v_t = D\Delta v + g(u, v) \qquad \text{in } \Omega \times (0, T)$$
$$\left.\frac{\partial}{\partial \nu}(u, v)\right|_{\partial\Omega} = 0, \qquad (u, v)|_{t=0} = (u_0(x), v_0(x)) > 0,$$

where the quasi-positivity (5.65) is reduced to

$$f(0, v) \geq 0, \ g(u, 0) \geq 0, \quad \forall u, v \geq 0.$$

It forms the prey–predator system if

$$f(u, v) = u(a - bv), \quad g(u, v) = v(-c + du),$$

where $a, b, c, d > 0$ are constants. The spatially homogeneous part

$$\frac{du}{dt} = f(u, v), \quad \tau\frac{dv}{dt} = g(u, v)$$

takes the unique stationary state $u_* = c/d$, $v_* = a/b$, and otherwise, the orbit becomes periodic in time. This fact follows because

$$\frac{du}{dt} = u(a - bv), \ \tau\frac{dv}{dt} = v(-c + du), \quad (u, v)|_{t=0} = (u_0, v_0) > 0 \tag{5.77}$$

has a first integral

$$a \log v - bv + c \log u - du = \text{constant}, \tag{5.78}$$

and (5.78) determines a Jordan curve in the first quadrant in uv space except for the stationary solution.

To derive (5.78), we use the growth rates

$$\xi = \log u, \quad \eta = \log v \tag{5.79}$$

to write (5.77) as

$$\frac{d\xi}{dt} = a - be^{\eta}, \quad \frac{d\eta}{dt} = -\tau^{-1}c + \tau^{-1}de^{\xi}. \tag{5.80}$$

The right-hand side of (5.80) does not depend on the variables in the left-hand side explicitly, and therefore, it takes the form of a Hamilton system

$$\xi_t = -H_\eta, \quad \eta_t = H_\xi \tag{5.81}$$

with the Hamiltonian

$$H(\xi, \eta) = -a\eta + be^{\eta} - \tau^{-1}c\xi + \tau^{-1}de^{\xi}. \tag{5.82}$$

Then we obtain (5.78) by

$$\frac{dH}{dt} = 0.$$

Under (5.79), the prey–predator system with diffusion

$$u_t = \varepsilon^2\Delta u + u(a - bv), \quad \tau v_t = D\Delta v + v(-c + du)$$

$$\frac{\partial}{\partial \nu}(u, v)\bigg|_{\partial\Omega} = 0, \quad (u, v)|_{t=0} = (u_0(x), v_0(x)) > 0 \tag{5.83}$$

takes the form

$$\xi_t = \varepsilon^2 e^{-\xi} \Delta e^{\xi} - H_\eta, \ \eta_t = \tau^{-1} D e^{-\eta} \Delta e^{\eta} + H_\xi, \ \frac{\partial}{\partial \nu}(\xi, \eta)\bigg|_{\partial\Omega} = 0.$$

Then we obtain

$$\begin{aligned}\frac{\partial}{\partial t} H(\xi, \eta) &= H_\xi \varepsilon^2 e^{-\xi} \Delta e^{\xi} + H_\eta \tau^{-1} D e^{-\eta} \Delta e^{\eta} \\ &= -\tau^{-1}(c\varepsilon^2 e^{-\xi} \Delta e^{\xi} + aDe^{-\eta} \Delta e^{\eta}) + \tau^{-1}(\varepsilon^2 d \Delta e^{\xi} + Db \Delta e^{\eta}),\end{aligned}$$

and therefore,

$$\frac{d}{dt} \int_\Omega H(\xi, \eta) = -\tau^{-1} \int_\Omega c\varepsilon^2 |\nabla \xi|^2 + aD|\nabla \eta|^2 \, dx \le 0. \tag{5.84}$$

Thus

$$\int_\Omega H(\xi, \eta)$$

gives the Lyapunov function to (5.83), and then (5.84) implies the a priori estimate

$$\int_\Omega e^{\xi} + e^{\eta} \, dx = \|u(\cdot, t)\|_1 + \|v(\cdot, t)\|_1 \le C. \tag{5.85}$$

We combine (5.85) and

$$u_t \le \varepsilon^2 \Delta u + au, \ u > 0, \quad \frac{\partial u}{\partial \nu}\bigg|_{\partial\Omega} = 0$$

derived from the first equation of (5.83). Repeating the semigroup estimates (5.76), we obtain

$$\limsup_{t \uparrow T} \|u(\cdot, t)\|_\infty < +\infty. \tag{5.86}$$

Then we write

$$\tau v_t = D\Delta v + v(-c + du), \quad \frac{\partial v}{\partial \nu}\bigg|_{\partial\Omega} = 0,$$

for the second equation of (5.83). By (5.86) and the linear theory, it holds that

$$\limsup_{t\uparrow T} \|v(\cdot,t)\|_\infty < +\infty$$

and hence $T = +\infty$. Then the parabolic regularity guarantees that the orbit

$$\mathcal{O} = \{(u(\cdot,t), v(\cdot,t)\}_{t\geq 0}$$

is compact in $C(\overline{\Omega})^2$.

From the theory of dynamical systems, the ω-limit set

$$\omega(u_0, v_0) = \{(u_\infty, v_\infty) \mid \exists t_k \uparrow +\infty, \lim_{k\to\infty} \|u(\cdot,t_k) - u_0, v(\cdot,t_k) - v_0\|_{C^2} = 0\}$$

induced by the initial value $(u_0, v_0) = (u, v)|_{t=0}$ is a non-empty, connected, and compact set. This $\omega(u_0, v_0)$ is invariant under the flow induced by (5.83), and so is the Lyapunov function

$$\int_\Omega H(\xi,\eta) = \int_\Omega -a\log v + bv - \tau^{-1}c\log u + \tau^{-1}du\ dx \tag{5.87}$$

on it.

Each $(u_\infty, v_\infty) \in \omega(u_0, v_0)$ is non-negative, but inequality (5.84) implies

$$\int_\Omega \log u_\infty + \log v_\infty\ dx > -\infty$$

by Fatou's lemma. Combining this inequality with (5.85), we see that the right-hand side of (5.87) converges absolutely as $t \uparrow +\infty$. It holds also that u_∞ and v_∞ are not identically zero, and hence the strong maximum principle of the parabolic equation is applicable. Thus, the solution $(\tilde{u}(\cdot,t), \tilde{v}(\cdot,t))$ to (5.83) with the initial value (u_∞, v_∞) is positive definite for $t > 0$, and the quantity defined by the right-hand side of (5.87),

$$L(\tilde{u}(\cdot,t), \tilde{v}(\cdot,t)) = \int_\Omega -a\log\tilde{v} + b\tilde{v} - \tau^{-1}c\log\tilde{u} + \tau^{-1}d\tilde{u}\ dx$$

is differentiable in $t > 0$. Putting $\tilde{\xi} = \log\tilde{u}$ and $\tilde{\eta} = \log\tilde{v}$, therefore, we obtain

$$\frac{d}{dt}\int_\Omega H(\tilde{\xi},\tilde{\eta}) = -\tau^{-1}\int_\Omega c\varepsilon^2|\nabla\tilde{\xi}|^2 + aD|\nabla\tilde{\eta}|^2\ dx = 0, \quad t > 0.$$

It holds that $\nabla\tilde{\xi}(\cdot,t) = \nabla\tilde{\eta}(\cdot,t) = 0$ for $t > 0$, and hence any $(u_\infty, v_\infty) \in \omega(u_0, v_0)$ is spatially homogeneous.

By the invariance of $\omega(u_0, v_0)$ under the flow (5.83) and the above-described classification of the spatially homogeneous orbit we obtain the following theorem.

Theorem 5.3 *It holds always that $T = +\infty$ in (5.83). Each initial value (u_0, v_0) admits a spatially homogeneous orbit $\mathscr{O} \subset \mathbf{R}^2_+$ such that*

$$\lim_{t\uparrow+\infty} dist_{C^2}((u(\cdot,t), v(\cdot,t)), \mathscr{O}) = 0.$$

If this $\mathscr{O}$ is not a singleton, therefore, there is $\ell > 0$ such that

$$\lim_{t\uparrow+\infty} \|u(\cdot, t+\ell) - u(\cdot,t), v(\cdot,t+\ell) - v(\cdot,t)\|_{C^2} = 0.$$

Inequality (5.85) is independent of D and hence so is (5.86). Then we obtain

$$\sup_{0\leq t<T} \|v(\cdot,t)\|_\infty \leq C_T$$

for each $T > 0$ by the second equation of (5.83), with $C_T > 0$ independent of D. The parabolic regularity now guarantees the convergence of $(u, v) = (u_D(\cdot,t), v_D(\cdot,t))$, the solution to (5.83), to $(U, V) = (U(\cdot,t), V(t))$, the solution to

$$U_t = \varepsilon^2 \Delta U + U(a - bV), \quad \left.\frac{\partial u}{\partial \nu}\right|_{\partial\Omega} = 0, \quad U|_{t=0} = u_0(x)$$
$$\tau \frac{dV}{dt} = \frac{V}{|\Omega|}\int_\Omega -c + dU\, dx, \quad V|_{t=0} = \overline{v}_0 \equiv \frac{1}{|\Omega|}\int_\Omega v_0(x) \tag{5.88}$$

locally uniformly in $\overline{\Omega} \times [0,\infty)$ as $D \uparrow +\infty$. System (5.88) is called the *shadow system* of (5.83).

The Hamiltonian $H(\xi, \eta)$ in (5.82) is independent of D, which induces

$$\frac{d}{dt}\int_\Omega H(\hat{\xi}, \hat{\eta})\, dx = -\tau^{-1}\int_\Omega c\varepsilon^2 |\nabla\hat{\xi}|^2\, dx \leq 0$$

for (5.84). The above argument now guarantees the convergence of $(U, V) = (U(\cdot,t), V(t))$ to a solution to the ODE part as $t \uparrow +\infty$. However, the periods and phases of these limits of (u, v) and (U, V) may be different. The shadow system (5.88) actually reproduces the transient profile of (5.83) for $D \gg 1$.

The spatially homogeneous part of the Gierer–Meinhardt system in morphogenesis is described by

$$\frac{du}{dt} = -u + \frac{u^p}{v^q}, \quad \tau\frac{dv}{dt} = -v + \frac{u^r}{v^s}. \tag{5.89}$$

It has the same property as that of the prey–predator system (5.83), if

$$\tau = \frac{s+1}{p-1}. \tag{5.90}$$

Thus there is a unique stationary solution and the other orbit is periodic-in-time in the first quadrant in the uv plane. There is actually a first integral of (5.89) for (5.90). Here we derive this fact from the Hamilton formalism as in (5.83).

For this purpose we rewrite (5.89) as

$$u^{-p}(u_t + u) = v^{-q}, \quad v^s(v_t + \tau^{-1}v) = \tau^{-1}u^r. \tag{5.91}$$

Then, regarding the first terms of the left-hand side of (5.91), we put

$$\xi = \frac{u^{-p+1}}{p-1}, \quad \eta = \frac{v^{s+1}}{s+1},$$

to obtain

$$\begin{aligned} \xi_t &= u^{-p+1} - v^{-q} = (p-1)\xi - \{(s+1)\eta\}^{-\frac{q}{s+1}} \\ \eta_t &= -\frac{v^{s+1}}{\tau} + \frac{u^r}{\tau} = -\tau^{-1}(s+1)\eta + \tau^{-1}\{(p-1)\xi\}^{-\frac{r}{p-1}}. \end{aligned} \tag{5.92}$$

The cross terms of the right-hand side of (5.92) are linear in ξ and η, which are contained in the Hamiltonian under the assumption of (5.90). Then, (5.92) is reduced to

$$\frac{d\xi}{dt} = H_\eta, \quad \frac{d\eta}{dt} = -H_\xi \tag{5.93}$$

for

$$\begin{aligned} H(\xi,\eta) &= (p-1)\xi\eta + \left(\frac{r}{p-1} - 1\right)^{-1} A(\xi) + \left(\frac{q}{s+1} - 1\right)^{-1} B(\eta) \\ A(\xi) &= \tau^{-1}(p-1)^{-\frac{r}{p-1}}\xi^{1-\frac{r}{p-1}}, \quad B(\eta) = (s+1)^{-\frac{q}{s+1}}\eta^{1-\frac{q}{s+1}}. \end{aligned} \tag{5.94}$$

Then a result similar to Theorem 5.3 is obtained under some more restrictions on the parameter. The other reduction of spatially inhomogeneous Gierer–Meinhardt system with diffusion is a parabolic equation with non-local term [36].

Even in the case that the Hamilton structure is not expected, some algebraic structures can make the system integrable. For example, a Lotka–Volterra system different from (5.77),

$$\tau_1\frac{du_1}{dt} = (u_2 - u_3)u_1, \ \tau_2\frac{du_2}{dt} = (u_3 - u_1)u_2, \ \tau_3\frac{du_3}{dt} = (u_1 - u_2)u_3 \tag{5.95}$$

can take the form

$$\tau_1 \frac{d\xi_1}{dt} = e^{\xi_2} - e^{\xi_3}, \ \tau_2 \frac{d\xi_2}{dt} = e^{\xi_3} - e^{\xi_1}, \ \tau_3 \frac{d\xi_3}{dt} = e^{\xi_1} - e^{\xi_2}, \tag{5.96}$$

in the variables $\xi_i = \log u_i$ for $i = 1, 2, 3$. If $\times$ denotes the outer product in $\mathbf{R}^3$, this (5.96) means

$$\mathcal{M} \frac{d\xi}{dt} = H(\xi) \times a, \tag{5.97}$$

where

$$\xi = \begin{pmatrix} \xi_1 \\ \xi_2 \\ \xi_3 \end{pmatrix}, \ a = \begin{pmatrix} 1 \\ 1 \\ 1 \end{pmatrix}, \ H(\xi) = \begin{pmatrix} e^{\xi_1} \\ e^{\xi_2} \\ e^{\xi_3} \end{pmatrix} \tag{5.98}$$

and $\mathcal{M} = \mathrm{diag}(\tau_1, \tau_2, \tau_3)$. This Poisson structure induces the conservation laws for (5.97),

$$\begin{aligned} 0 &= \frac{d}{dt}\mathcal{M}\xi \cdot a = \frac{d}{dt}(\tau_1\xi_1 + \tau_2\xi_2 + \tau_3\xi_3) \\ 0 &= \frac{d}{dt}\mathcal{M}\xi \cdot H(\xi) = \frac{d}{dt}(\tau_1 e^{\xi_1} + \tau_2 e^{\xi_2} + \tau_3 e^{\xi_3}). \end{aligned} \tag{5.99}$$

In the reaction–diffusion system

$$\begin{aligned} \tau_1 \frac{\partial u_1}{\partial t} &= d_1 \Delta u_1 + (u_2 - u_3)u_1, & \left.\frac{\partial u_1}{\partial \nu}\right|_{\partial\Omega} &= 0 \\ \tau_2 \frac{\partial u_2}{\partial t} &= d_2 \Delta u_2 + (u_3 - u_1)u_2, & \left.\frac{\partial u_2}{\partial \nu}\right|_{\partial\Omega} &= 0 \\ \tau_3 \frac{\partial u_3}{\partial t} &= d_3 \Delta u_3 + (u_1 - u_2)u_3, & \left.\frac{\partial u_3}{\partial \nu}\right|_{\partial\Omega} &= 0, \end{aligned} \tag{5.100}$$

the quantities

$$\int_\Omega \sum_i \tau_i \log u_i \, dx, \quad \int_\Omega \sum_i \tau_i u_i \, dx$$

associated with (5.99) form the total entropy and the total mass of the system, which reflects the structure of the solution set. Thus, a result analogous to Theorem 5.3 is valid for (5.100) if $n \leq 2$, where n denotes the space dimension.

5.3.4 Skew-Symmetric Lotka–Volterra Systems

This subsection is devoted to the spatially homogeneous Lotka–Volterra system

$$\tau_j \frac{dv_j}{dt} = (-e_j + \sum_{k=1}^{N} a_{jk} v_k) v_j, \quad v_j|_{t=0} = v_{j0} > 0, \quad 1 \leq j \leq N, \tag{5.101}$$

where $v_j = v_j(t) > 0$ stands for the population of the j-th species, $A = (a_{jk})$ is a square matrix of order N, and $\tau_j > 0$, $e_j \in \mathbf{R}$ are constants. Since the publications of Lotka [48] and Volterra [91], this model is used in several areas including ecology, economics, and chemistry, and has been studied extensively [13, 21, 22].

Any solution $v = (v_j(t))$ to (5.101) remains in the positive cone

$$\mathbf{R}^N_+ \equiv \{(v_j) \in \mathbf{R}^N \mid v_j > 0, 1 \leq \forall j \leq N\}$$

as far as it exists, because each coordinate plane in $\mathbf{R}^N$ denoted by $v_j = 0$ for some $j = 1, \cdots, N$ is an invariant set of (5.101), that is, quasi-positivity of the model. One of major questions in population dynamics, then, is whether some species go to extinction or not. Here, extinction means approaching zero as $t \to \infty$. Finding criteria for the long term coexistence of multiple species, referred to as *permanence* or uniform persistence, is thus an important issue [23, 30, 31, 39].

One approach to the question of permanence may be finding periodic-in-time orbits. The predator–prey system is a typical case that any orbit is periodic in time except for the unique equilibrium. This structure leads to asymptotic spatial homogenization corresponding to a partial differential equation provided with a diffusion term [62]. The present subsection formulates a class of systems provided with such a property via conserved quantities. This class is explicitly given by a set of algebraic conditions on $A = (a_{jk})$ and $e = (e_j)$. Then, if the system takes N components, we have $2N - 3$ and $2N - 1$ degrees of freedom without and with linear terms, respectively.

Thus we say that the Lotka–Volterra system (5.101) is provided with a property (P) if it satisfies the following conditions:

(P1) The set of stationary solutions (equilibria) E is the intersection of L, an affine space of co-dimension 2, and the positive cone $\mathbf{R}^N_+$, so that $E = L \cap \mathbf{R}^N_+$.

(P2) Any non-stationary solution is periodic in time, with the orbit $\mathcal{O} \cong S^1$ contractible to an equilibrium, that is, a point belonging to E, in $\mathbf{R}^N_+ \setminus E$.

(P3) Any two distinct orbits $\mathcal{O}_1, \mathcal{O}_2 \cong S^1$ do not link in $\mathbf{R}^N_+$. The set $\mathcal{O}_1 \cup \mathcal{O}_2$, in other words, shrinks into two distinct points through a continuous deformation.

Henceforth, we assume the following conditions for the matrix $A = (a_{jk})$:

(a1) A is irreducible.

(a2) A is skew-symmetric,

$$ {}^{t}A + A = 0. \tag{5.102} $$

($a3$) A has both positive and negative components in any row.

We recall that the square matrix A is called reducible if there is a permutation matrix P such that

$$ {}^{t}PAP = \begin{pmatrix} A_{11} & A_{12} \\ 0 & A_{22} \end{pmatrix}, $$

where A_{11} and A_{22} are non-trivial square matrices. Although this assumption is not essential in later arguments, since $A = (a_{jk})$ is skew-symmetric, system (5.101) does not have any non-trivial proper sub-systems under ($a3$).

Our results are divided into two cases, $e = 0$ and $e \neq 0$, where $e = (e_j)$.

Theorem 5.4 ([38]) *Let $A = (a_{jk})$ satisfy (a1), (a2), and (a3). Let $e = (e_j) = 0$, $N \geq 3$, and*

$$ a_{ij}a_{kl} + a_{il}a_{jk} - a_{ik}a_{jl} = 0, \quad \forall\, i, j, k, l \in \{1, \ldots, N\}. \tag{5.103} $$

Then, system (5.101) is provided with property (P).

Condition (5.103) is void if $N = 3$. Therefore, when $N = 3$, $e_1 = e_2 = e_3 = 0$, and

$$ A = \begin{pmatrix} 0 & c_3 & -c_2 \\ -c_3 & 0 & c_1 \\ c_2 & -c_1 & 0 \end{pmatrix}, \quad c_1, c_2, c_3 \text{ have the same sign}, \tag{5.104} $$

any non-stationary solution to (5.101) is periodic-in-time. This result is well-known. Here we confirm that the total set of orbits takes the topological feature described in ($P1$), ($P2$), and ($P3$). Hence the set of equilibria E is a half line in $\mathbf{R}^3_+$, any non-stationary orbit $\mathcal{O}$ is homeomorphic to S^1, surrounding E in $\mathbf{R}^3_+$, and any distinct two non-stationary orbits $\mathcal{O}_1$ and $\mathcal{O}_2$ do not link in $\mathbf{R}^3_+$.

Let $N \geq 4$. We first note that the set of skew-symmetric matrices, denoted by X, is identified with $\mathbf{R}^{N(N-1)/2}$. Second, if i, j, k, l are not distinct, Eq. (5.103) for $A = (a_{ij}) \in X$ is obvious. Third, condition (5.103) is invariant under the change of order on i, j, k, l for $A = (a_{jk}) \in X$. Therefore, system (5.103) for $N \geq 4$ is reduced to that of ${}_{N}C_4$ relations

$$ a_{ij}a_{kl} + a_{il}a_{jk} - a_{ik}a_{jl} = 0, \quad 1 \leq i < j < k < l \leq N. \tag{5.105} $$

The set of irreducible skew-symmetric matrices satisfying (5.105), denoted by Y, however, is identified with a dense set in $\mathbf{R}^{2N-3}$.

To see this remarkable property, let $a_{12} \neq 0$ without loss of generality. Then the entries $a_{34}, \ldots, a_{N-1N}$ of $A = (a_{jk})$ are represented by $a_{12}, \ldots, a_{2N}$ from the first

$(N-2)(N-3)/2$ relations of (5.105), that is,

$$a_{kl} = \frac{a_{1k}a_{2l} - a_{1l}a_{2k}}{a_{12}}, \quad 3 \leq k < l \leq N. \tag{5.106}$$

Then the rest of relations in (5.105) are satisfied for a_{kl} defined above. In other words, (5.105) for $i = 1$, $j = 2$, $3 \leq k < l \leq N$, assures all the other relations of (5.105), provided that $a_{12} \neq 0$. The proof of this reduction is elementary [38]. Since irreducibility is a generic property of matrices, the above Y is identified with a dense set in $\mathbf{R}^{2N-3}$.

We also note that if $\mathrm{sign}(a_{ij}) = (-1)^{i+j}$ is the case for $i = 1, 2$ and $i < j \leq N$, the matrix $A = (a_{jk}) \in Y$ satisfies the sign condition, *(a3)*. Thus, there are $(2N-3)$ degrees of freedom for system (5.101) with $e = (e_j) = 0$, provided with property (P).

Condition *(a3)* is associated with the persistence of (5.101). Let $A = (a_{jk}) \in Y$ and $e = (e_j) = 0$. Then, if property (P) does not arise to (5.101) any solution $v = (v_j(t))$ is global in time but satisfies

$$\lim_{t \uparrow +\infty} v_j(t) = 0, \quad 1 \leq \exists j \leq N \tag{5.107}$$

if it is non-stationary. In particular, any $v_* \in E$ is unstable. See Remark 5.1 below for the proof of this fact. A simple example of this case is $N = 3$ of (5.104), with the sign condition of c_1, c_2, c_3 violated.

By a result of [88], for $A = (a_{jk}) \in Y$ satisfying *(a3)*, there are at least $(2N-1)$ degrees of freedom of $A = (a_{jk})$, such that the ω-limit set of the solution to the associated reaction diffusion system of (5.101) is a Jordan curve or singleton contained in $\mathbf{R}^N_+$.

We turn to the case $e = (e_j) \neq 0$. When $N = 2$, $e_1 \cdot e_2 < 0$, and $a_{12} \cdot a_{21} < 0$, we call (5.101) the prey–predator system. It is classical that this system is provided with property (P). Thus, in this case there is a unique equilibrium $v_* \in \mathbf{R}^2_+$, and the total set of orbits denoted by $\mathscr{F}$ is composed of curves homeomorphic to S^1 and the equilibrium v_*. A general form of (5.101) provided with this property is given by the following theorem.

Theorem 5.5 (**[38]**) *Let $N \geq 3$ and $A = (a_{jk})$ satisfy (a1), (a2), and (a3). Let $e = (e_j)$ have both positive and negative components in (5.101). Assume, furthermore, that*

$$a_{jk}e_i - a_{ik}e_j + a_{ij}e_k = 0, \quad \forall i, j, k \in \{1, \ldots, N\}. \tag{5.108}$$

Then, system (5.101) is provided with property (P).

Condition (5.108) concerning $A = (a_{jk}) \in X$ and $e = (e_j) \in \mathbf{R}^N$ has the following reduction similar to (5.103). First, this condition is obvious if i, j, k are not distinct. Second, this relation is invariant under the change of order on i, j, k.

Hence (5.108) is reduced to the system of ${}_N C_3$ relations

$$a_{jk}e_i - a_{ik}e_j + a_{ij}e_k = 0, \quad 1 \le i < j < k \le N. \tag{5.109}$$

Finally, the degree of freedom of $\{a_{ij}, e_k\}$ satisfying (5.109) is $(2N-1)$.

To see the last fact, we assume $a_{1N} \cdot e_1 \neq 0$ without loss of generality, and define $a_{23}, \ldots, a_{2N}$ by $a_{12}, \ldots, a_{1N}$ and $e_1, \ldots, e_N$, using the first $(N-1)(N-2)/2$ relations of (5.109),

$$a_{jk} = \frac{a_{1k}e_j - a_{1j}e_k}{e_1}, \quad 2 \le j < k \le N. \tag{5.110}$$

Then the other relations of (5.109) are satisfied automatically. The proof of this property is given in [38].

Let $\hat{X}$ be the set of systems (5.101) with (5.102). This time, since both $A = (a_{ij})$ and $e = (e_k)$ are regarded as parameters, the set $\hat{X}$ is identified with $\mathbf{R}^{N(N+1)/2}$. Let $\hat{Y}$ be the set of $\{a_{ij}, e_k\}$ satisfying $A = (a_{ij}) \in X$ and (5.108). By the above reduction, this $\hat{Y}$ is identified with $\mathbf{R}^{2N-1}$.

To detect a class of $\{a_{ij}, e_k\} \in \hat{Y}$ with $A = (a_{ij})$ satisfying (*a3*), let

$$\text{sign}(a_{1j}) = (-1)^j, \ 2 \le j \le N, \quad \text{sign}(e_k) = (-1)^k, \ 3 \le k \le N,$$

and $e_1 > 0 > e_2$. Then it holds that

$$\text{sign}(a_{2j}) = (-1)^{j+1}, \quad 3 \le j \le N,$$

and therefore, $A = (a_{ij})$ actually satisfies (*a3*). Thus, there are $(2N-1)$ degrees of freedom of $\{a_{ij}, e_k\}$ such that (5.101) is provided with property (*P*).

The proof of Theorems 5.4 and 5.5 relies on the facts on the sign of the entries of the skew-symmetric matrix A.

Lemma 5.1 *If $N \ge 4$, one of the following cases arises for $A = (a_{jk})$ satisfying (a2) and (a3):*
Case 1. *There exist distinct $i, j, k \in \{1, 2, \cdots, N\}$ such that $a_{ij} > 0$, $a_{ik} < 0$, $a_{jk} > 0$.*
Case 2. *There exist distinct $i, j, k, l \in \{1, 2, \cdots, N\}$ such that $a_{ij} > 0$, $a_{ik} < 0$, $a_{il} = 0$, $a_{jk} = 0$, $a_{jl} > 0$, $a_{kl} < 0$.*

Lemma 5.2 *Under the assumptions of Theorem 5.5, either the first or the second case of Lemma 5.1 arises with $e_i \neq 0$, that is:*
Case 1. *There exist distinct $i, j, k \in \{1, 2, \cdots, N\}$ such that $a_{ij} > 0$, $a_{ik} < 0$, $a_{jk} > 0$, and $e_i \neq 0$.*
Case 2. *There exist distinct $i, j, k, l \in \{1, 2, \cdots, N\}$ such that $a_{ij} > 0$, $a_{ik} < 0$, $a_{il} = 0$, $a_{jk} = 0$, $a_{jl} > 0$, $a_{kl} < 0$, and $e_i \neq 0$.*

If the matrix A is given, we can easily check the above sign conditions of its entries. The point here, however, is that these conditions are the consequences of the assumptions of the theorems. An elementary but complicated proof is given in [38].

5.3.5 *The Case Without Linear Term*

To show Theorem 5.4, let $N \geq 3$ and $e = (e_j) = 0$ in (5.101):

$$\tau_j \frac{dv_j}{dt} = \sum_{k=1}^{N} a_{jk} v_k v_j, \quad v_j\big|_{t=0} = v_{j0} > 0, \quad 1 \leq j \leq N. \tag{5.111}$$

First, equality (5.102) implies the total mass conservation

$$\frac{dM}{dt} = 0, \tag{5.112}$$

where $M = \tau \cdot v$. Second, by $v = (v_j) > 0$ and (5.112) we obtain $T = +\infty$ with

$$\sup_{t\geq 0} \max_{1\leq j\leq N} v_j(t) < +\infty. \tag{5.113}$$

Here and henceforth, $v = (v_j) > 0$ indicates $v_j > 0$ for any $1 \leq j \leq N$. Let

$$\mathscr{O} = \{v(t)\}_{t\geq 0} \subset \mathbf{R}_+^N$$

be the corresponding orbit.

Boltzmann's H-function or entropy, on the other hand, is defined under the assumption that

$$\exists b \in \mathbf{R}^N \setminus \{0\}, \quad {}^t\widetilde{A}b = 0, \quad \widetilde{A} = (\tau_j^{-1} a_{jk}), \tag{5.114}$$

that is,

$$H = b \cdot \log v = \sum_{j=1}^{N} b_j \log v_j,$$

where $b = (b_j)$ and

$$\log v = (\log v_j), \quad v = (v_j) \in \mathbf{R}_+^N.$$

Thus we obtain

$$\frac{dH}{dt} = \sum_{j=1}^{N} b_j v_j^{-1} \frac{dv_j}{dt} = \sum_{j,k=1}^{N} b_j \tau_j^{-1} a_{jk} v_k = 0$$

by (5.114) for the solution $v = (v_j(t))$ to (5.101). This H actually induces an increasing quantity for the reaction–diffusion system [50, 62, 88].

Summarizing, we reach the following lemma.

Lemma 5.3 *Assume (5.102) and the existence of linearly independent r vectors*

$$b^i \in Ker\, {}^t\widetilde{A}, \quad 1 \leq i \leq r,$$

where $r \geq 1$ *and* $\tilde{A} = (\tau_i^{-1} a_{ij})$. *Put*

$$\beta_0 = \tau \cdot v_0, \quad \beta_i = b^i \cdot \log v_0, \ 1 \leq i \leq r,$$

where $v_0 = (v_{j0}) > 0$. *Then it holds that* $\mathcal{O} \subset \mathcal{O}_*$, *where*

$$\mathcal{O}_* = \{v \in \mathbf{R}_+^N \mid \tau \cdot v = \beta_0, \ b^i \cdot \log v = \beta_i, \ 1 \leq i \leq r\}. \tag{5.115}$$

Now we show the following lemma.

Lemma 5.4 *If* $r = N - 2$ *and* $b^i > 0$ *for some* $i \in \{1, \cdots, r\}$ *in Lemma 5.3, the set* $\mathcal{O}_*$ *defined by (5.115) is a Jordan curve or singleton.*

Proof Under the transformation $\xi = \log v \in \mathbf{R}^N$, the set $\mathcal{O}_*$ is homeomorphic to

$$\widetilde{\mathcal{O}}_* = \{\xi \in \mathbf{R}^N \mid \tau \cdot e^{\xi} = \beta_0, \ b^i \cdot \xi = \beta_i, \ 1 \leq i \leq N-2\}. \tag{5.116}$$

Let

$$M = \{\xi \in \mathbf{R}^N \mid \tau \cdot e^{\xi} \leq \beta_0\}.$$

This set is strictly convex, and $\widetilde{\mathcal{O}}_*$ is the intersection of its boundary

$$\partial M = \{\xi \in \mathbf{R}^N \mid \tau \cdot e^{\xi} = \beta_0\}$$

and linearly independent $(N-2)$ hyperplanes

$$P_i = \{\xi \in \mathbf{R}^N \mid b^i \cdot \xi = \beta_i\}, \quad 1 \leq i \leq N-2.$$

Hence $\widetilde{\mathcal{O}}_* \subset \mathbf{R}^N$ is a non-intersecting curve or singleton.

Now we show the boundedness of $\widetilde{\mathcal{O}}_*$. First, there arises

$$\sup_{1 \leq j \leq N} \xi_j \leq C, \quad \forall \xi = (\xi_j) \in M.$$

Second, since $b^i > 0$ for some $1 \le i \le N-2$ we have

$$\inf_{1\le j\le N} \xi_j \ge -C, \quad \forall \xi = (\xi_j) \in M \cap P_i.$$

Therefore, $\widetilde{\mathscr{O}}_* \subset \mathbf{R}^N$ is bounded. Then we see that $\mathscr{O}_*$ is a Jordan curve or singleton by the inverse transformation $v_j = e^{\xi_j}$, $1 \le j \le N$. □

Remark 5.1 If $\widetilde{\mathscr{O}}_*$ is not bounded, the set $\mathscr{O}_*$ is an open curve in $\mathbf{R}^N_+$. Then any non-stationary solution $v = (v_j(t)) > 0$ satisfies (5.107) by (5.112), because any stationary solution does not lie on $\mathscr{O}_*$ in this case. See Lemma 5.6 below.

Now we show the following lemma, using Lemma 5.1.

Lemma 5.5 *The requirements of Lemma 5.4 are fulfilled under the assumption of Theorem 5.4.*

Proof If $N = 3$ the assumption implies (5.104). Since the relation

$$b = \begin{pmatrix} b_1 \\ b_2 \\ b_3) \end{pmatrix} \in \operatorname{Ker} {}^t\widetilde{A}, \quad b_1 b_2 b_3 \neq 0$$

means

$$\frac{b_3}{b_2} = \frac{c_3}{c_2} \cdot \frac{\tau_3}{\tau_2}, \quad \frac{b_1}{b_3} = \frac{c_1}{c_3} \cdot \frac{\tau_1}{\tau_3}, \quad \frac{b_2}{b_1} = \frac{c_2}{c_1} \cdot \frac{\tau_2}{\tau_1},$$

we can take $0 < b = {}^t(b_1, b_2, b_3) \in \operatorname{Ker} {}^t\widetilde{A}$.

Let $N \ge 4$. First, we examine the condition for the existence of linearly independent $(N-2)$ vectors, denoted by

$$b^i = (b^i_j) \in \operatorname{Ker} {}^t\tilde{A}, \quad 1 \le i \le N-2.$$

This condition is equivalent to the linear independence of $B^i = (B^i_j)$, $1 \le i \le N-2$, for $B^i_j = \tau_j^{-1} b^i_j$, which means rank $A = 2$.

To confirm this property under the assumption of Theorem 5.4, let $A = [\mathbf{a_1} \cdots \mathbf{a_N}]$. Since A is irreducible, we have $\mathbf{a_k} \neq 0$ for $1 \le k \le N$. Also, since A is skew-symmetric, the vectors $\{\mathbf{a_k}, \mathbf{a_l}\}$, $k \neq l$, are linearly independent if $a_{kl} \neq 0$. For such k, l, which actually exists, condition (5.103) implies

$$\mathbf{a_i} - \frac{a_{il}}{a_{kl}} \mathbf{a_k} - \frac{a_{ik}}{a_{lk}} \mathbf{a_l} = 0, \quad \forall i \neq k, l, \tag{5.117}$$

and hence rank $A = 2$. Thus we have actually linearly independent $(N-2)$ vectors in $\operatorname{Ker} {}^t\tilde{A}$.

Now we shall show the existence of $0 < b \in \text{Ker}\, {}^t\widetilde{A}$, or equivalently, that of $0 < B \in \text{Ker}\, {}^tA = \text{Ker}\, A$. In fact, in the first case of Lemma 5.1 we have $i, j, k \in \{1, 2, \cdots, N\}$ such that $a_{ij} > 0$, $a_{ik} < 0$, and $a_{jk} > 0$. Here, the condition $B = {}^t(B_1, \cdots, B_N) \in \text{Ker}\, {}^tA$ is equivalent to

$$B_k = -\frac{1}{a_{jk}} \sum_{n \neq j,k} a_{jn} B_n, \quad B_j = \frac{1}{a_{jk}} \sum_{n \neq j,k} a_{kn} B_n.$$

Consequently, we obtain

$$\text{Ker}\, {}^tA = \langle \mathbf{c_n} \mid n \neq j, k \rangle$$

for $\mathbf{c_n} = {}^t(c_n^1, \cdots, c_n^N) \in \mathbf{R}^N$ defined by

$$c_n^m = \begin{cases} a_{jk}, & m = n \\ a_{kn}, & m = j \\ -a_{jn}, & m = k \\ 0, & \text{otherwise.} \end{cases}$$

Hence $\mathbf{c_i}$ is a vector of which i-th, j-th, and k-th components are positive and the others are zero. Recalling $c_n^n = a_{jk} > 0$, we obtain

$$0 < \tilde{B} \equiv \sum_{n \neq i,j,k} \mathbf{c_n} + s\mathbf{c_i} \in \text{Ker}\, {}^tA, \quad s \gg 1.$$

We turn to the second case of Lemma 5.1. For such a set (i, j, k, l) of four (distinct) elements in $\{1, \cdots, N\}$, the condition $B = {}^t(B_1, \cdots, B_N) \in \text{Ker}\, {}^tA$ means

$$B_j = -\frac{1}{a_{ij}} \sum_{n \neq i,j} a_{in} B_n, \quad B_i = \frac{1}{a_{ij}} \sum_{n \neq i,j} a_{jn} B_n.$$

Consequently, it holds that $\text{Ker}\, {}^tA = \langle \mathbf{c_n} \mid n \neq i, j \rangle$ for $\mathbf{c_n} = {}^t(c_n^1, \cdots, c_n^N) \in \mathbf{R}^N$ defined by

$$c_n^m = \begin{cases} a_{ij}, & m = n \\ a_{jn}, & m = i \\ -a_{in}, & m = j \\ 0, & \text{otherwise.} \end{cases}$$

We observe that $\mathbf{c_k}$ is a vector of which the j-th and k-th components are positive and the others are zero. Also, $\mathbf{c_l}$ is a vector of which the i-th and l-th components are positive and the others are zero. Recalling $c_n^n = a_{ij} > 0$, we obtain

$$0 < b \equiv \sum_{n \neq i,j,k,l} \mathbf{c_n} + s(\mathbf{c_k} + \mathbf{c_l}) \in \mathrm{Ker}\,{}^t A, \quad s \gg 1.$$

The proof is complete. □

Remark 5.2 We see that in (5.117), the pair $(c_1, c_2) = (a_{il}/a_{kl}, a_{ik}/a_{lk})$ is the unique choice for

$$\mathbf{a_i} - c_1 \mathbf{a_k} - c_2 \mathbf{a_l} = 0$$

to hold, regarding the l-th and the k-th components on the left-hand side. Consequently, the condition

$$a_{ij}a_{kl} + a_{il}a_{jk} - a_{ik}a_{jl} = 0, \quad \forall i, j, k, l \in \{1, \ldots, N\} \text{ such that } a_{kl} \neq 0, \tag{5.118}$$

slightly weaker than (5.103), is sufficient for rank $A = 2$ to hold, if $A = [\mathbf{a_1} \cdots \mathbf{a_N}]$ is irreducible and skew-symmetric. Condition (5.118) is also necessary for rank $A = 2$, if, furthermore, $\{\mathbf{a_k}, \mathbf{a_l}\}$ are linearly dependent whenever $a_{kl} = 0$. By this reason it is natural that the solution set to (5.103) has $(2N-3)$ degrees of freedom, because two vectors $\mathbf{a_k}$, $\mathbf{a_l}$, $k \neq l$, can be free in $A = [\mathbf{a_1} \cdots \mathbf{a_N}]$ if rank $A = 2$.

The following lemma excludes either hetero-clinic or homo-clinic orbits of (5.101), $e = (e_j) = 0$, when $A = (a_{ij})$ satisfies $(a1)$, $(a2)$, and (5.103).

Lemma 5.6 *Let $A \in X$ satisfy rank $A = 2$. Let $b^i \in \mathrm{Ker}\,{}^t\tilde{A}$, $1 \leq i \leq N-2$, be linearly independent $(N-2)$ vectors. Then the property $\mathscr{O}_* \cap \mathrm{Ker}\ A \neq \emptyset$ arises only if $\sharp\mathscr{O}_* = 1$, where $\mathscr{O}_*$ is the set defined by (5.115) with $r = N-2$.*

Proof We use the sets $\widetilde{\mathscr{O}_*}$, M, and P_i defined in the proof of Lemma 5.4. Let $\mathscr{M} = \partial M$ and $\tau e^{\xi} = (\tau_j e^{\xi_j})$ for $\tau = (\tau_j)$ and $\xi = (\xi_j)$. First, since

$$\mathscr{M} = \{\xi \in \mathbf{R}^N \mid \tau \cdot \mathrm{e}^{\xi} = \beta_0\},$$

it holds that

$$T_{\xi}\mathscr{M} = \left\langle \tau e^{\xi} \right\rangle^{\perp} \equiv \{\eta \in \mathbf{R}^N \mid \tau \mathrm{e}^{\xi} \cdot \eta = 0\}, \quad \xi \in \mathscr{M}.$$

Second, the space $\mathrm{Ker}\,{}^t\widetilde{A}$ is generated by b^i, $1 \leq i \leq N-2$, and therefore, it holds that

$$\mathrm{Ker}\ A = \mathrm{Ker}\,{}^t A = \left\langle \tau^{-1} b^i \mid 1 \leq i \leq N-2 \right\rangle$$

for $\tau^{-1}b^i = (\tau_j^{-1}b^i_j)$, where $b^i = (b^i_j)$ and $\langle \tau^{-1}b^i \mid 1 \leq i \leq N-2 \rangle$ denotes the vector space generated by $\tau^{-1}b^i$, $1 \leq i \leq N-2$.

If $v \in \mathscr{O}_* \cap \mathrm{Ker}\ A$, there is $\xi \in \widetilde{\mathscr{O}_*}$ such that $v = e^{\xi} \in \mathrm{Ker}\ A$. Then, we obtain

$$\tau e^{\xi} \in \langle b^i \mid 1 \leq i \leq N-2 \rangle. \tag{5.119}$$

Since

$$P \equiv \{\zeta \in \mathbf{R}^N \mid b^i \cdot \zeta = \beta_i,\ 1 \leq i \leq N-2\} = \{\xi\} + \bigcap_{i=1}^{N-2} \langle b^i \rangle^{\perp}, \tag{5.120}$$

condition (5.119) implies

$$\tau e^{\xi} \in (P - \{\xi\})^{\perp}.$$

Therefore, it holds that $P \subset T_{\xi}\mathscr{M} + \{\xi\}$, and we end up with

$$\widetilde{\mathscr{O}}_* = P \cap \mathscr{M} \subset T_{\xi}\mathscr{M} + \{\xi\}. \tag{5.121}$$

Equation (5.121) implies $\widetilde{\mathscr{O}}_* = \{\xi\}$ because $\mathscr{M}$ is the boundary of the strictly convex set

$$M = \{\xi \in \mathbf{R}^N \mid \tau \cdot \mathrm{e}^{\xi} \leq \beta_0\}.$$

□

Proof *(Theorem 5.4)* By Lemmas 5.3–5.5, it holds that $\mathscr{O} \subset \mathscr{O}_*$, where $\mathscr{O}_*$ is a Jordan curve or singleton. If $\mathscr{O}_*$ is a Jordan curve, we have $\mathscr{O} = \mathscr{O}_*$ by Lemma 5.6, which means that the solution $v = v(t)$ is periodic in time. If $\mathscr{O}_*$ is singleton, on the other hand, the orbit $\mathscr{O}$ is composed of an equilibrium. Hence the solution is stationary.

The set of equilibria to (5.111) is given by $E = \mathbf{R}^N_+ \cap \mathrm{Ker}\, A$. This set is non-empty because of the existence of $0 < B \in \mathrm{Ker}\, A$. Since rank $A = 2$, it is the intersection of the vector space Ker A of co-dimension two and the positive cone $\mathbf{R}^N_+$.

We recall that $\mathscr{M}$ is the boundary of the strictly convex space

$$M = \{\xi \in \mathbf{R}^N \mid \tau \cdot e^{\xi} \leq \beta_0\}$$

determined by β_0, that is, $\mathscr{M} = \partial M$. As $\beta_0 > 0$ varies, $\mathscr{F} = \{\mathscr{M}\}$ forms a nested family of hypersurfaces of co-dimension one, covering $\mathbf{R}^N \setminus \{0\}$. Furthermore, each orbit $\mathscr{O}$ of (5.111) is homeomorphic to the intersection of some $\mathscr{M} \in \mathscr{F}$ and an affine space P of dimension two defined by (5.120). Since M with $\partial M = \mathscr{M}$ is strictly convex and $\mathscr{O}$ is either a Jordan curve or singleton composed of an equilibrium, it is contractible in $\mathbf{R}^N_+ \setminus E$ to an equilibrium point as P moves in $\mathbf{R}^N$. Finally, if two distinct Jordan orbits $\mathscr{O}_1, \mathscr{O}_2 \subset \mathbf{R}^N_+$ are realized on the same hypersurface $\mathscr{M} = \partial M \in \mathscr{F}$, they do not link because of the strict convexity of

M. Therefore, the same property arises to any distinct two non-stationary orbits in $\mathbf{R}^N_+$. □

5.3.6 *The Case with Linear Term*

Here we study the case $N \geq 3$ of (5.101). To begin with, we have

$$\frac{d}{dt}\tau \cdot v = -e \cdot v \leq a\tau \cdot v, \quad a = \frac{\max_j(-e_j)}{\min_j \tau_j} \tag{5.122}$$

and hence $\tau \cdot v \leq (\tau \cdot v_0)e^{at}$, $v_0 = (v_{0j})$. Then it holds that $T = +\infty$ with

$$0 < v_j(t) \leq \tau_j^{-1}(\tau \cdot v_0)e^{at}.$$

Transform (5.101) to

$$\tau_j \frac{d\xi_j}{dt} = -e_j + \sum_{k=1}^{N} a_{jk}e^{\xi_k}, \quad \xi_j\big|_{t=0} = \xi_{j0} \in \mathbf{R}, \quad 1 \leq j \leq N, \tag{5.123}$$

using $\xi_j = \log v_j$, and let $\tau = (\tau_j) > 0$, $e = (e_j)$, $\xi = (\xi_j)$, and $\xi_0 = (\xi_{j0})$. Henceforth, without loss of generality we assume

$$(i, j, k) = (1, j, N), \quad j \in \{2, \ldots, N-1\}$$

and

$$(i, j, k, l) = (1, j, N, l), \quad j, l \in \{2, \ldots, N-1\}, \ j \neq l,$$

in the first and the second cases of Lemma 5.2, respectively. Thus we have $e_1 \neq 0$ and either

$$a_{1j} > 0, \ a_{1N} < 0, \ a_{jN} > 0, \quad 2 \leq \exists j \leq N-2 \tag{5.124}$$

or

$$\begin{aligned} &a_{1j} > 0, \ a_{1l} = 0, \ a_{1N} < 0, \ a_{jN} = 0, \ a_{jl} > 0, \ a_{N\ell} < 0 \\ &2 \leq \exists j \neq \exists l \leq N-2. \end{aligned} \tag{5.125}$$

In both cases it holds that $a_{1N} < 0$.

Lemma 5.7 *Under the above assumptions, define* $V_j(\xi)$, $1 \leq j \leq N-1$, *by*

$$V_1(\xi) = a_{1N}\sum_{k=1}^{N}\tau_k e^{\xi_k} + e_N\tau_1\xi_1 - e_1\tau_N\xi_N$$

$$V_j(\xi) = a_{jN}\tau_1\xi_1 - a_{1N}\tau_j\xi_j + a_{1j}\tau_N\xi_N, \quad 2 \leq j \leq N-1. \tag{5.126}$$

Then $V_j(\xi)$, $1 \leq j \leq N-1$, are invariant with respect to (5.123). Hence it holds that

$$\widetilde{\mathcal{O}} \equiv \{\xi(t) \mid t \geq 0\} \subset \widetilde{\mathcal{O}}_* \equiv \{\xi \in \mathbf{R}^N \mid V_j(\xi) = c_j, \ 1 \leq j \leq N-1\} \tag{5.127}$$

for $c_j = V_j(\xi_0)$. Moreover, $\widetilde{\mathcal{O}}_*$ is a Jordan curve or singleton.

Proof We shall confirm

$$\frac{d}{dt}V_j(\xi(t)) = 0, \quad 1 \leq j \leq N-1 \tag{5.128}$$

for the solution $\xi = (\xi_j(t))$ to (5.123). First, (5.102) implies

$$\frac{d}{dt}\sum_{k=1}^{N}\tau_k e^{\xi_k} = -\sum_{k=1}^{N}e_k e^{\xi_k},$$

while

$$e_N\tau_1\frac{d\xi_1}{dt} - e_1\tau_N\frac{d\xi_N}{dt} = \sum_{k=1}^{N}(e_N a_{1k} - e_1 a_{Nk})e^{\xi_k}$$

is obvious. Then, (5.112) for $j = 1$ is a consequence of (5.108) with $(i, j, k) = (1, N, k)$:

$$\frac{dV_1}{dt} = \sum_{k=1}^{N}(-a_{1N}e_k + e_N a_{1k} - e_1 a_{Nk})e^{\xi_k} = 0.$$

Here we use (5.108) for $(i, j, k) = (1, j, N)$, $(i, j, k) = (1, j, k)$, and $(i, j, k) = (1, N, k)$ to obtain

$$a_{jN}e_1 - a_{1N}e_j + a_{1j}e_N = 0$$
$$a_{jk}e_1 - a_{1k}e_j + a_{1j}e_k = 0$$
$$a_{Nk}e_1 - a_{1k}e_N + a_{1N}e_k = 0.$$

Then it follows that

$$(a_{jN}a_{1k} + a_{1j}a_{Nk} - a_{1N}a_{jk})e_1 = 0$$

and hence (5.103) for $(i, j, k, l) = (1, j, k, N)$, that is,

$$a_{jN}a_{1k} + a_{1j}a_{Nk} - a_{1N}a_{jk} = 0, \quad 2 \le j \ne k \le N - 1. \tag{5.129}$$

From (5.126) and (5.129) we obtain

$$\begin{aligned}\frac{dV_j}{dt} &= \sum_{k=1}^{N}(a_{jN}a_{1k} - a_{1N}a_{jk} + a_{1j}a_{Nk})e^{\xi_k} + \left(-a_{jN}e_1 + a_{1N}e_j - a_{1j}e_N\right) \\ &= -a_{jN}e_1 + a_{1N}e_j - a_{1j}e_N, \quad 2 \le j \le N - 1,\end{aligned} \tag{5.130}$$

which implies (5.112) for $2 \le j \le N - 1$ by (5.108) with $(i, j, k) = (1, j, N)$.

We show that $\widetilde{\mathcal{O}}_*$ is a Jordan curve or singleton. In fact, it is the intersection of the boundary of

$$M = \{\xi \in \mathbf{R}^N \mid V_1(\xi) \ge c_1\},$$

which is strictly convex by $a_{1N} < 0$, and the $N-2$ linearly independent hyperplanes

$$P_j = \{\xi \in \mathbf{R}^N \mid V_j(\xi) = c_j\}, \quad 2 \le j \le N - 1.$$

Thus we have only to show that $\widetilde{\mathcal{O}}_*$ defined by (5.127) is bounded. In fact, since $a_{1N} < 0$ implies

$$c_1 = V_1(\xi) \le a_{1N}(\tau_N e^{\xi_N} + \tau_1 e^{\xi_1}) + e_N\tau_1\xi_1 - e_N\tau_N\xi_N,$$

the components ξ_1, ξ_N are bounded above on $\widetilde{\mathcal{O}}_*$. From these bounds the first equality of (5.126) implies

$$-C \le a_{1N}\sum_{k=2}^{N-1}\tau_k e^{\xi_k}$$

with a constant $C > 0$, Therefore, the components $\xi_2, \cdots, \xi_{N-1}$ are also bounded above on $\widetilde{\mathcal{O}}_*$ by $a_{1N} < 0$.

Thus we have only to derive the boundedness below of $\xi_1, \cdots, \xi_N$ for $\xi = (\xi_1, \cdots, \xi_N) \in \widetilde{\mathcal{O}}_*$. For this purpose, we distinguish two cases of Lemma 5.2.

Case 1

We have (5.124). Hence it holds that $a_{1j} > 0$, $a_{1N} < 0$, $a_{jN} > 0$, and $e_1 \ne 0$ for some $2 \le j \le N - 1$. Then, V_j is a linear function of (ξ_1, ξ_j, ξ_N), the coefficients of which are positive. Since $\widetilde{\mathcal{O}}_*$ is bounded above, the coordinates ξ_1, ξ_j, ξ_N are bounded below on $\widetilde{\mathcal{O}}_*$. From this property and $V_i(\xi) = c_i$, so is ξ_i for $i \ne 1, j, N$. Hence $\widetilde{\mathcal{O}}_*$ is bounded below.

Case 2
We have (5.125). Hence there are $j \neq l$ satisfying $2 \leq j, l \leq N-1$, $a_{1j} > 0$, $a_{1N} < 0$, $a_{1l} = 0$, $a_{jN} = 0$, $a_{jl} > 0$, $a_{Nl} < 0$, and $e_1 \neq 0$. Then, V_j and V_l are linear functions of (ξ_j, ξ_N) and (ξ_1, ξ_l), respectively, and all of their coefficients are positive. Hence the coordinates $\xi_1, \xi_j, \xi_N, \xi_l$ are bounded below on $\widetilde{\mathscr{O}}_*$ because they are bounded above. From this property and $V_i(\xi) = c_i$, so is ξ_i for $i \neq 1, j, N, l$. Hence $\widetilde{\mathscr{O}}_*$ is bounded below. □

Lemma 5.8 *Under the assumptions of Theorem 5.5, the equilibrium set of (5.101) is given by* $E = \mathbf{R}^N_+ \cap E_*$ *for*

$$E_* = \{v_*\} + \langle \mathbf{b_j} \mid 2 \leq j \leq N-1 \rangle,$$

where $v_* = \frac{1}{a_{1N}}{}^t(-e_N, 0, \cdots, 0, e_1)$, $\mathbf{b_j} = (b_j^m)$, $2 \leq j \leq N-1$, *and*

$$b_j^m = \begin{cases} a_{jN}, & m = 1 \\ a_{1j}, & m = N \\ -a_{1N}, & m = j \\ 0, & \textit{otherwise}. \end{cases}$$

Proof We have $v = (v_k) \in E$ if and only if $v > 0$ and

$$I_j \equiv -e_j + \sum_{k=1}^{N} a_{jk} v_k = 0, \quad 1 \leq j \leq N. \tag{5.131}$$

Here it should be noted that the definition of I_j, $1 \leq j \leq N$, and (5.129) imply

$$a_{1j} I_N - a_{Nj} I_1 = a_{1N} I_j, \quad 2 \leq j \leq N-1$$

by (5.108) with $(i, j, k) = (1, j, N)$.

Since $a_{1N} \neq 0$, therefore, the system of equations (5.131) is reduced to $I_1 = I_N = 0$, that is,

$$v_1 = \frac{1}{a_{1N}} \left(-e_N - \sum_{k=2}^{N-1} a_{kN} v_k \right), \quad v_N = \frac{1}{a_{1N}} \left(e_1 - \sum_{k=2}^{N-1} a_{1k} v_k \right). \tag{5.132}$$

Then the result follows from (5.132). □

Let

$$\widetilde{E} = \{\xi \in \mathbf{R}^N \mid \xi = \log v, \ v \in E\}.$$

Lemma 5.9 *Under the above assumptions, the property* $\widetilde{\mathscr{O}}_* \cap \widetilde{E} \neq \emptyset$ *arises only if* $\sharp \widetilde{\mathscr{O}}_* = 1$.

Proof The hypersurface

$$\mathcal{M} = \{\xi \in \mathbf{R}^N \mid V_1(\xi) = c_1\}$$

is the boundary of a strictly convex set $M = \{\xi \in \mathbf{R}^N \mid V_1(\xi) \geq c_1\}$, and

$$P = \{\xi \in \mathbf{R}^N \mid V_j(\xi) = c_j, \ 2 \leq j \leq N-1\}$$

is a two-dimensional affine space.

Given $\xi \in \mathcal{M}$, we have

$$T_\xi \mathcal{M} = \{\eta = (\eta_j) \in \mathbf{R}^N \mid (a_{1N}\tau_1 e^{\xi_1} + e_N\tau_1)\eta_1 + \sum_{i=2}^{N-1} a_{1N}\tau_i e^{\xi_i}\eta_i + (a_{1N}\tau_N e^{\xi_N} - e_1\tau_N)\eta_N = 0\}.$$

In addition, for $\xi \in \widetilde{\mathcal{O}}_* = \mathcal{M} \cap P$, the plane P is parallel to $T_\xi \mathcal{M}$ if and only if

$$a_{1N}\tau_1 e^{\xi_1} + e_N\tau_1 = \sum_{i=2}^{N-1} k_i a_{iN}\tau_1$$
$$a_{1N}\tau_j e^{\xi_j} = -k_j a_{1N}\tau_j, \quad 2 \leq j \leq N-1$$
$$a_{1N}\tau_N e^{\xi_N} - e_1\tau_N = \sum_{i=2}^{N-1} k_i a_{1i}\tau_N$$

for some $k_2, \cdots, k_{N-1} \in \mathbf{R}$, which is equivalent to

$${}^t(e^{\xi_1}, e^{\xi_2}, e^{\xi_3}, \cdots, e^{\xi_{N-1}}, e^{\xi_N}) \in E,$$

or $\xi \in \widetilde{E}$. Therefore, if $\xi \in \widetilde{\mathcal{O}}_* \cap \widetilde{E}$, the plane P is parallel to $T_\xi \mathcal{M}$. Then $\widetilde{\mathcal{O}}_* = \{\xi\}$ follows because M is strictly convex. □

Proof *(Theorem 5.5)* To begin with, we show the existence of $v = (v_k) > 0$ such that (5.132) holds. In fact, in the case (5.124) we have $j \in \{2, \cdots, N-1\}$ such that $a_{1j} > 0$, $a_{jN} > 0$. Then the requirement $v = (v_k) > 0$ holds in (5.132) for $v_j > 0$ sufficiently large by $a_{1N} < 0$. In the case of (5.125), on the other hand, we have $j \in \{2, \cdots, N-1\}$ such that $a_{1j} > 0 = a_{jN}$. Here we find $i \in \{2, \cdots, N-1\}$ with $a_{iN} > 0$, using $(a1)$, $(a3)$, and $a_{1N} < 0$. Then $v = (v_k) > 0$ holds in (5.132) for sufficiently large $v_j > 0$ and $v_i > 0$.

Since (5.131) is equivalent to (5.132), we have $E \neq \emptyset$. Then the rest of the proof is similar to that of Theorem 5.4, using Lemmas 5.7–5.9. □

Remark 5.3 Persistence of (5.101) means that any solution $v = (v_j(t))$ exists global in time and it holds that $\liminf_{t\uparrow+\infty} v_j(t) > 0$ for any j. From the above method we can show the following theorem.

Theorem 5.6 *System (5.101), $N \geq 3$, exhibits persistence if ${}^t A + A \leq 0$ and there exists $i, j, l \in \{1, \cdots, N\}$ such that $a_{jl}, a_{ij} > 0 > a_{il}$ and*

$$a_{ij}a_{kl} + a_{il}a_{jk} - a_{ik}a_{jl} \leq 0, \quad a_{jk}e_i - a_{ik}e_j + a_{ij}e_k \leq 0, \qquad 1 \leq k \leq N. \tag{5.133}$$

Proof Since ${}^t A + A \leq 0$ we have

$$\frac{d}{dt}\tau \cdot v \leq -e \cdot v$$

for (5.112). Then it follows that $T = +\infty$ from $v = (v_k(t)) > 0$. We assume $i = 1$, $l = N$ without loss of generality, and take $V = V_j(\xi)$ for $2 \leq j \leq N-1$ in (5.126), recalling $\xi = (\log v_j)$. By the first equality of (5.130) and the assumption (5.133), it holds that

$$\frac{dV}{dt} \geq 0.$$

Here, the set

$$\hat{\mathscr{O}}_* = \{\xi \in \mathbf{R}^N \mid V(\xi) \geq V(\xi_0)\}$$

is bounded below for $\xi_0 = (\log v_{j0})$, because each coefficient of the linear function $V_j(\xi)$ is positive from the assumption of $a_{jl}, a_{ij} > 0 > a_{il}$ with $i = 1, l = N$. □

5.4 Reaction–Diffusion Systems with Mass Dissipation

5.4.1 Global Existence via Total Mass Control

Fundamental reactions are mostly modeled by quadratic nonlinearity. The general theory assures uniformly bounded, global-in-time orbit for the reaction–diffusion system of any dimension in this case, under the presence of total mass dissipation [16, 17]. Existence of entropy, then, induces asymptotic homogenization in space variables [62]. We begin this section with the Lotka–Volterra system with diffusion term.

Recall that the Lotka–Volterra system is concerned with population dynamics in ecology. For the case of N species with diffusion, it is described by

$$\tau_j u_{jt} = d_j \Delta u_j + (-e_j + \sum_k a_{jk} u_k) u_j \qquad \text{in } \Omega \times (0, T)$$

$$\left.\frac{\partial u_j}{\partial \nu}\right|_{\partial\Omega} = 0, \quad u_j\big|_{t=0} = u_{j0}(x) \geq 0, \qquad 1 \leq j \leq N, \tag{5.134}$$

where $\Omega \subset \mathbf{R}^n$ denotes a bounded domain with smooth boundary $\partial\Omega$, ν is the outer unit normal vector, $\tau_j, d_j > 0$, $e_j, a_{jk} \in \mathbf{R}$ are constants, and $u_0 = (u_{j0})$ is a smooth initial value. There arises a local-in-time solution, positive in $\overline{\Omega} \times (0, T)$ unless $u_{j0} \equiv 0$. System (5.100) is the case of $N = 3$, $e_1 = e_2 = e_3 = 0$, and

$$A \equiv (a_{jk}) = \begin{pmatrix} 0 & 1 & -1 \\ -1 & 0 & 1 \\ 1 & -1 & 0 \end{pmatrix}.$$

Let $A = (a_{jk})_{j:1\downarrow N; k:1\to N} \in M_N(\mathbf{R})$, $u = (u_j)_{j:1\downarrow N} \in \mathbf{R}^N$, $e = (e_j)_{j:1\downarrow N} \in \mathbf{R}^N$, $\tau = (\tau_j)_{j:1\downarrow N} \in \mathbf{R}^N$, $d = (d_j)_{j:1\downarrow N} \in \mathbf{R}^N$, and

$$|u| = (\sum_{i=1}^{N} u_j^2)^{1/2}. \tag{5.135}$$

We assume

$${}^tA + A \leq 0, \quad e \geq 0. \tag{5.136}$$

Here, ${}^tA + A \leq 0$ and $e = (e_j)_{j:1\downarrow N} \geq 0$ mean the non-positive definiteness of the symmetric matrix ${}^tA + A$ and $e_j \geq 0$ for any $1 \leq j \leq N$, respectively.

To examine the thermodynamical structure derived from (5.136), we take the spatially homogeneous part of (5.134),

$$\tau_j \frac{dv_j}{dt} = (-e_j + \sum_k a_{jk} v_k) v_j, \quad v_j\big|_{t=0} = v_{j0} > 0. \tag{5.137}$$

It holds that $v_j(t) > 0$ as far as the solution exists. Then

$$M = \tau \cdot v = \sum_j \tau_j v_j$$

stands for the total mass of the species, where $v = (v_j)_{j:1\downarrow N}$. We obtain

$$\frac{dM}{dt} = \tau \cdot \frac{dv}{dt} = \sum_j (-e_j + \sum_k a_{jk} v_k) v_j \leq -e \cdot v \leq 0 \tag{5.138}$$

by (5.136). Since M is bounded, it holds that $T = +\infty$ in (5.137), and furthermore,

$$0 \le v_j(t) \le C, \quad t \ge 0,\ 1 \le j \le N. \tag{5.139}$$

We assume, without loss of generality,

$$e_1, \cdots, e_\ell = 0 < e_{\ell+1}, \cdots, e_N \tag{5.140}$$

for some $\ell = 0, \cdots, N$, and put

$$c_0 = \frac{\min_{j=\ell+1,\cdots,N} e_j}{\max_{j=\ell+1,N} \tau_j} > 0, \quad \tilde{M} = \sum_{j=\ell+1}^{N} \tau_j v_j$$

if $\ell \le N-1$.

Inequality (5.138) implies

$$M(t) \le M(0) - \int_0^t e \cdot v(t')dt'$$

and hence

$$\tilde{M}(t) \le M(0) - c_0 \int_0^t \tilde{M}(t')dt', \quad t \ge 0. \tag{5.141}$$

This (5.141) means

$$f'(t) \le M(0) - c_0 f(t), \quad t \ge 0$$

for $f(t) = \displaystyle\int_0^t \tilde{M}(t')dt'$. Therefore, it follows that $f(t) \le C$, that is,

$$\int_0^\infty \tilde{M}(t')dt' < +\infty.$$

Then there is $t_k \uparrow +\infty$ such that $\tilde{M}(t_k) \to 0$. We have, furthermore,

$$\frac{d\tilde{M}}{dt} \le \sum_{j=\ell+1}^{N} \sum_{k=1}^{N} a_{jk} v_k v_j \le C\tilde{M}$$

by (5.139), which implies

$$\tilde{M}(t) \le \tilde{M}(t_k) + C \int_{t_k}^t \tilde{M}(t')dt', \quad t \ge t_k. \tag{5.142}$$

Making $t \uparrow +\infty$ and then $k \to \infty$ in (5.142), we obtain

$$\limsup_{t\uparrow+\infty} \tilde{M}(t) \le 0, \tag{5.143}$$

which means

$$\lim_{t\uparrow+\infty} v_j(t) = 0, \quad \ell+1 \le j \le N.$$

Here, the following structure of (5.137) assures increasing entropy, that is, the existence of $0 < b \in \mathbf{R}^\ell$ such that

$$^t\tilde{A}b \ge 0, \quad \tilde{A} = (\tau_j^{-1} a_{jk})_{j:1\downarrow\ell;\ k:1\to\ell}. \tag{5.144}$$

In fact, then,

$$H = \sum_{j=1}^{\ell} b_j \log v_j, \quad b = (b_j)_{j:1\downarrow\ell}$$

gives the entropy in (5.140) as

$$\frac{dH}{dt} = \sum_{j=1}^{\ell} b_j v_j^{-1} \frac{dv_j}{dt} = \sum_{j=1}^{\ell}\sum_{k=1}^{N} b_j \tau_j^{-1} a_{jk} v_k \ge 0. \tag{5.145}$$

Inequality (5.145) implies $\liminf_{t\uparrow+\infty} v_j(t) > 0$ for $1 \le j \le \ell$. The ω-limit set

$$\omega(v_0) = \{v_* = (v_{*j})_{j:1\downarrow N} \in \mathbf{R}^N \mid \exists t_k \uparrow +\infty \text{ s.t. } v(t_k) \to v_*\}$$

defined for the initial value $v_0 = (v_{0j})_{j:1\downarrow N}$ of (5.137) is non-empty, connected, and compact. Furthermore, it holds that

$$v_{\ell+1}^* = \cdots = v_N^* = 0 < v_1^*, \cdots, v_\ell^*, \quad b \cdot (\log v_j^*)_{j:1\downarrow\ell} \ge H(0)$$

for any $v_* = (v_{*j})_{j:1\downarrow N}$.

These structures of the system of ordinary differential equations induce those for (5.134) together with the *homogenization* in space [88]. Also, condition (5.144) will be clarified according to the algebraic structure of $A = (a_{jk})$ in the next chapter.

Here we make a preliminary study of (5.134) with (5.136). The fundamental property is

$$\frac{dM}{dt} = \int_\Omega \tau \cdot u_t \, dx \le -\int_\Omega e \cdot u \, dx \le 0 \tag{5.146}$$

for

$$M = \int_\Omega \tau \cdot u \, dx. \tag{5.147}$$

The a priori estimate

$$\sup_{t \geq 0} \|u(\cdot, t)\|_1 \leq C \tag{5.148}$$

follows from (5.146), and then there arises

$$\sup_{t \geq 0} \|u(\cdot, t)\|_\infty \leq C \tag{5.149}$$

and hence $T = +\infty$ in the case of $n = 1$. This is shown by the proof of the following theorem, global-in- time ε regularity for the case of $n = 2$.

Theorem 5.7 *If $n = 2$ and (5.136) in (5.134), there is $\varepsilon_0 > 0$ such that*

$$\lambda \equiv \int_\Omega \tau \cdot u_0 \, dx = \sum_j \int_\Omega \tau_j u_{j0} \, dx < \varepsilon_0 \tag{5.150}$$

implies

$$T = +\infty, \quad \sup_{t \geq 0} \|u(\cdot, t)\|_\infty < +\infty. \tag{5.151}$$

Proof First, we have (5.148) by (5.146). Second, it holds that

$$\tau_j u_{jt} \leq d_j \Delta u_j + a_0 \sum_k u_k u_j, \ u_j \geq 0 \ \left. \frac{\partial u_j}{\partial \nu} \right|_{\partial \Omega} = 0, \quad 1 \leq j \leq N \tag{5.152}$$

for $a_0 = \max_{j,k}(a_{jk})_+$ and $(a_{jk})_+ = \max\{a_{jk}, 0\}$. Then, Young's inequality implies

$$\begin{aligned} \frac{\tau_j}{2} \frac{d}{dt} \|u_j\|_2^2 + d_j \|\nabla u_j\|_2^2 &\leq a_0 \sum_k \int_\Omega u_k u_j^2 \, dx \\ &\leq \frac{2N a_0}{3} \|u_j\|_3^3 + \frac{a_0}{3} \sum_k \|u_k\|_3^3. \end{aligned} \tag{5.153}$$

For the right-hand side of (5.153) we apply the Gagliardo–Nirenberg inequality valid for $n = 2$,

$$\|w\|_3^3 \leq C \|w\|_{H^1}^2 \|w\|_1, \quad w \in H^1(\Omega), \tag{5.154}$$

and also the Poincaré–Wirtinger inequality

$$\mu_2 \|w - \overline{w}\|_2^2 \leq \|\nabla w\|_2^2, \quad \overline{w} = \frac{1}{|\Omega|} \int_\Omega w \, dx \tag{5.155}$$

for $w = u_j$. From (5.148) it holds that

$$\frac{\tau_j}{2} \frac{d}{dt} \|u_j\|_2^2 + d_j \|\nabla u_j\|_2^2 \leq C\lambda \|\nabla u\|_2^2 + C,$$

and hence

$$\frac{1}{2} \frac{d}{dt} \sum_j \tau_j \|u_j\|_2^2 + d_0 \|\nabla u\|_2^2 \leq NC\lambda \|\nabla u\|_2^2 + NC \tag{5.156}$$

for $d_0 = \min_j d_j$, where $|\nabla u|^2 = \sum_j |\nabla u_j|^2$.

If

$$\lambda < d_0/(2NC_2) \equiv \varepsilon_0$$

in (5.156) we have

$$\frac{d}{dt} \sum_j \tau_j \|u_j\|_2^2 + d_0 \|\nabla u\|_2^2 \leq C$$

and then it holds that

$$\frac{d}{dt} [u]_2^2 + \mu_2 \tau_0^{-1} d_0 [u]_2^2 \leq C$$

for (5.155), (5.148), $\tau_0 = \max_j \tau_j$, and

$$[u]_2^2 = \sum_j \tau_j \|u_j\|_2^2.$$

We thus obtain

$$\|u(\cdot, t)\|_2 \leq C. \tag{5.157}$$

Based on (5.157), we apply the comparison theorem. Regarding

$$\tau_j u_{jt} \leq d_j \Delta u_j + \frac{a_0}{2} (|u|^2 + N u_j^2), \ u_j \geq 0, \ \left. \frac{\partial u_j}{\partial \nu} \right|_{\partial \Omega} = 0, \quad 1 \leq j \leq N,$$

we take

$$g_j = \mu u_j + \frac{a_0}{2}(|u|^2 + Nu_j^2)$$

for $\mu \gg 1$. Let

$$\overline{u}_j = \overline{u}_j(x,t), \quad 1 \leq j \leq N,$$

be the solution to

$$\tau_j \overline{u}_{jt} = d_j \Delta \overline{u}_j - \mu \overline{u}_j + g_j(\cdot, t)$$
$$\left.\frac{\partial \overline{u}_j}{\partial \nu}\right|_{\partial\Omega} = 0, \quad \overline{u}_j\big|_{t=0} = u_{j0}(x) \geq 0. \tag{5.158}$$

Then we obtain

$$0 \leq u_j \leq \overline{u}_j \tag{5.159}$$

by the comparison theorem. Here we have

$$\|g_j(\cdot, t)\|_1 \leq C$$

by (5.157). Now we use the semigroup estimate (5.76). Then,

$$L_j = -d_j \Delta + \mu$$

satisfies

$$\|e^{-tL_j}\|_{L^q \to L^r} \leq C(q,r)e^{-\mu t} \max\{1, t^{-\frac{n}{2}\left(\frac{1}{q}-\frac{1}{r}\right)}\}, \quad 1 \leq q \leq r \leq \infty. \tag{5.160}$$

We thus apply (5.160) to

$$\overline{u}_j(\cdot, t) = e^{-t\tau_j^{-1}L_j} u_{j0} + \int_0^t e^{-(t-s)\tau_j^{-1}L_j} g_j(\cdot, s) ds \tag{5.161}$$

derived from (5.158), repeatedly. In fact, assuming

$$\|u(\cdot, t)\|_q \leq C, \quad \exists q \geq 2, \tag{5.162}$$

we obtain

$$\|g_j(\cdot, t)\|_{q/2} \leq C.$$

Then, we use (5.160)–(5.161) to get

$$\|\overline{u}_j(\cdot,t)\|_r \le C, \quad 1 \le j \le N \tag{5.163}$$

for $\frac{n}{2}(\frac{2}{q} - \frac{1}{r}) < 1$, satisfying $r \in [q, \infty]$. Then, inequality (5.159) implies

$$\|u_j(\cdot,t)\|_r \le C \tag{5.164}$$

for $1 \le r < \frac{nq}{2(n-q)_+}$ and $r = \infty$ if $q > n$ and $q \le n$, respectively.

If $n = 2$, we have (5.163) for $1 \le r < \infty$ by (5.162), $q = 2$. Then we apply (5.162) for $q > 2 = n$, to obtain (5.164). The proof is complete. □

Let $\ell = N$ in (5.140) for the moment. This assumption means $e_j = 0$ for $j = 1, \cdots, N$, and then, (5.134) admits the self-similar transformation

$$u_j^{\mu}(x,t) = \mu^2 u_j(\mu x, \mu^2 t), \quad \mu > 0, \ 1 \le j \le N. \tag{5.165}$$

In this case, the quantity λ in (5.150) is invariant in $\mu > 0$, if and only if $n = 2$. We find the thermodynamical structure of total mass and entropy, and also that local mass is controlled in time. Blowup analysis combined with the notion of a weak solution [62, 88] removes the assumption (5.150) in Theorem 5.7 [88].

Remark 5.4 The above proof guarantees (5.151) for $n = 1$ in (5.134) with (5.136). This case was studied by [50]. If the nonlinearity is quadratic growth in (5.134), furthermore, condition (5.157) implies (5.149) for $n \le 3$ [88]. Under the total mass control with quadratic growth rate of the nonlinearity, however, $n = 2$ is the critical dimension for global-in-time existence and pre-compactness of the orbit, as we have confirmed above; there is actually blowup in finite time of the solution to the Smolchowski–Poisson equation described in the next section for $n = 2$, where total mass control is achieved under the quadratic growth rate.

Remark 5.5 Pointwise estimates of the solution valid for the reaction–diffusion systems, however, extend this well-posedness to higher dimensions, with aid of fine parabolic estimates and duality argument. It has been established so far that quadratic nonlinearity guarantees uniform boundedness of the solution and hence global in time existence of the compact orbit [16, 17]. Below, we describe several approaches to confirm the distinguished role of the comparison property of the solution [81].

5.4.2 L^2 Control in Space and Time

We take the reaction–diffusion system

$$\tau_j \frac{\partial u_j}{\partial t} - d_j \Delta u_j = f_j(u) \quad \text{in } \Omega \times (0,T)$$

$$\left.\frac{\partial u_j}{\partial \nu}\right|_{\partial\Omega} = 0, \quad u_j\big|_{t=0} = u_{j0}(x) > 0, \tag{5.166}$$

where $\Omega \subset \mathbf{R}^n$ is a bounded domain with smooth boundary $\partial\Omega$, $\tau = (\tau_j) > 0$, $d = (d_j) > 0$, $1 \leq j \leq N$, are constants, and $u_{j0} = u_{j0}(x) \in C(\overline{\Omega})$. The nonlinearity

$$f_j = f_j(u) : \mathbf{R}^N \to \mathbf{R}, \quad u = (u_j),$$

is locally Lipschitz continuous to guarantee local-in-time well-posedness of the system. Let $T \in (0, +\infty]$ be the maximal existence time of the classical solution. Assuming quasi-positivity,

$$f_j(u_1, \cdots, u_{j-1}, 0, u_{j+1}, \cdots, u_N) \geq 0, \quad u = (u_j) \geq 0, \quad 1 \leq j \leq N, \tag{5.167}$$

we reach $u = (u_j(x,t)) > 0$ by the theory of *invariant region* [15, 71]. Henceforth, we assume these conditions together with mass dissipation

$$\sum_j f_j(u) \leq 0, \quad u = (u_j) \geq 0. \tag{5.168}$$

Inequality (5.168) ensures the total mass control

$$\|\tau \cdot u(t)\|_1 \leq \|\tau \cdot u_0\|_1, \quad 0 \leq t < T.$$

Under the quadratic growth,

$$|\nabla f_j(u)| \leq C(1 + |u|), \ 1 \leq j \leq N, \tag{5.169}$$

then, (5.166) takes the same scaling invariance as that of the Smoluchowski–Poisson equation described in the next section, which makes $n = 2$ the critical dimension for the uniform boundedness of the solution

$$T = +\infty, \ \|u(t)\|_\infty \leq C \tag{5.170}$$

to hold. We actually have ε-regularity on the initial total mass $M = \|\tau \cdot u_0\|_1$ for (5.170) if $n = 2$ in the previous subsection. In fact, we have (5.170) for $n = 1$, $n = 2$ by [50, 88], taking the Lotka–Volterra system as a model. Note that (5.168) is reduced to the equality by inducing a suspended system.

More precisely, let

$$\tilde{f}_{N+1}(u) = -\sum_{i=1}^N f_j(u)$$

and take $\tau_{N+1},\ d_{N+1} > 0$ arbitrarily, to introduce the extended system for $1 \leq j \leq N+1$,

$$\tau_j \frac{\partial \tilde{u}_j}{\partial t} - d_j \Delta \tilde{u}_j = \tilde{f}_j(\tilde{u}) \quad \text{in } \Omega \times (0,T)$$
$$\left.\frac{\partial \tilde{u}_j}{\partial \nu}\right|_{\partial\Omega} = 0, \quad \tilde{u}_j\big|_{t=0} = \tilde{u}_{j0}(x) > 0, \tag{5.171}$$

where

$$\tilde{u}_{j0} = u_{j0},\ 1 \leq j \leq N, \quad \tilde{u}_{N+1\,0} = 1, \quad \tilde{f}_j(\tilde{u}) = f_j(u), \quad 1 \leq j \leq N.$$

This $\tilde{f} = (\tilde{f}_j(\tilde{u}))$ is then locally Lipschitz continuous, quasi-positive, quadratic growth indicated by

$$|\nabla \tilde{f}_j(\tilde{u})| \leq C(1 + |\tilde{u}|), \quad \tilde{u} = (\tilde{u}_j) \geq 0,$$

and mass conservative,

$$\sum_{j=1}^{N+1} \tilde{f}_j(\tilde{u}) = 0, \quad \tilde{u} = (\tilde{u}_j) \geq 0.$$

Unique existence of the classical solution to (5.166), finally, assures $\tilde{u}_j(\cdot,t) = u_j(\cdot,t)$, $1 \leq j \leq N$, for $t \in [0,T)$. The question (5.170) for (5.166) is thus reduced to that for the mass conservative case in (5.168),

$$\sum_j f_j(u) = 0, \quad u = (u_j) \geq 0, \tag{5.172}$$

although additional assumptions on $f = (f_j(u))$, if requested later, have to be examined individually.

Some arguments are valid to induce (5.170) for $n = 2$, different from the case of the Smoluchowski–Poisson equation [80], where (5.170) is violated. One of them is the L^2 estimate in space and time derived from a duality argument [60]:

Lemma 5.10 *If $0 \leq u = (u_j(x,t))$ is smooth in $\overline{\Omega} \times [0,T)$ and satisfies*

$$\frac{\partial}{\partial t}(\tau \cdot u) - \Delta(d \cdot u) \leq 0, \ \textit{in } \Omega \times (0,T), \quad \left.\frac{\partial}{\partial \nu}(d \cdot u)\right|_{\partial\Omega} \leq 0, \tag{5.173}$$

then it follows that

$$\|u\|_{L^2(Q_T)} \leq CT^{1/2}\|u_0\|_2, \quad u|_{t=0} = u_0. \tag{5.174}$$

Proof Let $u_0 = u|_{t=0}$. By (5.112) we have

$$\tau \cdot u(\cdot, t) - \tau \cdot u_0 \leq \int_0^t \Delta(d \cdot u(\cdot, s))\, ds,$$

and hence

$$\begin{aligned} &(\tau \cdot u(\cdot, t), d \cdot u(\cdot, t)) - (\tau \cdot u_0, d \cdot u(\cdot, t)) \\ &\leq -(\nabla d \cdot u(\cdot, t), \nabla \int_0^t d \cdot u(\cdot, s)\, ds) \\ &= -\frac{1}{2}\frac{d}{dt}\|\nabla \int_0^t d \cdot u(\cdot, s)\, ds\|_2^2, \end{aligned} \tag{5.175}$$

where (,) denotes the L^2-inner product. Integration of (5.175) over $(0, T)$ implies

$$\begin{aligned} \int_0^T (\tau \cdot u(\cdot, t), d \cdot u(\cdot, t))\, dt &\leq \|\tau \cdot u_0\|_2 \cdot \int_0^T \|d \cdot u(\cdot, t)\|_2\, dt \\ &\leq T^{1/2}\|\tau \cdot u_0\|_2 \cdot \left(\int_0^T \|d \cdot u(\cdot, t)\|_2^2\, dt\right)^{1/2}, \end{aligned}$$

and hence (5.174) holds by $u = (u_j(\cdot, t)) \geq 0$. □

Remark 5.6 In [88] it is shown that inequalities (5.169) and (5.170) imply a slightly stronger estimate on the growth rate of $f = (f_j(u))$,

$$\sum_j f_j(u) \log u_j \leq C(1 + |u|^2),$$

which implies $T = +\infty$ by the argument in the previous subsection. For the exclusion of blowup in infinite time, a blowup argument is applicable, to reach (5.170).

We may use the following theorem to ensure (5.169) for $n \leq 2$.

Theorem 5.8 ([62]) *There exists a weak solution global in time, with the orbit* $\mathcal{O} = \{u(t)\}$ *pre-compact in* $L^1(\Omega)$ *for any* n.

The proof of this theorem is based on (5.174) and the property of compactness concerning

$$\frac{\partial w}{\partial t} = \Delta w + H \text{ in } Q_T, \quad \left.\frac{\partial w}{\partial \nu}\right|_{\partial\Omega} = 0, \quad w|_{t=0} = w_0(x) \tag{5.176}$$

for

$$w_0 \in L^1(\Omega), \quad H \in L^1(Q_T). \tag{5.177}$$

We rely on the following lemma [4] for the well-posedness of (5.176)–(5.177).

Lemma 5.11 (Baras–Pierre) *Given $w_0 \in L^1(\Omega)$ and $H \in L^1(Q_T)$, let*

$$w = w(\cdot, t) \in L^\infty(0, T; L^1(\Omega)) \cap L^1_{loc}(0, T; W^{1,1}(\Omega))$$

be the solution to (5.176). More precisely, for any $\varphi \in W^{1,\infty}(\Omega)$ it holds that

$$\frac{d}{dt} \int_\Omega w\varphi \, dx + \int_\Omega \nabla w \cdot \nabla \varphi \, dx = \int_\Omega H\varphi \, dx$$

in the sense of distributions with respect to t and

$$\lim_{t \downarrow 0} w(\cdot, t) = w_0$$

in the sense of measures on $\overline{\Omega}$. Then it follows that

$$w(\cdot, t) = e^{t\Delta} w_0 + \int_0^t e^{(t-s)\Delta} H(\cdot, s) \, ds, \quad 0 \le t \le T. \tag{5.178}$$

In particular, $w \in C([0, T], L^1(\Omega))$ and this solution exists uniquely.

The existence of the solution in the above lemma can be proven by the duality argument [4]. Then the following fact is obtained.

Lemma 5.12 *The mapping*

$$\mathscr{F} : (w_0, H) \in L^1(\Omega) \times L^1(Q_T) \mapsto w \in C([0, T], L^1(\Omega))$$

is continuous, where $w = w(x, t)$ is the solution to (5.176) in Lemma 5.11.

Under these preparations, we show a key fact for the proof of Theorem 5.8.

Lemma 5.13 *The mapping*

$$\mathscr{F} : (w_0, H) \in L^1(\Omega) \times L^1(Q_T) \mapsto w \in L^1(Q_T)$$

is compact, where $w = w(x, t)$ is the solution to (5.176) in Lemma 5.11. In other words, the image of each bounded set in $L^1(\Omega) \times L^1(Q_T)$ by $\mathscr{F}$ is relatively compact in $L^1(Q_T)$.

Proof By (5.178), the dual operator

$$\mathscr{F}^* : L^\infty(Q_T) \to L^\infty(\Omega) \times L^\infty(Q_T)$$

is realized as $\mathscr{F}^*(h) = (\theta|_{t=0}, \theta)$, where $\theta = \theta(\cdot, t)$ is the solution to the backward heat equation

$$\frac{\partial \theta}{\partial t} + \Delta\theta = h \text{ in } Q_T, \quad \left.\frac{\partial \theta}{\partial \nu}\right|_{\partial\Omega} = 0, \quad \theta|_{t=T} = 0. \tag{5.179}$$

Then the assertion follows because $\mathscr{F}^*$ is compact by the parabolic regularity. □

We describe roughly an alternative proof of (5.170) for $n = 2$ based on Theorem 5.8. This proof uses control in time in contrast with [88], which approaches from the control of space variables.

First, the semigroup estimate guarantees

$$\limsup_{t\uparrow T} \|u(\cdot,t)\|_2 < +\infty \quad \Rightarrow \quad \limsup_{t\uparrow T} \|u(\cdot,t)\|_\infty < +\infty$$

because of the quadratic growth [44]. Second, we have

$$\frac{d}{dt}\|u\|_2^2 + \delta\|u\|_{H^1}^2 \le C(\|u\|_3^3 + 1)$$

with $\delta > 0$, and then apply Gagliardo–Nirenberg inequality

$$\|u\|_3 \le C\|u\|_2^{1-a}\|u\|_{H^1}^a, \quad a = n(\frac{1}{2} - \frac{1}{3}) = \frac{n}{6}$$

valid for $n \le 5$, to reach

$$\frac{d}{dt}\|u\|_2^2 \le C(\|u\|_2^2 + 1)^{\frac{6-n}{4-n}}$$

for $n \le 3$ by Young's inequality. This inequality implies

$$-\frac{d}{dt}(\|u\|_2^2 + 1)^{-\frac{2}{4-n}} \le C. \tag{5.180}$$

We assume the contrary to (5.170), and obtain $t_k \uparrow T$ such that

$$\lim_{k\to\infty} \|u(\cdot, t_k)\|_2 = +\infty. \tag{5.181}$$

Letting $u^k(\cdot,t) = u(\cdot, t + t_k)$, we have

$$-\frac{d}{dt}(\|u^k\|_2^2 + 1)^{-\frac{2}{4-n}} \le C$$

by (5.180) with C independent of k which implies

$$(\|u^k(\cdot,-t)\|_2^2+1)^{-\frac{2}{4-n}} \le (\|u^k(\cdot,0)\|_2^2+1)^{-\frac{2}{4-n}} + Ct, \quad 0<t<t_k. \tag{5.182}$$

By a result in [62], there is a subsequence denoted by the same symbol converging in $C_{loc}((-T,0],L^1(\Omega)) \cap L^2_{loc}(\overline{\Omega}\times(-T,0])$.

Let u^∞ be the limit function:

$$u^k \to u^\infty \quad \text{in } C_{loc}((-T,0],L^1(\Omega)) \cap L^2_{loc}(\overline{\Omega}\times(-T,0]). \tag{5.183}$$

Then we obtain

$$(\|u^\infty(\cdot,-t)\|_2^2+1)^{-\frac{2}{4-n}} \le Ct \quad \text{a.e. } t\in(0,T)$$

by (5.181)–(5.183) which implies

$$\|u^\infty(\cdot,-t)\|_2^2+1 \ge C^{-1}t^{-\frac{4-n}{2}} = \delta t^{-1} \quad \text{a.e. } t\in(0,T)$$

for $n=2$ with $\delta = C^{-1}>0$. Hence there arises

$$u^\infty \notin L^2_{loc}(\overline{\Omega}\times(-T,0]),$$

a contradiction.

5.4.3 A Duality Argument

Here we employ the method of [10] that ensures (5.170) in $\Omega=\mathbf{R}^n$ under the entropy inequality

$$\sum_j f_j(u)\log u_j \le 0,\ u=(u_j)\ge 0. \tag{5.184}$$

The result covers chemical reactions of four species, $A_1+A_2 \leftrightarrow A_3+A_4$, satisfying (5.184) with (5.169). See [80] for other cases of chemical reactions. We take the system (5.166), where $\Omega\subset\mathbf{R}^n$ is a bounded domain with smooth boundary $\partial\Omega$. Theorem 5.12 below excludes the assumption (5.184) in the following theorem [10].

Theorem 5.9 (Caputo–Goudon–Vasseur) *The property (5.170) arises for any n, provided that (5.167)–(5.169), and (5.184) hold.*

To describe the proof [10, 81], assume the contrary,

$$\exists x_k \to x_0 \in \overline{\Omega},\ \exists t_k \uparrow T,\ |u(x_k,t_k)| \to \infty.$$

Take $0<r\ll 1$, and put

$$\tilde{u}^k = \tilde{u}^k(x,t) = r^2 u(rx + x_k, r^2 t + t_k), \quad \tilde{f}_j(u) = r^4 f_j(r^{-2}u).$$

It holds that

$$\tau_j \frac{\partial \tilde{u}_j^k}{\partial t} - d_j \Delta \tilde{u}_j^k = \tilde{f}_j(u^k), \ \tilde{u}^k = (\tilde{u}_j^k) \geq 0 \ \text{in} \ \tilde{Q}_8^k, \quad \left. \frac{\partial \tilde{u}_j^k}{\partial \nu} \right|_{\gamma^k \cap B_8} = 0, \tag{5.185}$$

where

$$\tilde{Q}_\rho^k = Q_\rho \cap Q_T^k, \quad \gamma^k = \partial \Omega^k, \quad Q_\rho = B(0,\rho) \times (-\rho^2, 0)$$

for

$$\begin{aligned} Q_T^k &= r^{-1}(\Omega - \{x_k\}) \times r^{-2}((0,T) - \{t_k\}) \\ \partial \Omega^k &= r^{-1}(\partial \Omega - \{x_k\}). \end{aligned}$$

We confirm that this $\tilde{f} = (\tilde{f}_j(u)) : \mathbf{R}^N \to \mathbf{R}$ is provided with local Lipschitz continuity, quasi-positivity, mass dissipation, quadratic growth, and inequality (5.184) uniformly in $0 < r \leq 1$ and k. Then we show uniform estimates on $(\tilde{u}_j^k)$ in k, recalling that γ^k in (5.185) is a scaling of a portion of $\partial\Omega$.

Henceforth, we drop the suffix k for simplicity. The first observation is an ε-regularity for (5.185) derived from the iteration scheme in [9].

Lemma 5.14 *Under the assumption of the previous theorem, any $p > 1$ admits ε_0 such that*

$$\|\tilde{u}\|_{L^p(\tilde{Q}_1)} < \varepsilon_0 \quad \Rightarrow \quad 0 \leq \tilde{u}_j(0,0) \leq 1, \ 1 \leq j \leq N. \tag{5.186}$$

Assuming mass conservation is justified as in Sect. 5.4.1. Then, second, we have

$$\frac{\partial \tilde{M}}{\partial t} - \Delta(\tilde{d}\tilde{M}) = 0 \ \text{in} \ \tilde{Q}_4, \quad \left. \frac{\partial}{\partial \nu}(\tilde{d}\tilde{M}) \right|_{\tilde{B}_4 \cap \gamma} = 0 \tag{5.187}$$

for $\tilde{M} = \tau \cdot \tilde{u}$ with

$$0 < d_* \leq \tilde{d} = \tilde{d}(x,t) \leq d^* < +\infty,$$

which controls

$$\|\tilde{u}\|_{L^p(\tilde{Q}_1)}.$$

For this purpose we apply the *Alexandroff–Bakelman–Pucci inequality* [1, 3, 63] in dual form to ensure the following lemma [10, 16, 90].

Lemma 5.15 *It holds that*

$$\|\tilde{M}\|_{L^{1+\frac{1}{n}}(\tilde{Q}_1)} \leq C \sup_{-4<t<0} \|\tilde{M}(t)\|_{L^1(\tilde{B}_2)} \tag{5.188}$$

in (5.187).

Control of the right-hand side of (5.188), on the other hand, is achieved by a global-in-space estimate on

$$\frac{\partial M}{\partial t} - \Delta(DM) = 0 \text{ in } Q_T \equiv \Omega \times (0, T), \quad \frac{\partial}{\partial \nu}(DM)\Big|_{\partial\Omega} = 0 \tag{5.189}$$

for $M = \tau \cdot u$ with

$$d_* \leq D = D(x,t) \leq d^*.$$

Here we assume $T \geq 2$ without loss of generality. Then we show the following lemma [81].

Lemma 5.16 *There is $\Phi = \Phi(x,t)$ in (5.189) such that*

$$-\Delta\Phi = M - \overline{M} \text{ in } \Omega, \quad \frac{\partial \Phi}{\partial \nu}\Big|_{\partial\Omega} = 0, \quad \overline{M} = \frac{1}{|\Omega|}\int_\Omega M\, dx \tag{5.190}$$

and

$$\begin{aligned} &[\Phi]^{\theta,\theta/2}_{\Omega\times(t_0,t_0+1)} \\ &\equiv \sup_{x,x'\in\Omega,\ x\neq x',\ t,t'\in(t_0,t_0+1),\ t\neq t'} \frac{|\Phi(x,t)-\Phi(x',t')|}{|x-x'|^\theta + |t-t'|^{\theta/2}} \\ &\leq C_\delta(\|\Phi\|_{L^\infty(\Omega\times(0,T))} + 1), \quad 0 < \delta < t_0 < t_0 + 1 < T, \end{aligned} \tag{5.191}$$

where $0 < \theta < 1$ and $\overline{M}$ is a constant independent of t.

Proof The constant $\overline{M}$ is independent of t by (5.193). There exists a unique $f_* = f_*(x,t)$ such that

$$-\Delta f_* = M - \overline{M}, \quad \frac{\partial f_*}{\partial \nu}\Big|_{\partial\Omega} = 0, \quad \overline{f_*} = 0. \tag{5.192}$$

Then we obtain

$$\Delta(\frac{\partial f_*}{\partial t} + DM) = 0, \quad \frac{\partial}{\partial \nu}(\frac{\partial f_*}{\partial t} + DM)\Big|_{\partial\Omega} = 0$$

and hence there is $d(t)$ such that

$$\frac{\partial f_*}{\partial t} = -DM + d(t).$$

Taking $c = c(t)$ in $c' = d$ and letting this $\Phi = f_* + c(t)$, we obtain (5.190) and

$$\frac{\partial \Phi}{\partial t} = D\Delta\Phi - D\overline{M} \text{ in } \Omega \times (0, T), \quad \left.\frac{\partial \Phi}{\partial \nu}\right|_{\partial\Omega} = 0, \tag{5.193}$$

which implies (5.191) by the *Krylov–Safonov inequality* [16, 40, 41] and standard parabolic regularity based on $D \in L^\infty(Q_T)$ [45]. □

Proof *(Theorem 5.9)* First, since $c(\cdot, 0)$ may be arbitrary in the proof of the previous lemma, we prescribe

$$\overline{f(\cdot, 0)} = 0$$

in (5.192). Then the elliptic estimate ensures

$$\|\Phi(\cdot, 0)\|_\infty \le C\|M(\cdot, 0)\|_\infty.$$

Second, the maximum principle for (5.193) implies

$$\|\Phi\|_{L^\infty(\Omega\times(0,T))} \le \|\Phi(\cdot, 0)\|_\infty.$$

We thus end up with

$$|\Phi(x, t) - \Phi(x', t)| \le C|x - x'|^\theta, \quad x, x' \in \Omega, \ t \in (t_0, t_0 + 1). \tag{5.194}$$

Putting $\Phi^k(x, t) = \Phi(rx + x_k, r^2t + t_k)$, we obtain

$$-\Delta\Phi^k(x, t) - r^2\overline{M} = -r^2M(rx + x_k, r^2t + t_k) = M^k(x, t), \quad \left.\frac{\partial \Phi^k}{\partial \nu}\right|_{\gamma^k\cap B_4} = 0.$$

Henceforth we drop the suffix k, because the following estimates are uniform in k.

Take $\zeta \in C_0^\infty(\mathbf{R}^n)$ such that

$$0 \le \zeta \le 1, \quad \text{supp}\, \zeta \subset B_4, \quad \zeta = 1 \text{ in } B_2.$$

Then we obtain

$$\|M^k(t)\|_{L^1(\tilde{B}_2)}$$
$$\le \int_{\tilde{B}_4} M^k(t)\zeta\, dx = \int_{\tilde{B}_4} (-\Delta\Phi^k)\zeta\, dx + O(r^2)$$

$$
\begin{aligned}
&= \int_{\tilde{B}_4} [\Phi^k(x,t) - \Phi^k(0,t)](-\Delta\zeta(x))\, dx + O(r^2) \\
&\leq \|\Delta\zeta\|_\infty \int_{\tilde{B}_4} |\Phi(rx + x_k, r^2 t + t_k) - \Phi(x_k, r^2 t + t_k)|\, dx + O(r^2) \\
&\leq Cr^\theta, \quad -4 < t < 0 \qquad (5.195)
\end{aligned}
$$

by Lemma 5.16. We thus end up with

$$
\|u^k\|_{L^{1+\frac{1}{n}}(\tilde{Q}_1^k)} \leq Cr^\theta, \quad 0 < r \ll 1
$$

of which the right-hand side may be smaller than ε_0 in Lemma 5.14 for $0 < r \ll 1$. Then we obtain

$$
0 \leq u(x_k, t_k) \leq Cr^{-2}, \quad k = 1, 2, \cdots,
$$

a contradiction. □

Remark 5.7 We confirm again that a parabolic regularity interpolation [35] makes it possible to exclude (5.184) and relax (5.169), if system (5.166) is dissipative and quasi-positive [17]. An explicit bound of the growth rate of nonlinearity, however, is prescribed under the entropy inequality, a generalization of (5.184) [81].

5.4.4 3D Case

By the above argument we obtain the following result.

Theorem 5.10 ([81]) *Let $n \leq 3$, and assume (5.167), (5.168),*

$$
|\nabla f_j(u)| \leq C(1 + |u|^{q-1}), \ 1 \leq j \leq N, \quad 1 < q < 2 + \frac{2}{n}, \qquad (5.196)
$$

and

$$
\sum_j f_j(u) \log u_j \leq C(1 + |u|^{1+\frac{2}{n}}), \quad u = (u_j) \geq 0. \qquad (5.197)
$$

Then it holds that (5.170).

Remark 5.8 Since inequalities (5.196) and (5.168) imply

$$
\sum_j f_j(u) \log u_j \leq C(1 + |u|^q), \quad u = (u_j) \geq 0
$$

as in [88], the case $n = 2$ is again critical for the critical growth to ensure (5.170) in the above formulation.

Remark 5.9 In [81] it is also shown that uniform boundedness of the solution is achieved for any n and $1 < q < \infty$ in (5.196) without (5.197), if uniform parabolic L^p-maximal regularity is achieved for

$$M_t - \Delta(DM) = 0$$

for

$$0 < d_* \leq D = D(x,t) = \frac{d \cdot u}{\tau \cdot u} \leq d^*.$$

Although this property is valid if $D(\cdot,t)$ remains in a compact set in VMO as $t \uparrow T$ [92], there is a counter example of the blowup of the solution for q large even for $n = 1$ [61].

Here we show a result weaker than Theorem 5.12 below using L^2-maximal regularity, to illustrate the argument of duality applied to rescaled systems.

Theorem 5.11 *It holds that (5.170) for $n = 3$ if (5.167), (5.168), and (5.196) are valid for $1 < q < 9/5$.*

We begin with the following lemma.

Lemma 5.17 *Under the assumption of Theorem 5.11 any $p > (\frac{n}{2}+1)(q-1)$ admits ε_0 such that (5.186) holds in (5.185).*

Proof Continue to drop the suffix k. From the assumption we can apply Lemma 5.14 to

$$v_j = (u_j+1)^r - 1, \ g_j(v) = rf_j(u)(u_j+1)^{r-1}, \quad r > \frac{n}{2}(q-1).$$

Then we obtain the result as in [9]. □

We continue to drop the suffix k.

Lemma 5.18 *Let $\omega = \tilde{B}_4$. Given $\theta = \theta(x,t) \geq 0$, define $w = w(x,t) \geq 0$ by*

$$w_t + \tilde{d}\Delta w = -\theta \ in\ \omega \times (-16,0), \quad w|_{t=0} = 0, \quad \left.\frac{\partial w}{\partial \nu}\right|_{\partial\omega} = 0. \tag{5.198}$$

Then it holds that

$$\sup_{-16<t<0} \|w(t)\|^2_{L^{\frac{2n}{n-2}}(\omega)} + \int_{-16}^0 \|\nabla w(t)\|^2_{L^{\frac{2n}{n-2}}(\omega)}\, dt \leq C\|\theta\|^2_{L^2(\omega\times(-16,0))}. \tag{5.199}$$

Proof This assertion is reduced to the case of smooth coefficients and solutions. Multiplying Δw to (5.198), we have

$$-\frac{1}{2}\frac{d}{dt}\|\nabla w\|_2^2 + d_*\|\Delta w\|_2^2 \le \|\theta\|_2\|\Delta w\|_2$$
$$\le \frac{d_*}{2}\|\Delta w\|_2^2 + \frac{8}{d_*}\|\theta\|_2^2$$

and hence

$$\sup_{t\in(-4,0)} \|\nabla w(t)\|_2^2 + \int_{-4}^{0} \|\Delta w\|_2^2\, dt \le C\|\theta\|^2_{L^2(\omega\times(-4,0))}.$$

Multiplying $-w$ to (5.198), on the other hand, ensures

$$-\frac{1}{2}\frac{d}{dt}\|w\|_2^2 = (\tilde{d}\Delta w, w) + (w, \theta) \le \|w\|_2(d^*\|\Delta w\|_2 + \|\theta\|_2)$$
$$\le \frac{1}{2}\|w\|_2^2 + C(\|\Delta w\|_2^2 + \|\theta\|_2^2),$$

which implies

$$-\frac{d}{dt}(e^t\|w\|_2^2) \le 2Ce^t(\|\Delta w\|_2^2 + \|\theta\|_2^2).$$

Thus we obtain

$$\sup_{-16<t<0} \|w(t)\|_2^2 \le C\int_{-16}^{0} \|\Delta w\|_2^2 + \|\theta\|_2^2\, dt \le C\|\theta\|^2_{L^2(\omega^\times(-4,0))}.$$

We end up with

$$\sup_{-16<t<0} \|w(t)\|_{H^1}^2 + \int_{-16}^{0} \|\Delta w\|_2^2\, dt \le C\|\theta\|^2_{L^2(\omega\times(-16,0))},$$

which implies (5.199) by Sobolev's embedding. □

Lemma 5.19 *If $n = 3$ it holds that*

$$\|\tilde{M}\|_{L^2(\tilde{Q}_1)} \le C \sup_{-4<t<0} \|\tilde{M}(t)\|^{1/3}_{L^1(\tilde{B}_2)} \cdot \|\tilde{M}\|^{2/3}_{L^{4/3}(\tilde{Q}_2^k)}. \tag{5.200}$$

in (5.187).

Proof We take $\varphi = \varphi(x) \in C_0^\infty(B_2)$ and $\eta = \eta(t) \in C_0^\infty(-4, 0]$ satisfying

$$0 \le \varphi \le 1, \quad \varphi = 1 \text{ in } B_1, \quad \left.\frac{\partial \varphi}{\partial \nu}\right|_{\gamma} = 0, \quad 0 \le \eta \le 1, \quad \eta = 1 \text{ on } [-1, 0].$$

We put $\zeta = \varphi(x)\eta(t)$ and $v = \tilde{M}\zeta$, to obtain

$$\frac{\partial v}{\partial t} - \Delta(\tilde{d}v) = f \text{ in } \omega \times (-4, 0), \qquad \left.\frac{\partial}{\partial \nu}(\tilde{d}v)\right|_{\partial\omega} = 0$$

for

$$\begin{aligned} f &= \tilde{M}\zeta_t - 2\nabla\zeta \cdot \nabla(\tilde{d}\tilde{M}) - \tilde{d}\tilde{M}\Delta\zeta \\ &= \tilde{M}\zeta_t - 2\nabla \cdot (\tilde{d}\tilde{M}\nabla\zeta) + \tilde{d}\tilde{M}\Delta\zeta. \end{aligned}$$

Then it holds that

$$\begin{aligned} \frac{d}{dt}(v, w) + (v, \theta) &= (v_t, w) + (v, w_t) + (v, \theta) \\ &= (\Delta(\tilde{d}v) + f, w) + (v, -\theta - \tilde{d}\Delta w) + (v, \theta) = (f, w) \end{aligned}$$

with $(v, w)|_{t=-4,0} = 0$. In

$$\begin{aligned} (f, w) &= (\zeta_t \tilde{M}, w) + 2(\tilde{d}\tilde{M}\nabla\zeta, \nabla w) + (\tilde{d}\tilde{M}\Delta\zeta, w) \\ &= \eta_t(\varphi\tilde{M}, w) + \eta\{2(\tilde{d}\tilde{M}\nabla\varphi, \nabla w) + (\tilde{d}\tilde{M}\Delta\varphi, w)\} \end{aligned}$$

we obtain

$$|(\varphi\tilde{M}, w)| + |(\tilde{d}\tilde{M}\Delta\varphi, w)| \le C\|w\|_6\|\tilde{M}\|_{6/5}, \quad -16 < t < 0.$$

It holds also that

$$2|(\tilde{d}\tilde{M}\nabla\zeta, \nabla w)| \le C\|\nabla w\|_6\|\tilde{M}\|_{6/5},$$

which implies

$$\begin{aligned} \int_{-4}^{0} (v, \theta)\, dt &= \int_{-4}^{0} (f, w)\, dt \\ &\le C \int_{-4}^{0} \|w(t)\|_{W^{1,6}} \|\tilde{M}(t)\|_{6/5} dt \\ &\le C\|\theta\|_{L^2(\omega\times(-4,0))} \left(\int_{-4}^{0} \|\tilde{M}(t)\|_{6/5}^2 dt \right)^{1/2} \end{aligned}$$

by (5.199). Hence we obtain

$$\|\tilde{M}\|_{L^2(\tilde{Q}_1)} \leq C \int_{-4}^{0} \left(\|\tilde{M}(t)\|_{6/5}^2 dt \right)^{1/2}$$

and the result follows from

$$\|\tilde{M}(t)\|_{6/5} \leq \|\tilde{M}(t)\|_1^{1/3} \|\tilde{M}(t)\|_{4/3}^{2/3}.$$

□

Proof *(Theorem 5.11)* First, we have

$$\|\tilde{M}\|_{L^{4/3}(\tilde{Q}_2)} \leq C \sup_{-16<t<0} \|\tilde{M}(t)\|_{L^1(\tilde{B}_4)}$$

similarly to (5.188), recalling $n = 3$, which implies

$$\|\tilde{M}\|_{L^2(\tilde{Q}_1)} \leq C \sup_{-16<t<0} \|\tilde{M}(t)\|_{L^1(\tilde{B}_4)}$$

by (5.200).

Second, we use

$$\|\tilde{M}(t)\|_{L^1(\tilde{B}^4)} \leq C r^{\theta}, \quad -16 < t < 0$$

derived similarly to (5.195). We thus end up with

$$\|\tilde{M}\|_{L^2(\tilde{Q}_1)} \leq C r^{\theta}$$

for $0 < r \ll 1$ which leads to a contradiction by Lemma 5.17. □

5.4.5 Quadratic Growth Rate

Below we give a rough description of the proof for the following result. Henceforth it is always assumed that $f_i = f_i(u) : \overline{\mathbf{R}_+^N} \to \mathbf{R}^N$, $1 \leq i \leq N$, is locally Lipschitz continuous. We assume also $n \geq 3$, because the result proven below [17] is known for $n = 2$ from [62].

Theorem 5.12 (Fellner–Morgan–Tang) *System (5.166) has a unique uniformly bounded global-in-time classical solution for any space dimension n if $f = (f_i(u)) : \overline{\mathbf{R}_+^N} \to \mathbf{R}^N$ is quasi-positive (5.167), mass dissipative (5.168), and with up to quadratic growth rate (5.169).*

First, uniqueness is a standard consequence of the linearization. For the existence, we construct a family of approximate solutions global in time, denoted by $u^k = u^k(x,t)$, $k = 1, 2, \cdots$, replacing $f_i(u)$ with $f_i^k(u) = f_i(u \wedge k)$. Then Theorem 5.12 is reduced to the uniform estimate

$$\sup_{t \geq 0} \|u^k(\cdot, t)\|_\infty \leq C.$$

Then we write $f(u) = (f_i(u))$ for $f_k(u) = (f_i^k(u))$, dropping suffix k for simplicity.

Second, we reduce (5.167) to (5.172) by the above argument. The fundamental property is now given by

$$\|\tau \cdot u(\cdot, t)\|_1 = \|\tau \cdot u_0\|_1 \equiv M.$$

The facts used for the proof of Theorem 5.12 are the following.

1. Sobolev inequality [43]:

$$\|f\|_{L^q(Q_T)} \leq C(q, \Omega, T)\|f\|_{Q_T}^{2,1}, \quad Q_T = \Omega \times (0, T)$$

for $1 \leq q < \frac{2(n+2)}{n-2}$, where

$$\|f\|_{Q_T}^{2,1} = \sum_{2r+|\alpha| \leq 2} \|\partial_t^r \partial_x^\alpha f\|_{L^2(Q_T)}.$$

2. Hölder regularity of (5.179), or equivalently,

$$w_t - d\Delta w = \theta \text{ in } Q_T, \quad \left.\frac{\partial w}{\partial \nu}\right|_{\partial\Omega} = 0, \quad w|_{t=0} = 0, \tag{5.201}$$

provided with

$$0 < d_* \leq d = d(x,t) \leq d^* < +\infty. \tag{5.202}$$

Hence there is $0 < \alpha < 1$ such that

$$[w]_{Q_T}^{\alpha,\alpha/2} \leq C(\Omega, T)(1 + \|w\|_{L^\infty(Q_T)} + \|\theta\|_{L^\infty(Q_T)}),$$

where

$$[w]_{Q_T}^{\alpha,\alpha/2} = \sup_{x \neq x' \in \Omega,\ t \neq t' \in (0,T)} \frac{|w(x,t) - w(x',t')|}{|x - x'|^\alpha + |t - t'|^{\alpha/2}}.$$

This estimate is a parabolic version of the Giorgi–Nash–Moser inequality. It is shown in [17] based on the interior estimate of [43].

3. *Regularity interpolation* for a counter part of (5.201),

$$u_t - \Delta u = \psi \text{ in } Q_T, \quad \left.\frac{\partial u}{\partial \nu}\right|_{\partial\Omega} = 0, \quad u|_{t=0} = u_0(x), \tag{5.203}$$

provided with

$$|u(x,t) - u(x',t)| \leq H|x - x'|^{\delta}, \quad \|\psi\|_{L^\infty(Q_T)} \leq F$$

for $0 \leq \delta < 1$. It follows that

$$\|\nabla_x u\|_{L^\infty(Q_T)} \leq C(\Omega, \gamma)\|u_0\|_{W^{1,\infty}} + B(\Omega, \delta) H^{\frac{1}{2-\delta}} F^{\frac{1-\delta}{2-\delta}}.$$

This property was noticed first in [35] for the case of $\Omega = \mathbf{R}^n$. The proof for the above form is given in [16].

4. *Souplet's trick* [72] valid for (5.203) with $u_0 = 0$, that is,

$$\|\Delta u\|_{L^\infty(Q_T)} \leq C\|u\|^{1/2}_{W^{1,\infty}(Q_T)} \|\psi\|^{1/2}_{W^{1,\infty}(Q_T)}.$$

It is obtained by the semigroup estimate.

We conclude this subsection with a remark on the maximal regularity of (5.201)–(5.202), that is, a refinement of Lemma 5.18.

Lemma 5.20 *It holds that*

$$\|\Delta w\|_{L^2(Q_T)} \leq \frac{1}{d_*}\|\theta\|_{L^2(Q_T)}, \quad \sup_{t\in(0,T)} \|\nabla w(\cdot,t)\|_2^2 \leq \frac{2}{d_*}\|\theta\|^2_{L^2(Q_T)} \tag{5.204}$$

in (5.201)–(5.202).

Proof Since

$$\frac{1}{2}\frac{d}{dt}\|\nabla w\|_2^2 + d_*\|\Delta w\|_2^2 = (\theta, \Delta w) \leq \|\theta\|_2 \|\Delta w\|_2$$

we obtain

$$\frac{1}{2}\|\nabla w(\cdot,t)\|_2^2 + d_*\|\Delta w\|^2_{L^2(Q_T)} \leq \|\theta\|_2 \|\Delta w\|_{L^2(Q_T)}, \quad 0 \leq t \leq T,$$

which implies

$$\|\Delta w\|_{L^2(Q_T)} \leq \frac{1}{d_*}\|\theta\|_{L^2(Q_T)}$$

and hence

$$\frac{1}{2}\sup_{t\in[0,T]}\|\nabla w(\cdot,t)\|_2^2\leq\frac{1}{d_*}\|\theta\|_{L^2(Q_T)}^2.$$

□

Lemma 5.21 *It holds that*

$$\|w\|_{Q_T}^{2,1}\leq C(\Omega,T)\|\theta\|_{L^2(Q_T)}$$

in (5.201).

Proof Noting

$$\frac{1}{2}\frac{d}{dt}\|w\|_2^2=(d\Delta w,w)+(\theta,w)\leq d^*\|\Delta w\|_2\|w\|_2+\|\theta\|_2\|w\|_2,$$

we apply the argument used in the proof of Lemma 5.18, to reach

$$\sup_{t\in(0,T)}\|w(\cdot,t)\|_2\leq C(T)\|\theta\|_{L^2(Q_T)}$$

by (5.204). Then the result follows from the previous lemma and

$$\|w_t\|_2\leq d^*\|\Delta w\|_2+\|\theta\|_2.$$

□

5.4.6 Mass Conservative Systems

Here we show several estimates derived only from mass conservation without any growth rate of the nonlinearity. Recall $n\geq 3$.

Lemma 5.22 *It holds that*

$$\|\tau\cdot u\|_{L^2(\Omega\times(t,t+1))}\leq C\|\tau\cdot u\|_{L^2(\Omega\times(t,t+2))}^{\frac{n-2}{n+1}},\quad t\geq 0. \tag{5.205}$$

in (5.166) if $n\geq 3$, (5.167), and (5.172) are satisfied.

Proof We put

$$d^*=\max_i \tau_i^{-1}d_i,\quad d_*=\min_i \tau_i^{-1}d_i,\quad Q=\Omega\times(t,t+2). \tag{5.206}$$

Given $0\leq\theta\in L^2(Q)$, we define $w=w(x,s)\geq 0$ by

$$w_s + d^* \Delta w = -\theta \text{ in } Q, \quad \left.\frac{\partial w}{\partial \nu}\right|_{\partial\Omega} = 0, \quad w|_{s=t+2} = 0.$$

It holds that

$$\|w\|_Q^{2,1} \le C(\Omega)\|\theta\|_{L^2(Q)}, \quad \|\Delta w\|_{L^2(Q)} \le \frac{1}{d^*}\|\theta\|_{L^2(Q)} \tag{5.207}$$

by Lemmas 5.20 and 5.21 for $d = d^*$.

Take smooth $\varphi : \mathbf{R} \to [0, 1]$ satisfying

$$\varphi(s) = \begin{cases} 0, & s \le 0 \\ 1, & s \ge 1, \end{cases} \quad 0 \le \varphi' \le 2,$$

and put $\varphi^t = \varphi(\cdot - t)$. Since

$$\begin{aligned} &\tau_i \frac{\partial}{\partial s}(\varphi^t u_i) - d_i \Delta(\varphi^t u_i) = \tau_i (\varphi^t)' u_i + \varphi^t f_i(u) \\ &\left.\frac{\partial}{\partial \nu}(\varphi^t u_i)\right|_{\partial\Omega} = 0, \quad \left.\varphi^t u_i\right|_{s=t} = 0 \end{aligned} \tag{5.208}$$

there arises

$$\begin{aligned} &\int\int_Q (\varphi^t u_i)\theta \, dxds = \int\int_Q \varphi^t u_i(-w_s - d^* \Delta w)\, dxds \\ &= \int\int_Q \tau_i^{-1} w \left(\tau_i \frac{\partial}{\partial s}(\varphi^t u_i) - d_i \Delta(\varphi^t u_i)\right) + (\tau_i^{-1} d_i - d^*)(\varphi^t u_i)\Delta w \, dxds \\ &= \int\int_Q \tau_i^{-1} w((\varphi^t)' \tau_i u_i + \varphi^t f_i(u)) + (\tau_i^{-1} d_i - d^*)(\varphi^t u_i)\Delta w \, dxds, \end{aligned}$$

and hence

$$\begin{aligned} &\int\int_Q \varphi^t \tau \cdot u\theta \, dxds \\ &= \int\int_Q w(\varphi^t)' \tau \cdot u + \sum_i \tau_i(\tau_i^{-1} d_i - d^*)(\varphi^t u_i)\Delta w \, dxds \\ &\le 2\int_Q w\tau \cdot u \, dxds + (d^* - d_*)\varphi_t \tau \cdot u|\Delta w| \, dxds \\ &\le 2\|w\|_{L^q(Q)}\|\tau \cdot u\|_{L^{q'}(Q)} + (d^* - d_*)\|\Delta w\|_{L^2(Q)}\|\varphi_t \tau \cdot u\|_{L^2(Q)} \end{aligned}$$

for

$$q = \frac{2(n+1)}{n-2}, \quad \frac{1}{q'} + \frac{1}{q} = 1.$$

We have

$$1 < q = \frac{2(n+1)}{n-2} < \frac{2(n+2)}{n-2}, \quad q' = \frac{2n+2}{n+4}$$

and hence

$$\|w\|_{L^2(Q)} \le C(n, \Omega)\|w\|_Q^{2,1} \le C'\|\theta\|_{L^2(Q)}$$

from Sobolev's inequality and Lemma 5.21. Using

$$\begin{aligned}
\|\tau \cdot u\|_{L^{q'}(Q)} = \|\tau \cdot u\|_{L^{\frac{2n+2}{n+4}}(Q)} &\le \|\tau \cdot u\|_{L^1(Q)}^{\frac{3}{n+1}} \|\tau \cdot u\|_{L^{\frac{3}{n+1}}(Q)}^{\frac{n-2}{n+1}} \\
&= (2M)^{\frac{3}{n+1}} \|\tau \cdot u\|_{L^{\frac{n-2}{n+1}}(Q)}^{\frac{n-2}{n+1}}
\end{aligned}$$

and (5.207), we now obtain

$$\int\int_Q \varphi_t \tau \cdot u\theta \, dxds \le C\|\tau \cdot u\|_{L^2(Q)}^{\frac{n-2}{n+1}} \|\theta\|_{L^2(Q)} + (1 - \frac{d_*}{d^*})\|\theta\|_{L^2(Q)} \|\varphi_t \tau \cdot u\|_{L^2(Q)},$$

which implies

$$\|\varphi_t \tau \cdot u\|_{L^2(Q)} \le \frac{d^*}{d_*} \cdot C\|\tau \cdot u\|_{L^2(Q)}^{\frac{n-2}{n+1}}$$

and hence (5.205). □

Lemma 5.23 *It holds that*

$$\|u\|_{L^2(\Omega \times (t,t+2))} \le C, \quad t \ge 0 \tag{5.209}$$

in (5.166) under the assumption of the previous lemma.

Proof If

$$\|\tau \cdot u\|_{L^2(\Omega \times (t,t+1))} \le \|\tau \cdot u\|_{L^2(\Omega \times (t+1,t+2))}$$

it holds that

$$\|\tau \cdot u\|_{L^2(\Omega \times (t,t+2))} \le 2\|\tau \cdot u\|_{L^2(\Omega(t+1,t+2))} \le 2C\|\tau \cdot u\|_{L^2(\Omega \times (t,t+2))}^{\frac{n-2}{n+1}}$$

by Lemma 5.22 and hence

$$\|\tau \cdot u\|_{L^2(\Omega\times(t,t+2))} \leq C.$$

Otherwise, we have

$$\|\tau \cdot u\|_{L^2(\Omega\times(t+1,t+2))} \leq \|\tau \cdot u\|_{L^2(\Omega\times(t,t+1))},$$

and consequently,

$$\|\tau \cdot u\|_{L^2(\Omega\times(n,n+1))} \leq C, \quad n = 1, 2, \cdots, \tag{5.210}$$

in both cases. Inequality (5.210) now implies (5.209). □

Lemma 5.24 *It holds that*

$$\left\| \int_t^{t+1} u(\cdot, s) ds \right\|_\infty \leq C, \quad t \geq 0 \tag{5.211}$$

under the assumption of the previous lemma, where $u = (u_i(\cdot, s)) \geq 0$.

Proof Equality (5.208) implies

$$\begin{aligned} &\frac{\partial}{\partial s}(\varphi^t \tau \cdot u) - \Delta(\varphi^t d \cdot u) = (\varphi^t)' \tau \cdot u \\ &\left.\frac{\partial}{\partial \nu}\varphi^t \tau \cdot u\right|_{\partial\Omega} = 0, \quad \varphi^t \tau \cdot u\big|_{s=t} = 0 \end{aligned} \tag{5.212}$$

and hence

$$\tau \cdot u(\cdot, t+2) - \Delta \int_t^{t+2} \varphi^t d \cdot u \, ds = \int_t^{t+2} \varphi^t \tau \cdot u \, ds.$$

We thus obtain

$$(-\Delta + 1) \int_t^{t+2} \varphi^t d \cdot u \, ds \leq H^t, \quad \left.\frac{\partial}{\partial \nu} \int_t^{t+2} \varphi^t d \cdot u \, ds\right|_{\partial\Omega} = 0 \tag{5.213}$$

for

$$H^t = \int_t^{t+2} \varphi^t d \cdot u \, ds + \int_t^{t+2} (\varphi^t)' \tau \cdot u \, ds$$

with

$$\|H^t\|_{L^2(\Omega)} \leq C \tag{5.214}$$

by Lemma 5.23.

A comparison theorem for (5.213) now ensures

$$0 \leq \int_t^{t+2} \varphi^t d \cdot u \, ds \leq \Psi,$$

for $\Psi = \Psi(x, t)$ defined by

$$(-\Delta + 1)\Psi = H^t, \quad \left. \frac{\partial \Psi}{\partial \nu} \right|_{\partial \Omega} = 0,$$

and then we obtain

$$\|\Psi\|_{H^2(\Omega)} \leq C$$

by (5.214) and the elliptic estimate. The Sobolev inequality then implies

$$\left\| \int_{t+1}^{t+2} u(\cdot, s) \, ds \right\|_p \leq C \left\| \int_t^{t+2} \varphi^t d \cdot u(\cdot, s) \, ds \right\|_p \leq C'$$

for $p = \infty$, $1 < p < \infty$, and $p = \frac{2n}{n-4}$ according to $n = 3$, $n = 4$, and $n \geq 5$, respectively.

We thus obtain

$$\left\| \int_t^{t+2} u(\cdot, s) \, ds \right\|_p \leq C$$

by a translation in t, and therefore, (5.214) is improved as

$$\|H^t\|_{L^p(\Omega)} \leq C.$$

A bootstrap argument now ensures (5.211). □

Let

$$U(t) = \|u\|_{L^\infty(\Omega \times (t, t+2))}, \quad v(x, s) = \int_t^s \varphi^t d \cdot u(x, t') \, dt'.$$

Lemma 5.25 *It holds that*

$$\|\nabla_x v\|_{L^\infty(\Omega \times (t, t+2))} \leq C \left(1 + U(t)^{\frac{1-\delta}{2-\delta}} \right), \quad t \geq 0 \tag{5.215}$$

under the assumption of the previous lemma.

Proof By (5.212) we have

$$b(x,s)\frac{\partial v}{\partial s} - \Delta v = \int_t^s (\varphi^t)'\tau \cdot u \, dt', \quad \left.\frac{\partial v}{\partial \nu}\right|_{\partial\Omega} = 0, \quad v|_{s=t} = 0$$

with

$$\frac{1}{d^*} \leq b(x,s) = \frac{\varphi_t \tau \cdot u}{\varphi_t d \cdot u} \leq \frac{1}{d_*},$$

for $0 < d* \leq d^* < \infty$ defined by (5.206). Then the Hölder regularity for parabolic equations assures

$$|v(x,s) - v(x',s)| \leq C|x - x'|^\delta, \quad x, x' \in \Omega, \ t < s < t+2, \tag{5.216}$$

where $0 < \delta < 1$.

Equation (5.212), on the other hand, implies

$$v_s - \Delta v = \psi, \quad \left.\frac{\partial v}{\partial \nu}\right|_{\partial\Omega} = 0, \quad v|_{s=t} = 0, \tag{5.217}$$

for

$$\psi(\cdot,s) = \int_t^s (\varphi^t)'\tau \cdot u(\cdot,t')dt' + \varphi^t(d-\tau)\cdot u$$

satisfying

$$\|\psi\|_{L^\infty(\Omega\times(t,t+2))} \leq C(1+U(t)). \tag{5.218}$$

by Lemma 5.24. By the regularity interpolation to (5.217), therefore, we obtain

$$\|\nabla_x v\|_{L^\infty(\Omega\times(t,t+2))} \leq C\|\psi\|_{L^\infty(\Omega\times(t,t+2))}^{\frac{1-\delta}{2-\delta}}$$

and hence (5.215) by (5.218). □

5.4.7 Uniform Boundedness of the Solution

We complete the proof of Theorem 5.12 under the additional assumption on the growth rate, (5.169). This quadratic rate may be so relaxed as

$$|f_j(u)| \leq C(1+|u|^{2+\varepsilon}), \quad 1 \leq j \leq N \tag{5.219}$$

for $\varepsilon > 0$ satisfying

$$\gamma \equiv \frac{3+\varepsilon}{4} + \frac{1-\delta}{2(2-\delta)} < 1. \tag{5.220}$$

Let

$$z_i(x,s) = \int_t^s (\varphi^t)' u_i(x,t')\, dt'.$$

Lemma 5.26 *It holds that*

$$\|\nabla z_i\|_{L^\infty(\Omega\times(t,t+2))} \leq C\left(1 + U(t)^{\frac{1+\varepsilon}{2}}\right), \quad t \geq 0 \tag{5.221}$$

in (5.166), if $n \geq 3$, (5.167), (5.172), and (5.219) are satisfied.

Proof We have

$$\tau_i \frac{\partial z_i}{\partial s} - d_i \Delta z_i = \int_t^s \tau_i (\varphi^t)'' u_i + (\varphi^t)' f_i(u)\, dt'$$
$$\left.\frac{\partial z_i}{\partial \nu}\right|_{\partial\Omega} = 0, \quad z_i|_{s=t} = 0. \tag{5.222}$$

Let

$$F_i(\cdot,s) = \int_t^s (\varphi^t)' f_i(u)\, dt', \quad e = \begin{pmatrix} 1 \\ \cdot \\ \cdot \\ 1 \end{pmatrix}.$$

It holds that

$$\begin{aligned} \|F_i(\cdot,s)\|_\infty &\leq C \left\| \int_t^{t+2} (1 + e\cdot u(\cdot,t'))^{2+\varepsilon}\, dt' \right\|_\infty \\ &\leq C(1+U(t))^{1+\varepsilon} \left\| \int_t^{t+2} 1 + e\cdot u(\cdot,t')\, dt' \right\|_\infty \\ &\leq C(1+U(t))^{1+\varepsilon}, \quad t \leq s \leq t+2 \end{aligned} \tag{5.223}$$

by Lemma 5.24. Now, apply interpolation regularity for $\delta = 0$ to (5.222), recalling Q in (5.206). We obtain

$$\begin{aligned} \|\nabla_x z_i\|_{L^\infty(Q)} &\leq C\|z_i\|_{L^\infty(Q)}^{1/2} (1 + \|F_i\|_{L^\infty(Q)}^{1/2}) \\ &\leq C(1+U(t)^{1+\varepsilon})^{1/2}, \end{aligned}$$

using Lemma 5.24 and (5.223). Then (5.221) follows. □

Lemma 5.27 *It holds that*

$$\|\nabla\varphi^t u_i\|_{L^\infty(\Omega\times(t,t+2))} \leq C\left(1+U(t)^{\frac{3+\varepsilon}{2}}\right), \quad t\geq 0 \tag{5.224}$$

under the assumption of the previous lemma.

Proof Applying interpolation regularity to (5.208) for $\delta = 0$, we obtain

$$\begin{aligned}\|\nabla\varphi^t u_i\|_{L^\infty(Q)} &\leq C\|\varphi^t u_i\|_{L^\infty(Q)}^{1/2}\|\tau_i(\varphi^t)'u_i + \varphi^t f_i(u)\|_{L^\infty(Q)}^{1/2} \\ &\leq CU(t)^{1/2}(1+U(t)^{2+\varepsilon})^{1/2}\end{aligned}$$

for Q defined by (5.206). Then (5.224) follows. □

Lemma 5.28 *It holds that*

$$\|\Delta v\|_{L^\infty(\Omega\times(t,t+2))} \leq C(1+U(t)^\gamma) \tag{5.225}$$

for $0<\gamma<1$ *defined by (5.220) under the assumption of the previous lemma.*

Proof We continue to use $Q = \Omega\times(t,t+2)$. By Souplet's trick to (5.217) we have

$$\|\Delta v\|_{L^\infty(Q)} \leq C\|v\|_{W^{1,\infty}(Q)}\|\psi\|_{W^{1,\infty}(Q)} \tag{5.226}$$

First, Lemmas 5.24 and 5.25 imply

$$\|v\|_{W^{1,\infty}(Q)} \leq C\left(1+U(t)^{\frac{1-\gamma}{2-\gamma}}\right). \tag{5.227}$$

Second, Lemmas 5.24, 5.26, 5.27 imply

$$\|\psi\|_{W^{1,\infty}(Q)} \leq C\left(1+U^{\frac{3+\varepsilon}{2}}\right). \tag{5.228}$$

Inequality (5.225) is now a consequence of (5.226)–(5.228). □

Proof *(Theorem 5.12)* We have

$$\varphi^t\tau\cdot u(\cdot,s) - \Delta v(\cdot,s) = \int_t^s (\varphi^t)'\tau\cdot u\, dt'$$

by (5.212). Then Lemmas 5.24 and 5.28 imply

$$\|\varphi^t\tau\cdot u\|_{L^\infty(Q)} \leq C\left(1+\|\tau\cdot u\|_{L^\infty(Q)}^\gamma\right), \quad t\geq 0.$$

The inequality

$$\|\tau \cdot u\|_{L^\infty(\Omega\times(t,t+1))} \leq C, \quad t \geq 0$$

now follows as in the proof of Lemma 5.23, and hence

$$\sup_{t\geq 0} \|u(\cdot,t)\|_\infty \leq C.$$

□

5.5 Smoluchowski–Poisson Equation

5.5.1 *System of Chemotaxis*

The Keller–Segel system

$$\begin{aligned} u_t &= \nabla\cdot(d_1(u,v)\nabla u) - \nabla\cdot(d_2(u,v)\nabla v) \\ v_t &= d_v\Delta v - k_1 vw + k_{-1}p + f(v)u \\ w_t &= d_w\Delta w - k_v w + (k\_1 + k_2)p + g(v,w)u \\ p_t &= d_p\Delta p + k_1 vw - (k_{-1}+k_2)p, \end{aligned} \tag{5.229}$$

models the chemotactic features of cellular slime molds at the scale of several hours in time and cell populations in space, where $u = u(x,t)$, $v = v(x,t)$, $w = w(x,t)$, and $p = p(x,t)$ stand for the density of cellular slime molds, the concentrations of chemical substances, enzymes, and complexes, respectively. It is a multi-scale model where several factors are taken into account [37].

The first concept involved in this model is the gradient. Given the scalar field φ, its gradient $\nabla\varphi$ stands for the vector field, the gradient of which maximizes the growth of ϕ and the length of which is its rate. The gradient operator ∇ is then defined, which leads to the divergence $\nabla\cdot j$ of the vector field j. If $\rho = \rho(x,t)$ stands for the mass density, equation of conservation (5.68) arises with the divergence theorem of Gauss. Einstein's formula ensures that j is proportional to $\nabla\rho$ from the mean field theory, while $j = v\rho$ follows from the transport theory, which means that the flux of the density of particles is a product of the mass ρ and the velocity v as is noticed in (4.61) and (5.70).

The terms in (5.229) are thus associated with material transport, composed of diffusion and chemotaxis,

$$\begin{aligned} u_t &= \nabla\cdot(d_1(u,v)\nabla u) - \nabla\cdot(d_2(u,v)\nabla v) \\ v_t &= d_v\Delta v, \ w_t = d_w\Delta w, \ p_t = d_p\Delta p. \end{aligned}$$

Concentrations of chemical material v, enzyme w, and complexes p are subject to the diffusion, while the density of slime molds u is under the influence of fluctuations of environments. Hence the diffusion coefficient $d_1 = d_1(u, v)$ and the chemotactic velocity $d_2/u = d_2(u, v)/u$ are under the control of (u, v).

The other part of (5.229) is a spatially homogeneous model, composed of the productions of u and v, the rates of which depend on v and (v, w),

$$v_t = f(v)u, \quad w_t = g(v, w)u,$$

and the chemical reaction

$$V + W \to P\ (k_1), \quad P \to V + W\ (k_{-1}), \quad P \to W\ (k_2),$$

described by the law of mass action,

$$\begin{aligned} v_t &= -k_1 vw + k_{-1}p \\ w_t &= -k_1 vw + (k_{-1} + k_2)p \\ p_t &= k_1 vw - (k_{-1} + k_2)p. \end{aligned} \tag{5.230}$$

A reduction is made by the Michaelis–Menten relation [57], valid in the quasi-stationary state of (5.230), where $p_t = 0$. Using the mass conservation $(w+p)_t = 0$, we reach

$$\begin{aligned} u_t &= \nabla \cdot (d_1(u, v)\nabla u) - \nabla \cdot (d_2(u, v)\nabla v) \\ v_t &= d_v \Delta v - k(v)v + f(v)u \end{aligned} \tag{5.231}$$

by (5.229), where

$$k(v) = \frac{ck_1k_2}{k_{-1} + k_2 + k_1 v}.$$

Then linear approximation of (4.17),

$$\begin{aligned} u_t &= d_u \Delta u - \chi \nabla \cdot (u\nabla v) \\ v_t &= d_v \Delta v - b_1 v + b_2 u, \end{aligned}$$

and its reduction to the elliptic-parabolic system, e.g.,

$$\begin{aligned} &u_t = \nabla \cdot (\nabla u - u\nabla v), \quad -\Delta v = u - \frac{1}{|\Omega|}\int_\Omega u\, dx \quad \text{in } \Omega \times (0, T) \\ &\frac{\partial u}{\partial \nu} - u\frac{\partial v}{\partial \nu} = \frac{\partial v}{\partial \nu} = 0 \quad \text{on } \partial\Omega \times (0, T), \end{aligned} \tag{5.232}$$

are used as simplified systems of chemotaxis. Then aggregation of the solution is suspected and studied for the case of two space dimensions [5, 7, 14, 19, 27, 34, 54, 55].

5.5.2 Non-equilibrium Statistical Mechanics

System (5.232) is nothing but the Smoluchowski–Poisson equation describing the motion of a mean field of many self-gravitating Brownian particles, and its stationary state is formulated by the Boltzmann–Poisson equation studied in Sect. 2.4.

We recall that non-equilibrium statistical mechanics is used to describe the motion of a mean field of many particles. The Kramers–Moyal expansion and the principle of maximal entropy production are used in semiconductor physics and astronomy, respectively. In both cases the first equation of (5.232) arises as an adiabatic limit, and then it is called the Smoluchowski equation. This equation is derived also by various methods to describe the biased motion of particles in polymer chemistry and biology [59].

The Kramers–Moyal expansion uses momentum expansion of the master equation and then applies the Langevin equation for the particle transport. This process ensures the Kramers equation, provided with time, position, and wave number as individual variables. Then the Fokker–Planck equation and the Smoluchowski equation arise as two different adiabatic limits of this equation [77].

More precisely, let $P(x, t \mid x_1, t_1)$ be the existence probability of a particle at the position x at the time t, which was at the position x_1 at the time t_1. The master equation for the $P(x, t \mid x_1, t_1)$ takes the form

$$\begin{aligned} \frac{\partial}{\partial t} P(x, t \mid x_1, t_1) &= -\int dx' \, W(x \to x') P(x, t \mid x_1, t_1) \\ &\quad + \int dx' \, W(x' \to x) P(x', t \mid x_1, t_1), \end{aligned} \tag{5.233}$$

where W denotes the transient probability. This equation induces the Kramers–Moyal expansion

$$\frac{\partial}{\partial t} P(x, t \mid x_1, t_1) = \sum_{k=1}^{\infty} \frac{1}{k!} (-\partial_x)^k \left[C_k(x) P(x, t \mid x_1, t_1) \right] \tag{5.234}$$

for

$$C_k(x) = \lim_{\Delta t \to 0} \frac{1}{\Delta t} \left\langle [x(t + \Delta t) - x(t)]^k \right\rangle_{x(t)=x}, \tag{5.235}$$

the mean k-th moment of the existence probability conditioned by $x(t) = x$.

We use the Langevin equation

$$\frac{dx}{dt} = v, \quad m\frac{dv}{dt} = -m\gamma v + R(t) + mF(x) \tag{5.236}$$

describing the motion of particles under fluctuation and diffusion, to seek the k-th moment in (5.235), where m, γ, $R(t)$, and $F(x)$ describe mass, friction coefficient, fluctuation, and outer force. For the moment, the position x and the wave number v are thought to be independent variables. Then the conditioned existence probability of particles takes the form $P(x, v, t \mid x_1, v_1, t_1)$ and (5.234)–(5.235) is replaced by

$$\begin{aligned}\frac{\partial}{\partial t}P(x, v, t \mid x_1, v_1, t_1) &= \sum_{k=1}^{\infty}\frac{1}{k!}[(-\partial_x)^k, (-\partial_v)^k]\\ &\quad\cdot[C_k(x, v)P(x, v, t \mid x_1, v_1, t_1)]\\ C_k(x, v) &= (\lim_{\Delta t\downarrow 0}\frac{1}{\Delta t}\langle|x(t+\Delta t) - x(t)|^k\rangle,\\ &\quad \lim_{\Delta t\downarrow 0}\frac{1}{\Delta t}\langle|v(t+\Delta t) - v(t)|^k\rangle).\end{aligned} \tag{5.237}$$

Then (5.236) implies

$$C_1(x, v) = (v, -\gamma v + F(x)), \quad C_2(x, v) = \left(0, 2D/m^2\right), \tag{5.238}$$

where $D > 0$ is the diffusion coefficient: $\langle R(t_1), R(t_2)\rangle = 2D\delta(t_1 - t_2)$.

Using (5.238) and neglecting (5.237) for $k \geq 3$, we obtain the Kramers equation

$$\begin{aligned}&\frac{\partial}{\partial t}P(x, v, t \mid x_1, v_1, t_1)\\ &= [-\frac{\partial}{\partial x}v + \frac{\partial}{\partial v}(-F(x) + \gamma v) + \frac{D}{m^2}\frac{\partial^2}{\partial v^2}]P(x, v, t \mid x_1, v_1, t_1).\end{aligned} \tag{5.239}$$

If P is uniform in x, we reach the Fokker–Planck equation

$$\frac{\partial P}{\partial t} = \frac{\partial}{\partial v}[-F + \gamma v + \frac{D}{m^2}\frac{\partial}{\partial v}]P.$$

The other reduction is the Smoluchowski equation where the temperature T and the potential force acting to the particle K are induced with the Boltzmann constant k. It thus holds that $F = \frac{K}{m}$ and $D = mkT$, which results in

$$\frac{\partial P}{\partial t} = -\frac{K}{m}\frac{\partial P}{\partial v} - v\frac{\partial P}{\partial x} + \gamma\frac{\partial}{\partial v}(vP + \frac{kT}{m}\frac{\partial P}{\partial v})$$

$$
\begin{aligned}
&= \gamma \frac{\partial}{\partial v}(vP + \frac{kT}{m}\frac{\partial P}{\partial v} - \frac{K}{\gamma m}P + \frac{kT}{\gamma m}\frac{\partial P}{\partial x}) - \frac{kT}{m}\frac{\partial^2 P}{\partial v \partial x} - v\frac{\partial P}{\partial x} \\
&= \gamma(\frac{\partial}{\partial v} - \frac{1}{\gamma}\frac{\partial}{\partial x})(vP + \frac{kT}{m}\frac{\partial P}{\partial v} - \frac{K}{\gamma m}P + \frac{kT}{\gamma m}\frac{\partial P}{\partial x}) \\
&\quad - \frac{\partial}{\partial x}(\frac{K}{\gamma m}P - \frac{kT}{\gamma m}\frac{\partial P}{\partial x})
\end{aligned}
\tag{5.240}
$$

by (5.239). Using $q = x + \frac{v}{\gamma}$ and $p = v$, we obtain

$$
\frac{\partial}{\partial v} = \frac{1}{\gamma}\frac{\partial}{\partial q} - \frac{\partial}{\partial p}, \quad \frac{\partial}{\partial x} = \frac{\partial}{\partial q}
$$

and hence

$$
\begin{aligned}
\frac{\partial P}{\partial t} &= -\gamma\frac{\partial A}{\partial p} - \frac{\partial}{\partial q}(\frac{K}{\gamma m}P - \frac{kT}{\gamma m}\frac{\partial P}{\partial q}) \\
A &= vP + \frac{kT}{m}\frac{\partial P}{\partial v} - \frac{K}{\gamma m}P + \frac{kT}{\gamma m}\frac{\partial P}{\partial x}
\end{aligned}
\tag{5.241}
$$

by (5.240). If

$$
\begin{aligned}
&\int P(q - \frac{p}{\gamma}, p, t)dp \sim f(x,t)|_{x=q-\frac{p}{\gamma}} \\
&\int K(q - \frac{p}{\gamma})P(q - \frac{p}{\gamma}, p, t)dp \sim K(x)f(x,t)|_{x=q-\frac{p}{\gamma}}
\end{aligned}
$$

for $f(x,t) = \displaystyle\int P(x,v,t)dv$ in (5.241), there arises the Smoluchowski equation

$$
\frac{\partial f}{\partial t} = \frac{1}{\gamma m}\frac{\partial}{\partial x}(-Kf + kT\frac{\partial f}{\partial x}). \tag{5.242}
$$

5.5.3 *Maximal Entropy Production*

The other derivation of (5.242) uses the principle of maximal entropy production [11] for the transport equation

$$
f_t + v \cdot \nabla_x f - \nabla_x \varphi \cdot \nabla_v f = -\nabla_v \cdot j, \tag{5.243}
$$

where $x, v, t, \varphi = \varphi(x,t)$, and $f = f(x,v,t) \geq 0$ are position, velocity, time, field potential, the particle density, respectively. Equation (5.243) arises if $\gamma = 0$ and

$F = \nabla_x \varphi$ in the Kramers equation (5.239), with $D = 0$ compensated by the general dissipation term $-\nabla_v \cdot j$ in the right-hand side. Maximal entropy production is used to prescribe this term.

Let $n = 3$. Total mass and total entropy are given by

$$M = \int \rho \, dx, \quad \rho = \int f \, dv \tag{5.244}$$

and

$$E = \frac{1}{2} \int \int f|v|^2 \, dxdv + \frac{1}{2} \int \rho\varphi \, dx = \frac{1}{2} \int \int f(|v|^2 + \varphi) \, dxdv, \tag{5.245}$$

respectively, while

$$S = \int \int s(f) \, dxdv \tag{5.246}$$

stands for the total entropy, where $s = s(f)$ is the entropy density. By (5.245), energy production is given by

$$\begin{aligned}
\dot{E} = \frac{dE}{dt} &= \frac{1}{2} \int \int f_t(|v|^2 + \varphi) + f\varphi_t \, dxdv \\
&= \frac{1}{2} \int \int (-\nabla_v \cdot j)(|v|^2 + \varphi) \, dxdv - \frac{1}{2} \int \int (U \cdot \nabla f)(|v|^2 + \varphi) - f\varphi_t \, dxdv \\
&= \frac{1}{2} \int \int (-\nabla_v \cdot j)|v|^2 - (v \cdot \nabla_x f)\varphi + (\nabla_x \varphi \cdot \nabla_v f)|v|^2 + f\varphi_t \, dxdv
\end{aligned}$$

for

$$U = \begin{pmatrix} v \\ -\nabla\varphi \end{pmatrix}, \quad \nabla = \begin{pmatrix} \nabla_x \\ \nabla_v \end{pmatrix}.$$

Since

$$\begin{aligned}
(v \cdot \nabla_x f)\varphi - (\nabla_x \varphi \cdot \nabla_v f)|v|^2 &= \nabla_x \cdot (vf\varphi) - \nabla_x \varphi \cdot (vf + |v|^2 \nabla_v f) \\
&= \nabla_x \cdot (vf\varphi) - \nabla_x \varphi \cdot \nabla_v \cdot ((v \otimes v) f).
\end{aligned}$$

it holds that

$$\dot{E} = \int \int j \cdot v + \frac{1}{2} f\varphi_t \, dxdv$$

Entropy production, on the other hand, is equal to

$$\dot{S} = \frac{dS}{dt} = \iint s'(f) f_t = \iint s'(f)[-v \cdot \nabla_x f + \nabla_x \varphi \cdot \nabla_v f - \nabla_v \cdot j]$$
$$= \iint s''(f) j \cdot f - v \cdot \nabla_x s(f) + \nabla_x \varphi \cdot \nabla_v s(f) \, dx dv = \iint s''(f) j \cdot f$$

by (5.246).

By the maximum entropy production, the general dissipation term j maximizes $\dot{S}$ under the constraint of $\dot{E} = 0$, where the boundedness of $|j|^2/(2f)$ is required. This formulation is used in [12] for the system of point vortices. Taking variation in j, we obtain

$$\delta \dot{S} - \beta(t) \delta \dot{E} - \iint \frac{1}{D} \cdot \delta\left(\frac{|j|^2}{2f}\right) = 0 \tag{5.247}$$

with the Lagrangian multipliers $\beta(t)$ and D. Equality (5.247) implies

$$j = D\left[f s''(f) \nabla_v f - \beta(t) f v\right]$$

and hence

$$f_t + v \cdot \nabla_x f - \nabla_x \varphi \cdot \nabla_v f = \nabla_v \cdot D[-f s''(f) \nabla_v f + \beta(t) f v] \tag{5.248}$$

by (5.243).

Viewing the mean field of particles via canonical ensembles, $\beta(t)$ in (5.248) is a constant. In the case of micro-canonical ensembles, we prescribe the constant E to couple (5.248) and (5.245). The Kramers equation (5.240) arises with the combination of canonical statistics and Boltzmann entropy

$$s(f) = -f(\log f - 1) \tag{5.249}$$

in (5.248). Fundamental equations in fluid dynamics arise by taking several moments of the general Kramers equation (5.248) with the canonical ensemble of constant temperature.

In taking the zero-th moment we integrate with respect to v in (5.248), which results in the equation of continuity

$$\frac{\partial \rho}{\partial t} + \nabla \cdot \rho u = 0, \tag{5.250}$$

where ρ is the particle density defined by (5.244) and

$$u = \frac{1}{\rho} \int f v \, dv$$

is the particle velocity. Proceeding to the first moment, let $w = u - v$ and

$$P_{ij} = \int f w_i w_j \, dv$$

be the relative velocity and stress tensor, respectively. Then the equation of motion is obtained by multiplying (5.248) by v, which results in

$$\frac{\partial}{\partial t}(\rho u_i) + \frac{\partial}{\partial x_j}(\rho u_i u_j) + \frac{\partial}{\partial x_j} P_{ij} + \rho \frac{\partial \varphi}{\partial x_i} = D \int f s''(f) \frac{\partial f}{\partial v_i} + \beta f v_i \, dv.$$

If $A(f)$ denotes a primitive function of $f s''(f)$ it holds that

$$\int f s''(f) \frac{\partial f}{\partial v_i} \, dv = \int \frac{\partial}{\partial v_i} A(f) \, dv = 0,$$

and hence the equation of motion takes the form of

$$\frac{\partial}{\partial t}(\rho u_i) + \frac{\partial}{\partial x_j}(\rho u_i u_j) + \frac{\partial}{\partial x_j} P_{ij} + \rho \frac{\partial \varphi}{\partial x_i} = -D\beta\rho u_i. \tag{5.251}$$

Finally, the energy equation arises with the second moment [12].

Fluid motion under thermal equilibrium causes vanishing of the right-hand side of (5.251), and if this fluid is incompressible and isotropic the density ρ is constant and it holds also that

$$P_{ij} = p\delta_{ij}. \tag{5.252}$$

Hence we obtain Euler's equation of motion

$$v_t + (v \cdot \nabla)v = -\nabla p, \ \nabla \cdot v = 0 \quad \text{in } \mathbf{R}^3 \times (0, T) \tag{5.253}$$

by (5.250).

In case of unknown ρ, we apply the principle of minimal free energy besides (5.250) and (5.251). Since this process arises at each time and for the moment we refrain from writing t explicitly. The Helmholtz free energy for the canonical ensemble is defined by

$$F = E - TS = \int\int \frac{1}{2}(|v|^2 + \varphi) f - T s(f) \, dx dv, \tag{5.254}$$

where T is the temperature. Then $\beta = T^{-1}$ is regarded as the inverse temperature by (5.247). In the local thermal equilibrium there arises the minimal (5.254) under the constraint of $\int\int f \, dx dv = \lambda$. Hence this f satisfies

$$s'(f) = \beta(\frac{|v|^2}{2} + \lambda(x)) \tag{5.255}$$

with the Lagrangian multiplier $\lambda(x)$. Equality (5.255) implies

$$f = A(\beta(\frac{|v|^2}{2} + \lambda(x))), \tag{5.256}$$

where $f = A(\sigma)$ denotes the inverse function of $\sigma = s'(f)$.

The density ρ and the pressure p are defined by

$$\rho = \int f \, dv \tag{5.257}$$

and

$$\begin{aligned} 3p &= \int f|w|^2 \, dv = \int f|v|^2 \, dv - \int fu^2 \, dv \\ &= \int fv^2 \, dv - \frac{1}{\rho}(\int fv \, dv)^2 \end{aligned} \tag{5.258}$$

for $w = u - v$, respectively. If equalities (5.256)–(5.258) ensure a functional relation

$$p = p(\rho), \tag{5.259}$$

this fluid is said to be barotropic, and Eq. (5.259) is called the state equation.

Assuming (5.252), we obtain

$$\frac{\partial \rho}{\partial t} + \nabla \cdot \rho u = 0, \quad \rho \frac{Du}{Dt} = -\nabla p - \rho \nabla \varphi - \xi \rho u \tag{5.260}$$

by (5.250) and (5.251), where $\xi = D\beta$ and

$$\frac{D}{Dt} = \frac{\partial}{\partial t} + u \cdot \nabla.$$

Then (5.259)–(5.260) imply the compressible Euler equation,

$$\frac{\partial \rho}{\partial t} + \nabla \cdot \rho u = 0, \quad \rho \frac{Du}{Dt} = -\nabla p - \rho \nabla \varphi, \quad p = p(\rho)$$

the fundamental equation of a gas. In the high friction limit, the second term on the right-hand side of (5.260) is zero, and then there arises the general Smoluchowski equation

$$\frac{\partial \rho}{\partial t} = \nabla[\frac{1}{\xi}(\nabla p + \rho \nabla \varphi)], \tag{5.261}$$

where the H-theorem and viral identity hold [70].

The state equation (5.259) is determined by the entropy density $s = s(f)$, (5.256), and (5.258). For the Boltzmann entropy (5.249), equality (5.256) becomes

$$f = e^{-\beta|v|^2/(2k)} \cdot e^{-\lambda(x)/k}$$

which implies the Boyle–Charles law

$$p = k\rho T. \tag{5.262}$$

Then the classical Smoluchowski equation arises with (5.262) and (5.261). If the physical constants are one we obtain

$$\frac{\partial \rho}{\partial t} = \nabla \cdot (\nabla \rho + \rho \nabla \varphi). \tag{5.263}$$

For the self-gravitating particles in astrophysics, the above $\varphi = \varphi(x, t)$ is the gravitational potential made by the particle density ρ. Hence the state equation (5.259) and generalized Smoluchowski equation (5.261) are coupled with the Poisson equation

$$\Delta\varphi = \rho. \tag{5.264}$$

If $n = 2$ the Boltzmann entropy (5.249) is associated with the critical exponent. For the case of total space $\mathbf{R}^2$, a form of the Smoluchowski–Poisson equation (5.263)–(5.264) is given by

$$u_t = \Delta u - \nabla \cdot (u \nabla \Gamma * u),\ u \geq 0 \quad \text{in } \mathbf{R}^2 \times (0, T),$$

where

$$\Gamma(x) = \frac{1}{2\pi} \log \frac{1}{|x|} \tag{5.265}$$

denotes the fundamental solution to $-\Delta$: $-\Delta\Gamma = \delta$.

Total mass conservation and the H-theorem take the form

$$\frac{d}{dt}\int_{\mathbf{R}^2} u = 0, \quad \frac{d}{dt}\mathscr{F}(u) = -\int_{\mathbf{R}^2} u|\nabla(\log u - v)|^2 \leq 0$$

for

$$\mathscr{F}(u) = \int_{\mathbf{R}^2} u(\log u - 1) - \frac{1}{2}\langle \Gamma * u, u \rangle.$$

The viral theorem, on the other hand, is represented by

$$\frac{dI}{dt} = 4\lambda - \frac{1}{2\pi}\lambda^2, \quad I = \int_{\mathbf{R}^2} |x|^2 u, \quad \lambda = \int_{\mathbf{R}^2} u.$$

Then Eq. (3.78) describes the stationary state, that is, the Liouville equation

$$-\Delta v = e^v \text{ in } \mathbf{R}^2, \quad \int_{\mathbf{R}^2} e^v = \lambda.$$

5.5.4 Blowup in Finite Time

There is a quantized blowup mechanism to (5.232) in stationary states, in finite time, and in infinite time, where $\Omega \subset \mathbf{R}^2$ is a bounded domain with smooth boundary [75, 77, 80, 83]. We note that system (5.232) is divided into the Smoluchowski part

$$u_t = \nabla \cdot (\nabla u - u \nabla v) \text{ in } \Omega \times (0, T), \quad \left. \frac{\partial u}{\partial \nu} - u \frac{\partial v}{\partial \nu} \right|_{\partial \Omega} = 0 \tag{5.266}$$

and the Poisson part

$$-\Delta v = u - \frac{1}{|\Omega|} \int_\Omega u \, dx, \quad \left. \frac{\partial v}{\partial \nu} \right|_{\partial \Omega} = 0, \quad \int_\Omega v \, dx = 0. \tag{5.267}$$

The initial condition is provided with

$$u|_{t=0} = u_0(x). \tag{5.268}$$

To make this model well-posed, here we added the last condition to (5.267), because otherwise, the solution is not definite up to an additional constant. This modification, however, has no effect on (5.266), where only ∇v is involved in the first equation.

The above stated quantized blowup mechanism is valid even for the other Smoluchowski–Poisson equation where (5.267) is replaced by

$$-\Delta v = u, \quad v|_{\partial \Omega} = 0. \tag{5.269}$$

In more detail, boundary blowup points are excluded for both blowup in finite time and blowup in infinite time for (5.266) with (5.269), while boundary blowup points actually exist for (5.266) with (5.267) where the collapse masses are reduced to half of those of interior blowup points.

Description of the kinetic blowup mechanism, that is, blowup in finite time and in infinite time, becomes simpler if we take the system on the whole space $\mathbf{R}^2$,

$$u_t = \Delta u - \nabla \cdot (u \nabla \Gamma * u) \text{ in } \mathbf{R}^2 \times (0, T), \quad u|_{t=0} = u_0(x) \geq 0, \tag{5.270}$$

where

$$(\Gamma * u)(x, t) = \int_{\mathbf{R}^2} \Gamma(x - x') u(x', t) \, dx'$$

for $\Gamma = \Gamma(x)$ defined by (5.265). If

$$u_0 \in X \equiv L^1(\mathbf{R}^2) \cap L^\infty(\mathbf{R}^2)$$

the semigroup theory guarantees a unique local-in-time solution $0 \leq u \in C([0, T), X)$ smooth in (x, t) for $t > 0$, provided with the total mass conservation

$$\|u(\cdot, t)\|_1 = \lambda \equiv \|u_0\|_1. \tag{5.271}$$

If the maximal existence time denoted by $T = T_{\max}$ is finite, it follows that

$$\lim_{t \uparrow T} \|u(\cdot, t)\|_\infty = +\infty, \tag{5.272}$$

because the existence time of the solution is estimated below by $\|u_0\|_\infty$.

Given $\varphi \in C_0^2(\mathbf{R}^2)$, we have the weak form of (5.270),

$$\begin{aligned} \frac{d}{dt} \int_{\mathbf{R}^2} u(x, t) \varphi(x) \, dx &= \int_{\mathbf{R}^2} u(x, t) \Delta \varphi(x) \, dx \\ &+ \frac{1}{2} \iint_{\mathbf{R}^2 \times \mathbf{R}^2} \rho_\varphi^0(x, x') u(x, t) u(x', t) \, dx dx' \end{aligned} \tag{5.273}$$

for

$$\rho_\varphi^0(x, x') = -\frac{x - x'}{2\pi |x - x'|^2} \cdot (\nabla \varphi(x) - \nabla \varphi(x')) \in L^\infty(\mathbf{R}^2 \times \mathbf{R}^2),$$

where $C_0^2(\mathbf{R}^2)$ denotes the set of C^2 functions in $\mathbf{R}^2$ with compact support. The process of deriving (5.273) is called *symmetrization*, because it uses the symmetry of the potential kernel for the Poisson part, the origin of which is in the action–reaction law valid for self-interacting particles.

Then, taking $\varphi = |x|^2$ is justified in (5.273) if we assume

$$I_0 \equiv \int_{\mathbf{R}^2} |x|^2 u_0(x) \, dx < +\infty, \tag{5.274}$$

which results in

$$\frac{d}{dt}\int_{\mathbf{R}^2} |x|^2 u(x,t)\, dx = 4\lambda - \frac{\lambda^2}{2\pi}. \tag{5.275}$$

Equality (5.275) implies

$$\lambda > 8\pi \quad \Rightarrow \quad T < +\infty \tag{5.276}$$

from

$$\sup_{0\le t<T} \int_{\mathbf{R}^2} |x|^2 u(x,t)\, dx \le T(4\lambda - \frac{\lambda^2}{4\pi}) + I_0. \tag{5.277}$$

Remark 5.10 As a consequence of Theorem 5.13 below, there arises

$$\lambda \le 8\pi \quad \Rightarrow \quad T = +\infty, \tag{5.278}$$

conversely to (5.276). In more detail, we have a blowup in finite time, blowup in infinite time, and uniformly bounded global-in-time solution to (5.270) with (5.271) for $u_0 \in L^\infty(\mathbf{R}^2)$ satisfying (5.274), if $\lambda > 8\pi$, $\lambda = 8\pi$, and $\lambda < 8\pi$, respectively.

The other property than (5.276) derived from equality (5.275) is the boundedness of the blowup set defined by

$$\mathscr{S} = \{x_0 \in \mathbf{R}^2 \cup \{\infty\} \mid \exists x_k \to x_0,\ \exists t_k \uparrow T,\ \lim_{k\to\infty} u(x_k, t_k) = +\infty\}, \tag{5.279}$$

where $\mathbf{R}^2 \cup \{\infty\}$ denotes the one-point compactification of $\mathbf{R}^2$. In fact, using the following lemma derived from the Gagliardo–Nirenberg inequality, we obtain $\mathscr{S} \subset \mathbf{R}^2$ by (5.271), (5.273), and Chebyshev's inequality applied to (5.277). This lemma is called an *ε-regularity* (Chapter 11 of [75]).

Lemma 5.29 *There is $\varepsilon_0 > 0$ such that*

$$\lim_{R\downarrow 0} \limsup_{t\uparrow T} \|u(\cdot,t)\|_{L^1(B(x_0,R))} < \varepsilon_0 \quad \Rightarrow \quad x_0 \notin \mathscr{S}.$$

Proof The proof is divided into two parts. First, we show the existence of $\varepsilon_0 > 0$ such that

$$\limsup_{t\uparrow T} \|u(\cdot,t)\|_{L^1(\Omega\cap B(x_0,4R))} < \varepsilon_0$$

$$\Rightarrow \quad \limsup_{t\uparrow T} \int_{\Omega\cap B(x_0,2R)} [u(\log u - 1)](x,t)\, dx < +\infty \tag{5.280}$$

for $R > 0$. Second, it follows that

$$\limsup_{t\uparrow T} \int_{\Omega\cap B(x_0,2R)} [u(\log u - 1)](x,t)\, dx < +\infty$$
$$\Rightarrow \quad \limsup_{t\uparrow T} \|u(\cdot,t)\|_{L^\infty(B(x_0,R))} < +\infty. \tag{5.281}$$

Here, a nice cut-off function is used for the proof. It is $\varphi = \varphi_{x_0,R}(x) \in C_0^2(\mathbf{R}^2)$ with the support radius $0 < R \leq 1$ satisfying

$$0 \leq \varphi = \varphi_{x_0,R}(x) = \begin{cases} 1, & x \in B(x_0, R/2) \\ 0, & x \in B(x_0, R)^c, \end{cases} \tag{5.282}$$

and

$$|\nabla \varphi_{x_0,R}| \leq CR^{-1}\varphi^{5/6}, \quad |\nabla^2 \varphi_{x_0,R}| \leq CR^{-2}\varphi^{2/3}. \tag{5.283}$$

This φ is obtained as $\varphi = \psi^6$ for $\psi = \psi_{x_0,R}$ satisfying (5.282).

Property (5.280) follows from the use of a local entropy, while (5.281) is a consequence of a standard argument of parabolic regularity; Gagliardo–Nirenberg inequality and Moser's iteration as in Chapter 11 of [75]. □

Writing

$$u_t = \nabla \cdot (u\nabla(\log u - \Gamma * u))$$

for (5.270) is justified under additional assumption on u_0, to control the behavior of $u(x,t)$ for $|x| \gg 1$. Then the *free energy*

$$\mathcal{F}(u) = \int_{\mathbf{R}^2} u(\log u - 1)\, dx - \frac{1}{2}\int\int_{\mathbf{R}^2\times\mathbf{R}^2} \Gamma(x - x')u \otimes u$$

defined for $u \otimes u = u(x,t)u(x',t)\, dxdx'$ satisfies

$$\frac{d}{dt}\mathcal{F}(u) = -\int_{\mathbf{R}^2} u|\nabla(\log u - v)|^2\, dx \leq 0. \tag{5.284}$$

The *dual Trudinger–Moser inequality*

$$\inf\{\mathcal{F}(u) \mid u \geq 0,\ \|u\|_1 = 8\pi\} > -\infty \tag{5.285}$$

now guarantees global-in-time existence of uniformly bounded solution to (5.270) for the case of $\lambda < 8\pi$. Hence $\lambda = 8\pi$ is the threshold for the existence of the solution global in time as is described in Remark 5.10.

To avoid technical difficulties in controlling the behavior of the solution as $|x| \to +\infty$, however, we restrict the use of this free energy for problems on bounded domain. This case is treated later in Sects. 5.5.8–5.5.11. Here, we show the following theorem on (5.270), which assures (5.278), without using $\mathscr{F}(u)$.

Theorem 5.13 *Given*

$$0 \le u_0 \in L^1(\mathbf{R}^2) \cap L^\infty(\mathbf{R}^2), \quad |x|^2 u_0 \in L^1(\mathbf{R}^2),$$

let $u = u(\cdot, t)$ be the solution to (5.270). Assume blowup in finite time indicated by $T < +\infty$. Then $u(x,t)dx = \mu(dx,t)$ is extended up to $t = T$ as

$$\mu(dx,t) \in C_*([0,T], \mathscr{M}(\mathbf{R}^2 \cup \{\infty\})), \tag{5.286}$$

where $\mathscr{M}(\mathbf{R}^2 \cup \{\infty\}) = C(\mathbf{R}^2 \cup \{\infty\})'$. It holds that $\sharp\mathscr{S} < +\infty$, $\mathscr{S} \subset \mathbf{R}^2$, and

$$\mu(dx,T) = \sum_{x_0 \in \mathscr{S}} m(x_0)\delta_{x_0}(dx) + f(x)dx, \tag{5.287}$$

where $0 < f = f(x) \in L^1(\mathbf{R}^2) \cap C(\mathbf{R}^2 \setminus \mathscr{S})$ and $m(x_0) \in 8\pi\mathbf{N}$.

Proof We have readily shown $\mathscr{S} \subset \mathbf{R}^2$ by inequality (5.277) and Lemma 5.29. Let $C_b^2(\mathbf{R}^2)$ be the set of C^2 functions in $\mathbf{R}^2$, bounded up to their second derivatives.

First, the weak form (5.273) implies

$$\left| \frac{d}{dt} \int_{\mathbf{R}^2} u(x,t)\varphi(x)\, dx \right| \le C \|\nabla\varphi\|_{C^1}(\lambda + \lambda^2) \tag{5.288}$$

for each $\varphi \in C_b^2(\mathbf{R}^2)$. Inequality (5.288) is called the *monotonicity formula.* Since this inequality implies

$$\int_0^T \left| \frac{d}{dt} \int_{\mathbf{R}^2} u(x,t)\varphi(x)\, dx \right| dt < +\infty,$$

the extension of $u(x,t)dx$ to

$$\mu(dx,t) \in C_*([0,T], \mathscr{M}(\mathbf{R}^2 \cup \{\infty\}))$$

is achieved.

Second, the ε regularity, Lemma 5.29, implies

$$\mu(\{x_0\}, T) \ge \varepsilon_0, \quad \forall x_0 \in \mathscr{S}.$$

Hence $\mathscr{S} < +\infty$ follows from the total mass conservation, $\mu(\mathbf{R}^2, T) = \lambda$. It holds also that (5.287) with

$$m(x_0) \geq \varepsilon_0, \quad 0 \leq f = f(x) \in L^1(\mathbf{R}^2).$$

The property $f \in L^\infty_{loc}(\mathbf{R}^2 \setminus \mathscr{S})$ also follows from Lemma 5.29 and (5.286). Therefore, $f = f(x)$ is smooth in $\mathbf{R}^2 \setminus \mathscr{S}$ by the elliptic–parabolic regularity of

$$u_t = \Delta u - \nabla \cdot (u \nabla v), \; -\Delta v = u \quad \text{in } \mathbf{R}^2 \times (0, T) \tag{5.289}$$

derived from (5.270). Then the strong maximum principle to (5.289) guarantees

$$f = f(x) > 0, \quad \forall x \in \mathbf{R}^2 \setminus \mathscr{S}$$

unless $f \equiv 0$. This case of $f \equiv 0$ is not consistent with $T < +\infty$ because it implies $u_0 \equiv 0$ and hence $u \equiv 0$.

We have thus (5.286)–(5.287) with

$$m(x_0) \geq \varepsilon_0, \quad 0 < f = f(x) \in L^1(\mathbf{R}^2) \cap C(\mathbf{R}^2 \setminus \mathscr{S}),$$

and therefore, only *quantization of the collapse mass*,

$$m(x_0) \in 8\pi \mathbf{N}, \quad \forall x_0 \in \mathscr{S}$$

is left to complete the proof of Theorem 5.13 below. □

Now we show $m(x_0) \in 8\pi\mathbf{N}$ in (5.287) for any $x_0 \in \mathscr{S}$. The next lemma is a refinement of Lemma 5.29, based on the smoothing effect of several norms of the solution (Chapter 12 of [77]).

Lemma 5.30 ([68]) *There is $\varepsilon_0 > 0$ such that if u is a classical solution to (5.270) and if*

$$\|u_0\|_{L^1(B(x_0,2R))} < \varepsilon_0, \; u_0 = u|_{t=0}, \quad x_0 \in \mathbf{R}^2, \; R > 0,$$

then it follows that

$$\|u\|_{L^\infty(B(x_0,R)\times(\tau,t_0))} \leq C, \quad 0 < \forall \tau < t_0, \tag{5.290}$$

for $0 < t_0 \ll 1$ determined by R.

Proof We note the following issues in advance. First, by (5.288) there is $t_1 \in (0, T)$ such that

$$\|u_0\|_{L^1(B(x_0,4R))} < \varepsilon_0/2 \Rightarrow \sup_{t \in (0,t_1)} \|u(\cdot, t)\|_{L^1(B(x_0,2R))} < \varepsilon_0. \tag{5.291}$$

Second, property (5.280) relies on the inequality derived from (5.270),

$$\frac{d}{dt}\int_{\mathbf{R}^2} u(\log u - 1)\varphi\, dx + \frac{1}{4}\int_{\mathbf{R}^2} u^{-1}|\nabla u|^2\varphi\, dx \le 2\int_{\mathbf{R}^2} u^2\varphi\, dx + C_\varphi \tag{5.292}$$

applied to $\varphi = \varphi_{x_0,R}$ in (5.282)–(5.283) (Chapter 11 of [75]).

Here the first term on the right-hand side of (5.292) is absorbed into the second term on the left-hand side by the Gagliardo–Nirenberg inequality, under the cost of smallness of the local L^1 norm of the solution as in the conclusion of (5.291). At this stage, however, the estimate of

$$\int_{\mathbf{R}^2} u(\log u - 1)\varphi\, dx$$

at $t > 0$ involves that of $t = 0$. This term, equivalent to a local Zygmund norm [77] of u_0, is stronger than $\|u_0\|_{L^1(B(x_0,2R))}$. Hence parabolic smoothing of this norm is necessary to get (5.290), differently from the proof of (5.280).

For this purpose, we derive

$$\frac{dJ}{dt} + 2\int_{\mathbf{R}^2} u^2\varphi\, dx \le C_R$$

from (5.292) for $\varphi = \varphi_{x_0,R}$ and

$$J = \int_{\mathbf{R}^2} (u\log u + e^{-1})\varphi\, dx,$$

recalling $s\log s + e^{-1} \ge 0$ for $s \ge 0$. Then we use

$$\int_{\mathbf{R}^2} (u\log u + e^{-1})\varphi \le \{\int_{\mathbf{R}^2} [(u\log u + e^{-1})\varphi]^{3/2}\}^{2/3}|B_R|^{1/3}$$

to infer

$$\frac{dJ}{dt} + 3J^{3/2} \le C_R + \int_{\mathbf{R}^2} [-2u^2 + C_R(u\log u + e^{-1})^{3/2}]\varphi\, dx.$$

An elementary inequality on the right-hand side then guarantees

$$\frac{dJ}{dt} + 3J^{3/2} \le C_R'. \tag{5.293}$$

Here we use

$$\frac{d}{dt}t^{-2} + 3(t^{-2})^{3/2} = t^{-3}$$

to derive

$$J(t) \leq t^{-2}, \quad 0 < t \leq \min\{t_1, t_0\}, \tag{5.294}$$

where $t_0^{-3} = C'_R$. Regarding (5.294) as a parabolic smoothing, we apply an iteration scheme to reach the conclusion (Chapter 12 of [75]). □

By Lemma 5.30 and the monotonicity formula (5.288) applied to direct and reverse directions of t, there exist $0 < \varepsilon_0 \ll 1$, $0 < \sigma_0 \ll 1$, and $C > 0$ such that

$$\|u_0\|_1 < \varepsilon_0 \quad \Rightarrow \quad \sup_{|t|<\sigma_0} \|u(\cdot, t)\|_\infty \leq C,$$

if $u = u(x, t)$ is a classical solution to

$$u_t = \Delta u - \nabla \cdot (u\nabla\Gamma * u) \text{ in } \mathbf{R}^2 \times (-T, T), \quad u_0 = u|_{t=0} \tag{5.295}$$

for $T > 0$. There is also a scaling invariance of (5.295) described by

$$u_\mu(x, t) = \mu^2 u(\mu x, \mu^2 t), \quad \mu > 0. \tag{5.296}$$

Thus if $u = u(x, t)$ is a solution then $u_\mu = u_\mu(x, t)$ also solves (5.295). Hence we can make Lemma 5.30 up to the following scaling invariant form [80].

Lemma 5.31 *There are positive constants ε_0, σ_0, and C such that*

$$\begin{aligned} &\|u_0\|_{L^1(B(x_0,2R))} < \varepsilon_0, \quad u_0 = u|_{t=0} \\ &\Rightarrow \sup_{t\in[-\sigma_0 R^2, \sigma_0 R^2]\cap(-T,T)} \|u(\cdot, t)\|_{L^\infty(B(x_0,R))} \leq CR^{-2} \end{aligned} \tag{5.297}$$

for any $x_0 \in \mathbf{R}^2$ and $R > 0$, if $u = u(x, t) \geq 0$ is a classical solution to (5.295).

Remark 5.11 Given $x_0 \in \mathscr{S}$, take $0 < R \ll 1$ satisfying

$$\|f\|_{L^1(B(x_0,4R))} < \varepsilon_0/2$$

for $0 < f = f(x) \in L^1(\mathbf{R}^2) \cap C(\mathbf{R}^2 \setminus \mathscr{S})$ in (5.287), and apply Lemma 5.31 for $B = B(x_1, R)$ and $x_1 \in \partial B(x_0, 2R)$. Then it follows that

$$\limsup_{t\uparrow T} \sup_{x\in\partial B(x_0,R)} u(x, t) \leq CR^{-2}, \quad 0 < R \ll 1 \tag{5.298}$$

from (5.288) and (5.297). Inequality (5.298), however, does not imply $f \in L^\infty_{loc}(\mathbf{R}^2)$, although $f \in L^\infty(\mathbf{R}^2 \setminus B(0, R))$ for $R \gg 1$ holds by (5.277) and (5.298). An open question here is $f \in L \log L(\Omega)$.

5.5.5 Scaling Limit

To prove $m(x_0) \in 8\pi\mathbf{N}$, the mass quantization in Theorem 5.13, we use the backward self-similar transformation

$$z(y,s) = (T-t)u(x,t), \quad y = (x-x_0)/(T-t)^{1/2}, \quad s = -\log(T-t), \tag{5.299}$$

to reach

$$z_s = \nabla\cdot(\nabla z - z\nabla(\Gamma * z + |y|^2/4)) \quad \text{in } \mathbf{R}^2 \times (-\log T, +\infty) \tag{5.300}$$

together with $z(y,s) \geq 0$ and $\|z(\cdot,s)\|_1 = \lambda$. Similarly to (5.288) it holds also that

$$\left|\frac{d}{dt}\int_{\mathbf{R}^2} z(y,s)\varphi(y)\,dy\right| \leq C_\varphi, \quad \varphi \in C_0^2(\mathbf{R}^2). \tag{5.301}$$

Inequality (5.301) is sufficient to guarantee the weak scaling limit. Thus any $s_k \uparrow +\infty$ admits a subsequence denoted by the same symbol such that

$$z(y, s+s_k)dy \rightharpoonup \zeta(dy,s) \quad \text{in } C_*(-\infty,+\infty;\mathscr{M}(\mathbf{R}^2)), \tag{5.302}$$

where $\mathscr{M}(\mathbf{R}^2) = C_\infty(\mathbf{R}^2)'$ for

$$C_\infty(\mathbf{R}^2) = \{\varphi \in C(\mathbf{R}^2 \cup \{\infty\}) \mid \varphi(\infty) = 0\}.$$

This $\zeta(dy,s) \geq 0$ is a weak solution $z = \zeta(dy,s)$ to

$$z_s = \nabla\cdot(\nabla z - z\nabla(\Gamma * z + |y|^2/4)) \quad \text{in } \mathbf{R}^2 \times (-\infty,+\infty). \tag{5.303}$$

To define this notion, let $\mathscr{E}$ be the closure in $L^\infty(\mathbf{R}^2 \times \mathbf{R}^2)$ of

$$\mathscr{E}_0 = \{\rho_\varphi^0 + \psi \mid \varphi \in C_0^2(\mathbf{R}^2),\ \psi \in C_0(\mathbf{R}^2 \times \mathbf{R}^2)\}$$

for

$$\rho_\varphi^0(y,y') = -\frac{y-y'}{2\pi|y-y'|^2}\cdot(\nabla\varphi(y) - \nabla\varphi(y')) \in L^\infty(\mathbf{R}^2 \times \mathbf{R}^2),$$

where $C_0(\mathbf{R}^2 \times \mathbf{R}^2)$ denotes the set of continuous functions in $\mathbf{R}^2 \times \mathbf{R}^2$ with compact support. Although $\rho_\varphi(y,y')$ is not continuous at $y = y'$, this $\mathscr{E}$ is separable. Thanks to this property, there is $0 \leq \mathscr{K} \in L_*^\infty(0,T;\mathscr{E}')$, called a multiplicated operator, in accordance with $\zeta(dy,s)$ generated by (5.302). It satisfies

$$\|\mathscr{K}(\cdot,s)\|_{\mathscr{E}'} \leq \lambda^2, \quad \mathscr{K}(\cdot,s)|_{C_0(\mathbf{R}^2\times\mathbf{R}^2)} = \zeta(dy,s)\otimes\zeta(dy',s), \text{ a.e. } s,$$

and

$$\frac{d}{ds}\langle\varphi, \zeta(dy,s)\rangle = \left\langle \Delta\varphi + \frac{y}{2}\cdot\nabla\varphi, \zeta(dy,s)\right\rangle + \frac{1}{2}\langle\rho_\varphi^0, \mathscr{K}(\cdot,s)\rangle \tag{5.304}$$

in the sense of distributions in $s \in (-\infty, +\infty)$ for each $\varphi \in C_0^2(\mathbf{R}^2)$, where

$$C_0(\mathbf{R}^2 \times \mathbf{R}^2)$$

denotes the set of continuous functions in $\mathbf{R}^2 \times \mathbf{R}^2$ with compact support.

Remark 5.12 Generally, given a family of weak solutions of which initial values take uniformly bounded total measures, and the associated multiplicated operators denoted by $\mathscr{K}$ above are uniformly bounded in $L_*^\infty(0,T;\mathscr{E}')$. Hence we can subtract a sequence converging weakly to a weak solution. These conditions are always satisfied when we generate a weak solution from their sequences provided with the above properties, although we do not mention it explicitly below.

Let $\varphi = \varphi_{x_0,R}(x) \in C^2(\overline{\Omega})$ be the cut-off function satisfying (5.282)–(5.283). Since

$$\|\nabla\varphi_{x_0,R}\|_{C^1} \le CR^{-2}, \quad 0 < R \ll 1$$

it holds that

$$\left|\int_{\mathbf{R}^2} \varphi_{x_0,R}(x)u(x,t)\,dx - \langle\varphi_{x_0,R}, \mu(dx,T)\rangle\right| \le C_\lambda R^{-2}(T-t),$$

which implies

$$\lim_{b\uparrow+\infty}\limsup_{t\uparrow T}\left|\langle\varphi_{x_0,b(T-t)^{1/2}}, \mu(\cdot,t)\rangle - m(x_0)\right| = 0 \tag{5.305}$$

by (5.287). We thus obtain

$$m(x_0) = \zeta(\mathbf{R}^2, s), \quad -\infty < s < +\infty. \tag{5.306}$$

Equality (5.306) is called the *first parabolic envelope* [75]. At the above process of the scaling limit, we thus pick up collapse mass encountered inside the parabolic envelope.

Similarly, there is the *second parabolic envelope* described by

$$\left\langle |y|^2, \zeta(dy,s)\right\rangle \le C, \quad -\infty < s < +\infty \tag{5.307}$$

which implies the *tightness* of the measures $\{\zeta(dy,s)\}$.

The *scaling back* $A(dy,s)$ of $\zeta(dy,s)$ is now defined by

$$\zeta(dy, s) = e^{-s} A(dy', s'), \quad y' = e^{-s/2} y, \quad s' = -e^{-s}. \tag{5.308}$$

It is a weak solution to

$$A_{s'} = \nabla \cdot (\nabla A - A \nabla \Gamma * A), \ A \geq 0 \quad \text{in } \mathbf{R}^2 \times (-\infty, 0) \tag{5.309}$$

defined similarly to the weak solution to (5.303), with the term $\frac{y}{2} \cdot \nabla \varphi$ on the right-hand side eliminated in (5.304). It satisfies

$$A(\mathbf{R}^2, s') = m(x_0), \quad -\infty < s' < 0.$$

The following lemma is concerned with the full orbit of (5.309) which may be called the *weak Liouville property*.

Lemma 5.32 *If* $a = a(dy, s) \in C_*(-\infty, +\infty; \mathcal{M}(\mathbf{R}^2))$ *is a weak solution to*

$$a_s = \nabla \cdot (\nabla a - a \nabla \Gamma * a), \ a \geq 0 \quad \text{in } \mathbf{R}^2 \times (-\infty, +\infty) \tag{5.310}$$

satisfying $a(\mathbf{R}^2, 0) = M$, *then it follows that either* $M = 0$ *or* $M = 8\pi$.

Proof The proof is similar to the case of classical solution in [42]. First, we have

$$a(\mathbf{R}^2, s) = M, \quad -\infty < s < +\infty.$$

Second, we use the local second moment to infer that $M > 8\pi$ and $M < 8\pi$ contradict (5.310) for $s \geq 0$ and $s \leq 0$, respectively, if the initial mass $a(dy, 0)$ is sufficiently concentrated at the origin.

Then, the scaling invariance of (5.310),

$$a^{\mu}(dy, s) = \mu^2 a(dy', s'), \quad y' = \mu y, \quad s' = \mu^2 s, \tag{5.311}$$

excludes this condition of concentration. □

Remark 5.13 For (5.309) concerning the half orbit $-\infty < s < 0$, the above proof assures only $m(x_0) \geq 8\pi$. Then there is an argument taking $s'_\ell \uparrow +\infty$ and applying the *concentration compactness principle* [46] to the sequence of probability measures $\{A_\ell(dy')\}$ defined by

$$\tilde{A}_\ell(dy') = A(dy', -s'_\ell)/m(x_0)$$

[77]. This argument visualizes the formation of collapses with normalized mass 8π at $s' = -\infty$ to $A(dy', s')$, which concentrates to the origin as $s' \uparrow 0$ with possible collision. The other argument guarantees the formation of collapses with normalized mass to $\zeta(dy, s)$ by the use of the Schwarz symmetrization ([69], Chapter 14 of [75]). There is a different argument in Sect. 5.1.3 from these ones for the proof

of formation of collapses with normalized mass to $a(dy', s')$ in $s' < 0$; the weak Liouville property for the *translation* and *dilation* limits.

To approach (5.303) directly, we confirm that the scale-free ε regularity for rescaled equation, Lemma 5.31, is valid for the weak solution $A(dy', s')$ to (5.309) generated by a family of classical solutions. Hence we have

$$\begin{aligned} & A(B(y_0', 2R), s') < \varepsilon_0 \\ & \Rightarrow \|A(\cdot, s')\|_{L^\infty(B(y_0', R))} \leq CR^{-2}, \quad \forall s' < 0,\ y_0' \in \mathbf{R}^2,\ R > 0. \end{aligned}$$

Then we use the scaling back (5.308) to apply this result to (5.309):

$$\begin{aligned} & \zeta(B(e^{s/2}y_0', 2Re^{s/2}), s) < \varepsilon_0 \\ & \Rightarrow \|\zeta(\cdot, s)\|_{L^\infty(B(e^{s/2}y_0, Re^{s/2})} \leq CR^{-2}e^{-s}, \quad \forall s \in \mathbf{R},\ y_0' \in \mathbf{R}^2,\ R > 0, \end{aligned}$$

which reads

$$\begin{aligned} & \zeta(B(y_0, 2r), s) < \varepsilon_0 \\ & \Rightarrow \|\zeta(\cdot, s)\|_{L^\infty(B(y_0, r))} \leq Cr^{-2}, \quad \forall s \in \mathbf{R},\ y_0 \in \mathbf{R}^2,\ r > 0. \end{aligned} \tag{5.312}$$

This property, combined with the second parabolic envelope (5.307), assures

$$\|\zeta(dy, s)\|_{L^\infty(\mathbf{R}^2 \setminus B(0,R))} \leq C \quad -\infty < s < +\infty \tag{5.313}$$

for $R > 0$ sufficiently large. Writing

$$g(y, s)dy = \zeta^{ac}(dy, s),$$

we thus obtain

$$\|g(\cdot, s)\|_1 + \|g(\cdot, s)\|_{L^\infty(\mathbf{R}^2 \setminus B(0,R))} \leq C, \tag{5.314}$$

which plays a fundamental role in controlling the chemotaxis term $\Gamma * z$ in (5.303).

It holds also that the singular part of $\zeta(dy, s)$ denoted by $\zeta^s(dy, s)$ is composed of a finite sum of delta functions again by the ε regularity. The coefficients of these delta functions are quantized as 8π,

$$\zeta^s(dy, s) = 8\pi \sum_{y_0 \in \mathscr{S}(s)} \delta_{y_0}(dy), \quad \sharp\mathscr{S}(s) < +\infty, \quad \mathscr{S}(s) \subset B(0, 2R)$$

by a result of [69]. This quantization may be proven in a different way, by suitable weak limits and the weak Liouville property through scaling back (5.308)–(5.309), as is described in Sect. 5.1.3.

We thus end up with

$$\zeta(dy,s) = 8\pi \sum_{j=1}^{m(s)} \delta_{y_j(s)}(dy) + g(y,s)dy \tag{5.315}$$

in (5.302), where $m(s) \in \mathbf{N}$, $y_j(s) \in B(0,2R)$, and $g = g(y,s) \geq 0$ satisfies (5.314).

5.5.6 *Residual Vanishing*

Proof *(Theorem 5.13, Completion)* If $m(x_0) \not\in 8\pi\mathbf{N}$ in (5.306) there arises $g(\cdot,s) \not\equiv 0$ and $m(x_0) > 8\pi m(s)$ in (5.315) for any $s \in \mathbf{R}$. Hence we have $\delta > 0$ such that

$$\|g(\cdot,s)\|_1 \geq \delta \quad \text{a.e. } s. \tag{5.316}$$

We show that (5.316) gives a contradiction, because (a) the regular part of $\zeta(dy,s)$ is swept away to $y = \infty$ as $s \uparrow +\infty$ by the $|y|^2/4$ term in (5.303), (b) uniform L^p bound, $1 \leq p \leq \infty$, arises in the outer region of the chemotaxis term by (5.314), and (c) the second parabolic envelope (5.307) implies that $\zeta(dy,s)$ in $\mathbf{R}^2 \setminus B(0,R)$ with $R \gg 1$ is uniformly thin.

First, we use

$$\int_{\mathbf{R}^2} |y|^2 g(y,s)\, dy \leq C \tag{5.317}$$

derived from (c). It holds also that

$$g > 0 \quad \text{in } H = \bigcup\nolimits_{s>0} (\mathbf{R}^2 \setminus \mathscr{S}(s)) \times \{s\} \tag{5.318}$$

from the maximum principle for parabolic equations. By

$$\zeta(dy,s) \in C_*(0,+\infty; \mathscr{M}(\mathbf{R}^2)) \tag{5.319}$$

and (5.315) this set H is open in $\mathbf{R}^2 \times (-\infty,+\infty)$, and furthermore, the elliptic–parabolic regularity implies the smoothness of $g = g(y,s)$ in H. Then there arises

$$\begin{aligned} g_s &= \Delta g - \nabla \cdot g \nabla w \quad \text{in } H \\ w(y,s) &= \frac{|y|^2}{4} + 4\sum_{j=1}^{m(s)} \log \frac{1}{|y - y_j(s)|} + (\Gamma * g)(y,s). \end{aligned} \tag{5.320}$$

Given $s_0 \in \mathbf{R}$, let $\mathscr{S}_\varepsilon(s_0) = \bigcup_{j=1}^m B(y_j(s_0), \varepsilon)$ be the ε-neighborhood of $\mathscr{S}(s_0)$. Then we obtain

$$y_i(s) \in B(y_j(s_0), \varepsilon) \Rightarrow (y - y_i(s)) \cdot \nu_y \geq 0, \ \forall y \in \partial B(y_j(s_0), \varepsilon) \tag{5.321}$$

for $1 \leq j \leq m(s_0)$ and $|s - s_0| \ll 1$ by (5.319), where ν denotes the outer unit normal vector. From (5.320), then it follows that

$$\begin{aligned} \frac{d}{ds} \int_{B(0,r) \setminus \mathscr{S}_\varepsilon(s_0)} g \, dy &= \int_{\partial(B(0,r) \setminus \mathscr{S}_\varepsilon(s_0))} \frac{\partial g}{\partial \nu} - g \frac{\partial w}{\partial \nu} \, dS \\ &\leq \int_{\partial(B(0,r) \setminus \mathscr{S}_\varepsilon(s_0))} \frac{\partial g}{\partial \nu} - g \frac{\partial v}{\partial \nu} \, dS \end{aligned} \tag{5.322}$$

for $r > 2R$, $|s - s_0| \ll 1$, and

$$v = \frac{|y|^2}{4} + (\Gamma * g)(y, s).$$

Here we take $0 < h \ll 1$ and operate $\frac{1}{h} \int_s^{s+h} \cdot \, ds'$ on both sides of (5.320). Then, making $\varepsilon \downarrow 0$ results in

$$\frac{1}{h} \left[\int_{B(0,r)} g(y, s') \, dy \right]_{s'=s}^{s'=s+h} \leq \frac{1}{h} \int_s^{s+h} ds' \int_{\partial B(0,r)} (g_r - g v_r)(\cdot, s') \, dS$$

by (5.321). Since inequality (5.314) implies

$$\|\nabla(\Gamma * g)(\cdot, s)\|_{L^\infty(\mathbf{R}^2 \setminus B(0,2R))} \leq C_0,$$

it holds that

$$\begin{aligned} \int_{\partial B(0,r)} g_r - g v_r \, dS &\leq \int_{\partial B(0,r)} g_r - \frac{r}{2} g + C_0 g \, dS \\ &= \frac{d}{dr}(r \int_0^{2\pi} g \, d\theta) - \int_0^{2\pi} (1 + \frac{r^2}{2}) g \, d\theta + C_0 r \int_0^{2\pi} g \, d\theta. \end{aligned}$$

Then we obtain

$$\begin{aligned} &\frac{1}{h} \left[\int_0^r (r\overline{g})(r, s') \, dr' \right]_{s'=s}^{s'=s+h} \leq \frac{1}{h} \int_s^{s+h} \left[\frac{d}{dr}(r\overline{g}) - (\frac{r^2}{2} - C_1 r + 1)\overline{g} \right] (r, s') \, ds' \\ &\overline{g} = \frac{1}{2\pi} \int_0^{2\pi} g \, d\theta. \end{aligned} \tag{5.323}$$

This inequality means

$$\begin{aligned}\frac{1}{h}\left[B(r,s')\right]_{s'=s}^{s'=s+h} &\leq \frac{1}{h}\int_s^{s+h}(B_{rr}-a(r)B_r)(r,s')\,ds' \\ &= \frac{1}{h}\int_s^{s+h}[e^A(e^{-A}B_r)_r](r,s')\,ds'\end{aligned}$$

in $(\mathbf{R}^2 \setminus B(0,2R)) \times (-\infty,+\infty)$, where

$$\begin{aligned}B = B(r,s) &= \int_0^r r\overline{g}(r,s)\,dr \\ a = a(r) &= \frac{r^2}{2} - C_0 r + \frac{1}{r} \\ A = A(r) &= \frac{r^2}{4} - C_0 r + \log r.\end{aligned}$$

Hence it follows that

$$\frac{1}{h}\left[(e^{-A}B)(r,s')\right]_{s'=s}^{s'=s+h} \leq \frac{1}{h}\int_s^{s+h}(e^{-A}B_r)_r(r,s')\,ds' \tag{5.324}$$

for $r > 2R$ and $0 < s < +\infty$. By (5.316)–(5.317) we obtain

$$B(r,s) = \int_0^r r\overline{g}(r,s)\,dr \geq \frac{\delta}{2} - \frac{C_1}{r^2+1}, \quad r > 0,\ s > 0. \tag{5.325}$$

It holds also that

$$B(r,s) > 0, \quad r > 0,\ s \in \mathbf{R}$$

by (5.318). Now we take $r_1 > 2R$ such that

$$a(r) \geq \frac{r}{4} + 1, \quad \frac{\delta}{2} - \frac{C_1}{r^2+1} \geq \frac{\delta}{4} \qquad \forall r \geq r_1, \tag{5.326}$$

to put

$$\varphi(r) = \begin{cases} c_1 r + c_2, & r \geq r_2 \\ \sin\beta(r-r_1), & r_1 \leq r_2 = r_1 + \frac{\pi}{4\beta}, \end{cases} \tag{5.327}$$

using $c_2 = \frac{1}{\sqrt{2}}(1-\frac{\pi}{4}) - r_1 c_1$ and $\beta = \sqrt{2}c_1$ defined for $0 < c_1 \ll 1$. It holds that

$$0 \leq \varphi = \varphi(r) \in C^1[r_1,\infty), \quad \varphi(r_1) = 0, \quad \varphi_r > 0 \text{ in } (r_1,+\infty) \tag{5.328}$$

and

$$\int_{r_1}^{\infty} e^{-A}\varphi\, dr < +\infty, \qquad \lim_{r\uparrow+\infty} e^{-A}(\varphi+\varphi_1) = 0. \tag{5.329}$$

We have also $r_2 \uparrow +\infty$ as $c_1 \downarrow 0$, and then

$$\varphi_{rr} - a\varphi_r \le -\mu\varphi \tag{5.330}$$

is valid for $r \ge r_2$ with $0 < \mu \ll 1$ by (5.325)–(5.326). Since

$$\varphi_r > 0,\ \varphi_{rr} = -\beta^2\varphi, \quad r_1 < r < r_2,$$

inequality (5.330) is extended to $r \ge r_1$ for $0 < \mu \ll 1$.

Now we apply (5.329)–(5.330) to (5.324). First, this inequality holds for $r \ge r_1$, $-\infty < s < +\infty$, and therefore, we obtain

$$\frac{1}{h}\int_{r_1}^{\infty} \left[(e^{-A}B)(r,s')\varphi(r)\right]_{s'=s}^{s'=s+h} dr \le \int_s^{s+h}\int_{r_1}^{\infty} [(e^{-A}B_r)_r\varphi](r,s')\, dr.$$

Second, by

$$B(\cdot,s') > 0, \quad B_r(\cdot,s') > 0, \quad B(\infty,s') \le 2\pi C$$

there is $r_j \uparrow +\infty$ such that $B_r(s',r_j) \to 0$, which guarantees

$$\begin{aligned}\int_{r_1}^{\infty} (e^{-A}B_r)_r(s',r)\, dr &= -\int_{r_1}^{\infty} (e^{-A}B_r)(r,s')\varphi_r(r)\, dr \\ &\le \int_{r_1}^{\infty} [(e^{-A}\varphi_r)_r B](r,s')\, dr \\ &= \int_{r_1}^{\infty} [(\varphi_{rr} - a\varphi_r)e^{-A}B](r,s')\, dr\end{aligned}$$

by (5.328). Then it follows that

$$\frac{1}{h}\left[\int_{r_1}^{\infty} (e^{-A}B\varphi)(r,s')\, dr\right]_{s'=s}^{s'=s+h} \le -\mu\cdot\frac{1}{h}\int_s^{s+h} ds' \int_{r_1}^{\infty} (e^{-A}B\varphi)(r,s')\, dr$$

and hence

$$\frac{d^+}{ds}\int_{r_1}^{\infty} e^{-A}B\varphi\, dr \le -\mu\int_{r_1}^{\infty} e^{-A}B\varphi\, dr, \quad 0 < s < +\infty, \tag{5.331}$$

where

$$\frac{d^+}{ds}H(s) = \limsup_{h\downarrow 0} \frac{1}{h}\left(H(s+h) - H(s)\right).$$

Since (5.331) implies

$$\frac{d^+}{ds}\left(e^{\mu s/2}\int_{r_1}^{\infty} e^{-A}B\varphi\, dr\right) \leq -\frac{\mu}{2}e^{\mu s/2}\int_{r_1}^{\infty} e^{-A}B\varphi\, dr < 0$$

the mapping $s \mapsto e^{\mu s/2}\int_{r_1}^{\infty} e^{-A}B\varphi\, dr$ is monotone decreasing. It holds that

$$e^{\mu s/2}\int_{r_1}^{\infty}[(e^{-A}B\varphi](r,s)\, dr \leq \int_{r_1}^{\infty}[(e^{-A}B\varphi](r,0)\, dr, \quad s>0$$

and hence

$$\lim_{s\uparrow+\infty}\int_{r_1}^{\infty}(e^{-A}B\varphi)(r,s)\, dr = 0.$$

This equality is a contradiction because

$$\int_{r_1}^{\infty}(e^{-A}B\varphi)(r,s)dr \geq \frac{\delta}{4}\int_{r_1}^{\infty}(e^{-A}\varphi)(r)\, dr > 0$$

follows from (5.323) and (5.325).

Hence we have $g \equiv 0$, which implies $m(x_0) \in 8\pi\mathbf{N}$ and

$$\zeta(dy,s) = 8\pi\sum_{j=1}^{m}\delta_{y_j(s)}(dy) \tag{5.332}$$

with $m = m(x_0)/(8\pi)$. The proof of Theorem 5.13 is thus complete. □

Remark 5.14 The method of *outer second moment* induces an alternative proof of residual vanishing, which is more consistent with the proof of blowup in infinite time below, that is, Lemma 5.38 in Sect. 5.5.11. Using (5.307), we have thus readily shown in (5.313) the existence of $R \gg 1$ such that

$$\begin{aligned}&\zeta(dy,s) = \zeta^{ac}(dy,s), \qquad |y| > \frac{R}{2}\\ &\|\zeta(\cdot,s)\|_{L^\infty(\mathbf{R}^2\setminus B_{R/2})} \leq C, \quad -\infty < s < +\infty\end{aligned} \tag{5.333}$$

with $\zeta^{ac}(dy,s) = g(y,s)dy$, which implies

$$\|\nabla\Gamma * \zeta(\cdot,s)\|_{L^\infty(\mathbf{R}^2\setminus B_R)} \leq C, \quad -\infty < s < +\infty. \tag{5.334}$$

It holds also that

$$\|z_k(\cdot, s)\|_{L^\infty(\mathbf{R}^2\setminus B_{R/2})} \leq C, \quad -\log T - s_k < s < +\infty$$

for $z_k = z(\cdot, s + s_k)$, $k \gg 1$, similarly by Lemma 5.31, because (5.302) and (5.313) imply

$$\|z_k(\cdot, s)\|_{L^1(\mathbf{R}^2\setminus B_{R/4})} < \varepsilon_0$$

for $R \gg 1$. Hence we obtain

$$\|\nabla \Gamma * z_k(\cdot, s)\|_{L^\infty(\mathbf{R}^2\setminus B_R)} \leq C, \quad -\log T - s_k < s < +\infty. \tag{5.335}$$

Equation (5.300) now implies

$$\frac{d}{ds}\int_{|y|\geq R} z_k\varphi \geq \int_{|y|\geq R} z_k \Delta\varphi - C z_k \varphi_r + z_k \frac{y}{2}\cdot\nabla\varphi \, dy \tag{5.336}$$

for $\varphi = \varphi(|y|) \in C^2(|y| \geq R)$ satisfying

$$0 \leq \varphi(|y|) \leq C(1 + |y|^2), \ \varphi_r \geq 0, \quad r = |y| \geq R; \qquad \varphi|_{r=R} = 0. \tag{5.337}$$

We now put

$$\varphi(y) = \xi(|y|/R), \quad \xi(r) = r^2 - 1,$$

which satisfies (5.337). Here, we confirm

$$\Delta\varphi + \frac{y}{2}\cdot\nabla\varphi \geq C\varphi_r, \quad |y| \geq R \gg 1$$

for this φ, which is equivalent to

$$\frac{1}{R^2}\Delta\xi + (\frac{r}{2} - \frac{C}{R})\,\xi_r \geq 0, \quad r \geq 1. \tag{5.338}$$

In fact, since $\Delta\xi = 4$ and $\xi_r = 2r$, inequality (5.338) is achieved for $R \gg 1$. More precisely, we have

$$\min_{r\geq 1} H(r) = H(1)$$

if $R > C$ for

$$H(r) \equiv \frac{1}{R^2}\Delta\xi + (\frac{r}{2} - \frac{C}{R})\,\xi_r = \frac{4}{R^2} + r^2 - \frac{C}{R}2r$$

$$= (r - \frac{C}{R})^2 + \frac{4 - C^2}{R^2},$$

and then it follows that

$$H(1) = \frac{4}{R^2} + 1 - \frac{2C}{R} \geq 0, \quad R \gg 1.$$

We thus deduce

$$\frac{d}{ds} \int_{|y| \geq R} z_k \varphi \, dy \geq 0, \quad -\log T - s_k < s < +\infty$$

and hence

$$\frac{d}{ds} \int_{\mathbf{R}^2} (\frac{|y|^2}{R^2} - 1)_+ \zeta(dy, s) \geq 0, \quad -\infty < s < +\infty \tag{5.339}$$

in the sense of distributions by the second parabolic envelope (5.307). Inequality (5.339) implies

$$\zeta(y, s) = 0, \quad |y| \geq R, \ -\infty < s < +\infty$$

as in Lemma 5.38 in Sect. 5.5.11, again by (5.307). Then the residual vanishing

$$\zeta^{ac}(dy, s) \equiv 0$$

holds by the unique continuation theorem for parabolic equations as in Theorem 5.13 in Sect. 5.5.4.

Remark 5.15 Putting $\varphi = |x|^2$ in (5.304) is justified by (5.306)–(5.307), which results in

$$\frac{dI}{ds} = m(x_0) - \frac{m(x_0)^2}{2\pi} + I, \quad \text{a.e. } s \in \mathbf{R}$$

for the total second moment

$$I = \langle |y|^2, \zeta(dy, s) \rangle$$

[64, 76]. If $m(x_0) = 8\pi$, therefore, it holds that $I(s) \equiv 0$ by (5.307), and hence

$$\zeta(dy, s) = 8\pi \delta_0(dy), \quad -\infty < s < +\infty. \tag{5.340}$$

Remark 5.16 We say that the blowup point $x_0 \in \mathscr{S}$ is *simple* if $m(x_0) = 8\pi$. Then it holds that

$$A(dy', s') = 8\pi\delta_0(dy'), \quad -\infty < s' < 0 \tag{5.341}$$

by (5.340). Otherwise, if $m(x_0) = 8\pi m$ for $m \geq 2$, we say that *collision of sub-collapses* occurs at $x_0 \in \mathscr{S}$. Equality (5.332) then implies

$$A(dy', s') = 8\pi \sum_{j=1}^{m} \delta_{y'_j(s')}(dy'), \quad -\infty < s' < 0. \tag{5.342}$$

In this case we obtain the sub-collapse dynamics,

$$\frac{dy'_j}{ds'} = 8\pi \nabla_{y'_j} H_m(y'_1, \cdots, y'_m), \ 1 \leq j \leq m, \quad \text{a.e. } s' < 0, \tag{5.343}$$

where

$$H_m(y'_1, \cdots, y'_m) = \sum_{1 \leq i < j \leq m} \Gamma(y'_i - y'_j), \quad \Gamma(y') = \frac{1}{2\pi} \log \frac{1}{|y'|} \tag{5.344}$$

is the point vortex Hamiltonian on the whole space. Then the second equality in (5.44),

$$\lim_{s' \uparrow 0} y'_j(s') = 0, \quad 1 \leq j \leq m$$

follows from

$$|y_j(s)| \leq C, \quad -\infty < s < +\infty \tag{5.345}$$

in (5.332).

Remark 5.17 Equation (5.343) with (5.344) is derived from the weak form of (5.309), that is,

$$\frac{d}{ds'}\langle \varphi, A \rangle = \langle \Delta\varphi, A \rangle + \frac{1}{2}\langle \rho^0_\varphi, \mathscr{K}(\cdot, s') \rangle, \quad \text{a.e. } s' < 0$$
$$\rho^0_\varphi(y, y') = -\frac{y - y'}{2\pi |y - y'|^2} \cdot (\nabla\varphi(y) - \nabla\varphi(y')) \in L^\infty(\mathbf{R}^2 \times \mathbf{R}^2). \tag{5.346}$$

Here, $\varphi \in C^2_0(\mathbf{R}^2)$, $0 \leq \mathscr{K} = \mathscr{K}(\cdot, s') \in L^\infty_*(-\infty, 0; \mathscr{E}')$ is the associated multiplicated operator satisfying

$$\mathscr{K}(\cdot, s')\big|_{C_0(\mathbf{R}^2 \times \mathbf{R}^2)} = A(dy, s') \otimes A(dy', s'),$$

and $\mathscr{E}$ is the closure of

$$\mathscr{E}_0 = \{\psi + \rho^0_\varphi \mid \psi \in C_0(\mathbf{R}^2 \times \mathbf{R}^2),\ \varphi \in C^2_0(\mathbf{R}^2)\}$$

in $L^\infty(\mathbf{R}^2 \times \mathbf{R}^2)$. The local second moment applied to (5.346), more precisely, traces the movement of sub-collapses, $y'_j(s')$, $1 \leq j \leq m$, as in the proof of Lemma 5.34 in Sect. 5.5.10, which induces (5.343)–(5.344) [77].

Remark 5.18 We have readily shown (5.332) with (5.345). Inequality (5.307), on the other hand, is uniform with respect to the selection of the sequence $t_k \uparrow T$, and hence so is (5.313). Inequality (5.345), therefore, is also uniform. We thus obtain the type II blowup rate,

$$\limsup_{t\uparrow T}(T-t)\|u(\cdot,t)\|_{L^\infty(B(x_0,b(T-t)^{1/2}))} = +\infty, \quad \exists b > 0,\ \forall x_0 \in \mathscr{S}. \tag{5.347}$$

Remark 5.19 If the blowup point $x_0 \in \mathscr{S}$ is simple as in $m(x_0) = 8\pi$, we obtain (5.340), regardless of the choice of $s_k \uparrow +\infty$. Hence there arises

$$z(y,s)dy \rightharpoonup m_*(x_0)\delta_0(dy) \text{ in } \mathscr{M}(\mathbf{R}^2), \quad s \uparrow +\infty, \tag{5.348}$$

and therefore, the type II blowup rate, (5.347), is refined as

$$\lim_{t\uparrow T}(T-t)\|u(\cdot,t)\|_{L^\infty(\Omega\cap B(x_0,b(T-t)^{1/2})} = +\infty, \quad \forall b > 0. \tag{5.349}$$

Remark 5.20 For the radially symmetric solution $u = u(|x|,t)$, it holds always that (5.341), while the exact blowup rate is known to be the one set for the inner region in the matched asymptotic expansion [27],

$$L(t) = (T-t)^{1/2}e^{-\sqrt{|\log(T-t)|/2}}.$$

If $u = u(|x|,t)$ in $\mathbf{R}^2 \times (0,T)$ with $T < +\infty$, therefore, it holds that

$$u(x,t) = \frac{K}{L(t)^2}u_*(\sqrt{K}x/L(t))(1+o(1)) \quad \text{locally uniformly in } x \in \mathbf{R}^2,$$

where $K = 4^{-1}\exp(e^2+\gamma)$, γ is the Euler constant,

$$u_*(x) = \frac{8}{(1+|x|^2)^2}$$

is a stationary solution derived from the Liouville equation

$$-\Delta v = e^v,\ v \leq v(0) = 0 \text{ in } \mathbf{R}^2, \quad \int_{\mathbf{R}^2} e^v < +\infty$$

through $u_* = e^v$, which results in $\|u_*\|_1 = 8\pi$ [53].

Remark 5.21 There is a formal solution with $m(x_0) = 2$ to (5.270), constructed by the method of matched asymptotic expansion [49].

5.5.7 *Bounded Domains*

For the case of bounded domains, free energy is easier to handle. If $\Omega \subset \mathbf{R}^2$ is a bounded domain with smooth boundary $\partial\Omega$, the Poisson equation (5.267) or (5.269) thus takes Green's function denoted by $G = G(x, x')$. Then it holds that

$$v(x,t) = \int_\Omega G(x,x')u(x',t)\,dx'.$$

This $G = G(x, x')$ is smooth in $\overline{\Omega} \times \overline{\Omega} \setminus \overline{D}$ for $D = \{(x, x) \mid x \in \Omega\}$. The singularity on the diagonal D is distinguished according to the interior and the boundary.

First, interior singularity is given by

$$G(x,x') = \Gamma(x-x') + K(x,x'), \tag{5.350}$$

where

$$K = K(x,x') \in C^{1+\theta,\theta}(\Omega \times \overline{\Omega}) \cap C^{\theta,1+\theta}(\overline{\Omega} \times \Omega)$$

for $0 < \theta < 1$. Second, boundary singularity around $x_0 \in \partial\Omega$ is described by the conformal diffeomorphism

$$X : \overline{\Omega \times B(x_0, 2R)} \to \overline{\mathbf{R}^2_+}$$

defined for $0 < R \ll 1$, where $\mathbf{R}^2_+ = \{(X_1, X_2) \mid X_2 > 0\}$. Using the reflection

$$X = \begin{pmatrix} X_1 \\ X_2 \end{pmatrix} \quad \mapsto \quad X_* = \begin{pmatrix} X_1 \\ -X_2 \end{pmatrix},$$

parametrices of the Poisson part, (5.267) and (5.269), are given by

$$E(x,x') = \Gamma(X-X') + \Gamma(X-X'_*) \tag{5.351}$$

and

$$E(x,x') = \Gamma(X-X') - \Gamma(X-X'_*), \tag{5.352}$$

respectively. Then it follows that

$$G(x, x') = E(X, X') + K(x, x') \tag{5.353}$$

with

$$K = K(x, x') \in C^{1+\theta,\theta} \cap C^{\theta,1+\theta}(\overline{\Omega \cap B(x_0, R)} \times \overline{\Omega \cap B(x_0, R)})$$

[75].

Using these properties of Green's function, we obtain Theorems 5.14–5.15 below, analogous to Theorem 5.13.

Theorem 5.14 *Let $u = u(x, t)$ be the classical solution to (5.266)–(5.268). Assume the blowup in finite time, $T < +\infty$, and define the blowup set by*

$$\mathcal{S} = \{x_0 \in \overline{\Omega} \mid \exists x_k \to x_0,\ \exists t_k \uparrow T,\ \lim_{k\to\infty} u(x_k, t_k) = +\infty\}. \tag{5.354}$$

Then $\sharp\mathcal{S} < +\infty$ and it holds that

$$u(x, t)dx \rightharpoonup \sum_{x_0\in\mathcal{S}} m(x_0)\delta_{x_0}(dx) + f(x)\,dx \quad in\ \mathcal{M}(\overline{\Omega}) \tag{5.355}$$

as $t \uparrow T$, where

$$0 < f = f(x) \in L^1(\Omega) \cap C(\overline{\Omega} \setminus \mathcal{S}), \quad m(x_0) \in m_*(x_0)\mathbf{N}$$

for

$$m_*(x_0) = \begin{cases} 8\pi, & x_0 \in \Omega \\ 4\pi, & x_0 \in \partial\Omega. \end{cases} \tag{5.356}$$

Remark 5.22 For $x_0 \in \Omega \cap \mathcal{S}$ we can define $A(dy', s')$ by (5.302) and (5.308). Then it holds that (5.341) if $m(x_0) = m_*(x_0) = 8\pi$. In the other case of

$$m = m(x_0)/m_*(x_0) \geq 2,$$

we have collapse dynamics (5.342)–(5.343).

Remark 5.23 If $x_0 \in \partial\Omega \cap \mathcal{S}$, we take the zero extension of $z = z(y, s)$ of (5.299) to the region where it is not defined. We get (5.302) and this $\zeta(dy, s)$ has the support included in a closed half space independent of s. Then we take its even reflection denoted by $\tilde{\zeta}(dy, s)$ to define $\tilde{A}(dy', s')$ by (5.308):

$$\tilde{\zeta}(dy, s) = e^{-s}\tilde{A}(dy', s'), \quad y' = e^{-s/2}y, \quad s' = -e^{-s}.$$

The same property as in (5.341) and (5.342)–(5.343) arises to $\tilde{A}(dy', s')$, according to $m = 1$ and $m \geq 2$, respectively, where $m = m(x_0)/m_*(x_0)$.

In the next subsection, we show the exclusion of boundary blowup to (5.266), (5.268), and (5.269) to complete the proof of the result stated in Sect. 5.1.3.

Theorem 5.15 *Let $u = u(x,t)$ be the classical solution to (5.266), (5.268), and (5.269). Assume the blowup in finite time, $T < +\infty$, and define the blowup set $\mathcal{S}$ by (5.354). Then it holds that $\mathcal{S} \subset \Omega$ and $\sharp\mathcal{S} < +\infty$. There arises also (5.355) as $t \uparrow T$, where*

$$0 < f = f(x) \in L^1(\Omega) \cap C(\overline{\Omega} \setminus \mathcal{S}), \quad m(x_0) \in 8\pi\mathbf{N}.$$

Remark 5.24 The same conclusion as in Remark 5.22 for interior blowup points occurs in the above theorem at each $x_0 \in \mathcal{S}$, for $A(dy', s')$ defined by (5.302) and (5.308).

We conclude this subsection with a remark on the free energy. In fact, associated with system (5.266) and (5.267), or (5.266) and (5.269), this free energy is defined with Green's function $G = G(x, x')$, that is,

$$\mathcal{F}(u) = \int_\Omega u(\log u - 1)\, dx - \frac{1}{2} \iint_{\Omega\times\Omega} G(x,x')u \otimes u, \tag{5.357}$$

where $u \otimes u = u(x,t)u(x',t)\, dxdx'$. Then it follows that

$$\frac{d}{dt}\mathcal{F}(u) = -\int_\Omega u|\nabla(\log u - v)|^2\, dx \leq 0 \tag{5.358}$$

for the classical solution $u = u(\cdot, t) > 0$. In fact, we have

$$u = u(x,t) > 0 \quad \text{in } \overline{\Omega} \times (0, T)$$

unless $u_0 \equiv 0$ by the strong maximum principle.

Theorem 5.16 (**[80]**) *Any $x_0 \in \mathcal{S}$ is simple if*

$$T < +\infty, \quad \lim_{t\uparrow T} \mathcal{F}(u(\cdot,t)) > -\infty \tag{5.359}$$

in (5.266) and (5.267), or (5.266) and (5.269).

Proof We take the case of (5.266) with (5.269). By (5.357) and (5.359) it holds that

$$\int_0^T \int_\Omega u|\nabla(\log u - v)|^2\, dxdt < +\infty. \tag{5.360}$$

Writing (5.266) as

$$u_t = \nabla \cdot u\nabla(\log u - v), \quad u\frac{\partial}{\partial \nu}(\log u - v)\Big|_{\partial\Omega} = 0,$$

we obtain

$$\left|\frac{d}{dt}\int_\Omega \varphi u \, dx\right| \leq \left|\int_\Omega u\nabla(\log u - v)\cdot\nabla\varphi \, dx\right|$$
$$\leq \|\nabla\varphi\|_\infty \lambda^{1/2}\|u^{1/2}\nabla(\log u - v)\|_2, \quad \lambda = \|u_0\|_1.$$

With $\varphi = \varphi_{x_0,R}$ satisfying (5.282)–(5.283), therefore, it follows that

$$\begin{aligned}
&\left|\langle\varphi_{x_0,R}, \mu(dx,t)\rangle - \langle\varphi_{x_0,R}, \mu(dx,T)\rangle\right| \\
&\leq \int_t^T \left|\frac{d}{dt}\int_\Omega \varphi_{x_0,R}\cdot u \, dx\right| dt \\
&\leq \|\nabla\varphi_{x_0,R}\|_\infty \lambda^{1/2}\int_t^T \|u^{1/2}\nabla(\log u - v)\|_2 \, dt \\
&\leq C\lambda^{1/2}R^{-1}(T-t)^{1/2}\{\int_t^T\int_\Omega u|\nabla(\log u - v)|^2 \, dxdt\}^{1/2}
\end{aligned}$$

for $x_0 \in \mathscr{S}$.

Putting $R = b(T-t)^{1/2}$, we thus obtain

$$\begin{aligned}
&\left|\langle\varphi_{x_0,b(T-t)^{1/2}}, \mu(dx,t)\rangle - \langle\varphi_{x_0,b(T-t)^{1/2}}, \mu(dx,T)\rangle\right| \\
&\leq C\lambda^{1/2}b^{-1}\{\int_t^T\int_\Omega u|\nabla(\log u - v)|^2 \, dxdt\}^{1/2}
\end{aligned}$$

for any $b > 0$. Equality (5.305) is thus improved as

$$\limsup_{t\uparrow T} |\langle\varphi_{x_0,b(T-t)^{1/2}}, \mu(dx,t)\rangle - m(x_0)| = 0, \quad \forall b > 0 \tag{5.361}$$

by (5.360). Then it follows that

$$\zeta(dy,s) = m(x_0)\delta_0(dy), \quad -\infty < s < +\infty$$

in (4.83), and hence

$$A(dy',s') = m(x_0)\delta_0(dy'), \quad s' < 0. \tag{5.362}$$

The case $m \geq 2$ thus does not occur, and therefore it holds that $m(x_0) = 8\pi$. □

Remark 5.25 If $x_0 \in \mathscr{S}$ is simple, $m(x_0) = m_*(x_0)$, there arises also the *free energy transmission*

$$\lim_{t\uparrow T} \mathscr{F}_{x_0, b(T-t)^{1/2}}(u(\cdot,t)) = +\infty, \quad \forall b > 0, \tag{5.363}$$

where

$$\mathscr{F}_{x_0,R}(u) = \int_{\Omega\cap B(x_0,R))} u(\log u - 1)\, dx$$
$$- \frac{1}{2} \iint_{(\Omega\cap B(x_0,R))\times(\Omega\cap B(x_0,R)))} u \otimes u \, dxdx'$$

stands for the local free energy (Chapter 16 of [75]).

5.5.8 *Exclusion of Boundary Blowup*

Here we show $\mathscr{S} \subset \Omega$ in Theorem 5.15, following [76]. Let $u = u(x,t) > 0$ be the classical solution to (5.266) with (5.269) and (5.268), satisfying $T < +\infty$, and define the blowup set $\mathscr{S}$ by (5.343).

Let

$$\mathscr{Y}_0 = \{\varphi \in C_0^2(\overline{\mathbf{R}^2_+}) \mid \left.\frac{\partial \varphi}{\partial \nu}\right|_{\partial \mathbf{R}^2_+} = 0\}$$

$$\rho^0_\varphi(y,y') = \nabla\varphi(y)\cdot\nabla_y E(y,y') + \nabla\varphi(y')\cdot\nabla_{y'}E(y,y'),$$

and $\mathscr{E}$ be the closure of

$$\mathscr{E}_0 = \{\psi + \rho^0_\varphi \mid \psi \in C_0(\overline{\mathbf{R}^2_+}\times\overline{\mathbf{R}^2_+}),\ \varphi \in \mathscr{Y}_0\}$$

in $L^\infty(\mathbf{R}^2\times\mathbf{R}^2)$, for $E = E(y,y')$ defined by (5.34).

If $x_0 \in \partial\Omega \cap \mathscr{S}$ exists, there is

$$\zeta = \zeta(dy,s) \in C_*(-\infty,+\infty; \mathscr{M}(\overline{\mathbf{R}^2_+}))$$

satisfying

$$\zeta(\overline{\mathbf{R}^2_+}, s) = m(x_0) > 0,\ \langle |y|^2, \zeta(dy,s) \le C, \quad -\infty < s < +\infty$$

and

$$\frac{d}{ds}\langle \varphi, \zeta(dy,s)\rangle = \langle \Delta\varphi + \frac{y}{2}\cdot\nabla\varphi, \zeta(dy,s)\rangle + \frac{1}{2}\langle \rho_\varphi^0, \mathscr{K}(\cdot,s)\rangle_{\mathscr{E},\mathscr{E}'}, \quad \text{a.e. } s \in \mathbf{R}$$

for $\varphi \in \mathscr{Y}_0$, where

$$\begin{aligned} \mathscr{K} &= \mathscr{K}(\cdot,s) \in L_*^\infty(-\infty,+\infty;\mathscr{E}') \\ \mathscr{K}(\cdot,s)|_{C_0(\mathbf{R}^2\times\mathbf{R}^2)} &= \zeta(dy,s)\otimes\zeta(dy',s). \end{aligned} \tag{5.364}$$

Take $\xi = \xi(|y|) \in C_0^2(\mathbf{R}^2)$ in

$$0 \le \xi \le 1, \quad \xi = 1 \text{on } B(0,1), \quad \xi = 0 \text{ on } B(0,2)^c,$$

and put

$$\varphi(y) = |y|^2\xi_R(y), \quad \xi_R(y) = \xi(y/R).$$

It holds that $\varphi \in \mathscr{Y}_0$, and therefore,

$$\begin{aligned} &\int_{s_1}^{s_2} \langle \Delta\varphi + \frac{y}{2}\cdot\nabla\varphi, \zeta(dy,s)\rangle + \frac{1}{2}\langle \rho_\varphi^0, \mathscr{K}(\cdot,s)\rangle_{\mathscr{E},\mathscr{E}'}\, ds \\ &\quad = [\langle \varphi, \zeta(dy,s)\rangle]_{s=s_1}^{s=s_2}, \quad -\infty < s_1 < s_2 < +\infty. \end{aligned} \tag{5.365}$$

Letting $R \uparrow +\infty$ in (5.365), we apply the dominated convergence theorem to the left-hand side of (5.365), regarding (5.306)–(5.307). Since

$$\Delta|y|^2 = 4, \quad y\cdot\nabla|y|^2 = 2|y|^2$$

we obtain

$$\lim_{R\uparrow+\infty}\int_{s_1}^{s_2}\langle \Delta\varphi + \frac{y}{2}\cdot\nabla\varphi, \zeta(dy,s)\rangle ds = 4(s_2-s_1)m(x_0) + \int_{s_1}^{s_2} I(s)\, ds, \tag{5.366}$$

where

$$I(s) = \langle |y|^2, \zeta(dy,s)\rangle.$$

The second term of the left-hand side of (5.365) is treated by (5.364). Since

$$\begin{aligned} (y-y_*')\cdot y + (y'-y_*)\cdot y' &= |y|^2 - 2y\cdot y_*' + |y_*|^2 \\ &= |y-y_*'|^2 = |y'-y_*|^2 \end{aligned}$$

we obtain

$$\nabla\Gamma(y-y')\cdot(\nabla|y|^2-\nabla|y'|^2)=-\frac{1}{\pi}$$

$$\nabla\Gamma(y-y'_*)\cdot\nabla|y|^2+\nabla\Gamma(y'-y_*)\cdot\nabla|y'|^2=-\frac{1}{\pi},$$

and therefore,

$$\rho^0_{|y|^2}(y,y')=0. \tag{5.367}$$

Then it holds that

$$\begin{aligned}\rho^0_\varphi(y,y') &= \nabla(|y|^2\xi_R(y))\cdot\nabla_y E(y,y')+\nabla(|y'|^2\xi_R(y'))\cdot\nabla_{y'}E(y,y')\\ &= \left\{|y|^2\nabla\xi_R(y)\cdot\nabla_y E(y,y')+|y'|^2\nabla\xi_R(y')\cdot\nabla_{y'}E(y,y')\right\}\\ &\quad+\left\{\xi_R(y)\nabla|y|^2\cdot\nabla_y E(y,y')+\xi_R(y')\nabla|y'|^2\cdot\nabla_{y'}E(y,y')\right\}\\ &= I+II \end{aligned} \tag{5.368}$$

for $\varphi=|y|^2\xi_R$.

For the term II in the right-hand side of (5.368), we divide $\mathbf{R}^2_+\times\mathbf{R}^2_+$ in the yy' plane into three parts. First, for $|y|<2R$, $y\in\mathbf{R}^2_+$ we have

$$II=(\xi_R(y)-\xi_R(y'))\nabla|y|^2\cdot\nabla_y E(y,y') \tag{5.369}$$

by (5.367). Then we obtain

$$\begin{aligned}|II| &\le \|\nabla\xi\|_\infty\cdot\frac{|y-y'|}{R}\cdot 2|y|\cdot\frac{1}{\pi|y-y'|}\\ &\le \frac{2}{\pi}\|\nabla\xi\|_\infty\varphi_{0,4R}(y)\cdot\frac{|y|}{R}\end{aligned}$$

by

$$\frac{|y-y'|}{|y-y'_*|}\le 1,\quad y,y'\in\mathbf{R}^2_+.$$

A similar estimate is valid for $|y'|<2R$ by the symmetry, while there arises

$$|y|\ge 2R,\ |y'|\ge 2R \quad\Rightarrow\quad II=0.$$

We thus obtain

$$|II|\le C(\varphi_{0,4R}(y)\frac{|y|}{R}+\varphi_{0,4R}(y')\frac{|y'|}{R}),\quad y,y'\in\mathbf{R}^2_+. \tag{5.370}$$

Inequality (5.370) implies

$$
\begin{aligned}
&|\langle II, \mathcal{K}(\cdot, s)\rangle_{\mathcal{E},\mathcal{E}'}| \\
&\leq \langle C(\varphi_{0,4R}(y)\frac{|y|}{R} + \varphi_{0,4R}(y')\frac{|y'|}{R}), \mathcal{K}(\cdot, s)\rangle_{\mathcal{E},\mathcal{E}'} \\
&= C\langle \varphi_{0,4R}(y)\frac{|y|}{R} + \varphi_{0,4R}(y')\frac{|y'|}{R}, \zeta(dy, s) \otimes \zeta(dy', s)\rangle_{\mathcal{E},\mathcal{E}'}
\end{aligned}
$$

by (5.364), while the right-hand side of (5.370) is uniformly bounded. The dominated convergence theorem again, therefore, assures

$$
\lim_{R\uparrow+\infty} \int_{s_1}^{s_2} \langle II, \mathcal{K}(\cdot, s)\rangle_{\mathcal{E},\mathcal{E}'} ds = 0. \tag{5.371}
$$

For the first term on the right-hand side of (5.368), denoted by I, we use

$$
\begin{aligned}
I &= (|y|^2 - |y'|^2)\nabla\xi_R(y) \cdot \nabla_y E(y, y') + |y'|^2(\nabla\xi_R(y) - \nabla\xi_R(y')) \cdot \nabla_{y'} E(y, y') \\
&= III + IV
\end{aligned}
$$

and divide $\mathbf{R}^2_+ \times \mathbf{R}^2_+$ into four parts. First, for $|y| < 4R$ and $|y'| < 4R$ it holds that

$$
\begin{aligned}
|III| &\leq C\left(\varphi_{0,8R}(y)|y| + \varphi_{0,8R}(y')|y'|\right) \cdot \frac{|y|}{R} \\
|IV| &\leq C\left(\varphi_{0,8R}(y') + \varphi_{0,8R}(y)\right) \cdot \frac{|y'|^2}{R^2}.
\end{aligned}
$$

Second, there arises

$$
|y| \geq 2R, \ |y'| \geq 2R \quad \Rightarrow \quad III = IV = 0.
$$

Third, for $|y| < 2R$ and $|y'| \geq 4R$, we return to

$$
III + IV = I = |y|^2\nabla\xi_R(y) \cdot \nabla_y E(y, y') + |y'|^2\nabla\xi_R(y') \cdot \nabla_{y'} E(y, y')
$$

and use $|y - y'| \geq 2R$, to obtain

$$
\begin{aligned}
&\left||y|^2\nabla\xi_R(y) \cdot \nabla_y E(y, y')\right| \leq C\varphi_{0,4R}(y)\frac{|y|^2}{R^2} \\
&|y'|^2\nabla\xi_R(y') \cdot \nabla_{y'} E(y, y') = 0.
\end{aligned}
$$

For $|y| \geq 4R$ and $|y'| < 2R$, finally, there arises

$$
|y|^2\nabla\xi_R(y) \cdot \nabla_y E(y, y') = 0
$$

$$\left||y'|^2\nabla\xi_R(y')\cdot\nabla_{y'}E(y,y')\right|\leq C\varphi_{0,4R}(y')\frac{|y'|^2}{R^2}$$

similarly.

We end up with

$$|I|\leq C\left(\varphi_{0,8R}(y)(1+|y|)+\varphi_{0,8R}(y')(1+|y'|)\right)$$
$$\cdot(\frac{|y|}{R}+\frac{|y|^2}{R^2}+\frac{|y'|}{R}+\frac{|y'|^2}{R^2}),\quad y,y'\in\mathbf{R}^2_+$$

and then the dominated convergence theorem implies

$$\lim_{R\uparrow+\infty}\int_{s_1}^{s_2}\langle I,\mathscr{K}(\cdot,s)\rangle_{\mathscr{E},\mathscr{E}'}ds=0 \tag{5.372}$$

by (5.306)–(5.307).

From (5.371)–(5.372) it follows that

$$\lim_{R\uparrow+\infty}\int_{s_1}^{s_2}\langle\rho^0_\varphi,\mathscr{K}(\cdot,s)\rangle_{\mathscr{E},\mathscr{E}'}ds=0, \tag{5.373}$$

and hence we obtain

$$I(s_2)-I(s_1)=\int_{s_1}^{s_2}4m(x_0)+I(s)\,ds,\quad -\infty<s_1<s_2<\infty$$

by (5.365), (5.366), and (5.373). The mapping

$$s\in(-\infty,+\infty)\quad\mapsto\quad I(s)=\langle|y|^2,\zeta(dy,s)\rangle \tag{5.374}$$

is locally absolutely continuous, and there arises

$$\frac{dI}{ds}=4m(x_0)+I\quad\text{a.e. } s. \tag{5.375}$$

Then it follows that

$$\lim_{s\uparrow+\infty}I(s)=+\infty$$

from $m(x_0)\geq\varepsilon_0$, which contradicts (5.307). Thus we obtain $\partial\Omega\cap\mathscr{S}=\emptyset$. □

5.5.9 Blowup in Infinite Time

Blowup in infinite time indicates

$$T = +\infty, \quad \limsup_{t \uparrow +\infty} \|u(\cdot, t)\|_\infty = +\infty. \tag{5.376}$$

In the other case of

$$T = +\infty, \quad \sup_{t \geq 0} \|u(\cdot, t)\|_\infty < +\infty, \tag{5.377}$$

the orbit $\mathscr{O} = \{u(\cdot, t)\}$ is pre-compact in $C(\overline{\Omega})$, and then, the ω-limit set defined by

$$\omega(u_0) = \{u_\infty \in C^2(\overline{\Omega}) \mid \exists t_k \uparrow +\infty, \ \lim_{k \to \infty} \|u(\cdot, t_k) - u_\infty\|_\infty = 0\} \tag{5.378}$$

is non-empty, compact, and connected from the parabolic regularity [26]. In accordance with the free energy, acting as a Lyapunov function to (5.266) and (5.267), or (5.266) and (5.269), this $\omega(u_0)$ is contained in the set of stationary solutions.

To describe this fact more precisely, we assume the contrary, and define the ω-limit set by (5.378). This set is invariant under the flow, which means that the solution $\tilde{u} = \tilde{u}(\cdot, t)$ to (5.266) and (5.267), or (5.266) and (5.269), taking initial value $u_\infty \in \omega(u_0)$, remains $\tilde{u}(\cdot, t) \in \omega(u_0)$, as far as it exists: $0 < t \leq t_0$.

Since

$$\|u_\infty\|_1 = \|\tilde{u}(\cdot, t)\|_1 = \|u_0\|_1 = \lambda > 0, \tag{5.379}$$

we obtain $\tilde{u}(x, t) > 0$ on $\overline{\Omega}$ for $0 < t \leq t_0$ by the strong maximum principle. Then $\mathscr{F}(\tilde{u}(\cdot, t))$ is well-defined by (5.357), and is independent of t from the LaSalle principle [77]. It thus follows that

$$\frac{d}{dt} \mathscr{F}(\tilde{u}(\cdot, t)) = 0, \quad t > 0 \tag{5.380}$$

in (5.358), and equalities (5.379)–(5.380) imply

$$\log \tilde{u}(\cdot, t) - \tilde{v}(\cdot, t) = \text{constant}, \quad \tilde{v}(\cdot, t) = (-\Delta)^{-1} \tilde{u}(\cdot, t)$$

for $(-\Delta)^{-1}$ defined by

$$((-\Delta)^{-1} u)(x) = \int_\Omega G(x, x') u(x') \, dx'.$$

Then we obtain

$$\tilde{u}(\cdot,t) = \frac{\lambda e^{\tilde{v}(\cdot,t)}}{\int_\Omega e^{\tilde{v}(x,t)}dx}$$

by $\|\tilde{u}(\cdot,t)\|_1 = \lambda$. Sending $t \downarrow 0$, we get $v_\infty \equiv (-\Delta)^{-1}u_\infty$, where E_λ denotes the set of stationary solutions.

Thus $v \in E_\lambda$ if and only if

$$-\Delta v = \lambda\left(\frac{e^v}{\int_\Omega e^v dx} - \frac{1}{|\Omega|}\right), \quad \left.\frac{\partial v}{\partial \nu}\right|_{\partial\Omega} = 0, \quad \int_\Omega v\, dx = 0 \tag{5.381}$$

for the case of (5.266) and (5.267), or

$$-\Delta v = \frac{\lambda e^v}{\int_\Omega e^v\, dx}, \quad v|_{\partial\Omega} = 0 \tag{5.382}$$

for the case of (5.266) and (5.269), where $\lambda = \|u_0\|_1$.

Remark 5.26 By Theorems 5.14 and 5.15, blowup in finite time $T < +\infty$ does not arise in the case of $\lambda = 4\pi$ for (5.266) and (5.267), or $\lambda = 8\pi$ for (5.266) and (5.269). Therefore, if there is no classical stationary solution for this value of λ, the orbit $\mathcal{O}$ cannot be bounded. Blowup in infinite time, (5.376), thus occurs actually for (5.266) and (5.269) with $\lambda = 8\pi$ if Ω is close to a disc, because $E_{8\pi} = \emptyset$ in this case. As problem (5.381) admits the trivial solution $v = 0$, on the other hand, there is a compact orbit $\mathcal{O}$ to (5.266) and (5.267) global in time for any Ω and λ, that is, $u \equiv \lambda$. See [65] for more details on (5.381).

The problem (5.381) (or (5.382)) is the Boltzmann–Poisson equation studied in Sects. 2.4 and 3. It is provided with a quantized blowup mechanism for the family $\{\lambda_k, v_k\}$ of solutions.

Concerning (5.381), if

$$\lambda_k \to \lambda_0, \quad \|v_k\|_\infty \to +\infty \tag{5.383}$$

then $\lambda_0 \in 4\pi\mathbf{N}$, and passing to a subsequence, it holds that

$$v_k \to \sum_{x_0 \in \mathcal{S}} m_*(x_0)G(\cdot, x_0) \quad \text{locally uniformly in } \overline{\Omega} \setminus \mathcal{S} \tag{5.384}$$

with the blowup set defined by

$$\mathcal{S} = \{x_0 \in \overline{\Omega} \mid \exists x_k \to x_0,\ \lim_{k\to\infty} v_k(x_k) = +\infty\}. \tag{5.385}$$

Here we recall (5.356):

$$m_*(x_0) = \begin{cases} 8\pi, \ x_0 \in \Omega \\ 4\pi, \ x_0 \in \partial\Omega. \end{cases}$$

The set $\mathscr{S}$ in (5.385) is finite, and it holds that

$$\nabla_x \left(m_*(x_0)K(x, x_0) + \sum_{x_0' \in \mathscr{S} \setminus \{x_0\}} m_*(x_0')G(x, x_0')\right)\Bigg|_{x=x_0} = 0, \quad \forall x_0 \in \mathscr{S}, \tag{5.386}$$

where

$$K(x, x') = G(x, x') + \begin{cases} \frac{1}{2\pi} \log |x - x'|, \ x \in \Omega \\ \frac{1}{\pi} \log |x - x'|, \ \ x \in \partial\Omega \end{cases} \tag{5.387}$$

(Chapter 7 of [75]).

Similar results hold for (5.382), where $\mathscr{S} \subset \Omega$ always occurs for the blowup set $\mathscr{S}$ defined by (5.385). Hence we obtain $\lambda_0 \in 8\pi \mathbf{N}$ in (5.383), and (5.384) is simplified to

$$v_k \to \sum_{x_0 \in \mathscr{S}} 8\pi G(\cdot, x_0) \quad \text{locally uniformly in } \overline{\Omega} \setminus \mathscr{S}. \tag{5.388}$$

If $\mathscr{S} = \{x_1^*, \cdots, x_\ell^*\}$, furthermore, then $(x_1^*, \cdots, x_\ell^*)$ is a critical point of the ℓ-th Hamiltonian defined by

$$H_\ell(x_1, \cdots, x_\ell) = \frac{1}{2} \sum_{j=1}^{\ell} R(x_j) + \sum_{1 \leq i < j \leq \ell} G(x_i, x_j), \tag{5.389}$$

where $R = R(x)$ is the Robin function $R(x) = K(x, x)$ defined through $K = K(x, x')$ in (5.350). The limit function in (5.388),

$$v_0(x) = \sum_{x_0 \in \mathscr{S}} 8\pi G(x, x_0),$$

finally, is a singular limit of stationary solutions. See Theorem 2.5 of Sect. 2.4.1.

Henceforth, we assume blowup in infinite time, (5.376), and

$$t_k \uparrow +\infty, \quad \lim_{k \to \infty} \|u(\cdot, t_k)\|_\infty = +\infty \tag{5.390}$$

for (5.266) with (5.267), or (5.266) with (5.269). The principle of the generation of a weak solution works for

$$u_k(\cdot, t) = u(\cdot, t + t_k), \quad k = 1, 2, \cdots,$$

and passing to a subsequence we obtain

$$u(x, t + t_k)dx \rightharpoonup \mu(dx, t) \quad \text{in } C_*(-\infty, +\infty; \mathscr{M}(\overline{\Omega})), \tag{5.391}$$

where $\mu = \mu(dx, t)$ is a weak solution.

We show the profile of this $\mu(dx, t)$ for (5.266) with (5.267) first, and then that with (5.269).

Theorem 5.17 *If (5.390) occurs for (5.266) with (5.267), it holds that*

$$\mu^s(dx, t) = \sum_{x_0 \in \mathscr{S}_t} m_*(x_0)\delta_{x_0}(dx), \quad -\infty < t < +\infty \tag{5.392}$$

in (5.391), where $\mathscr{S}_t$ is the blowup set defined by

$$\mathscr{S}_t = \{x_0 \in \overline{\Omega} \mid \exists x_k \to x_0, \ \lim_{k\to\infty} u(x_k, t + t_k) = +\infty\}.$$

The numbers of boundary and interior blowup points, $\sharp(\partial\Omega \bigcap \mathscr{S}_t)$ and $\sharp(\Omega \bigcap \mathscr{S}_t)$, respectively, are independent of t. Letting

$$\partial\Omega \cap \mathscr{S}_t = \{x_1(t), \cdots, x_{\ell_1}(t)\}, \quad \Omega \cap \mathscr{S}_t = \{x_{\ell_1+1}(t), \cdots, x_{\ell_1+\ell_2}(t)\},$$

there arises

$$\begin{aligned} \tau \cdot \frac{dx_i}{dt} &= \tau \cdot V_i(x_1, \cdots, x_\ell), \quad && 1 \leq i \leq \ell_1 \\ \frac{dx_i}{dt} &= V_i(x_1, \cdots, x_\ell), && \ell_1 + 1 \leq i \leq \ell = \ell_1 + \ell_2, \end{aligned} \tag{5.393}$$

for

$$V_i(x_1, \cdots, x_\ell) = m_*(x_i)\nabla_x K(x_i, x_i) + \sum_{j\neq i} m_*(x_j)\nabla_x G(x_i, x_j),$$

where τ denotes a unit tangential vector. The solution $x = (x_i(t))$, $-\infty < t < +\infty$, to (5.393) makes up a pre-compact orbit $\{x(t)\}$ in $\overline{\Omega}^\ell \setminus \overline{D}$, where

$$D = \{x = (x_1, \cdots, x_\ell) \in \Omega^\ell \mid \exists i \neq j, \ x_i = x_j\}. \tag{5.394}$$

The residual vanishing indicates $\mu^{ac}(dx, t) = 0$ in (5.392), where $\mu^{ac}(dx, t)$ denotes the absolutely continuous part of $\mu(dx, t)$ generated by (5.391). This property always arises.

Theorem 5.18 *It holds that $\mu^{ac}(dx, t) = 0, \ -\infty < t < +\infty$, in the previous theorem.*

Remark 5.27 By Theorems 5.17–5.18, blowup in infinite time, (5.376), does not occur for (5.266) with (5.267) if the initial mass is disquantized as $\lambda \equiv \|u_0\|_1 \notin 4\pi\mathbf{N}$. We have, more precisely,

$$4\pi\ell_1 + 8\pi\ell_2 = \lambda, \tag{5.395}$$

where $\ell_1 \geq 0$ and $\ell_2 \geq 0$ denote the numbers of boundary and interior blowup points in (5.390), respectively.

Similar but simpler properties are valid for (5.266) and (5.269). Let $\lambda = \|u_0\|_1$.

Theorem 5.19 *If (5.390) occurs for (5.266) with (5.269), it holds that* $\lambda \in 8\pi\mathbf{N}$ *and*

$$\mu(dx,t) = \sum_{i=1}^{\ell} 8\pi\delta_{x_i(t)}(dx), \quad -\infty < t < +\infty$$

in (5.391), for $\ell = \lambda/8\pi$ *and* $x = (x_1(t), \cdots, x_\ell(t)) \in \Omega^\ell \setminus D$ *satisfying*

$$\frac{dx_i}{dt} = 8\pi\nabla_{x_i} H_\ell(x_1, \cdots, x_\ell), \quad -\infty < t < +\infty, \quad 1 \leq i \leq \ell, \tag{5.396}$$

where $H_\ell(x)$, $x = (x_1, \cdots, x_\ell)$, *is the* ℓ*-th point vortex Hamiltonian defined by (5.389). The orbit* $\mathcal{O} = \{x(t)\}$ *made by (5.396), furthermore, is pre-compact in* $\Omega^\ell \setminus D$.

Remark 5.28 By Theorem 5.19, blowup in infinite time indicated by (5.376) does not occur for (5.266) with (5.269), if $\lambda = \|u_0\|_1 \notin 8\pi\mathbf{N}$. Since (5.396) is an anti-gradient system of H_ℓ and $\mathcal{O} \subset \Omega^\ell \setminus D$ is pre-compact, the property (5.376) is excluded even if $\lambda \in 8\pi\mathbf{N}$ in this system, if there is no critical point of H_ℓ for $\ell = \lambda/8\pi$, and hence no singular solution to (5.382). A similar property is valid for (5.266) with (5.267). Hence (5.376) does not arise for this system even for $\lambda = \|u_0\|_1 \in 4\pi\mathbf{N}$, if there is no critical point of

$$\tilde{H}_\ell(x_1, \cdots, x_\ell) = \sum_{i=1}^{\ell} m_*(x_i)^2 R(x_i) + \sum_{1\leq i<j\leq \ell} m_*(x_i)m_*(x_j)G(x_i, x_j)$$

for any $\ell = \ell_1 + \ell_2$ with $(\ell_1, \ell_2) \in (\mathbf{N} \cup \{0\})^2$ satisfying (5.395) and

$$\ell_1 = \sharp\{i \mid x_i \in \partial\Omega\}, \quad \ell_2 = \sharp\{i \mid x_i \in \Omega\},$$

and $R(x) = K(x, x)$ for $K = K(x, x')$ defined by (5.387).

Remark 5.29 Comparing Theorems 5.15 to Theorem 5.19, we see that blowup in infinite time obeys rather similar features to those of the family of stationary solutions:

1. Quantized blowup mechanism without collision.
2. Hamiltonian control (5.393) of the location of blowup points.

Remark 5.30 If Ω is convex, there is no critical point of $H_\ell = H_\ell(x_1, \cdots, x_\ell)$ for $\ell \geq 2$, and $H_1 = R(x)$ takes a unique critical point [24]. Therefore, blowup in infinite time occurs only to $\lambda = 8\pi$ for such domains, and furthermore,

$$u(x,t)dx \rightharpoonup 8\pi\delta_{x_0}(dx) \quad \text{in } \mathcal{M}(\overline{\Omega})$$

as $t \uparrow +\infty$, where $x_0 \in \Omega$ is the critical point of $R = R(x)$.

Remark 5.31 A class of convex domains admits stationary solutions even for $\lambda \geq 8\pi$. If Ω is close to a disc, on the contrary, $\lambda > 8\pi$ always implies blowup in finite time of the solution (Remark 5.34 in Sect. 5.5.11).

5.5.10 Collapse Dynamics

Henceforth, we concentrate on the proof of Theorem 5.19 concerning (5.266) and (5.269). Its proof is divided into several steps.

First, from the improved ε regularity, Lemma 5.30, the singular part of $\mu(dx,t)$ in (5.391), denoted by $\mu^s(dx,t)$, is a finite sum of delta functions:

$$\mu^s(dx,t) = \sum_{x_0 \in \mathcal{S}_t} m_t(x_0)\delta_{x_0}(dx), \quad m_t(x_0) \geq \varepsilon_0. \tag{5.397}$$

Second, we have collapse mass normalization. Here we provide a proof using the scaling limit [80], different from the original [68].

Lemma 5.33 *It holds that $\mathcal{S}_t \subset \Omega$ and $m_t(x_0) = 8\pi$ in (5.397).*

Proof The fact $\partial\Omega \cap \mathcal{S}_t = \emptyset$ for (5.266) with (5.269) is due to the form of parametrix $E(x,x')$ in (5.352), and the proof is similar to the case of blowup in finite time as is described in Sect. 5.1.4. In fact, assuming $x_0 \in \partial\Omega \cap \mathcal{S}_T$, we take

$$\zeta(dy,s) = (T-t)\mu(dx,t), \quad y = (x-x_0)/(T-t)^{1/2}, \quad s = -\log(T-t)$$

and put it to be zero where it is not defined. Then we can confirm the generation of the weak solution. Any $s_k \uparrow +\infty$ thus admits a subsequence denoted by the same symbol and the limit measure $\tilde{\zeta}(dy,s)$ such that

$$\zeta(dy, s+s_k) \rightharpoonup \tilde{\zeta}(dy,s) \quad \text{in } C_*(-\infty,+\infty; \mathcal{M}(\mathbf{R}^2)).$$

This $\tilde{\zeta}(dy,s)$ has the support included in a closed subspace independent of s, denoted by $\overline{L}$, and is a weak solution to

$$z_s = \Delta z - \nabla \cdot z \nabla (E * z + \frac{|y|^2}{4}) \quad \text{in } L \times (-\infty, +\infty)$$

$$\frac{\partial z}{\partial \nu} - z \frac{\partial}{\partial \nu}(E * z + \frac{|y|^2}{4}) = 0 \quad \text{on } \partial L \times (-\infty, +\infty).$$

It also holds that $\tilde{\zeta}(\overline{L}, s) = m(x_0) > 0$, similarly to the blowup in finite time of the classical solution. Then we get a contradiction, from the proof in Sect. 5.5.8 of exclusion of its boundary blowup.

Given $x_0 \in \Omega \cap \mathscr{S}_0$ in (5.266) with (5.269), we show $m_0(x_0) = 8\pi$ in (5.397). Then we get the conclusion by translating the time variable t. For this purpose we assume $x_0 = 0 \in \Omega$ to take

$$\mu_\beta(dx, t) = \beta^2 \mu(dx', dt'), \quad x' = \beta x, \ t' = \beta^2 t$$

for $\beta > 0$, which is a weak solution in $\beta^{-1}\Omega \times (-\infty, +\infty)$. Then each $\beta_k \downarrow 0$ admits a subsequence denoted by the same symbol and the limit measure $\tilde{\mu}(dx, t)$ such that .

$$\mu_{\beta_k}(dx, t) \rightharpoonup \tilde{\mu}(dx, t) \quad \text{in } C_*(-\infty, +\infty; \mathscr{M}(\mathbf{R}^2)).$$

This $\tilde{\mu}(dx, t)$ is a weak solution to (5.310). Then we obtain

$$\tilde{\mu}(\mathbf{R}^2, t) \leq \lambda, \quad \tilde{\mu}(\mathbf{R}^2, 0) = m(x_0),$$

which implies $m_0(x_0) = 8\pi$ by Lemma 5.32. □

It holds that

$$\mu(dx, t) = \sum_{x_0 \in \mathscr{S}_t} 8\pi \delta_{x_0}(dx) + f(x, t)dx \tag{5.398}$$

in (5.391) with $0 \leq f = f(\cdot, t) \in L^1(\Omega)$. Since

$$\mu(dx, t) \in C_*(-\infty, ; \infty; \mathscr{M}(\overline{\Omega})),$$

the set $Q = \bigcup_t (\Omega \setminus \mathscr{S}_t) \times \{t\}$ is open. From the elliptic–parabolic regularity this $f = f(x, t)$ is smooth in Q, and satisfies (5.266) there, with smooth v. Hence it is also positive everywhere unless identically zero by the strong maximum principle.

The third step is the following lemma.

Lemma 5.34 *The number $\sharp \mathscr{S}_t = \ell$ in (5.398) is independent of t. Writing*

$$\mu^s(dx, t) = \sum_{i=1}^{\ell} 8\pi \delta_{x_i(t)}(dx),$$

we have

$$\frac{dx_i}{dt} = 8\pi \nabla_{x_i} H_\ell(x_1, \cdots, x_\ell), \quad 1 \leq i \leq \ell \tag{5.399}$$

in the sense of distributions in $t \in \mathbf{R}$ *with* $H_\ell = H_\ell(x_1, \cdots, x_\ell)$ *defined by (5.389).*

Proof Since $\mu(dx,t) \in C_*(-\infty,+\infty; \mathscr{M}(\overline{\Omega}))$ defined by (5.391) is a weak solution, there is $0 \leq \nu \in L^\infty_*(-\infty,+\infty; \mathscr{E}')$ satisfying

$$\nu|_{C(\overline{\Omega}\times\overline{\Omega})} = \mu \otimes \mu$$

and

$$\frac{d}{dt}\langle \xi, \mu(dx,t)\rangle = \langle \Delta\xi, \mu(dx,t)\rangle + \frac{1}{2}\langle \rho_\xi, \nu(t)\rangle$$

in the sense of distributions for each $\xi \in C^2(\overline{\Omega})$ with $\left.\frac{\partial \xi}{\partial \nu}\right|_{\partial\Omega} = 0$, where

$$\rho_\xi(x,x') = \nabla\xi(x)\cdot\nabla_x G(x,x') + \nabla\xi(x')\cdot\nabla_{x'}G(x,x')$$

and $\mathscr{E} \subset L^\infty(\Omega\times\Omega)$ is the closure of

$$\mathscr{E}_0 = \{\rho_\xi + \psi \mid \xi \in C^2(\overline{\Omega}), \ \left.\frac{\partial\xi}{\partial\nu}\right|_{\partial\Omega} = 0, \ \psi \in C(\overline{\Omega}\times\overline{\Omega})\}.$$

Thus we obtain

$$-\int_{\mathbf{R}} \eta'(t)\langle \xi, \mu(dx,t)\rangle dt = \int_{\mathbf{R}} \eta(t)[\langle \Delta\xi, \mu(dx,t)\rangle + \frac{1}{2}\langle \rho_\xi, \nu(t)\rangle]dt \tag{5.400}$$

for any $\eta = \eta(t) \in C_0^1(\mathbf{R})$. We have

$$\mathscr{S}_t = \{x_i(t) \mid 1 \leq i \leq \ell(t)\} \tag{5.401}$$

in (5.398).

There arises the decomposition

$$\begin{aligned} &\nu(t) = \nu_0(t) + \nu_1(t) + \nu_2(t) + f \otimes f \ dx dx' \\ &0 \leq \nu_0(t), \ \nu_1(t), \nu_2(t) \in L^\infty_*(-\infty,+\infty; \mathscr{E}') \end{aligned} \tag{5.402}$$

with

$$\nu_0(t)|_{C(\overline{\Omega}\times\overline{\Omega})} = 64\pi^2 \sum_{j,k=1}^{\ell(t)} \delta_{x_j(t)}(dx) \otimes \delta_{x_k(t)}(dx')$$

$$\nu_1(t)|_{C(\overline{\Omega}\times\overline{\Omega})} = f(x,t)dx \otimes \sum_{k=1}^{\ell(t)} 8\pi\delta_{x_k(t)}(dx')$$

$$\nu_2(t)|_{C(\overline{\Omega}\times\overline{\Omega})} = \sum_{j=1}^{\ell(t)} 8\pi\delta_{x_j(t)}(dx) \otimes f(x',t)dx'. \tag{5.403}$$

Take $-\infty < t_0 < +\infty$, $1 \leq i \leq \ell(t_0)$, and put $x_0 = x_i(t_0)$. Then we have $0 < \delta \ll 1$ and $0 < r \ll 1$ such that

$$\sharp(\mathscr{S}_t \cap B(x_0, r)) \leq 1, \quad |t - t_0| < \delta$$

by $\mu(dx,t) \in C_*(-\infty, +\infty; \mathscr{M}(\overline{\Omega}))$.

We may assume $x_0 = 0$ without loss of generality. Let

$$m(t) = \begin{cases} 8\pi, & \sharp(\mathscr{S}_t \cap B(x_0, r)) = 1 \\ 0, & \text{otherwise} \end{cases}$$

and

$$\xi = |x|^2\varphi(x),$$

where $0 \leq \varphi = \varphi(x) \in C_0^\infty(\Omega)$ is a cut-off function taking the value 1 in $B(0, r/2)$ and 0 outside $B(0, r)$. We have also $0 < c_0 \ll 1$ such that

$$|x_j(t) - x_i(t)| \geq c_0, \quad j \neq i, \ |t - t_0| < \delta. \tag{5.404}$$

Then we obtain

$$\rho_\xi(x,x') = -\frac{1}{\pi} + 2x \cdot \nabla_x K(x,x') + 2x' \cdot \nabla_{x'} K(x,x')$$
$$(x,x') \in B(0,r/2) \times B(0,r/2),$$

using $K = K(x,x')$ defined by (5.350). Therefore, since

$$4m(t) - \frac{m(t)^2}{2\pi} = 0$$

it follows that

$$\langle \rho_\xi, \nu_0(t) \rangle$$

$$
\begin{aligned}
&= m(t)^2 x_i(t) \cdot (\nabla_x K(x_i(t), x_i(t)) + \nabla_{x'} K(x_i(t), x_i(t)) \\
&\quad + m(t)^2 x_i(t) \cdot \sum_{j \neq i} (\nabla_x G(x_i(t), x_j(t)) + \nabla_{x'} G(x_i(t), x_j(t)) \\
&= 2m(t)^2 x_i(t) \cdot \nabla_{x_i} H_{\ell(t)}(x_1(t), \cdots, x_{\ell(t)}).
\end{aligned}
\tag{5.405}
$$

By (5.400), (5.402), and (5.405) we end up with

$$
\begin{aligned}
&- \int_{\mathbf{R}} \eta'(t) \left\{ m(t)|x_i(t)|^2 + \int_{\Omega} \xi(x) f(x,t)\, dx \right\} dt \\
&= \int_{\mathbf{R}} \eta(t) \big\{ 2m(t)^2 x_i(t) \cdot \nabla_{x_i} H_{\ell(t)}(x_1(t), \cdots, x_{\ell(t)}(t)) \\
&\quad + \frac{1}{2} \langle \rho_\xi, f \otimes f \rangle + \frac{1}{2} \langle \rho_\xi, \nu_1(t) + \nu_2(t) \rangle \big\}\, dt
\end{aligned}
\tag{5.406}
$$

if $\eta = \eta(t) \in C_0^1(\mathbf{R})$ is supported in $|t - t_0| < \delta$.

Now we let $r \downarrow 0$ in (5.406). In fact we have

$$
\rho_\xi(x, x') = \rho_\xi^0(x, x') + \rho_\xi^1(x, x')
$$

with

$$
\begin{aligned}
\rho_\xi^0(x, x') &= -\frac{1}{2\pi} \frac{\nabla \xi(x) - \nabla \xi(x')}{|x - x'|^2} \cdot (x - x') \\
\rho_\xi^1(x, x') &= \nabla \xi(x) \cdot \nabla_x K(x, x') + \nabla \xi(x') \cdot \nabla_{x'} K(x, x').
\end{aligned}
$$

It holds also that

$$
|\nabla \xi(x)| \le C|x|\chi_B(x), \quad |\nabla^2 \xi(x)| \le C\chi_B(x)
$$

for $B = B(0, r)$ by

$$
r|\nabla \varphi(x)| + r^2 |\nabla^2 \varphi(x)| \le C, \tag{5.407}
$$

which implies

$$
\begin{aligned}
|\rho_\xi^0(x, x')| &\le C(\chi_B(x) + \chi_B(x')) \\
|\rho_\xi^1(x, x')| &\le C(|x|\chi_B(x) + |x'|\chi_B(x'))
\end{aligned}
$$

for $B = B(0, r)$.

First, we have

$$\langle \rho_\xi, f \otimes f \rangle = \langle \rho_\xi^0, f \otimes f \rangle + \langle \rho_\xi^1, f \otimes f \rangle \leq C \left(\int_B f \, dx \right)^2$$

and hence

$$\int_{\mathbf{R}} \eta(t) \langle \rho_\xi, f \otimes f \rangle \, dt = o(1), \quad r \downarrow 0. \tag{5.408}$$

It holds also that

$$\int_{\mathbf{R}} \eta(t) \langle \rho_\xi^1, \nu_1(t) + \nu_2(t) \rangle \, dt = o(1), \quad r \downarrow 0.$$

Second, we show

$$\int_{\mathbf{R}} \eta(t) \langle \rho_\xi^0, \nu_1(t) \rangle \, dt = o(1), \quad r \downarrow 0. \tag{5.409}$$

In fact, the other case

$$\int_{\mathbf{R}} \eta(t) \langle \rho_\xi^0, \nu_2(t) \rangle \, dt = o(1), \quad r \downarrow 0$$

is proven similarly to (5.408), and then it follows that

$$\int_{\mathbf{R}} \eta(t) \langle \rho_\xi^0, \nu_1(t) + \nu_2(t) \rangle \, dt = o(1).$$

To this end we note

$$\nabla \xi(x) = 2x\varphi(x) + |x|^2 \nabla \varphi(x),$$

to infer

$$\begin{aligned} &\nabla \xi(x) - \nabla \xi(x') \\ &\quad = 2(x - x')\varphi(x) + 2x'(\varphi(x) - \varphi(x')) \\ &\qquad + (|x|^2 - |x'|^2)\nabla \varphi(x) + |x' - a|^2 (\nabla \varphi(x) - \nabla \varphi(x')). \end{aligned} \tag{5.410}$$

Then it follows that

$$\begin{aligned} \rho_\xi^0(x, x') &= -\frac{1}{\pi}\varphi(x) - \frac{1}{\pi}\frac{x - x'}{|x - x'|^2} \cdot x'(\varphi(x) - \varphi(x')) \\ &- \frac{1}{2\pi}(|x|^2 - |x'|^2) \cdot \frac{\nabla \varphi(x)}{|x - x'|^2} \cdot (x - x') \end{aligned}$$

$$
- \frac{1}{2\pi} \frac{|x'|^2}{|x - x'|^2} (\nabla \varphi(x) - \nabla \varphi(x')) \cdot (x - x'). \tag{5.411}
$$

Here we use the decomposition

$$
\rho_\xi^0(x, x') = \rho_\xi^{0,0}(x, x') + \rho_\xi^{0,1}(x, x') \tag{5.412}
$$

for

$$
\rho_\xi^{0,0}(x, x') = \varphi(x')\rho_\xi^0(x, x'), \quad \rho_\xi^{0,1}(x, x') = (1 - \varphi(x'))\rho_\xi^0(x, x').
$$

First, equality (5.410) implies

$$
\rho_\varphi^{0,0}(x, x') + \frac{1}{\pi}\varphi(x)\varphi(x') = -I - II - III
$$

with

$$
\begin{aligned}
I &= \frac{1}{\pi} \frac{x - x'}{|x - x'|^2} \cdot x'(\varphi(x) - \varphi(x'))\varphi(x') \\
II &= \frac{1}{2\pi}(x - x') \cdot (x + x') \frac{\nabla \varphi(x)}{|x - x'|^2} \varphi(x') \cdot (x - x') \\
III &= \frac{1}{2\pi} \frac{|x'|^2}{|x - x'|^2} (\nabla \varphi(x) - \nabla \varphi(x')) \cdot (x - x')\varphi(x').
\end{aligned}
$$

Using (5.407), we obtain

$$
|I| \leq \frac{C}{r} |x'\varphi(x')| \leq C\chi_B(x')
$$

by the mean value theorem. Similarly, we have

$$
\begin{aligned}
|II| &\leq |x \cdot \nabla \varphi(x)|\varphi(x') + |x'||\nabla \varphi(x)|\varphi(x') \\
&\leq C(\varphi(x') + \frac{1}{r}|x'\varphi(x')|) \leq C\chi_B(x')
\end{aligned}
$$

and

$$
|III| \leq C|x'|^2 r^{-2} \varphi(x') \leq C\chi_B(x').
$$

It thus follows that

$$
\left| \rho_\xi^{0,0}(x, x') + \frac{1}{\pi}\varphi(x)\varphi(x') \right| \leq C\chi_{\hat{B}}(x'),
$$

and hence

$$\int_{\mathbf{R}} \eta(t)\langle \rho_\xi^{0,0}, \nu_1(t)\rangle\, dt = -\frac{1}{\pi}\cdot 8\pi \iint_{\Omega\times\mathbf{R}} \eta(t)\varphi(x)f(x,t)dxdt + O(r)$$
$$= o(1), \quad r \downarrow 0.$$

We have also

$$\langle \rho_\xi^{0,1}(t), \nu_1\rangle = -\frac{1}{\pi}\cdot 8\pi \sum_{j\neq i}\int_{\Omega}\Big[\frac{x-x_j(t)}{|x-x_j(t)|^2}\cdot x_j(t)\varphi(x)$$
$$+\frac{1}{2}\frac{|x|^2}{|x-x_j(t)|^2}\nabla\varphi(x)\cdot(x-x_j(t))\Big] f(x,t),\ dx$$

using

$$\rho_\xi^0(x,x') = -\frac{1}{\pi}\varphi(x) - \frac{1}{\pi}\frac{x-x'}{|x-x'|^2}\cdot x'(\varphi(x)-\varphi(x'))$$
$$-\frac{|x|^2}{2\pi}\frac{\nabla\varphi(x)}{|x-x'|^2}\cdot(x-x') + \frac{|x'|^2}{2\pi}\frac{\nabla\varphi(x')}{|x-x'|^2}\cdot(x-x')$$

for (5.411). Then we obtain

$$\int_{\mathbf{R}} \eta(t)\left|\langle \rho_\xi^{0,1}, \nu_1(t)\rangle\right|\, dt \le C\cdot c_0^{-1}\iint_{B\times\mathbf{R}} \eta(t)f(x,t)\, dxdt = o(1), \quad r\downarrow 0$$

by (5.404) and (5.407).

It thus follows that (5.409) and then equality (5.406) is reduced to

$$-\int_{\mathbf{R}} \eta'(t)m(t)|x_i(t)|^2\, dt$$
$$= 2\int_{\mathbf{R}} \eta(t)m(t)^2 x_i(t)\cdot\nabla_{x_i} H_{\ell(t)}(x_1(t),\cdots,x_{\ell(t)}(t)). \tag{5.413}$$

Therefore, we have

$$\frac{d}{dt}(m(t)|x_i(t)|^2) \in L^\infty_{loc}(\mathbf{R})$$

in the sense of distributions, which implies

$$m(t) = 8\pi, \quad |t-t_0| < \delta.$$

Since this property holds for any $x_i = x_i(t)$, $1 \leq i \leq \ell(t)$, the number $\ell(t) = \ell$ is locally constant in t, and hence is independent of t. Furthermore, in the sense of distributions, there arises

$$\frac{1}{2}\frac{d}{dt}|x_i|^2 = 8\pi x_i \cdot \nabla_{x_i} H_\ell(x_1, \ldots, x_\ell).$$

Replacing the above $B(x_0, r)$ with $x_0 = 0$ to $B(a, r)$ with $|a| \ll 1$, we obtain

$$\frac{1}{2}\frac{d}{dt}|x_i - a|^2 = 8\pi(x_i - a) \cdot \nabla_{x_i} H_\ell(x_1, \cdots, x_\ell),$$

and hence

$$(x_i - a) \cdot \frac{dx_i}{dt} = 8\pi(x_i - a) \cdot \nabla_{x_i} H_\ell(x_1, \cdots, x_\ell).$$

From the arbitrariness of $a \in \mathbf{R}^2$ in $|a| \ll 1$, it follows that (5.399). □

We conclude this section with the following lemma.

Lemma 5.35 *The closure of the orbit $\{x(t)\} \subset \Omega^\ell \setminus D$ made by (5.399) is compact for D defined by (5.394). The ω-limit and α-limit sets of this orbit are contained in the critical point of H_ℓ.*

Proof Recalling (5.401), we apply the scaling argument used for the proof of $\partial\Omega \cap \mathscr{S}_t = \emptyset$ in Lemma 5.33. Then it follows that

$$\liminf_{t \to \pm\infty} \mathrm{dist}(x_j(t), \partial\Omega) > 0, \quad 1 \leq j \leq \ell. \tag{5.414}$$

Similarly, assuming the existence $t_k \to \pm\infty$ and $i \neq j$ such that

$$\lim_{k \to \infty} |x_i(t_k) - x_j(t_k)| = 0,$$

we take a subsequence satisfying

$$\mu(dx, t + t_k) \rightharpoonup \tilde{\mu}(dx, t) \quad \text{in } C_*(-\infty, +\infty; \mathscr{M}(\overline{\Omega}))$$

and

$$\lim_{k \to \infty} x_k = x_0 \in \Omega.$$

This $\tilde{\mu}(dx, t)$, however, is a weak solution to (5.266) with (5.269) satisfying

$$\tilde{\mu}(\{x_0\}, 0) \geq 16\pi,$$

a contradiction to Lemma 5.33. Since closure of the orbit $\{x(t)\} \subset \Omega^\ell \setminus D$ is compact, the result follows from the LaSalle principle assured by (5.417). □

Remark 5.32 A part of the result in the above lemma is derived differently. In fact, by Lemma 5.35 we have $t_0 \gg 1$, $0 < r_0 \ll 1$, and $x_1^0, \cdots, x_\ell^0 \in \Omega$ such that

$$B(x_i^0, 4r_0) \cap \mathscr{S}_t = \{x_i(t)\}, \quad t \geq t_0, \ 1 \leq i \leq \ell, \tag{5.415}$$

which implies

$$\inf_{t \geq t_0, \ i \neq j} |x_i(t) - x_j(t)| \geq c_0 \tag{5.416}$$

with $c_0 > 0$. Concerning (5.414) in Lemma 5.35 we can prove

$$\liminf_{t \uparrow +\infty} \mathrm{dist}(x_j(t), \partial\Omega) > 0, \quad 1 \leq j \leq \ell$$

directly, using

$$\frac{d}{dt} H_\ell(x_1, \cdots, x_\ell) = 8\pi |\nabla H_\ell(x_1, \cdots, x_\ell)|^2 \geq 0, \quad -\infty < t < +\infty \tag{5.417}$$

derived from (5.399).

Remark 5.33 The theory of gradient inequality [47] clarifies the profile of $\mathscr{O} = \{x(t)\}$ in more detail as is described in Sect. 5.1.4. This $\mathscr{O}$ is thus a connected orbit of critical points of H_ℓ as in (5.51).

5.5.11 Initial Mass Quantization

Residual vanishing for radially symmetric solutions is proven through the use of defect measures [58, 78]. Here we apply the argument for the general case, using Liouville's first volume derivative [80]. Put

$$x_i = x_i(t), \ u_k(x,t) = u(x, t + t_k), \ v_k(x,t) = v(x, t + t_k) \tag{5.418}$$

in Lemma 5.34. By Lemma 5.35 each $1 \leq i \leq \ell$ admits $r_1 > 0$ such that

$$\mathscr{S}_t \cap B(x_i(t), 2r_1) = \{x_i(t)\}, \quad -\infty < t < +\infty.$$

Lemma 5.36 *It holds that*

$$\frac{d}{dt} \int_{B(x_i, r)} (|x - x_i|^2 - r^2) u_k dx$$

$$\leq 4 \int_{B(x_i,r)} u_k \, dx + 2 \int_{B(x_i,r)} (x - x_i) \cdot u_k \nabla v_k dx$$

$$- 2 \int_{B(x_i,r)} (x - x_i) \cdot \dot{x}_i u_k \, dx, \quad 0 < r \leq r_1. \tag{5.419}$$

Proof By Liouville's first volume derivative, we have

$$\begin{aligned}
&\frac{d}{dt} \int_{B(x_i,r)} |x - x_i|^2 u_k \, dx \\
&= \int_{B(x_i,r)} \frac{\partial}{\partial t}(|x - x_i|^2 u_k) + \dot{x}_i \cdot \nabla(|x - x_i|^2 u_k) \, dx \\
&= \int_{B(x_i,r)} |x - x_i|^2 u_{kt} - 2(x - x_i) \cdot \dot{x}_i u_k \\
&\quad +2\dot{x}_i \cdot (x - x_i) u_k + \dot{x}_i |x - x_i|^2 \cdot \nabla u_k dx \\
&= \int_{B(x_i,r)} |x - x_i|^2 u_{kt} + \dot{x}_i \cdot |x - x_i|^2 \nabla u_k \, dx.
\end{aligned} \tag{5.420}$$

Here, it holds that

$$\begin{aligned}
\int_{B(x_i,r)} |x - x_i|^2 u_{kt} \, dx &= \int_{B(x_i,r)} |x - x_i|^2 \nabla \cdot (\nabla u_k - u_k \nabla v_k) \, dx \\
&= \int_{\partial B(x_i,r)} |x - x_i|^2 (\frac{\partial u_k}{\partial \nu} - u_k \frac{\partial v_k}{\partial \nu}) dS \\
&\quad - \int_{B(x_i,r)} 2(x - x_i) \cdot (\nabla u_k - u_k \nabla v_k) \, dx \\
&= r^2 \int_{\partial B(x_i,r)} \frac{\partial u_k}{\partial \nu} - u_k \frac{\partial v_k}{\partial \nu} \, dS \\
&\quad -2 \int_{\partial B(x_i,r)} (x - x_i) \cdot \nu \, u_k \, dS \\
&\quad + 4 \int_{B(x_i,r)} u_k \, dx + 2 \int_{B(x_i,r)} (x - x_i) \cdot u_k \nabla v_k \, dx
\end{aligned}$$

and hence the first term of the right-hand side of (5.420) is estimated as

$$\begin{aligned}
\int_{B(x_i,r)} |x - x_i|^2 u_{kt} \, dx &\leq r^2 \int_{B(x_i,r)} \nabla \cdot (\nabla u_k - u_k \nabla v_k) \, dx \\
&\quad + 4 \int_{B(x_i,r)} u_k \, dx + \int_{B(x_i,r)} 2(x - x_i) \cdot u_k \nabla v_k \, dx
\end{aligned}$$

$$= r^2 \int_{B(x_i,r)} u_{kt}\, dx + 4 \int_{B(x_i,r)} u_k\, dx$$

$$+ 2 \int_{B(x_i,r)} (x - x_i) \cdot u_k \nabla v_k\, dx. \tag{5.421}$$

For the second term of the right-hand side of (5.420), we note

$$\int_{B(x_i,r)} \dot{x}_i \cdot |x - x_i|^2 \nabla u_k\, dx$$

$$= \int_{\partial B(x_i,r)} (\dot{x}_i \cdot \nu)|x - x_i|^2 u_k\, dS - \int_{B(x_i,r)} 2(x - x_i) \cdot \dot{x}_i u_k\, dx$$

$$= r^2 \int_{\partial B(x_i,r)} (\dot{x}_i \cdot \nu) u_k\, dx - \int_{B(x_i,r)} 2(x - x_i) \cdot \dot{x}_i u_k\, dx$$

$$= r^2 \int_{B(x_i,r)} \dot{x}_i \cdot \nabla u_k\, dx - \int_{B(x_i,r)} 2(x - x_i) \cdot \dot{x}_i u_k\, dx. \tag{5.422}$$

By (5.420)–(5.422), there arises (5.419) because

$$\frac{d}{dt} \int_{B(x_i,r)} u_k\, dx = \int_{B(x_i,r)} u_{kt} + \dot{x}_i \cdot \nabla u_k\, dx$$

is derived from Liouville's first volume derivative again. □

Lemma 5.37 *It holds that*

$$\frac{d}{dt} \int_{B(x_i,r)} (|x - x_i|^2 - r^2) u_k dx \le 4 \int_{B(x_i,r)} u_k\, dx - \frac{1}{2\pi} \Big(\int_{B(x_i,r)} u_k\, dx\Big)^2$$

$$+ C \int_{B(x_i,r)} |x - x_i| u_k\, dx, \quad 0 < r \le r_2,\ t \ge t_0,\ k \ge k_0. \tag{5.423}$$

for some r_2, t_0, and k_0.

Proof The last term on the right-hand side of (5.419) we use Lemma 5.35 to deduce

$$|\dot{x}| \le C.$$

For the second term, we divide v_k as

$$v_k(x,t) = \int_{B(x_i,r)} \Gamma(x - x') u_k(x',t)\, dx' + \int_{B(x_i,r)} K(x,x') u_k(x',t)\, dx'$$

$$+ \int_{\Omega \setminus B(x_i,r)} G(x,x') u_k(x',t)\, dx' \equiv v_k^0(x,t) + v_k^1(x,t) + v_k^2(x,t).$$

We use the method of symmetrization for the first term, to obtain

$$2\int_{B(x_i,r)} (x - x_i) \cdot u_k \nabla v_k^0 \, dx = -\frac{1}{2\pi}(\int_{B(x_i,r)} u_k \, dx)^2. \tag{5.424}$$

For the second term, we have

$$\|\nabla v_k^1(\cdot, t)\|_{L^\infty(B(x_i,r))} \le C,$$

because $K = K(x, x')$ is smooth in $\Omega \times \Omega$. We have also

$$\|\nabla v_k^2(\cdot, t)\|_{L^\infty(B(x_i,r))} \le C, \quad k \gg 1.$$

In fact, $u_k(\cdot, t)$ is locally uniformly bounded in $\Omega \setminus \mathscr{S}_t$ with respect to $k \gg 1$, while $B(x_i(t), 4r_i) \cap \mathscr{S}_t = \{x_i(t)\}$ holds for any $-\infty < t < +\infty$. Since

$$v_k^2(x,t) = \int_{\Omega \setminus \mathscr{S}_t^{2r}} G(x, x')u_k(x', t)dx' + \int_{\mathscr{S}_t^{2r} \setminus B(x_i,r)} G(x, x')u_k(x', t)dx',$$

the result follows from $\|u_k(\cdot, t)\|_1 = \lambda$ and the smoothness of $G = G(x, x')$ in $\Omega \times \Omega \setminus D$ for the first term on the right-hand side, and

$$\sup_x \int_\Omega |\nabla_x G(x, x')| \, dx' < +\infty$$

for the second term on the right-hand side. We thus end up with (5.423) for some r_2, t_0, and k_0. □

We are ready to complete the proof of residual vanishing.

Lemma 5.38 *It holds that $f \equiv 0$ in (5.398).*

Proof We continue to write $x_i = x_i(t)$. Making making $k \to \infty$ in (5.423), we obtain

$$\begin{aligned}
&\frac{d}{dt}\int_{B(x_i,r)} (|x - x_i|^2 - r^2) f \, dx \\
&\le 4(8\pi + \int_{B(x_i,r)} f \, dx) - \frac{1}{2\pi}(8\pi + \int_{B(x_i,r)} f \, dx)^2 + C\int_{B(x_i,r)} |x - x_i| f \, dx \\
&\le -4\int_{B(x_i,r)} f \, dx + C\int_{B(x_i,r)} |x - x_i| f \, dx
\end{aligned}$$

in the sense of distributions in t, and hence

$$\begin{aligned}
&\frac{d}{dt}\int_{B(x_i,r)}(|x-x_i|^2-r^2)f\,dx \\
&\leq -2\int_{B(x_i,r)} f\,dx + \frac{C^2}{8}\int_{B(x_i,r)}|x-x_i|^2 f\,dx \\
&= \frac{C^2}{8}\int_{B(x_i,r)}(|x-x_i|^2-r^2)f\,dx + (\frac{C^2}{8}r^2-2)\int_{B(x_i,r)} f\,dx \\
&\leq \frac{C^2}{8}\int_{B(x_i,r)}(|x-x_i|^2-r^2)f\,dx
\end{aligned} \tag{5.425}$$

for $0 < r \ll 1$.

Let

$$I(t) = \int_{B(x_i,r)}(|x-x_i|^2-r^2)f\,dx \leq 0.$$

By (5.425) if there is $t_1 \geq t_0$ such that $I(t_1) < 0$ then it holds that

$$\lim_{t\uparrow+\infty} I(t) = -\infty,$$

a contradiction. Hence we have $I(t) \equiv 0$, which implies

$$f(\cdot,t) = 0 \quad \text{in } B(x_i,r) = B(x_i(t),r).$$

Then we obtain $f \equiv 0$ from the strong maximum principle or the unique continuation theorem for parabolic equations as in the proof of $f = f(x) > 0$ in Theorem 5.13 in Sect. 5.5.4. □

Proof *(Theorem 5.19)* This theorem is a consequence of Lemmas 5.33–5.35, and 5.38. □

A result analogous to Theorem 5.16 is the following theorem. It is applicable if $\lambda = 8\pi$ and Ω is close to a disc, in which case there is no classical solution to (5.382).

Theorem 5.20 ([68, 80]) *If*

$$T = +\infty, \quad \lim_{t\uparrow+\infty}\|u(\cdot,t)\|_\infty = +\infty, \quad \lim_{t\uparrow+\infty}\mathscr{F}(u(\cdot,t)) > -\infty, \tag{5.426}$$

then $\mu(dx,t)$ generated in (5.391) is independent of t and hence takes the form of the singular limit of the Boltzmann–Poisson equation (5.382).

Proof Take the case of (5.266) with (5.269) for simplicity. We have readily shown

$$\lambda = \|u_0\|_1 \in 8\pi\mathbf{N}$$

and (5.391) up to a subsequence, for $t_k \uparrow +\infty$ satisfying (5.390). There arises also

$$\mu(dx,t) = 8\pi \sum_{i=1}^{\ell} \delta_{x_i(t)}(dx), \quad 8\pi\ell = \lambda, \quad x_i(t) \neq x_j(t), \ i \neq j \tag{5.427}$$

with $x_i = x_i(t) \in \Omega$ satisfying (5.399), where $H_\ell = H_\ell(x)$, $x = (x_1, \cdots, x_\ell)$, is the Hamiltonian defined by (5.389).

First, since

$$\|\nabla u - u\nabla v\|_1^2 \leq \lambda \int_\Omega u^{-1}|\nabla u - u\nabla v|^2 = \lambda \int_\Omega u|\nabla(\log u - v)|^2$$

it holds that

$$\int_0^\infty \|\nabla u - u\nabla v\|_1^2 \, dt < +\infty \tag{5.428}$$

by (5.365) and (5.426. The equation

$$-\Delta v = u,$$

on the other hand, is equivalent to

$$v_{z\overline{z}} = -\frac{u}{4}, \quad z = x_1 + \sqrt{-1}x_2, \ \overline{z} = x_1 - \sqrt{-1}x_2, \ x = (x_1, x_2)$$

and hence it holds that

$$s_{\overline{z}} = v_{zz\overline{z}} - v_z v_{z\overline{z}} = -\frac{1}{4}u_z + \frac{1}{4}uv_z = -\frac{1}{4}(u_z - uv_z) \tag{5.429}$$

for

$$s = v_{zz} - \frac{1}{2}v_z^2.$$

We thus obtain

$$\int_0^\infty \|s_{\overline{z}}\|_1^2 \, dt < +\infty \tag{5.430}$$

by (5.428).

Elliptic L^1 estimate [8], second, implies

$$v(\cdot, t + t_k) \rightharpoonup \tilde{v}(\cdot, t) \quad \text{in } W^{1,q}(\Omega), \quad 1 \leq q < 2,$$

in accordance with (5.391). Then it holds that

$$\tilde{v}(\cdot,t) = 8\pi \sum_{i=1}^{\ell} G(\cdot, x_i(t)) \tag{5.431}$$

by (5.427).

Given $t_k \uparrow +\infty$, we have a subsequence denoted by the same symbol such that $t_{k+1} > t_k + 2$. Then it follows that

$$\lim_{k\to\infty} \int_{t_k-1}^{t_k+1} \|s_{\overline{z}}\|_1^2 \, dt = 0,$$

from (5.430) and

$$\sum_k \int_{t_k-1}^{t_k+1} \|s_{\overline{z}}\|_1^2 \, dt \le \int_0^\infty \|s_{\overline{z}}\|_1^2 \, dt < +\infty,$$

which implies

$$\tilde{s}(\cdot,t)_{\overline{z}} = 0 \text{ in } \overline{\Omega} \setminus \mathscr{S}_t, \quad -1 < t < 1$$

for $\tilde{s} = \tilde{s}(\cdot,t)$ defined by

$$\tilde{s} = \tilde{v}_{zz} - \frac{1}{2}\tilde{v}_z^2.$$

Then the complex analysis developed for the stationary solutions in Chap. 2 guarantees

$$\nabla_{x_i} H_\ell(x_1(t), \cdot, x_\ell(t)) = 0, \ 1 \le i \le \ell, \quad -1 < t < 1. \tag{5.432}$$

Hence $x(t) = (x_1(t), \cdots, x_\ell(t))$ is a stationary solution to (5.399), and consequently, (5.432) is valid for any $-\infty < t < +\infty$. □

Remark 5.34 Theorems 5.15, 5.16, 5.19, and 5.20 control the dynamics of (5.266) and (5.269) as follows. First, if $0 < \lambda < 8\pi$, there is a uniformly bounded global-in-time solution. It converges to a stationary solution by the theory of gradient inequality. If $\lambda = 8\pi$, it holds always that $T = +\infty$. If blowup in infinite time occurs at this level of λ, the limit measure $\mu(dx,t)$ generated by (5.391) takes the form of $\mu(dx,t) = 8\pi\delta x_0(dx)$ with $x_0 \in \Omega$ satisfying $\nabla R(x_0) = 0$. This case actually arises either if Ω is close to a disc, or if u_0 is concentrated as $\mathscr{F}(u_0) \ll -1$. If Ω is simply connected and $\lambda \gg 1$, it holds always that $T < +\infty$. If Ω is convex and $\lambda > 8\pi$, blowup in infinite time does not occur. Hence there arises either blowup in finite time or uniformly bounded global in time solution. If Ω is convex and $\lambda = 8\pi$, it holds that

$$u(x,t)dx \rightharpoonup 8\pi\delta_{x_0}(dx), \quad \nabla R(x_0) = 0$$

as $t \uparrow +\infty$. Such $x_0 \in \Omega$ is unique if Ω is convex as is described in Remark 5.30. (See [82] for more quantitative statement.)

Remark 5.35 There is a model (B) equation in higher space dimensions, the stationary state of which is realized as (2.163). It is a *degenerate parabolic equation* studied by [84–87]. There are a critical mass for the blowup in finite time, ε-regularity, and finiteness of the type II blowup points, in spite of the lack of the monotonicity formula of this model. Normalized Ricci flow on compact Riemann surface is the other model (B) equation associated with Helmholtz's free energy. There is a global-in-time pre-compact orbit even for the critical mass [82].

5.6 Methods of Mathematical Oncology

5.6.1 Cancer Evolution

Cancer is a malignant tumor which possesses the properties of clonal proliferation, and motility, and the ability to invade normal tissues. Organs, tissues, cells, organelles, molecules, and genes are the biological hierarchy observed in living individuals. Cancer evolution is an event interacting through these levels.

Cancer cells are created by mutation of normal cells. A normal cell stops dividing after forty or fifty times, while cancer cells make a colony by uncontrollable proliferation. Then they secret VEGF (vascular endothelial growth factor) to cause angiogenesis of the existing capillary, composed of endothelial and basement cells. VEGF captured by the receptor of the EC (endothelial cell) induces degradation of basement cells as well as sprouting new ECs. The new ECs move inside the ECM (extracellular matrix) to create a network connecting the tumor cell colony and capillary. Then oxygen is delivered through the network, which causes degradation of molecules associated with cell adhesion. Cancer cells thus acquire the ability to invade normal tissues and achieve metastasis using existing blood vessels inside the body.

There are malignant signalling pathways inside the cell. The first oncogene to be discovered was Src. The Ras was found in human cells. Several anti-oncogenes are also known. A huge amount of data on these signal transmissions and inhibitions is available. System biology is used as a tool to determine biological laws from such data.

Mathematical oncology has two aspects, data science and mathematical modeling. In the former, biological data are taken, analyzed, and used to describe biological laws. It is a combination of bio-informatics, bio-experiments, and bio-statistics. In the latter, mathematical models are used to describe these laws. In systems biology, ODE descriptions are mostly used to compile complex pathways inside the cell, while PDE descriptions can efficiently describe the events at tissue level. Multi-scale models are used for interaction across the biological hierarchy, and numerical simulations involving fluctuations are effective for visualizing biodynamics. The ultimate purpose of mathematical oncology is the prediction and control of biological phenomena observed in cancer evolution [79].

Evolution of cells viewed on the tissue level is sometimes reduced to the Gompertz model, and its clinical applications have been described recently [20, 51]. First, we are concerned with the cell population rates $0 \leq p_i = p_i(x,t) \leq 1$, $1 \leq i \leq N$, of N-species with constant total population and common velocity denoted by v. It holds that

$$\sum_i p_i = 1 \tag{5.433}$$

and

$$\frac{\partial p_i}{\partial t} + \nabla \cdot (v p_i) = (\gamma_i(M) + \sum_j a_{ij}(M) p_j - \beta_i(M)) p_i, \quad 1 \leq i \leq N \tag{5.434}$$

by the transport theory (5.70)–(5.71), where $\gamma_i = \gamma_i(M)$, $a_{ij} = a_{ij}(M)$, and $\beta_i = \beta_i(M)$ stand for the growth, interaction, and decay rates, respectively, depending on the micro-environment factor M.

From (5.433) it follows that

$$\nabla \cdot v = \sum_i (\gamma_i(M) + \sum_j a_{ij}(M) p_j - \beta_i(M)) p_i.$$

In the other assumption the driving force of the movement of cells is due to a gradient which results in the irrotational velocity: $v = -\nabla \pi$. Hence we obtain

$$-\Delta \pi = \sum_i (\gamma_i(M) + \sum_j a_{ij}(M) p_j - \beta_i(M)) p_i \tag{5.435}$$

and provide a boundary condition to this π.

The first case of vascularization is for $N = 2$, where $p = p_1$, $S = p_2$, and M stand for the population of tumor cells, that outside tumor, and the oxygen concentration, respectively, and $\alpha > 0$ is a constant determined by clinical data:

$$p_t + \nabla \cdot vp = Mp, \quad S_t + \nabla \cdot vS = 0, \quad M_t = -\alpha M. \tag{5.436}$$

Assuming that $M_0 = M|_{t=0}$ is spatially homogeneous, we obtain

$$M = M_0 e^{-\alpha t},$$

and then the volume rate of total tumor cells, $V = \int_{\Omega_t} p$, satisfies

$$\frac{dV}{dt} = \int_{\Omega_t} p_t + \int_{\partial\Omega_t} pv \cdot \nu = \int_{\Omega_t} p_t + \nabla \cdot vp \, dx = \int_{\Omega_t} Mp$$
$$= M(t)V = M_0 e^{-\alpha t} V,$$

where Ω_t is the cancer region.

We thus obtain an explicit formula of $V(t) = V(t, M_0, V_0)$. The parameter identification is now made for (M_0, V_0) by $V(t_k)$, $k = 0, 1$. Having $M(t)$, we determine the tumor cell density $p = p(x, t)$ by

$$p_t + \nabla \cdot vp = M(t)p, \quad p|_{t=0} = p_0(x)$$
$$-\Delta\pi = M(t)p, \quad \left.\frac{\partial \pi}{\partial \nu}\right|_{\partial\Omega} = 0$$
$$v = -\nabla\pi,$$

with Ω standing for the region effective to prescribe v above. Adaptive methods are useful here both for the parameter identification and the initial density segmentation.

The other case concerns the lung metastasis. It is assumed that $N = 1$ with p and M denote tissue density and malignant factor, respectively:

$$p_t + \nabla \cdot vp = Mp, \quad M_t = -\alpha M, \quad p = 1.$$

Assuming similarly that $M_0 = M|_{t=0}$ is spatially homogeneous, we obtain $M = M_0 e^{-\alpha t}$ and $\nabla \cdot v = M$, and then the volume rate $V = \int_{\Omega_t} p$ is determined by

$$\frac{dV}{dt} = \int_{\Omega_t} \nabla \cdot v = \mu, \quad \mu = \int_{\Omega_t} M$$
$$\frac{d\mu}{dt} = \int_{\Omega_t} M_t + \nabla \cdot vM \, dx = -\alpha\mu + M\mu.$$

5.6.2 Multi-scale Models

Multi-scaleness is used in modeling interactions inside and outside cell. In this scope of observation, sensitivity, activation, and inhibition are written by functional relations, while chemical reactions and material transport are described by ODE (ordinary differential equations) and PDE (partial differential equations), respectively.

Here we describe a modification [89] of the Anderson–Chaplain model [2] for tumor-induced angiogenesis. It is concerned with the tip cell density, ECM, and VEGF concentration, denoted by $n = n(x, t)$, $f = f(x, t)$, and $c = c(x, t)$, respectively, depending on the position $x = (x_1, x_2, x_3)$ and the time t. The principal equation is for n,

$$n_t + \nabla \cdot v_m n = d_n \Delta n, \tag{5.437}$$

where $v_m = v_m(x,t)$ is a advection velocity composed of chemotaxis and haptotaxis,

$$v_m = \nabla \chi(c) + d_f \nabla f. \tag{5.438}$$

Here and henceforth, $\nabla = {}^t(\partial_{x_1}, \partial_{x_2}, \partial_{x_3})$ and

$$\Delta = \nabla \cdot \nabla = \frac{\partial^2}{\partial x_1^2} + \frac{\partial^2}{\partial x_2^2} + \frac{\partial^2}{\partial x_3^2}$$

denote the gradient operator the Laplacian, respectively.

The right-hand side of (5.437) is due to the diffusion with $d_n = D > 0$ standing for the diffusion coefficient. It is derived from the limit of the *master equation*

$$\frac{q(x,t+\Delta t) - q(x,t)}{\Delta t} = \int_{S^2} T_{-\omega}(x+\omega\Delta x, t) q(x+\omega\Delta x, t) - T_\omega(x,t) q(x,t)\, d\omega,$$

where $S^2 = \{\omega \in \mathbf{R}^3 \mid |\omega| = 1\}$ and $T = T_\omega(x,t)$ is the transient probability of q-particle at (x,t) in the direction ω. Then the diffusion equation

$$q_t = D\Delta q$$

arises as $\Delta x \downarrow 0$ and $\Delta t \downarrow 0$ by Einstein's formula

$$D = \frac{(\Delta x)^2}{2n\tau}, \quad n = 3$$

under the assumption that $T_\omega(x,t) = T$ is a constant, where τ denotes the mean waiting time defined by

$$\int_{S^2} T_\omega(x,t)\, d\omega = \tau^{-1}$$

[83]. Then the left-hand side of (5.437) is a consequence of the transport theory.

The sensitivity $\chi(c)$ in the right-hand side of (5.438) takes the form

$$\chi(c) = -\frac{\alpha_c}{\beta_c + c} + \text{constant} \tag{5.439}$$

because the tip cell detects the *ligand–receptor complex*. The ligand VEGF and the receptor denoted by S and E, respectively, create the product P through the process

$S + E \leftrightarrow ES \to E + P$. If this process is quasi-static, there arises

$$[ES] = \frac{e_T[S]}{\gamma^{-1} + [S]}$$

for the equilibrium constant γ and $\chi(c)$ is assumed to be proportional to this quantity.

The model for the ECM density f is based on the fact that *haptotaxis* arises as a consequence of ECM degradation and also the observation of the remodeling of ECM by the tip cell. Hence it holds that

$$f_t + \nabla \cdot vf = -\eta_f nf + \alpha_f n \tag{5.440}$$

for

$$v = -\frac{d_n}{n}\nabla n + v_m \tag{5.441}$$

standing for the velocity of the tip cell. Finally, the VEGF concentration is uptaken by the tip cell, and hence there arises

$$c_t + \nabla \cdot vc = -\delta_c nc. \tag{5.442}$$

The zero flux condition to n,

$$d_n \frac{\partial n}{\partial \nu} - \nu \cdot v_m n \bigg|_{\partial\Omega} = 0, \tag{5.443}$$

is imposed on the boundary, to make (5.437)–(5.442) well-posed, where ν denotes the outer unit normal vector.

5.6.3 Hybrid Simulations

Hybrid simulation is based on a combination of deterministic and stochastic discretizations. First, we introduce a model-faithful scheme, where non-negativity and total mass conservation of the main variable n is preserved. Second, this scheme then makes it possible to define the transient probability per unit time of the particle n. Third, we execute a Monte Carlo simulation, defining its velocity.

Here we apply a finite difference scheme for the first part. We set physical constants to be 1 and assume that the space dimension is 1 for the sake of simple explanations. Thus, let $\Omega = (0, 1)$ and $h = \Delta x > 0$ and $\tau = \Delta t > 0$ be the spatial and temporal meshes, respectively. Concerning the space discretization we use a uniform mesh, $hN = 1$, for a positive integer $N \gg 1$. In terms of

$$D_h f(x) = \frac{f(x+h) - f(x)}{h}, \quad \overline{D_h} f(x) = \frac{f(x) - f(x-h)}{h},$$

the derivatives $f'(x)$ and $f''(x)$ at $x = ih$, $1 \leq h \leq N-1$, are approximated by the mean,

$$\frac{1}{2}(D_h f(x) + \overline{D_h} f(x)) = \frac{f(x+h) - f(x-h)}{2h},$$

and the center,

$$D_h \overline{D_h} f(x) = \overline{D_h} D_h f(x) = \frac{f(x+h) + f(x-h) - 2f(x)}{h^2}, \tag{5.444}$$

respectively.

The other fundamental scheme is the upwind difference approximation applicable for the advection model

$$n_t = -(nG)_x,$$

where $(nG)_x$ is replaced by

$$G_+(x)(\overline{D_h} n)(x) - G_-(x)(D_h n)(x), \quad G_\pm = \{\pm G, 0\} \tag{5.445}$$

at $x = ih$, $1 \leq i \leq N-1$. Regarding (5.437) as an advection–diffusion model,

$$n_t = \Delta n - \nabla \cdot n v_m$$

we use the center (5.444) and the upwind (5.445) schemes at the interior nodes $x_i = ih$, $1 \leq i \leq N-1$, for the first and the second terms on the right-hand side, respectively. The boundary condition

$$\left. \frac{\partial n}{\partial \nu} - n \frac{\partial}{\partial \nu}(\chi(c) + d_f f) \right|_{\partial \Omega} = 0$$

is provided, with D_h and $\overline{D_h}$ for the normal derivative at $x = 0$ and $x = 1$, respectively. This idea is applicable if Ω is a polygonal domain in $\mathbf{R}^2$ with slight modifications.

We use the explicit Euler scheme for time discretization, hence $\frac{\partial}{\partial t}$ is replaced by D_τ, $\tau = \Delta t$. To realize the fundamental property of (5.437) with (5.443), non-negativity and total mass conservation of n,

$$n = n(x,t) \geq 0, \quad \frac{d}{dt}\|n(\cdot,t)\|_1 = 0$$

in this discretized scheme, we use a non-uniform adapting mesh $\tau_k >$ so that $\{t_k\}$, defined by $t_{k+1} = t_k + \tau_k$ and $t_0 = 0$, is the time sequence which defines the approximate solution at $x = ih$, $0 \leq i \leq N$, and $t = t_k$, that is,

$$n_i^k \approx n(ih, t_k).$$

This requirement is fulfilled if τ_k is small, but a qualitative criterion is given by Varga's theorem.

Thus, if an $\tilde{N} \times \tilde{N}$ matrix A for $\tilde{N} = N + 1$ is *irreducible*, *diagonally dominant*, and

$$a_{ij} \leq 0,\ i \neq j; \quad a_{ii} > 0,\ 1 \leq i \leq \tilde{N},$$

it is non-singular and $A^{-1} > 0$, that is, each entry is positive. This property assures

$$n_i^k \geq 0,\ 0 \leq i \leq N \quad \Rightarrow \quad n_i^{k+1} \geq 0,\ 0 \leq i \leq N.$$

Using proper discretized boundary conditions, we have also the total mass conservation in the form of

$$\|\hat{n}^{k+1}\|_1 = \|\hat{n}^k\|_1,$$

where $\hat{n}^k$ is the piecewise linear continuous function taking the value n_i^k at $x = x_i$, $0 \leq i \leq N$.

Recall that A is irreducible if there is no permutation matrix P such that

$$P^{-1}AP = \begin{pmatrix} A_1 & A_2 \\ 0 & A_3 \end{pmatrix},$$

where A_1, A_2, and A_3 are square matrices. We say also that $A = (a_{ij})$ is diagonally dominant if

$$|a_{ii}| \geq \sum_{j \neq i} |a_{ij}|, \quad 1 \leq i \leq \tilde{N}.$$

If the above inequality is always strict, then this A is irreducible.

The above-described scheme of n with positivity and total mass preservation for any initial value ensures its *detailed mass balance*. Hence writing

$$n_i^{k+1} = P_i^{k,0} n_i^k + p_{i+1}^{k,-1} n_{i+1}^k + P_{i-1}^{k,+1} n_{i-1}^k, \quad P_0^{k,-1} = P_N^{k,+1} = 0,$$

we have

$$P_i^{k,\ell} \geq 0, \quad 0 \leq i \leq N, \quad \ell = 0, \pm 1, \quad k = 0, 1, 2, \cdots,$$

as well as

$$P_i^{k,0} + P_i^{k,+1} + P_i^{k,-1} = 0, \quad 0 \le i \le N, \ k = 0, 1, \cdots .$$

Thus, $\{P_i^{k,\ell}\}_\ell$ casts the transient probability of n at x_i; right, neutral, and left according to $\ell = +1$, $\ell = 0$, and $\ell = -1$, respectively. From these probabilities the values n_i^{k+1}, $0 \le i \le N$, are calculated by n_i^k, $0 \le i \le N$, via *Monte Carlo simulation.*

To proceed to the next step, we need to prescribe the approximate values of c and f at $t = t_{k+1}$ for $x = ih$, $0 \le i \le N$. We use again the explicit Euler scheme for f_t in (5.440) and (5.442). The gradient operator ∇ is replaced by the *mean* and *unilateral* at the interior and boundary nodes, respectively, while the approximated velocity of v at $(x,t) = (ih, t_k)$ is defined by

$$v_i^k = \frac{h}{\tau_k}(P_i^{k,+1}e_{+1} + P_i^{k,-1}e_{-1}), \quad e_{\pm 1} = \pm 1.$$

Boolean reduction is sometimes used to compute f_i^{k+1} and c_i^{k+1} for $0 \le i \le N$. Namely, the value n at $x = ih$ on the right-hand side of the discrete equations for f and c is replaced by n_*, which takes the value 1 if the n-particle under the Monte Carlo simulation exists at $(x,t) = (ih, t_k)$ and 0 otherwise. Numerical simulations have been tried for the rectangle $\Omega \subset \mathbf{R}^2$, where the trajectories of tip cells are regarded as the new vessels.

5.6.4 Chemotactic Paradox

The Keller–Segel model in Sect. 5.5.1 is a multi-scale model for the self-assembly of cellular slime molds subject to a chemical gradient. These cells move toward the spot where the signal is produced. Since this signal is a spiral pulse, they sometimes move against the chemical gradient. For this phenomenon, called chemotactic paradox, it is suspected that the cell detects the velocity of signal and moves against this velocity [56]. In an in vivo experiment, patching an arterial graft into a vein induced angiogenesis, where VEGF remain around the vessels and tip cells move toward the direction where the VEGF gradient decreases [32].

This phenomenon is a problem of *cell polarization*, determine the head and tail of the cell, denoted by $x = \ell$ and $x = -\ell$, respectively, where activation of the RAS pathway is involved. The LEGI model is concerned with this process, where R and $R^* = R_0 - R$ denote active and inactive Ras concentrations, respectively. This process is described by

$$\frac{dR}{dt} = AF(R) - IG(R) \tag{5.446}$$

for

$$F(R) = k_A \cdot \frac{R_0 - R}{R_0 - R + K_A}, \quad G(R) = k_I \cdot \frac{R}{R + K_I}. \tag{5.447}$$

Equalities (5.447) are a modification of the simpler model

$$F(R) = k_A(R_0 - R), \quad G(R) = k_I R,$$

where relaxations are taken into account for Ras activation and inactivation. The coefficients A and I are concentration of activation and inactivation factors, respectively, subject to

$$\frac{dA}{dt} = k_a S - \gamma_a A, \quad \frac{dI}{dt} = k_i S - \gamma_i I \tag{5.448}$$

where S stands for the signal strength affecting this cell.

Writing (5.446) as

$$\frac{dR}{dt} = F(R)I(Q - H(R)), \quad Q = A/I, \ H(R) = G(R)/F(R),$$

we have $\pm \frac{dR}{dt} > 0$ according to $Q > H(R)$ and $Q < H(R)$, respectively. Given $S = S(t)$, therefore, the orbit in the RQ plane of the solution

$$(R, A, I) = (R(t), A(t), I(t))$$

to (5.446)–(5.448) makes a curve in $0 < R < R_0$. In numerical results for the pulse-like signal

$$S = \exp(-(t - a)^2/c)$$

with the initial value $(R, A, I) = (0, 0, 0)$, this trajectory approaches

$$\Gamma : Q = H(R),$$

with Q and R decreasing and increasing, respectively, and then crosses Γ to reach $(R, Q) = (0, 0)$ along and below Γ. Hence except for the initial layer, Q and R obeys a positive correspondence, where the decrease of Q means that of R. This property means that the endpoint with relatively high Q forms the head of the cell.

Here we take a reduction of (5.448) in the region of

$$\gamma_{ai} \gg 1, \quad \frac{k_{ai}}{\gamma_{ai}} \approx 1,$$

which causes

$$\varepsilon \frac{dB}{dt} = g - B \tag{5.449}$$

for $B = A, I$, $0 < \varepsilon \ll 1$, and $g = S$. Using

$$E(t) = B(t) - g(t - \varepsilon), \quad h_\varepsilon(t) = \frac{1}{\varepsilon}(g(t) - g(t - \varepsilon)) - g'(t - \varepsilon),$$

we have

$$\frac{dE}{dt} = h_\varepsilon - \frac{1}{\varepsilon} E$$

and hence

$$E(t) = e^{-t/\varepsilon} E(0) + \int_0^t e^{-(t-s)/\varepsilon} h_\varepsilon(s)\, ds.$$

Since $g_\varepsilon(t) = o(1)$ locally uniformly in t, we obtain

$$E(t) = e^{-t/\varepsilon} E(0) + o(1)\varepsilon(1 - e^{-t/\varepsilon}),$$

and hence (5.449) implies

$$B(t) = g(t - \varepsilon) + o(\varepsilon)$$

locally uniformly in $t > 0$.

Using this formula, we see how

$$Q_\pm = A_\pm / I_\pm$$

is detected at the head $x = \ell$ and the tail $x = -\ell$ of the cell, under the Turing paradigm. Thus we assume that the activator is localized as

$$A_\pm(t) = A(\pm\ell, t),$$

while the inhibitor is uniformized as

$$I(t) = \frac{1}{2}(I(-\ell, t) + I(\ell, t)),$$

which results in

$$A_\pm(t) \sim \gamma_a^{-1} k_a S(\pm\ell, t - \gamma_a^{-1}),$$
$$I(t) \sim \frac{\gamma_i^{-1} k_i}{2} \left(S(\ell, t - \gamma_i^{-1}) + S(-\ell, t - \gamma_i^{-1}) \right).$$

We thus obtain

$$Q_{\pm}(t) = \frac{A_{\pm}(t)}{I(t)} \sim Q_0\{1+\frac{S_t(\pm\ell,t)}{S(\pm\ell,t)}(\gamma_i^{-1}-\gamma_a^{-1})\pm\ell\frac{S_x(\pm\ell,t)-\gamma_i^{-1}S_{xt}(\pm\ell,t)}{S(\pm\ell,t)}\}$$

as $\gamma_{ai} \gg 1$, where

$$Q_0 = \frac{\gamma_i k_a}{\gamma_a k_i}.$$

The cell is polarized according to

$$Q_+ > Q_-.$$

Since $\gamma_i > \gamma_a$ is assumed implicitly, it coincides with the direction $v = -S_t/S_x < 0$ in the chemical velocity, $S_x > 0$ in the chemical gradient, and $S_{xt} < 0$ in the chemical gradient change.

Thus all factors observed above are realized in this range of parameters. A quantitative realization is achieved by hybrid simulation where the chemical gradient decreasing is taken as a driving force of cell movement, which is adjusted by real data [52].

References

1. A.D. Alexandrov, Majorization of solutions of second order linear equations. Vestnik Lenningrad Univ. Math. **21**, 5–25 (1966). English Transl. Am. Math. Soc. Transl. **68**, 120–143 (1968)
2. A.R.A. Anderson, M.A.J. Chaplain, Continuous and discrete mathematical models of tumor-induced angiogenesis. Bull. Math. Biol. **60**, 857–899 (1998)
3. I.Ya. Bakel'man, Theory of quasilinear elliptic equations. Siberian Math. J. **2**, 179–186 (1961)
4. P. Baras, M. Pierre, Problème paraboliques semi-linéaires avec donées measures. Appl. Anal. **18**, 111–149 (1984)
5. P. Biler, Local and global solvability of some parabolic system modelling chemotaxis. Adv. Math. Sci. Appl. **8**, 715–743 (1998)
6. P. Biler, D. Hilhorst, T. Nadzieja, Existence and nonexistence of solutions for a model of gravitational interaction of particles, II. Colloq. Math. **67**, 297–308 (1994)
7. A. Blanchet, J. Dolbeault, B. Perthame, Two-dimensional Keller-Segel model: optimal critical mass and qualitative properties of the solutions. Electron. J. Differ. Equ. **2006**(44), 1–13 (2006)
8. H. Brezis, W. Strauss, Semi-linear second-order elliptic equations in L^1. J. Math. Soc. Jpn. **25**, 565–590 (1973)
9. M.C. Caputo, A. Vasseur, Global regularity of solutions to systems of reaction-diffusion with sub-quadratic growth in any dimension. Commun. Partial Differ. Equ. **34**, 1228–1250 (2009)
10. M.C. Caputo, T. Goudon, A.F. Vasseur, Solutions of the 4-species quadratic reaction-diffusion system are bounded and C^∞-smooth, in any space dimension. Anal. Partial Differ. Equ. **12**, 1773–1804 (2019)
11. P.-H. Chavanis, Generalized thermodynamics and Fokker-Planck equation: applications to steller dynamics and two-dimensional turbulence. Phys. Rev. E **68**, 036108 (2003)

12. P.-H. Chavanis, J. Sommeria, R. Robert, Statistical mechanics of two-dimensional vortices and collisionless stellar systems. Astrophys. J. **471**, 385–399 (1996)
13. T. Cheon, Evolutionary stability of ecological hierarchy. Phys. Rev. Lett. **4**, 58–105 (2003)
14. S. Childress, J.K. Percus, Nonlinear aspects of chemotaxis. Math. Biosci. **56**, 217–237 (1981)
15. K. Chueh, C. Conley, J. Smoller, Positively invariant regions for systems of nonlinear diffusion equations. Indiana Univ. Math. J. **26**, 373–392 (1977)
16. K. Fellner, J. Morgan, B.Q. Tang, Global classical solutions to quadratic systems with mass control in arbitrary dimensions. Ann. Inst. H. Poincaré - Analyse non linéaire **37**, 281–307 (2020)
17. K. Fellner, J. Morgan, B.Q. Tang, Uniform-in-time bounds for quadratic reaction-diffusion systems with mass dissipation in higher dimensions. Discrete Contin. Dyn. Syst. S **12**, 635–651 (2021)
18. H. Fujita, N. Saito, T. Suzuki, *Operator Theory and Numerical Methods* (Elsevier Science Publishers, Amsterdam, 2001)
19. H. Gajewski, K. Zecharias, Global behaviour of a reaction-diffusion system modelling chemotaxis. Math. Nachr. **195**, 77–114 (1998)
20. O. Gallinato, T. Colin, O. Saut, C. Poignard, Tumor growth model for ductal carcinoma: from in situ phase to stroma invasion. J. Theor. Biol. **429**, 253–266 (2017)
21. P.Y. Gao, Hamiltonian structure and first integrals for the Lotka-Volterra systems. Phys. Lett. A. **273**, 85–96 (2000)
22. K. Geisshirt, E. Praestgaard, S. Toxvaerd, Oscillating chemical reactions and phase separation simulated by molecular dynamics. J. Chem. Phys. **107**, 9406–9412 (1997)
23. K. Gopalsamy, Persistence in periodic and almost periodic Lotka-Volterra systems. J. Math. Biol. **21**, 145–148 (1984)
24. M. Grossi, F. Takahashi, Nonexistence of multi-bubble solutions to some elliptic equations on convex domains. J. Funct. Anal. **259**, 904–917 (2010)
25. R. Hamilton, The Ricci flow on surface, in *Mathematics and General Relativity*. Contem. Math., vol. 71 (American Mathematical Society, Providence, 1988), pp. 237–262
26. D. Henry, *Geometric Theory of Semilinear Parabolic Equations*. Lecture Notes in Math., vol. 840 (Springer, Berlin, 1981)
27. M.A. Hererro, J.J.L. Velázquez, Singularity patterns in a chemotaxis model. Math. Ann. **306**, 583–623 (1996)
28. D.D. Holm, T. Schmah, C. Stoica, *Geometry Mechanics and Symmetry* (Oxford University Press, New York, 2009)
29. S.-Z. Huang, *Gradient Inequalities with Applications to Asymptotic Behavior and Stability of Gradient-Like Systems* (American Mathematical Society, Providence, 2006)
30. V. Huston, A theorem on average Liapunov functions. Monatsh. Math. **98**, 267–275 (1984)
31. V. Huston, K. Schmitt, Permanence and the dynamics of biological systems. Math. Biosci. **111**, 1–71 (1992)
32. Y. Ito, M. Yoshida, D. Maeda, M. Takahashi, H. Nanjo, H. Masuda, A. Goto, Neovasculature can be induced by patching an arterial graft into a vein: a novel in vivo modeled of spontaneous arteriovenous fistula formation. Sci. Rep. **8**, 3156 (2018)
33. T. Ito, Y. Kumagai, K. Itano, T. Maruyama, K. Tamura, S. Kawasaki, T. Suzuki, Y. Murakami, Mathematical analysis of gefitinib resistance of lung adenocarcinoma caused by MET amplification. Biochem. Biophys. Res. Commun. **511**(3), 544–550 (2019)
34. W. Jäger, S. Luckhaus, On explosions of solutions to a system of partial differential equations. Trans. Am. Math. Soc. **329**, 819–824 (1992)
35. J.I. Kanel, Solvability in the large of a system of reaction-diffusion equations with the balance condition. Differ. Equ. **26**, 331–339 (1990)
36. N.I. Kavallaris, T. Suzuki, On the dynamics of a non-local equation associated with Gierer-Meinhardt system. Nonlinearity **30**, 1734–1761 (2017)
37. E.F. Keller, L.A. Segel, Initiation of slime mold aggregation viewed as an instability. J. Theor. Biol. **26**, 399–415 (1970)

38. M. Kobayashi, T. Suzuki, Y. Yamada, Lotka-Volterra systems with periodic orbits. Funckial. Ekvac **62**, 129–155 (2019)
39. R. Kon, J. Hofbauer, Structure of n-dimensional Lotka-Volterra systems for qualitative permanence. Kokyuroku RIMS, Kyoto Univ. **1309**, 146–153 (2003)
40. N.V. Krylov, M.V. Safanov, An estimate for the probability of a diffusion process hitting a set of positive measure. Doklady Akademii Nauk SSSR **245**(1), 18–20 (1979)
41. N.V. Krylov, M.V. Safanov, A property of the solutions to parabolic equations with measurable coefficients. Izvesiya Akademii Nauk SSSR, Seriya Matematicheskaya **44**(1), 161–175 (1980)
42. M. Kurokiba, T. Ogawa, Finite time blow-up of the solution for a nonlinear parabolic equation of drift-diffusion type. Differ. Integral Equ. **16**, 427–452 (1993)
43. O. Ladyženskaja, V.A. Solonnikov, N.N. Ural'ceva, *Linear and Quasi-Linear Equations of Parabolic Type* (American Mathematical Society, Providence, 1968)
44. E. Latos, T. Suzuki, Y. Yamada, Transient and asymptotic dynamics of a prey-predator system with diffusion. Math. Methods Appl. Sci. **35**, 1101–1109 (2012)
45. G.M. Leberman, *Second Order Parabolic Eifferential Equations* (World Scientific, Singapore, 1996)
46. P.L. Lions, The concentration compactness principle in the calculus of variation. The locally compact case. Part I. Ann. Inst. H. Poincaré Anal. Non Linéaire **1**, 109–145 (1984)
47. S. Łojasiewicz, Une propriété topologique des sous-ensembles analytiques réels, Colloques internationaux du C.N.R.S #117, Les équations aux dérivées partielles (1963)
48. A.J. Lotka, *Elements of Physical Biology* (William and Wilkins, New York, 1925)
49. S. Luckhaus, Y. Sugiyama, J.J.L. Velázquez, Measure valued solutions for the $2D$ Keller-Segel system. Arch. Ration. Mech. Anal. **206**, 31–80 (2012)
50. K. Masuda, K. Takahashi, Asymptotic behavior of solutions of reaction-diffusion systems of Lotka-Volterra type. Differ. Integral Equ. **7**, 1041–1053 (1994)
51. T. Michel, J. Fehrenbach, V. Lobjois, J. Laurent, A. Gomes, T. Colin, C. Poignard, Mathematical modeling of the proliferation gradient in multicellular tumor spheroids. J. Theor. Biol. **458**, 133–147 (2018)
52. D. Minerva, S. Tasaki, Y. Ito, M. Yoshida, et al., New chemotactic mechanism governs long-range angiogeneis induced by patching an arterial graft into a vein, preprint
53. N. Mizoguchi, Refined asymptotic behavior of blowup solutions to a simplified chemotaxis system. Commun. Pure Appl. Math. **75**, 1870–1886 (2020)
54. T. Nagai, Blow-up of radially symmetric solutions to a chemotaxis system. Adv. Math. Sci. Appl. **5**, 581–601 (1995)
55. T. Nagai, T. Senba, K. Yoshida, Application of the Trudinger-Moser inequality to a parabolic system of chemotaxis. Funkcial Ekvac. **40**, 411–433 (1997)
56. A. Nakajima, S. Ishihara, D. Imoto, S. Sawai, Rectified directional sensing in long-range cell migration. Nat. Commun. **17**, 5:5367 (2014)
57. V. Nanjundiah, Chemotaxis, signal relaying, and aggregation morphology. J. Theor. Biol. **42**, 65–105 (1973)
58. H. Ohtsuka, T. Senba, T. Suzuki, Blowup in infinite time in the simplified system of chemotaxis. Adv. Math. Sci. Appl. **17**, 445–472 (2007)
59. C.S. Patlak, Random walk with persistence and external bias. Bull. Math. Biophys. **15**, 311–338 (1953)
60. M. Pierre, Global existence in reaction - diffusion systems with control of mass: a survey. Milan J. Math. **78**, 417–455 (2010)
61. M. Pierre, D. Schmitt, Blow up in reaction-diffusion systems with dissipation of mass. SIAM J. Math. Anal. **28**, 259–269 (1997)
62. M. Pierre, T. Suzuki, Y. Yamada, Dissipative reaction diffusion systems with quadratic growth. Indiana Univ. Math. J. **68**, 291–322 (2019)
63. C. Pucci, Limitazioni per soluzioni di equaziones ellittiche. Ann. Mat. Pura Appl. **161**, 15–30 (1987)
64. T. Senba, Type II blowup of solutions to a simplified Keller-Segel system in two dimensions. Nonlinear Anal. **66**, 1817–1839 (2007)

65. T. Senba, T. Suzuki, Some structures of the solution set for a stationary system of chemotaxis. Adv. Math. Sci. Appl. **10**, 191–224 (2000)
66. T. Senba, T. Suzuki, Chemotactic collapse in a parabolic-elliptic system of chemotaxis. Adv. Differ. Equ. **10**, 21–50 (2001)
67. T. Senba, T. Suzuki, Weak solutions to a parabolic-elliptic system of chemotaxis. J. Funct. Anal. **191**, 17–51 (2002)
68. T. Senba, T. Suzuki, Time global solutions to a parabolic-elliptic system modelling chemotaxis. Asympt. Anal. **32**, 63–89 (2002)
69. T. Senba, T. Suzuki, Blowup behavior of solutions to rescaled Jäger-Luckhaus system. Adv. Differ. Equ. **8**, 787–820 (2003)
70. C. Sire, P.H. Chavanis, Thermodynamics and collapse of self-gravitating Brownian particles in D dimensions. Phys. Rev. E **66**, 046133 (2002)
71. J. Smoller, *Shock Waves and Reaction Diffusion Systems* (Springer, New York, 1896)
72. P. Souplet, Global existence for reaction-diffusion systems with dissipation of mass and quadratic growth. J. Evol. Equ. **18**, 1713–1720 (2018)
73. T. Suzuki, Mass quantization of the non-equilibrium mean field to self-interacting particles. Dyn. Contin. Discrete Impulsive Syst. **10**, 693–720 (2003)
74. T. Suzuki, Mass normalization of collapses in the theory of self-interacting particles. Adv. Math. Sci. Appl. **13**, 611–623 (2003)
75. T. Suzuki, *Free Energy and Self-interacting Particles* (Birkhäuser, Boston, 2005)
76. T. Suzuki, Exclusion of boundary blowup for $2D$ chemotaxis system provided with Dirichlet boundary condition. J. Math. Pure Appl. **100**, 347–367 (2013)
77. T. Suzuki, *Mean Field Theories and Dual Variation*, 2nd edn. (Atlantis Press, Paris, 2015)
78. T. Suzuki, Blowup in infinite time for $2D$ Smoluchowski-Poisson equation. Differ. Integral Equ. **28**, 601–630 (2015)
79. T. Suzuki, *Mathematical Methods for Cancer Evolution*. Lecture Notes in Mathematical Modelling in the Life Science (Springer Nature, Singapore, 2017)
80. T. Suzuki, *Chemotaxis, Reaction, Network* (World Scientific, Singapore, 2018)
81. T. Suzuki, Dissipative reaction diffusion systems with polynomial growth. Adv. Differ. Equ. **25**, 81–104 (2020)
82. T. Suzuki, *Semilinear Elliptic Equations: Modern and Classical Theories*, 2nd edn. (Du Gruyter, Berlin, 2020)
83. T. Suzuki, T. Senba, *Applied Analysis, Mathematical Methods in Natural Science*, 2nd edn. (Imperial College Press, London, 2011)
84. T. Suzuki, R. Takahashi, Degenerate parabolic equation with critical exponent derived form the kinetic theory, I. Generation of the weak solution. Adv. Differ. Equ. **14**, 433–476 (2009)
85. T. Suzuki, R. Takahashi, Degenerate parabolic equation with critical exponent derived from the kinetic theory, II. Blowup threshold. Differ. Integral Equ. **22**, 1153–1172 (2009)
86. T. Susuki, R. Takahashi, Degenerate parabolic equation with critical exponent derived from the kinetic theory, IV. Structure of the blowup set. Adv. Differ. Equ. **15**, 853–892 (2010)
87. T. Suzuki, R. Takahashi, Degenerate parabolic equation with critical exponent derived from the kinetic theory, III. ε-regularity. Differ. Integral Equ. **25**, 223–250 (2012)
88. T. Suzuki, Y. Yamada, Global-in-time behavior of Lotka-Volterra system with diffusion - skew symmetric case. Indiana Univ. Math. J. **64**, 181–216 (2015)
89. T. Suzuki, D. Minerva, K. Nishiyama, N. Koshikawa, M. Chaplain, Study on the tumor-induced angiogenesis using mathematical models. Cancer Sci. **109**, 15–23 (2018)
90. K. Tso, On an Aleksandrov-Bakelman maximum principle for second-order parabolic equations. Commun. Partial Differ. Equ. **10**, 543–553 (1985)
91. V. Volterra, Variazioni e fluttuazioni del numero d'individui in specie animali conviventi. Mem. Acad. Lincei. **2**, 31–113 (1926)
92. P. Weidemaier, On L_p estimates of optimal type for the parabolic oblique derivative problem with VMO coefficients. Differ. Integral Equ. **18**, 935–946 (2005)

Index

T. Suzuki, *Liouville's Theory in Linear and Nonlinear Partial Differential Equations*, Springer Asia Pacific Mathematics Series 8,
https://doi.org/10.1007/978-981-95-1975-0

MIX
Papier aus verantwortungsvollen Quellen
Paper from responsible sources
FSC® C105338

If you have any concerns about our products,
you can contact us on
ProductSafety@springernature.com

In case Publisher is established outside the EU,
the EU authorized representative is:
Springer Nature Customer Service Center GmbH
Europaplatz 3, 69115 Heidelberg, Germany

Printed by Libri Plureos GmbH
in Hamburg, Germany